Differential Calculus

Differential Calculus

Hari Kishan

M.Sc. (Maths) Gold Medalist, Ph.D.

Senior Reader

Department of Mathematics

K.R. College, Mathura

ATLANTIC

PUBLISHERS & DISTRIBUTORS (P) LTD

ATLANTIC
PUBLISHERS & DISTRIBUTORS (P) LTD

7/22, Ansari Road, Darya Ganj, New Delhi
Tel.: +91-11-4077 5252, 2327 3880
E-mail: orders@atlanticbooks.com
Web: www.atlanticbooks.com

Reprint 2025, 2016

Published by Atlantic Publishers & Distributors (P) Ltd.

Printed & bound in India by Atlantic Print Services

Preface

This book is meant for graduate level students of different universities across India. The underlying object of the book is to provide the reader with the thorough understanding of the topics included in the UGC syllabus. The book exactly covers the prescribed syllabus. Nothing undesirable is included and nothing essential has been escaped. Its approach is explanatory and language is lucid and communicable. The explanation of the subject matter is systematic and the students are better prepared to solve the problems. All fundamental concepts of the included topics are explained with a micro-analysis. Sufficient number of well-graded solved examples have been given to let the students understand the various skills necessary to solve the problems. Unsolved practise exercises of multiple varieties included in the book will enable the students to have command of the subject. The figures incorporated herein are precise and accurate. These enable the students to grasp the subject matter in an easier way. All the answers have been thoroughly checked and verified.

Meticulous care has been taken to make the book error-free. However, constructive suggestions from the readers for the improvement of the book would be incorporated in the ensuing edition and will also be gratefully acknowledged.

I am thankful to my son Manish Goyal, Lecturer in Mathematics, GLAITM, Mathura, also an author, for his immense help in the preparation of this book.

Last but not least, I am thankful to Dr. K.R. Gupta, Chairman, Atlantic Publishers & Distributors (P) Ltd. for his encouragement and publishing the book in record time.

I wish my readers brilliant success and bright future.

Hari Kishan

Contents

1
Successive Differentiation and Leibnitz's Theorem

1.1. Definitions and Notations

Let y be a function of x. If we differentiate y with respect to x once, we get $\frac{dy}{dx}$; $\frac{dy}{dx}$ is known as *first differential coefficient* or *first derivative* or *first derived function* of y with respect to x. In general $\frac{dy}{dx}$ is a function of x and therefore can further be differentiated. The differential coefficient of dy/dx with respect to x is known as *second differential coefficient* or *second derivative* or *second derived function* of y with respect to x and is written as $\frac{d^2y}{dx^2}$. Similarly, the differential coefficient of $\frac{d^2y}{dx^2}$ is called the *third differential coefficient* of y and is written as $\frac{d^3y}{dx^3}$. Similarly fourth, fifth, ...etc. differential coefficients are obtained and are written as $\frac{d^4y}{dx^4}, \frac{d^5y}{dx^5}$, ... etc. respectively. *The nth differential coefficient of y with respect to x is denoted by* $\frac{d^ny}{dx^n}$.

If $y = f(x)$, then its successive derivatives are denoted by

$$\frac{dy}{dx}, \frac{d^2y}{dx^2}, \frac{d^3y}{dx^3}, \ldots, \frac{d^ny}{dx^n}, \ldots$$

or $$y_1, y_2, y_3, \ldots, y_n, \ldots$$

or $$y', y'', y''', \ldots y^{(n)}, \ldots$$

or $$f'(x), f''(x), f'''(x), \ldots, f^n(x), \ldots$$

or $$Dy, D^2y, D^3y, \ldots, D^ny, \ldots$$

or $$Df(x), D^2f(x), D^3f(x), \ldots, D^nf(x), \ldots$$

The process of differentiating the same function again and again is called *Successive Differentiation.*

The derivatives obtained during this process are called *successive derivatives.*

If $y = f(x)$, then the value of $\frac{d^n y}{dx^n}$ at $x = a$ is denoted by

$$\left(\frac{d^n y}{dx^n}\right)_{x=a} \text{ or } (y_n)_{x=a} \text{ or } (y_n)\, a \text{ or } f^n(a).$$

ILLUSTRATIVE EXAMPLES

Example 1. *Find the possible derivatives of the functions—*

$$f(x) = x^3 + 3x^2 - 4x + 2.$$

Solution: Here, $f(x) = x^3 + 3x^2 - 4x + 2,$

therefore $f'(x) = 3x^2 + 6x - 4;$

$$f''(x) = 6x + 6;$$

and $f'''(x) = 6.$

All higher order derivatives are evidently zero.

Example 2. *If $y = A \cos mx + B \sin mx$, show that* $\frac{d^2y}{dx^2} + m^2 y = 0.$

Solution : Here, $y = A \cos mx + B \sin mx$ ... (1)

Differentiating (1) both sides of w.r.t. x, we get

$$\frac{dy}{dx} = -Am \sin mx + Bm \cos mx \quad ...(2)$$

Again differentiating (2) both sides w.r.t. x, we get

$$\frac{d^2y}{dx^2} = -Am^2 \sin mx - Bm^2 \cos mx$$

$$\Rightarrow \quad \frac{d^2y}{dx^2} = -m^2 (A \sin mx + B \cos mx)$$

$$\Rightarrow \quad \frac{d^2y}{dx^2} = -m^2 y \quad \text{[By (1)]}$$

$$\Rightarrow \quad \frac{d^2y}{dx^2} + m^2 y = 0.$$

Example 3. *Find the value of* $\frac{dy}{dx}$ *if*

$$e^y = xy.$$

Solution : $e^y = xy$

Differentiating w.r.t. x, we get

$$e^y \frac{dy}{dx} = x\frac{dy}{dx} + y \cdot 1$$

$$\Rightarrow \quad \frac{dy}{dx}(e^y - x) = y$$

$$\Rightarrow \quad \frac{dy}{dx} = \frac{y}{e^y - x}$$

$$= \frac{y}{xy - x} \qquad [\because e^x = xy]$$

$$= \frac{y}{x(y-1)}.$$

Example 4. *If* $ax^2 + 2hxy + by^2 + 2gx + 2fy + c = 0$ *then show that*

$$D^2y = \frac{\Delta}{(hx + by + f)^3},$$

where, $\Delta = abc + 2fgh - af^2 - bg^2 - ch^2$.

Solution : $ax^2 + 2hxy + by^2 + 2gx + 2fy + c = 0 \qquad \ldots (1)$

Differentiating w.r.t. x, we get

$$2ax + 2hy + 2hx\frac{dy}{dx} + 2by\frac{dy}{dx} + 2g + 2f\frac{dy}{dx} = 0$$

$$\Rightarrow \quad (hx + by + f)\frac{dy}{dx} = -(ax + hy + g)$$

$$\Rightarrow \quad \frac{dy}{dx} = -\frac{ax + hy + g}{hx + by + f}.$$

Again differentiating w.r.t. x, we get

$$\frac{d^2y}{dx^2} = -\left[\frac{(hx + by + f)\left(a + h\frac{dy}{dx}\right) - (ax + hy + g)\left(h + b\frac{dy}{dx}\right)}{(hx + by + f)^2}\right]$$

$$= -\left[\frac{\{(ab - h^2)\,y + (af - gh)\} + (h^2x + fh - abx + bg)\dfrac{dy}{dx}}{(hx + by + f)^2}\right]$$

$$= -\left[\frac{\{(ab - h^2)y + (af - gh)\} + (h^2x + fh - abx + bg)\left(-\dfrac{ax + hy + g}{hx + by + f}\right)}{(hx + by + f)^2}\right]$$

$$= -\left[\frac{(aby - h^2y + af - gh)(hx + by + f) - (h^2x + fh - abx - bg)(ax + hy + g)}{(hx + by + f)^3}\right]$$

$$= -\left[\frac{ab(ax^2 + 2hxy + by^2 + 2gx + 2fy) \\ -h^2(ax^2 + 2hxy + by^2 + 2gx + 2fy) + (af^2 - 2fgh + bg^2)}{(hx + by + f)^3}\right]$$

$$= -\left[\frac{ab(-c) - h^2(-c) + af^2 - 2fgh + bg^2}{(hx + by + f)^2}\right]$$

$$= \frac{abc + 2fgh - af^2 - bg^2 - ch^2}{(hx + by + f)^2} = \frac{\Delta}{(hx + by + f)^3}$$

where, $\Delta = abc + 2fgh - af^2 - bg^2 - ch^2$.

Example 5. *If* $y = Ae^{-kt}\cos(pt + c)$, *then show that*

$$\frac{d^2y}{dt^2} + 2k\frac{dy}{dt} + n^2y = 0, \text{ where } n^2 = p^2 + k^2.$$

Solution :
$$y = \underset{\text{I}}{Ae^{-kt}}\ \underset{\text{II}}{\cos(pt + c)} \quad \text{... (1)}$$

Differentiating w.r.t. t, we get

$$\frac{dy}{dt} = A[e^{-kt}\{-\sin(pt + c)\cdot p\} + (-ke^{-kt})\cos(pt + c)]$$

$$\Rightarrow \quad \frac{dy}{dt} = -Ape^{-kt}\sin(pt + c) - ky$$

$$\Rightarrow \quad \frac{dy}{dt} + ky = \underset{\text{I}}{-Ape^{-kt}}\ \underset{\text{II}}{\sin(pt + c)} \quad \text{... (2)}$$

Again differentiating w.r.t. t, we get

$$\frac{d^2y}{dt^2} + k\frac{dy}{dt} = -Ap[e^{-kt}\{p\cdot\cos(pt + c)\} + (-ke^{-kt})\sin(pt + c)]$$

$$\Rightarrow \quad \frac{d^2y}{dt^2} + k\frac{dy}{dt} = -Ap^2e^{-kt}\cos(pt + c) + Apke^{-kt}\sin(pt + c)$$

$$= -p^2y + k\left[-\frac{dy}{dt} - ky\right] \quad \text{[From (1) and (2)]}$$

$$= -p^2y - k\frac{dy}{dt} - k^2y$$

$$\Rightarrow \quad \frac{d^2y}{dt^2} + 2k\frac{dy}{dt} + (p^2 + k^2)\,y = 0$$

$$\Rightarrow \quad \frac{d^2y}{dt^2} + 2k\frac{dy}{dt} + n^2y = 0, \text{ since } n^2 = p^2 + k^2.$$

Example 6. *If* $y = a\cos(\log x) + b\sin(\log x)$, *show that*

$$x^2y_2 + xy_1 + y = 0.$$

Solution :

$$y = a\cos(\log x) + b\sin(\log x) \qquad \ldots (1)$$

Differentiating (1) with respect to x, we get

$$\frac{dy}{dx} = -a\frac{\sin(\log x)}{x} + \frac{b\cos(\log x)}{x}$$

$$\Rightarrow \quad x\frac{dy}{dx} = -a\sin(\log x) + b\cos(\log x) \qquad \ldots (2)$$

Differentiating (2) with respect to x, we get

$$x\frac{d^2y}{dx^2} + \frac{dy}{dx} = \frac{-a\cos(\log x)}{x} - \frac{b\sin(\log x)}{x}$$

$$\Rightarrow \quad x^2\frac{d^2y}{dx^2} + x\frac{dy}{dx} = -[a\cos(\log x) + b\sin(\log x)]$$

$$\Rightarrow \quad x^2\frac{d^2y}{dx^2} + x\frac{dy}{dx} = -y \qquad \text{[By (1)]}$$

$$\Rightarrow \quad x^2\frac{d^2y}{dx^2} + x\frac{dy}{dx} + y = 0.$$

Hence Proved.

Example 7. *If* $y = x\log\left(\frac{x}{a+bx}\right)$, *then prove that*

$$x^3\frac{d^2y}{dx^2} = \left(x\frac{dy}{dx} - y\right)^2$$

Solution :

$$y = x\log\left(\frac{x}{a+bx}\right)$$

$$\Rightarrow \quad \frac{y}{x} = \log x - \log(a + bx).$$

Differentiating both sides w.r.t. x, we get

$$\frac{x\frac{dy}{dx} - y\cdot 1}{x^2} = \frac{1}{x} - \frac{b}{a+bx}$$

$$= \frac{a + bx - bx}{x(a+bx)}$$

$$\Rightarrow \quad \frac{1}{x^2}\left(x\frac{dy}{dx}-y\right)=\frac{a}{x(a+bx)}$$

$$\Rightarrow \quad x\frac{dy}{dx}-y=\frac{ax}{a+bx} \qquad \ldots (1)$$

Again differentiating w.r.t. x, we get

$$x\frac{d^2y}{dx^2}+\frac{dy}{dx}-\frac{dy}{dx}=a\left(\frac{(a+bx)\cdot 1-x\cdot b}{(a+bx)^2}\right)$$

$$=\frac{a^2}{(a+bx)^2}$$

$$\Rightarrow \quad x\frac{d^2y}{dx^2}=\frac{a^2}{(a+bx)^2}$$

$$\Rightarrow \quad x^3\frac{d^2y}{dx^2}=\frac{a^2x^2}{(a+bx)^2}$$

$$\Rightarrow \quad x^3\frac{d^2y}{dx^2}=\left(\frac{ax}{a+bx}\right)^2$$

$$\Rightarrow \quad x^3\frac{d^2y}{dx^2}=\left(x\frac{dy}{dx}-y\right)^2$$

[From (1)]

Hence Proved.

Example 8. *If* $y=\sin(m\sin^{-1}x)$ *then prove that*

$$(1-x^2)\frac{d^2y}{dx^2}-x\frac{dy}{dx}+m^2y=0.$$

Solution : $$y=\sin(m\sin^{-1}x) \qquad \ldots (1)$$

Differentiating (1) with respect to x, we get

$$\frac{dy}{dx}=\cos(m\sin^{-1}x)\frac{m}{\sqrt{1-x^2}}$$

$$\Rightarrow \quad \sqrt{1-x^2}\frac{dy}{dx}=m\cos(m\sin^{-1}x).$$

Squaring both sides, we get

$$(1-x^2)\left(\frac{dy}{dx}\right)^2=m^2\cos^2(m\sin^{-1}x)$$

$$\Rightarrow \quad (1-x^2)\left(\frac{dy}{dx}\right)^2=m^2[1-\sin^2(m\sin^{-1}x)]$$

$$\Rightarrow \quad (1-x^2)\left(\frac{dy}{dx}\right)^2=m^2[1-y^2].$$

Again differentiating w.r.t. x, we get

$$(1-x^2)\,2\frac{dy}{dx}\frac{d^2y}{dx^2}-2x\left(\frac{dy}{dx}\right)^2 = -2m^2y\frac{dy}{dx}$$

Dividing by $2\frac{dy}{dx}$ throughout,

$$(1-x^2)\frac{d^2y}{dx^2}-x\frac{dy}{dx}+m^2y=0.$$

Hence Proved.

Example 9. *If* $y=A(x+\sqrt{x^2-1})^n+B(x-\sqrt{x^2-1})^n$, *then prove that*

$$(x^2-1)\,y_2+xy_1=n^2y.$$

Solution : $y = A(x+\sqrt{x^2-1})^n+B\,(x-\sqrt{x^2-1})^n.$

Differentiating both sides with respect to x, we get

$$y_1 = A\cdot n\,(x+\sqrt{x^2-1})^{n-1}\left\{1+\frac{1}{2}\cdot\frac{2x}{\sqrt{x^2-1}}\right\} + B\cdot n\,(x-\sqrt{x^2-1})^{n-1}\left\{1-\frac{1}{2}\cdot\frac{2x}{\sqrt{x^2-1}}\right\}$$

$$= \frac{n}{\sqrt{x^2-1}}\,[A(x+\sqrt{x^2-1})^n-B\,(x-\sqrt{x^2-1})^n]$$

$$\Rightarrow y_1\sqrt{x^2-1} = n\,[A\,(x+\sqrt{x^2-1})^n-B\,(x-\sqrt{x^2-1})^n].$$

Squaring both sides, we get

$$y_1^2\,(x^2-1) = n^2\,[\{A\,(x+\sqrt{x^2-1})^n+B\,(x-\sqrt{x^2-1})^n\}^2 - 4\,AB\,(x+\sqrt{x^2-1})^n\,(x-\sqrt{x^2-1})^n]$$

$$\left[\because (\alpha-\beta)^2=(\alpha+\beta)^2-4\alpha\beta\right]$$

$$y_1^2\,(x^2-1) = n^2\,[y^2-4AB].$$

Again differentiating w.r.t. x, we get

$$2y_1y_2(x^2-1)+2xy_1^2=2n^2yy_1$$

Cancelling $2y_1$ throughout, we get

$$(x^2-1)\,y_2+xy_1=n^2y.$$

Hence Proved.

Example 10. *If* $x=f(t),\ y=F(t)$, *prove that*

$$\frac{d^2y}{dx^2}=\frac{x'y''-y'x''}{x'^3},$$

where, $x'=\frac{dx}{dt}$ and $y''=\frac{d^2y}{dt^2}$ *etc.*

Solution : $$\frac{dy}{dx} = \frac{dy/dt}{dx/dt} = \frac{y'}{x'} \qquad \text{... (1)}$$

$$\therefore \quad \frac{d^2y}{dx^2} = \frac{d}{dx}\left(\frac{dy}{dx}\right)$$

$$= \frac{d}{dx}\left(\frac{y'}{x'}\right) \qquad \text{[From (1)]}$$

$$= \frac{d}{dt}\left(\frac{y'}{x'}\right)\frac{dt}{dx}$$

$$= \frac{\frac{d}{dt}\left(\frac{y'}{x'}\right)}{dx/dt}$$

$$= \frac{\frac{x'y'' - y'x''}{x'^2}}{x'}$$

$$= \frac{x'y'' - y'x''}{x'^3}.$$

Hence Proved.

Example 11. *If* $x = q - \sin q$, $y = 1 - \cos q$, *then prove that* $\frac{d^2y}{dx^2}$ *at* $(\pi, 2)$ *is* $\frac{-1}{4}$.

Solution : $$x = \theta - \sin\theta \Rightarrow \frac{dx}{d\theta} = 1 - \cos\theta$$

$$y = 1 - \cos\theta \Rightarrow \frac{dy}{d\theta} = \sin\theta$$

$$\therefore \quad \frac{dy}{dx} = \frac{\frac{dy}{d\theta}}{\frac{dx}{d\theta}} = \frac{\sin\theta}{1-\cos\theta} = \frac{2\sin\frac{\theta}{2}\cos\frac{\theta}{2}}{2\sin^2\frac{\theta}{2}} = \cot\frac{\theta}{2}$$

$$\frac{d^2y}{dx^2} = \frac{d}{dx}\left(\cot\frac{\theta}{2}\right) = \frac{d}{d\theta}\left(\cot\frac{\theta}{2}\right)\frac{d\theta}{dx}$$

$$= -\text{cosec}^2\frac{\theta}{2}\cdot\frac{1}{2}\times\frac{1}{1-\cos\theta}$$

$$= \frac{-1}{2}\text{cosec}^2\frac{\theta}{2}\cdot\frac{1}{2\sin^2\frac{\theta}{2}}$$

$$= \frac{-1}{4}\text{cosec}^4\frac{\theta}{2} \qquad \text{... (1)}$$

We are to find $\frac{d^2y}{dx^2}$ at $(\pi, 2)$, that is $x = \pi, y = 2$.

$$\therefore \quad y = 1 - \cos\theta \Rightarrow 2 = 2\sin^2\frac{\theta}{2} \Rightarrow \sin\frac{\theta}{2} = 1$$

$$\Rightarrow \quad \frac{\theta}{2} = \frac{\pi}{2} \Rightarrow \theta = \pi$$

and $\quad x = \theta - \sin\theta, \quad$ when $\theta = \pi$

$$x = \pi - \sin\pi = \pi$$

Thus, we get $\quad \theta = \pi$ for point $(\pi, 2)$, therefore

$$\frac{d^2y}{dx^2} \text{ at } (\pi, 2), = \frac{1}{4}\ \text{cosec}^4\ \frac{\pi}{2} = -\frac{1}{4}$$

Example 12. *If* $p^2 = a^2\cos^2\theta + b^2\sin^2\theta$, *prove that*

$$p + \frac{d^2p}{d\theta^2} = \frac{a^2b^2}{p^3}.$$

Solution : $\quad p^2 = a^2\cos^2\theta + b^2\sin^2\theta \qquad \text{... (1)}$

Differentiating w.r.t. θ, we get

$$2p\frac{dp}{d\theta} = -2a^2\sin\theta\cos\theta + 2b^2\sin\theta\cos\theta$$

$$2p\frac{dp}{d\theta} = (b^2 - a^2)\sin 2\theta \qquad \text{... (2)}$$

Again differentiating (2) w.r.t. θ,

$$2p\frac{d^2p}{d\theta^2} + 2\left(\frac{dp}{d\theta}\right)^2 = 2(b^2 - a^2)\cos 2\theta$$

$$\Rightarrow p\frac{d^2p}{d\theta^2} + \left(\frac{dp}{d\theta}\right)^2 = (b^2 - a^2)\cos 2\theta \qquad \text{... (3)}$$

Multiplying (3) by p^2, we get

$$p^3\frac{d^2p}{d\theta^2} + p^2\left(\frac{dp}{d\theta}\right)^2 = p^2(b^2 - a^2)\cos 2\theta$$

$$\Rightarrow \quad p^2\frac{d^2p}{d\theta^2} = p^2(b^2 - a^2)\cos 2\theta - \left(p\frac{dp}{d\theta}\right)^2$$

$$= p^2(b^2 - a^2)(\cos^2\theta - \sin^2\theta) - \frac{(b^2 - a^2)^2}{4}\sin^2 2\theta$$

[Using (2)]

$$= p^2(b^2\cos^2\theta + a^2\sin^2\theta) - p^2(a^2\cos^2\theta + b^2\sin^2\theta)$$
$$- (a^4 + b^4 - 2a^2b^2)\sin^2\theta\cos^2\theta$$

$$= (a^2 \cos^2 \theta + b^2 \sin^2 \theta)(b^2 \cos^2 \theta + a^2 \sin^2 \theta)$$
$$- p^4 - (a^4 + b^4 - 2a^2b^2) \sin^2 \theta \cos^2 \theta$$
[Using (1)]

$$\Rightarrow \quad p^4 + p^3 \frac{d^2p}{d\theta^2} = a^2b^2(\cos^4 \theta + \sin^4 \theta + 2 \sin^2\theta \cos^2\theta)$$
$$= a^2b^2 (\cos^2 \theta + \sin^2 \theta)^2$$
$$= a^2b^2$$
$$\Rightarrow \quad p + \frac{d^2p}{d\theta^2} = \frac{a^2b^2}{p^3}.$$
Hence Proved.

EXERCISE 1 (A)

1. If $y = e^{ax} \sin bx$, show that $y_2 - 2ay_1 + (a^2 + b^2) y = 0$.
2. If $y = \sin (\sin x)$, show that $y_2 + y_1 \tan x + y \cos^2 x = 0$.
3. If $y = x^3 \log x$, show that $y_4 = \dfrac{6}{x}$.
4. If $y = e^{-x} \cos x$, show that $y_4 + 4y = 0$.
5. If $y = ae^{mx} + be^{-mx}$, show that $y_2 = m^2y$.
6. If $y = \tan x + \sec x$, show that $\dfrac{d^2y}{dx^2} = \dfrac{\cos x}{(1 - \sin x)^2}$.
7. If $y^3 - 3ax^2 + x^3 = 0$, show that $\dfrac{d^2y}{dx^2} + \dfrac{2a^2x^2}{y^5} = 0$.
8. If $x^3 + y^3 - 3axy = 0$, show that $\dfrac{d^2y}{dx^2} = \dfrac{2a^3xy}{(ax - y^2)^3}$.
9. If $y = \dfrac{\log x}{-x}$, show that $y_2 = \dfrac{3 - 2 \log x}{x^3}$.
10. If $y = x + \cot x$, show that $\sin^2 x \dfrac{d^2y}{dx^2} - 2y + 2x = 0$.
11. If $y = a \sin (\log x)$, show that $x^2y_2 + xy_1 + y = 0$.
12. If $x = 2 \cos t - \cos 2t$ and $y = 2 \sin t - \sin 2t$, show that the value of $\dfrac{d^2y}{dx^2}$ at $t = \dfrac{\pi}{2}$ is $-\dfrac{3}{2}$.

13. If $ax^2 + 2hxy + by^2 = 1$, show that $\dfrac{d^2y}{dx^2} = \dfrac{h^2 - ab}{(hx + by)^3}$.

1.2. Standard Results for Calculation of *n*th Derivatives

In this section we shall find the *n*th derivatives in some simple cases.

1. n^{th} derivative of x^n

Suppose $\quad y = x^n$,

therefore $\quad y_1 = nx^{n-1}$

$$y_2 = n(n-1)\,x^{n-2}$$

$$\ldots = \ldots \quad \ldots$$

$$y_r = n(n-1)\ldots(n-(r-1))\,x^{n-r},\ r < n$$

$$\ldots = \ldots \quad \ldots$$

Continuing this process *n* times, we get

$$\mathbf{y_n = n\,(n-1)(n-2)\ldots 3, 2, 1.}$$

i.e., $\quad D^n(x^n) = n\,!.$

COROLLARY 1. $y_{n+1} = D^{n+1}(x^n) = y_{n+2} = D^{n+2}(x^n) = \ldots = 0.$

i.e., if $m > n$, then

$$D^m(x^n) = 0.$$

COROLLARY 2. If $\quad y = x^n$,

then

$$y_{n-1} = n\,!\,\frac{x}{1\,!}$$

$$y_{n-2} = n\,!\,\frac{x^2}{2\,!}.$$

2. n^{th} derivative of $(ax + b)^m$

Let $\quad y = (ax + b)^m$,

then $\quad y_1 = m\,(ax + b)^{m-1}\,.\,a.$

$$y_2 = m\,(m-1)\,(ax+b)^{m-2}\,.\,a^2$$

$$y_3 = m\,(m-1)\,(m-2)\,(ax+b)^{m-3}\,.\,a^3$$

....

....

$$\therefore \quad \mathbf{y_n = m\,(m-1)\,(m-2)\ldots(m-n+1)\ (ax+b)^{m-n}\,.\,a^n}$$

COROLLARY 3. If *m* is a +ve integer and $m = n$, then

$$y_n = n\,(n-1)\,(n-2)\ldots(n-n+1)\ (ax+b)^{m-n}\ .\,a^n$$

$$\Rightarrow \quad \mathbf{y_n = n!a^n \text{ i.e., } D^n(ax+b)^n = n\,!\,a^n}$$

i.e., $D^n(ax+b)^m = m(m-1)(m-2)\ldots(m-n+1)(ax+b)^{m-n}$ where $m > n$

COROLLARY 4. If m is a +ve integer and $m < n$, then evidently $y_n = 0$.

3. n^{th} derivative of e^{ax}

Let $\quad y = e^{ax}$,

then
$$y_1 = ae^{ax}$$
$$y_2 = a^2e^{ax}$$
$$y_3 = a^3e^{ax}$$
$$\ldots\ldots \quad \ldots\ldots$$
$$\ldots\ldots \quad \ldots\ldots$$

$\therefore \quad y_n = a^ne^{ax}$. i.e., $D^n(e^{ax}) = a^ne^{ax}$.

COROLLARY 5. If $a = 1$, then $y_n = e^x$.

4. n^{th} derivative of a^x

If $\quad y = a^x$,

then
$$y = e^{\log a^x}$$
$$= e^{x \log a}$$

and
$$y_n = D^n(a^x)$$
$$= D^n(e^{x \log a})$$
$$= e^{x \log a} \cdot (\log a)^n$$
$$= (\log a)^n a^x.$$

i.e., $\quad D^n a^x = (\log a)^n a^x$.

5. n^{th} derivative of $(ax + b)^{-1}$

Let $\quad y = (ax + b)^{-1}$,

then
$$y_1 = (-1)\, a(ax + b)^{-2}$$
$$y_2 = (-1)(-2)\, a^2 (ax + b)^{-3}$$
$$= (-1)^2\, 1 \cdot 2a^2 (ax + b)^{-3}$$
$$y_3 = (-1)(-2)(-3)a^3 (ax + b)^{-3}$$
$$= (-1)^3\, 1 \cdot 2 \cdot 3\, a^3 (ax + b)^{-3}$$
$$= \frac{(-1)^3\, 3!\, a^3}{(ax + b)^3}$$
$$\ldots = \ldots$$
$$y_n = \frac{(-1)^n\, n!\, a^n}{(ax + b)^{n+1}}$$

i.e., $$D^n (ax+b)^{-1} = \frac{(-1)^n n!\, a^n}{(ax+b)^{n+1}}.$$

6. n^{th} derivative of log $(ax + b)$

Let $$y = \log (ax+b),$$

then $$y_1 = \frac{a}{ax+b}$$

$$= a(ax+b)^{-1}$$

$\therefore$ $$y_n = D^{n-1} y_1$$

$$= D^{n-1} a(ax+b)^{-1}.$$

$$= \frac{a\, a^{n-1} (-1)^{n-1} (n-1)!}{(ax+b)^n}$$ [By 5]

i.e., $$y_n = D^n y = \frac{a^n (-1)^{n-1} (n-1)!}{(ax+b)^n}$$

7. n^{th} derivative of sin $(ax + b)$

Let $$y = \sin (ax+b),$$

then $$y_1 = a\cos(ax+b)$$

$$= a\sin\left(ax+b+\frac{\pi}{2}\right)$$

$$y_2 = a^2\cos\left(ax+b+\frac{\pi}{2}\right)$$

$$= a^2\sin\left(ax+b+\frac{\pi}{2}+\frac{\pi}{2}\right)$$

$$= a^2\sin\left(ax+b+2\frac{\pi}{2}\right)$$

$$y_3 = a^3\cos\left(ax+b+\frac{2\pi}{2}\right)$$

$$= a^3\sin\left(ax+b+\frac{2\pi}{2}+\frac{\pi}{2}\right)$$

$$= a^3\sin\left(ax+b+\frac{3\pi}{2}\right)$$

....

....

$\therefore$ $$y_n = a^n\sin\left(ax+b+\frac{n\pi}{2}\right)$$

i.e., $$D^n \sin(ax+b) = a^n \sin\left(ax+b+\frac{n\pi}{2}\right)$$

COROLLARY : If $a = 1$ and $b = 0$, then

$$y = \sin x$$

$$\therefore \quad y^n = \sin\left(x+\frac{n\pi}{2}\right).$$

8. n^{th} derivative of cos $(ax + b)$

Let $y = \cos(ax+b)$,

then $y_1 = -a\sin(ax+b)$

$$\Rightarrow \quad y_1 = a\cos\left(ax+b+\frac{\pi}{2}\right)$$

$$\therefore \quad y_2 = -a^2\sin\left(ax+b+\frac{\pi}{2}\right)$$

$$= a^2\cos\left(ax+b+\frac{\pi}{2}+\frac{\pi}{2}\right)$$

$$= a^2\cos\left(ax+b+\frac{2\pi}{2}\right)$$

$$y_3 = -a^3\sin\left(ax+b+\frac{2\pi}{2}\right)$$

$$= a^3\cos\left(ax+b+\frac{2\pi}{2}+\frac{\pi}{2}\right)$$

$$= a^3\cos\left(ax+b+\frac{3\pi}{2}\right)$$

....

....

$$\therefore \quad y_n = a^n\cos\left(ax+b+\frac{n\pi}{2}\right)$$

i.e., $$D^n \cos(ax+b) = a^n \cos\left(ax+b+\frac{n\pi}{2}\right).$$

COROLLARY : If $a = 1$ and $b = 0$, then

$$y = \cos x$$

$$\therefore \quad y_n = \cos\left(x+\frac{n\pi}{2}\right)$$

9. n^{th} derivative of $e^{ax} \sin (bx + c)$

Let
$$y = e^{ax} \sin (bx + c)$$
$$\quad \text{I} \qquad \text{II}$$

$\therefore$
$$y_1 = e^{ax} \cos (bx + c) \cdot b + ae^{ax} \sin (bx + c)$$
$$= e^{ax} [a \sin (bx + c) + b \cos (bx + c)]$$

Put $a = r \cos \phi$ and $b = r \sin \phi$.

Squaring and adding,
$$r^2 = a^2 + b^2 \Rightarrow r = (a^2 + b^2)^{1/2}$$

Dividing,
$$\tan \phi = \frac{b}{a} \Rightarrow \phi = \tan^{-1}\left(\frac{b}{a}\right)$$

$\therefore$
$$y_1 = e^{ax} [r \cos \phi \sin (bx + c) + r \sin \phi \cos (bx + c)]$$
$$= re^{ax} \sin (bx + c + \phi).$$

Similarly,
$$y_2 = r^2 e^{ax} \sin (bx + c + 2\phi),$$
$$y_3 = r^3 e^{ax} \sin (bx + c + 3\phi),$$
$$\ldots\ldots \quad \ldots\ldots \ldots\ldots \ldots\ldots \ldots\ldots \ldots\ldots$$
$$\ldots\ldots \quad \ldots\ldots \ldots\ldots \ldots\ldots \ldots\ldots \ldots\ldots$$

$\therefore$
$$y_n = r^n e^{ax} \sin (bx + c + n\phi)$$

$\Rightarrow$
$$y_n = (a^2 + b^2)^{n/2} e^{ax} \sin\left(bx + c + n \tan^{-1} \frac{b}{a}\right)$$

i.e.,
$$D^n e^{ax} \sin (bx + c) = (a^2 + b^2)^{n/2} e^{ax} \sin\left(bx + c + n \tan^{-1} \frac{b}{a}\right).$$

10. n^{th} derivative of $e^{ax} \cos (bx + c)$

Proceeding exactly as in § 9 of 1.2 above, we can show that if
$$y = e^{ax} \cos (bx + c)$$

then
$$y_n = (a^2 + b^2)^{n/2} e^{ax} \cos\left(bx + c + n \tan^{-1} \frac{b}{a}\right)$$

i.e.,
$$D^n e^{ax} \cos (bx + c) = (a^2 + b^2)^{n/2} e^{ax} \cos\left(bx + c + n \tan^{-1} \frac{b}{a}\right).$$

1.3. Decomposition into Sum or Difference of Standard Forms

Sometimes it is not possible to find the *n*th derivative of the function in its given form. In such cases, we try to decompose them into sum or difference of suitable standard functions and then we find out the *n*th derivative.

1.4. Use of Partial Fractions

The nth differential coefficient of a rational algebraic fraction can be conveniently determined by first resolving them into partial fractions which reduce them to standard forms. Then we can conveniently apply the standard results to find the nth derivative of the given function.

1.5. Use of De-Moivre's Theorem

Sometimes the denominator of the algebraic fraction cannot be resolved into real linear factors. In such cases we resolve the denominator into imaginary factors. The factors which are imaginary are then put into the form $r(\cos\theta + i\sin\theta)$ and De-Moivre's theorem is applied to simplify the result.

1.6. Trigonometrical Transformations

A function of the form $\sin^n x\cos^n x$ can be expressed as sum of sines and cosines of multiple angles by trigonometrical substitutions. For this, we have if

$$z = \cos x + i\sin x, \qquad \frac{1}{z} = \cos x - i\sin x$$

$$z^n = \cos nx + i\sin nx, \qquad \frac{1}{z^n} = \cos nx - i\sin nx$$

so that $\quad z^n + \dfrac{1}{z^n} = 2\cos nx \quad$ and $\quad z^n - \dfrac{1}{z^n} = 2i\sin nx.$

ILLUSTRATIVE EXAMPLES

Example 1. *If* $f(x) = \dfrac{ax+b}{cx+d}$, *then find the value of* $f^n(x)$.

Solution : Here, $f(x) = \dfrac{ax+b}{cx+d}$, then by simple division,

$$f(x) = \frac{a}{c} + \frac{b-(ad/c)}{cx+d}.$$

Therefore, $\quad D^n f(x) = D^n\left(\dfrac{a}{c} + \dfrac{b-(ad/c)}{cx+d}\right)$

i.e., $\quad f^n(x) = (b - ad/c)\dfrac{(-1)^n n!\, c^n}{(cx+d)^{n+1}}$

$$= \frac{bc-ad}{c}\,\frac{(-1)^n n!\, c^n}{(cx+d)^{n+1}}.$$

Example 2. *If* $f(x) = \sin 2x\cos 3x$, *then find the value of* $f^n(x)$.

Solution : Here, $f(x) = \sin 2x\cos 3x$

$$= \frac{1}{2}[\sin 5x - \sin x].$$

Therefore,
$$f^n(x) = D^n f(x)$$
$$= \frac{1}{2}[D^n \sin 5x - D^n \sin x]$$
$$= \frac{1}{2}\left[5^n \sin\left(5x + \frac{n\pi}{2}\right) - \sin\left(x + \frac{n\pi}{2}\right)\right].$$

Example 3. *If* $f(x) = \dfrac{x}{(x-1)(x-2)}$, *then find the value of* $f^n(x)$.

Solution :
$$f(x) = \frac{x}{(x-1)(x-2)}.$$

By partial fractions,
$$\frac{x}{(x-1)(x-2)} = \frac{A}{x-1} + \frac{B}{x-2}$$
$$\Rightarrow \quad x = A(x-2) + B(x-1)$$
$$\Rightarrow \quad x = (A+B)x - 2A - B$$
$$\Rightarrow \quad \left.\begin{array}{r} A + B = 1 \\ -2A - B = 0 \end{array}\right\}$$

Solving,
$$A = -1, B = 2$$
$$\therefore \quad f(x) = \frac{-1}{x-1} + \frac{2}{x-2}$$
$$\Rightarrow \quad f^n(x) = -\frac{(-1)^n n!}{(x-1)^{n+1}} + 2\frac{(-1)^n n!}{(x-2)^{n+1}}$$
$$= (-1)^n n!\left[\frac{2}{(x-2)^{n+1}} - \frac{1}{(x-1)^{n+1}}\right].$$

Example 4. *Find the* n^{th} *derivative of* $\sin^2 x \sin 2x$.

Solution :
$$y = \sin^2 x \sin 2x$$
$$= \frac{1 - \cos 2x}{2}.\ \sin 2x$$
$$= \frac{1}{2}[\sin 2x - \sin 2x \cos 2x]$$
$$= \frac{1}{2}[\sin 2x - \frac{1}{2}\cdot 2 \sin 2x \cos 2x]$$
$$= \frac{1}{2}[\sin 2x - \frac{1}{2}\sin 4x]$$
$$= \frac{1}{2}\sin 2x - \frac{1}{4}\sin 4x$$
$$\Rightarrow \quad y_n = \frac{1}{2}\cdot 2^n \sin\left(2x + \frac{n\pi}{2}\right) - \frac{1}{4}\cdot 4^n \sin\left(4x + \frac{n\pi}{2}\right)$$

$$y_n = 2^{n-1} \sin\left(2x + \frac{n\pi}{2}\right) - 4^{n-1} \sin\left(4x + \frac{n\pi}{2}\right).$$

Example 5. *Find the nth derivative of $e^x \cos^3 x$.*

Solution : Let $y = e^x \cos^3 x.$

Now, $\cos 3x = 4\cos^3 x - 3\cos x$

$\therefore$ $4\cos^3 x = \cos 3x + 3\cos x$

$\Rightarrow$ $\cos^3 x = \dfrac{\cos 3x + 3\cos x}{4}$

$\therefore$
$$y = \frac{1}{4} e^x (\cos 3x + 3\cos x)$$
$$= \frac{1}{4} e^x \cos 3x + \frac{3}{4} e^x \cos x$$

$\therefore$
$$y_n = \frac{1}{4}(1^2 + 3^2)^{n/2} e^x \cos\left(3x + n\tan^{-1}\frac{3}{1}\right) + \frac{3}{4}(1^2 + 1^2)^{n/2} e^x \cos\left(x + n\tan^{-1}\frac{1}{1}\right)$$
$$= \frac{1}{4} \cdot 10^{n/2} e^x \cos(3x + n\tan^{-1} 3) + \frac{3}{4} \cdot 2^{n/2} e^x \cos\left(x + n\frac{\pi}{4}\right)$$
$$= \frac{1}{4} e^x \left[10^{n/2} \cos(3x + n\tan^{-1} 3) + 3 \cdot 2^{n/2} \cos\left(x + \frac{n\pi}{4}\right)\right].$$

Example 6. *Find the nth differential coefficient of* $\dfrac{x^3}{x^2 - 3x + 2}$.

Solution : The given function is improper function, therefore, by actual division and breaking into partial fractions,

$$\frac{x^3}{x^2 - 3x + 2} = x + 3 + \frac{7x - 6}{(x-1)(x-2)} = x + 3 - \frac{1}{x-1} + \frac{8}{x-2}.$$

If $n > 1$, then

$$D^n\left(\frac{x^3}{x^2 - 3x + 2}\right) = D^n(x) + D^n(3) - D^n\left(\frac{1}{x-1}\right) + 8D^n\left(\frac{1}{x-2}\right)$$
$$= 0 + 0 - \frac{(-1)^n n!}{(x-1)^{n+1}} + 8\frac{(-1)^n n!}{(x-2)^{n+1}}$$
$$= (-1)^n n! \left[\frac{8}{(x-2)^{n+1}} - \frac{1}{(x-1)^{n+1}}\right].$$

Example 7. *Find the* n^{th} *differential coefficient of* $\cos x \cos 2x \cos 3x$.

Solution : Let $y = \cos x \cos 2x \cos 3x$

$\Rightarrow$ $$y = \frac{1}{2}[2\cos 2x \cos x]\cos 3x$$

$$= \frac{1}{2}[\cos 3x + \cos x]\cos 3x$$

$[\because 2\cos A \cos B = \cos(A+B) + \cos(A-B)]$

$$= \frac{1}{2}[\cos^2 3x + \cos x \cos 3x]$$

$$= \frac{1}{4}[2\cos^2 3x + 2\cos x \cos 3x]$$

$$= \frac{1}{4}[1 + \cos 6x + \cos 2x + \cos 4x]$$

Now, differentiating each term n times, we get

$$y_n = \frac{1}{4}\left[0 + 6^n \cos\left(6x + \frac{n\pi}{2}\right) + 2^n \cos\left(2x + \frac{n\pi}{2}\right) + 4^n \cos\left(4x + \frac{n\pi}{2}\right)\right]$$

$$= 2^{n-2}\left[3^n \cos\left(6x + \frac{n\pi}{2}\right) + \cos\left(2x + \frac{n\pi}{2}\right) + 2^n \cos\left(4x + \frac{n\pi}{2}\right)\right].$$

Example 8. *If* $y = \sin mx + \cos mx$, *prove that*

$$y_n = m^n[1 + (-1)^n \sin 2mx]^{1/2}.$$

Solution : $y = \sin mx + \cos mx$

$\therefore$ $$y_n = m^n \sin\left(mx + \frac{n\pi}{2}\right) + m^n \cos\left(mx + \frac{n\pi}{2}\right)$$

$$= m^n\left[\left\{\sin\left(mx + \frac{n\pi}{2}\right) + \cos\left(mx + \frac{n\pi}{2}\right)\right\}^2\right]^{1/2}$$

[Squaring and taking square root]

$$= m^n\left[1 + 2\sin\left(mx + \frac{n\pi}{2}\right)\cos\left(mx + \frac{n\pi}{2}\right)\right]^{1/2}$$

$$= m^n [1 + \sin (2mx + n\pi)]^{1/2}$$
$$= m^n [1 + \sin 2mx \cos n\pi + \cos 2mx \sin n\pi]^{1/2}$$
$$= m^n [1 + (-1)^n \sin 2mx]^{1/2}$$
$$[\because \cos n\pi = (-1)^n \text{ and } \sin n\pi = 0]$$

Therefore, $y_n = m^n [1 + (-1)^n \sin 2mx]^{1/2}$.

Example 9. *Find the nth differential coefficient of*

$$\frac{1}{(x-1)^3 (x-2)}.$$

Solution : Let
$$y = \frac{1}{(x-1)^3 (x-2)}$$

To break into partial fractions, let $x - 1 = z \Rightarrow x = 1 + z$.

$$\therefore \quad y = \frac{1}{z^3 (1 + z - 2)}$$
$$= \frac{1}{z^3 (z-1)}$$
$$= -\frac{1}{z^3} \cdot \frac{1}{1-z}$$
$$= -\frac{1}{z^3}\left[1 + z + z^2 + \frac{z^3}{1-z}\right]$$
[On dividing 1 by (1 – z)]
$$= -\frac{1}{z^3} - \frac{1}{z^2} - \frac{1}{z} + \frac{1}{z-1}$$
$$y = -\frac{1}{(x-1)^3} - \frac{1}{(x-1)^2} - \frac{1}{x-1} + \frac{1}{x-2}$$
$$\therefore \quad y_n = \frac{(-1)^n (n+2)!}{2(x-1)^{n+3}} - \frac{(-1)^n (n+1)!}{1.(x-1)^{n+2}}$$
$$= -\frac{(-1)^n n!}{(x-1)^{n+1}} + \frac{(-1)^n n!}{(x-2)^{n+1}}$$
$$= (-1)^{n+1} n! \left[\frac{(n+2)(n+1)}{2(x-1)^{n+3}} + \frac{n}{(x-1)^{n+2}} + \frac{1}{(x-1)^{n+1}} - \frac{1}{(x-2)^{n+1}}\right].$$

Example 10. *If* $y = \dfrac{1}{x^2 + a^2}$, *find* y_n.

Solution :

$$y = \frac{1}{x^2 + a^2}$$

$$= \frac{1}{x^2 - i^2 a^2} \qquad [\because i^2 = -1]$$

$$= \frac{1}{(x - ia)(x + ia)}$$

$$= \frac{1}{2ai}\left[\frac{1}{x - ia} - \frac{1}{x + ia}\right]$$

[Resolving into partial fractions]

$$\therefore \quad y_n = \frac{1}{2ai}\left[\frac{(-1)^n n!}{(x - ia)^{n+1}} - \frac{(-1)^n n!}{(x + ia)^{n+1}}\right]$$

$$= \frac{(-1)^n n!}{2ai}\left[\frac{1}{(x - ia)^{n+1}} - \frac{1}{(x + ia)^{n+1}}\right]$$

Suppose $x = r\cos\theta$, $a = r\sin\theta$.

Then squaring and adding, $x^2 + a^2 = r^2$.

Dividing, we get

$$\tan\theta = \frac{a}{x} \Rightarrow \theta = \tan^{-1}\left(\frac{a}{x}\right)$$

$$\therefore \quad y_n = \frac{(-1)^n n!}{2ai\, r^{n+1}}[(\cos\theta - i\sin\theta)^{(n+1)} - (\cos\theta + i\sin\theta)^{-(n+1)}]$$

$$= \frac{(-1)^n n!}{2ai\, r^{n+1}}[\cos(n+1)\theta + i\sin(n+1)\theta - \cos(n+1)\theta + i\sin(n+1)\theta]$$

$$= \frac{(-1)^n n!}{2ai\, r^{n+1}} 2i\sin(n+1)\theta$$

$$= \frac{(-1)^n n!}{a r^{n+1}}\sin(n+1)\theta$$

$$= \frac{(-1)^n n!}{a}\,\frac{\sin^{n+1}\theta}{a^{n+1}}\sin(n+1)\theta$$

$$\left[\because a = r\sin\theta \Rightarrow r = \frac{a}{\sin\theta}\right]$$

$$= \frac{(-1)^n n!}{a^{n+2}} \sin^{n+1}\theta \sin(n+1)\theta,$$

where, $\theta = \tan^{-1}\left(\frac{a}{x}\right)$.

COROLLARY : If $y = \tan^{-1}\left(\frac{x}{a}\right)$, then

$$y_1 = \frac{1}{1+\frac{x^2}{a^2}} \frac{1}{a} = \frac{a}{x^2+a^2}.$$

$$\therefore \qquad y_n = aD^{n-1}\left(\frac{1}{x^2+a^2}\right)$$

$$= a\,\frac{(-1)^{n-1}(n-1)!\sin^n\theta\sin n\theta}{a^{n+1}}$$

$$= \frac{(-1)^{n-1}(n-1)!\sin^n\theta\sin n\theta}{a^n}, \text{ where, } \tan\theta = \frac{a}{x} = \cot y.$$

Example 11. *Find the nth differential coefficient of* $\cos^4 x$.

Solution : From Trigonometry,

$$\cos^4 x = \frac{1}{4}(2\cos^2 x)^2$$

$$= \frac{1}{4}(1+\cos 2x)^2$$

$$= \frac{1}{4}(1+2\cos 2x+\cos^2 2x)$$

$$= \frac{1}{8}(2+4\cos 2x+2\cos^2 2x)$$

$$= \frac{1}{8}(2+4\cos 2x+1+\cos 4x)$$

$$= \frac{1}{8}(3+4\cos 2x+\cos 4x)$$

$$\therefore \qquad D^n(\cos^4 x) = \frac{1}{8}D^n[3+4\cos 2x+\cos 4x]$$

$$= \frac{1}{8}\left[0+4\cdot 2^n\cos\left(2x+\frac{n\pi}{2}\right)+4^n\cos\left(4x+\frac{n\pi}{2}\right)\right]$$

$$= \frac{1}{8}\left[4\cdot 2^n\cos\left(2x+\frac{n\pi}{2}\right)+4^n\cos\left(4x+\frac{n\pi}{2}\right)\right].$$

Example 12. *Find the* n^{th} *derivative of* $\cos^2 x \sin^3 x$.

Solution : Let $\quad z = \cos x + i \sin x$, then $\quad z^{-1} = \cos x - i \sin x$

$\therefore \quad 2 \cos x = z + z^{-1} \quad$ and $\quad 2i \sin x = z - z^{-1}$.

Also, $\quad z^r = \cos rx + i \sin rx \quad$ and $\quad z^{-r} = \cos rx - i \sin rx$

$z^r + z^{-r} = 2 \cos rx \quad$ and $\quad z^r - z^{-r} = 2i \sin rx$

$$\therefore \quad 2^2 \cdot 2^3 \cdot i^3 \cos^2 x \sin^3 x = (z + z^{-1})^2 (z - z^{-1})^3$$

$$= (z^5 - z^{-5}) - (z^3 - z^{-2}) - 2(z - z^{-1})$$

$$= 2i \sin 5x - 2i \sin 3x - 4i \sin x$$

$$\therefore \quad -16 \cos^2 x \sin^3 x = \sin 5x - \sin 3x - 2 \sin x$$

$$\Rightarrow \quad \cos^2 x \sin^3 x = \frac{1}{16}(-\sin 5x + \sin 3x + 2 \sin x)$$

$$\therefore \quad D^n (\cos^2 x \sin^3 x) = \frac{1}{16}\left[-5^n \sin\left(5x + \frac{n\pi}{2}\right) + 3^n \sin\left(3x + \frac{n\pi}{2}\right) + 2 \sin\left(x + \frac{n\pi}{2}\right)\right].$$

Example 13. *If* $y = \tan^{-1} \dfrac{1+x}{1-x}$, *find* y_n.

Solution :

$$y = \tan^{-1}\left(\frac{1+x}{1-x}\right)$$

Put $x = \tan \theta$ so that $\theta = \tan^{-1} x$.

Then

$$y = \tan^{-1}\left(\frac{1+\tan\theta}{1-\tan\theta}\right)$$

$$= \tan^{-1}\left(\frac{\tan\frac{\pi}{4} + \tan\theta}{1 - \tan\frac{\pi}{4}\tan\theta}\right)$$

$$= \tan^{-1} \tan\left(\frac{\pi}{4} + \theta\right)$$

$$= \frac{\pi}{4} + \theta = \frac{\pi}{4} + \tan^{-1} x$$

$$\therefore \quad y_n = (-1)^n (n-1)! \sin^n \theta \sin n\theta, \text{ where } \tan\theta = \frac{1}{x}.$$

EXERCISE 1 (B)

Find the nth differential coefficient of the following functions :

1. $\tan^{-1} x$
2. $(ax+b)^{p/q}$
3. $\log (ax+b)^p$
4. $\log \{(ax+b)(cx+d)\}$
5. $\dfrac{x}{c+dx}$
6. $\dfrac{a+bx}{c+dx}$
7. $\sin^2 x$
8. $\cos^2 x$
9. $\sin^3 x$
10. $\cos^3 x$
11. $\sin^4 x$
12. $\dfrac{1}{2}\log\left(\dfrac{x+a}{x-a}\right)$
13. $\sin x \sin 3x$
14. $\sin x \cos 2x$
15. $\cos x \cos 2x$
16. $\sin x \sin 2x \sin 3x$
17. $\sin^4 x \cos^3 x$
18. $e^{ax}\sin^3 bx$
19. $e^{ax}\cos^3 bx$
20. $\dfrac{1}{x^2+5x+6}+\sin 2x \sin 3x$
21. $\dfrac{1}{x^2+x+1}$
22. $\dfrac{1}{a^2-x^2}$
23. $\dfrac{1}{x^4-a^4}$
24. $\tan^{-1}\dfrac{2x}{1-x^2}$ **[Hint : Put $x=\tan\theta$]**
25. $\dfrac{x}{x^2+a^2}$
26. If $y=(x^2-1)^m$, prove that $y_{2m}=2m\,!$.

ANSWERS

1. $(-1)^{n-1}(n-1)!\sin^n\theta\sin n\theta$, where $\theta=\cot^{-1}x$.
2. $\dfrac{p}{q}\left(\dfrac{p}{q}-1\right)\left(\dfrac{p}{q}-2\right)\cdots\left(\dfrac{p}{q}-n+1\right)a^n(ax+b)^{p/q-n}$
3. $\dfrac{p(-1)^{n-1}(n-1)!\,a^n}{(ax+b)^n}$.
4. $\dfrac{(-1)^{n-1}(n-1)!\,a^n}{(ax+b)^n}+\dfrac{(-1)^{n-1}(n-1)!\,c^n}{(cx+d)^n}$
5. $\dfrac{(-1)^{n+1}n!\,cd^{n-1}}{(c+dx)^{n+1}}$
6. $\dfrac{(-1)^n n!\,d^{n-1}(ad-bc)}{(c+dx)^{n+1}}$
7. $-2^{n-1}\cos\left(2x+n\dfrac{\pi}{2}\right)$
8. $2^{n-1}\cos\left(2x+n\dfrac{\pi}{2}\right)$

9. $\frac{3}{4}\sin\left(x+\frac{n\pi}{2}\right)-\frac{1}{4}\,.\,3^n\sin\left(3x+\frac{n\pi}{2}\right).$

10. $\frac{1}{4}3^n\cos\left(3x+\frac{n\pi}{2}\right)+\frac{3}{4}\cos\left(x+\frac{n\pi}{2}\right).$

11. $-2^{n-1}\cos\left(2x+\frac{n\pi}{2}\right)+2^{2n-3}\cos\left(4x+\frac{n\pi}{2}\right).$

12. $\frac{1}{2}\left[\frac{(-1)^{n-1}(n-1)!}{(x+a)^n}-\frac{(-1)^{n-1}(n-1)!}{(x-a)^n}\right].$

13. $\frac{1}{2}\left[2^n\cos\left(2x+\frac{n\pi}{2}\right)-4^n\cos\left(4x+\frac{n\pi}{2}\right)\right].$

14. $\frac{1}{2}\left[3^n\sin\left(3x+\frac{n\pi}{2}\right)-\sin\left(x+\frac{n\pi}{2}\right)\right].$

15. $\frac{1}{2}\left[3^n\cos\left(3x+\frac{n\pi}{2}\right)+\cos\left(x+\frac{n\pi}{2}\right)\right].$

16. $\frac{1}{4}\left[2^n\sin\left(2x+\frac{n\pi}{2}\right)+4^n\sin\left(4x+\frac{n\pi}{2}\right)-6^n\sin\left(6x+\frac{n\pi}{2}\right)\right].$

17. $\frac{1}{64}\left[7^n\cos\left(7x+\frac{n\pi}{2}\right)-5^n\cos\left(5x+\frac{n\pi}{2}\right)-3^{n+1}\cos\left(3x+\frac{n\pi}{2}\right)\right.$
$\left.+3\cos\left(x+\frac{n\pi}{2}\right)\right].$

18. $\frac{2}{3}(a^2+b^2)^{n/2}e^{ax}\sin\left(bx+n\tan^{-1}\frac{b}{a}\right)$
$-\frac{1}{4}(a^2+9b^2)^{n/2}e^{ax}\sin\left(3bx+n\tan^{-1}\frac{3b}{a}\right).$

19. $\frac{1}{4}(a^2+9b^2)^{n/2}e^{ax}\cos\left\{3bx+n\tan^{-1}\frac{3b}{a}\right\}$
$+\frac{3}{4}(a^2+b^2)^{n/2}e^{ax}\cos\left\{bx+n\tan^{-1}\frac{b}{a}\right\}.$

20. $(-1)^n n!\,[(x+2)^{-n-1}-(x+3)^{-n-1}]$
$+\frac{1}{2}\left[\cos\left(x+\frac{n\pi}{2}\right)-5^n\cos\left(5x+\frac{n\pi}{2}\right)\right].$

21. $\frac{(-1)^n n!}{(\sqrt{3}/2)^{n+2}}\sin\,(n+1)\,\theta\,\sin^{n+1}\theta$, where, $\theta=\tan^{-1}\left(\frac{\sqrt{3}}{2x+1}\right).$

22. $(-1)^n n! \left(\frac{1}{2a}\right) [(x+a)^{-n-1} - (x-a)^{-n-1}]$

23. $\frac{(-1)^n n!}{4a^3}\left[\frac{1}{(x-a)^{n+1}} - \frac{1}{(x+a)^{n+1} - 2a^{n+1}} \sin^{n+1}\theta \sin(n+1)\theta\right]$,

where, $\theta = \tan^{-1}\left(\frac{a}{x}\right)$.

24. $2(-1)^{n-1}(n-1)! \sin^n\theta \sin n\theta$, where, $\theta = \tan^{-1}\left(\frac{1}{x}\right)$.

25. $\frac{(-1)^n n!}{a^{n+1}} \sin^{n+1}\theta \cos(n+1)\theta$, where, $\theta = \tan^{-1}\left(\frac{a}{x}\right)$.

1.7. Leibnitz's Theorem

Leibnitz's Theorem help us to find the nth derivative of the product of two functions.

STATEMENT : ***If $y = uv$, where u and v are any functions of x, then***

$$y_n = u_n v + {}^nC_1 u_{n-1} v_1 + {}^nC_2 u_{n-2} v_2 + \dots + {}^nC_r u_{n-r} v_r + \dots + uv_n,$$

where, suffixes of u and v denote the number of times they are differentiated.

PROOF : We shall prove the theorem by Mathematical Induction.

$\because \quad y = uv$

$\therefore$ By actual differentiation,

$$y_1 = u_1 v + uv_1.$$

Hence, the theorem is true for $n = 1$.

$$y_2 = u_2 v + u_1 v_1 + u_1 v_1 + uv_2$$

$$= u_2 v + 2u_1 v_1 + uv_2.$$

Hence, the theorem is true for $n = 2$.

Let us suppose that the theorem is true for $n = m$. Then we have

$$y_m = u_m v + {}^mC_1 u_{m-1} v_1 + {}^mC_2 u_{m-2} v_2 + \dots + {}^mC_r u_{m-r} v_r$$
$$+ {}^mC_{r+1} u_{m-r-1} v_{r+1} + \dots + uv_m$$

Differentiating both sides, we get

$$y_{m+1} = u_{m+1} v + u_m v_1 + {}^mC_1 [u_m v_1 + u_{m-1} v_2]$$
$$+ {}^mC_3 [u_{m-1} v_2 + u_{m-2} v_3] + ... + {}^mC_r [u_{m-r+1} v_r + u_{m-r} v_{r+1}]$$
$$+ {}^mC_{r+1} [u_{m-r} v_{r+1} + u_{m-r-1} v_{r+2}] + ... + u_1 v_m + u v_{m+1}$$

$$= u_{m+1} v + ({}^mC_0 + {}^mC_1) u_m v_1 + ({}^mC_1 + {}^mC_2) u_{m-1} v_2 + ...$$
$$+ ({}^mC_r + {}^mC_{r+1}) u_{m-r} v_{r+1} + ... + u v_{m+1} \qquad [\because {}^mC_0 = 1]$$

$$= u_{m+1} v + {}^{m+1}C_1 u_m v_1 + {}^{m+1}C_2 u_{m-1} v_2$$
$$+ .... + {}^{m+1}C_{r+1} u_{m-r} v_{r+1} + ... + u v_{m+1}$$
$$[\because {}^mC_r + {}^mC_{r+1} = {}^{m+1}C_{r+1}]$$

Hence, the theorem is true for $n = m + 1$, if it is true for $n = m$.

But, we have already proved that the theorem is true for $n = 1, 2$. Hence, it is true for 3, 4 ... and so on.

Thus, the theorem is true for every positive integral value of n.

Note : If x^m, where, m is a positive integer, is one of the factors taking $v = x^m$ simplifies the process of writing the nth derivative.

ILLUSTRATIVE EXAMPLES

Example 1. *Find the nth derivative of $x^3 \cos x$.*

Solution : Let

$$u = \cos x \text{ and } \qquad v = x^3, \text{ then}$$
$$u_1 = -\sin x, \qquad v_1 = 3x^2$$
$$u_2 = -\cos x, \qquad v_2 = 6x$$
$$u_3 = \sin x, \qquad v_3 = 6$$
$$u_4 = \cos x, \qquad v_4 = 0$$
$$\ldots \quad \ldots \quad \ldots \quad \ldots \quad \ldots \quad \ldots \quad \ldots \quad \ldots \quad \ldots$$
$$u_n = \cos\left(x + \frac{n\pi}{2}\right).$$

By Leibnitz's Theorem

$$D^n (x^3 \cos x) = u_n + {}^nC_1 u_{n-1} v_1 + {}^nC_2 u_{n-2} v_2 + {}^nC_3 u_{n-3} v_3 + ... + v_n$$
$$= \cos\left(x + \frac{n\pi}{2}\right) + n \cos\left(x + \frac{(n-1)\pi}{2}\right) \cdot 3x^2$$
$$+ \frac{n(n-1)}{2 \cdot 1} \cos\left(x + \frac{(n-2)\pi}{2}\right) \cdot 6x$$

$$+\frac{n(n-1)(n-2)}{3\cdot2\cdot1}\cos\left(x+\frac{(n-3)}{2}\pi\right)6.$$

Example 2. *Find the nth derivative of $x^3 e^{ax}$.*

Solution : Let

$u = e^{ax}$	$v = x^3$
$u_n = a^n e^{ax}$	$v_1 = 3x^2$
$u_{n-1} = a^{n-1} e^{ax}$	$v_2 = 6x$
$u_{n-2} = a^{n-2} e^{ax}$	$v_3 = 6$
$u_{n-3} = a^{n-3} e^{ax}$	$v_4 = 0$

$\therefore$ By Leibnitz's Theorem,

$$\frac{d^n}{dx^n}(x^3 e^{ax}) = u_n v + {}^nC_1 u_{n-1} v_1 + {}^nC_2 u_{n-2} v_2 + {}^nC_3 u_{n-3} v_3$$

$$= a^n e^{ax}\cdot x^3 + n\cdot a^{n-1} e^{ax}\cdot 3x^2 + \frac{n(n-1)}{2!}\cdot a^{n-2} e^{ax}\cdot 6x$$

$$+\frac{n(n-1)(n-2)}{3!} a^{n-3} e^{ax}\cdot 6$$

$$= e^{ax}[a^n x^3 + 3na^{n-1}x^2 + 3n(n-1)a^{n-2}x$$

$$+ n(n-1)(n-2)a^{n-3}].$$

Example 3. *If $y = x^{n-1}\log x$, then prove that*

$$y_n = \frac{(n-1)!}{x}$$

Solution : $y = x^{n-1}\log x$

$$\therefore \quad \frac{dy}{dx} = x^{n-1}\cdot\frac{1}{x} + (n-1)x^{n-2}\log x$$

$$\Rightarrow \quad \frac{dy}{dx} = x^{n-2} + (n-1)x^{n-2}\log x$$

$$\Rightarrow \quad xy_1 = x^{n-1} + (n-1)x^{n-1}\log x$$

$$\Rightarrow \quad xy_1 = x^{n-1} + (n-1)y \qquad \text{...(1)}$$

Differentiate (1), $(n-1)$ times by Leibnitz's Theorem, we get

$$y_n\cdot x + (n-1)y_{n-1}\cdot 1 = (n-1)! + (n-1)y_{n-1}$$

$$\Rightarrow \quad y_n\cdot x = (n-1)!$$

$$\Rightarrow \quad y_n = \frac{(n-1)!}{x}.$$

Example 4. *If* $I_n = \frac{d^n}{dx^n}(x^n \log x)$, *prove that*

$$I_n = n\,I_{n-1} + (n-1)\,!$$

Solution :

$$\begin{aligned} I_n &= \frac{d^n}{dx^n}(x^n \log x) \\ &= \frac{d^{n-1}}{dx^{n-1}}\left[\frac{d}{dx}(x^n \log x)\right] \\ &= \frac{d^{n-1}}{dx^{n-1}}\left[x^n \cdot \frac{1}{x} + nx^{n-1} \log x\right] \\ &= \frac{d^{n-1}}{dx^{n-1}}\left[x^{n-1} + nx^{n-1} \log x\right] \\ &= \frac{d^{n-1}}{dx^{n-1}}(x^{n-1}) + n\frac{d^{n-1}}{dx^{n-1}}(x^{n-1} \log x) \\ &= (n-1)\,! + n\,I_{n-1} \\ &= nI_{n-1} + (n-1)\,!. \end{aligned}$$

Hence Proved.

Example 5. *Differentiate n times the equation*

$$x^2\frac{d^2y}{dx^2} + x\frac{dy}{dx} + y = 0.$$

Solution : Differentiating n times by Leibnitz's Theorem, we have

$$D^n(x^2 y_2) + D^n(xy_1) + D^n(y) = 0.$$

Now, taking each term separately.

$$\begin{aligned} D^n(x^2 \cdot y_2) &= y_{n+2} \cdot x^2 + {}^nC_1 \cdot y_{n+1} \cdot 2x + {}^nC_2 \cdot y_n \cdot 2 \\ D^n(x \cdot y_1) &= y_{n+1} \cdot x + {}^nC_1\, y_n \cdot 1 \\ D^n(y) &= y_n. \end{aligned}$$

Adding, we get

$$0 = x^2 y_{n+2} + (2n+1)\,xy_{n+1} + (n^2+1)\,y_n$$

$$\Rightarrow \quad x^2 y_{n+2} + (2n+1)\,xy_{n+1} + (n^2+1)\,y_n = 0.$$

Example 6. *If* $y = a \cos \log x + b \sin \log x$, *show that*

$$x^2 y_2 + xy_1 + y = 0$$

and that $x^2y_{n+2} + (2n+1)\,xy_{n+1} + (n^2+1)\,y_n = 0.$

Solution : $$y = a\cos\log x + b\sin\log x.$$

Differentiating w.r.t. x,

$$y_1 = -a\sin(\log x)\cdot\frac{1}{x} + b\cos(\log x)\cdot\frac{1}{x}$$

$\therefore$ $$xy_1 = -a\sin(\log x) + b\cos(\log x).$$

Again differentiating w.r.t. x, we get

$$xy_2 + y_1 = -a\cos(\log x)\frac{1}{x} - b\sin(\log x)\frac{1}{x}$$

$\Rightarrow$ $$x^2y_2 + xy_1 = -a\cos\log x - b\sin\log x$$

$\Rightarrow$ $$x^2y_2 + xy_1 = -y$$

$\Rightarrow$ $$x^2y_2 + xy_1 + y = 0.$$

Differentiating this equation n times by Leibnitz's Theorem, we get

$$[x^2y_{n+1} + {}^nC_1\cdot y_{n+1}\cdot 2x + {}^nC_2\cdot y_n\cdot 2] + [y_{n+1}\cdot x + {}^nC_1\cdot y_n\cdot 1] + y_n = 0$$

which on simplification gives $x^2y_{n+2} + (2n+1)xy_{n+1} + (n^2+1)\,y_n = 0.$

Example 7. *If* $\cos^{-1}\left(\frac{y}{b}\right) = \log\left(\frac{x}{n}\right)^n$, *then show that*

$$x^2y_{n+2} + (2n+1)\,xy_{n+1} + 2n^2y_n = 0.$$

Solution : $$\cos^{-1}\left(\frac{y}{b}\right) = \log\left(\frac{x}{n}\right)^n$$

$$= n\log\left(\frac{x}{n}\right)$$

$$\frac{y}{b} = \cos\left(n\log\frac{x}{n}\right)$$

$$y = b\cos\left(n\log\frac{x}{n}\right) \qquad \ldots(1)$$

Differentiating both sides w.r.t. x,

$$y_1 = -\,b\sin\left(n\log\frac{x}{n}\right)\cdot n\cdot\frac{n}{x}\cdot\frac{1}{n}$$

$$xy_1 = -nb\sin\left(n\log\frac{x}{n}\right).$$

Differentiating again w.r.t. x,

$$xy_2 + y_1 = -nb\cos\left(n\log\frac{x}{n}\right)\cdot n\frac{n}{x}\cdot\frac{1}{n}$$

$$\Rightarrow \quad x^2y_2 + y_1x = -n^2b\cos\left(n\log\frac{x}{n}\right)$$

$$\Rightarrow \quad x^2y_2 + y_1x = -n^2y, \qquad \text{[Using (1)]}$$

$$\Rightarrow \quad x^2y_2 + y_1x + n^2y = 0$$

Differentiate it n times more by Leibnitz's theorem,

$$x^2y_{n+2} + n\,.\,y_{n+1}\,2x + \frac{n(n-1)}{1\cdot 2}\cdot y_n^2 + xy_{n+1} + ny_{n+1} + n^2y_n = 0.$$

$$\Rightarrow \quad x^2y_{n+2} + (2n+1)\,xy_{n+1} + (n^2 - n + n + n^2)\,y_n = 0.$$

$$\Rightarrow \quad x^2y_{n+2} + (2n+1)\,xy_{n+1} + 2n^2y_n = 0.$$

Example 8. *If* $y^{1/m} + y^{-1/m} = 2x$, *then prove that*

$$(x^2-1)\,y_{n+2} + (2n+1)\,xy_n + (n^2 - m^2)\,y_n = 0.$$

Solution :

$$y^{1/m} + y^{-1/m} = 2x$$

$$\Rightarrow \quad y^{2/m} + 1 = 2xy^{1/m}$$

$$\Rightarrow \quad y^{2/m} - 2xy^{1/m} + 1 = 0$$

$$\therefore \quad y^{1/m} = \frac{2x \pm \sqrt{4x^2 - 4x}}{2} = (x \pm \sqrt{x^2-1})$$

$$\Rightarrow \quad y = [x \pm \sqrt{x^2-1}]^m \qquad \text{...(1)}$$

Taking positive sign, $\quad y = [x + \sqrt{x^2-1}]^m$.

Differentiating it w.r.t. x,

$$y_1 = m[x + \sqrt{x^2-1}]^{m-1}\left[1 + \frac{1}{2}\cdot\frac{1}{\sqrt{x^2-1}}\cdot 2x\right]$$

$$= m[x + \sqrt{x^2-1}]^{m-1}\left[1 + \frac{x}{\sqrt{x^2-1}}\right]$$

$$= m[x + \sqrt{x^2-1}]^{m-1}\left[\frac{\sqrt{x^2-1} + x}{\sqrt{x^2-1}}\right]$$

$$= m\,[x+\sqrt{x^2-1}\,]^m \frac{1}{\sqrt{x^2-1}}$$

$$\Rightarrow \quad \sqrt{x^2-1}\;y_1 = m\,[x+\sqrt{x^2-1}\,]^m$$

$$\Rightarrow \quad \sqrt{(x^2-1)}\;y_1 = my \qquad \text{[Using (1)]}$$

$$(x^2-1)\,y_1^2 = m^2y^2 \qquad \dots (2)$$

Now, taking negative sign,

$$y = (x-\sqrt{x^2-1}\,)^m$$

$$y_1 = m\,(x-\sqrt{x^2-1}\,)^{m-1}\left[1-\frac{2x}{2\sqrt{x^2-1}}\right]$$

$$= m\,(x-\sqrt{x^2-1}\,)^{m-1}\frac{(\sqrt{x^2-1}-x)}{\sqrt{x^2-1}}$$

$$= \frac{-m\,(x-\sqrt{x^2-1}\,)^m}{\sqrt{x^2-1}}$$

$$\Rightarrow \sqrt{x^2-1}\;y_1 = -m\,(x-\sqrt{x^2-1}\,)^m$$

$$\Rightarrow \sqrt{x^2-1}\;y_1 = -my$$

$$\Rightarrow (x^2-1)\,y_1^2 = m^2y^2.$$

Thus, we get the same value of y_1 for positive and negative signs of (1), thus,

$$(x^2-1)\,y_1^2 = m^2y^2.$$

Differentiate it again w.r.t. x,

$$(x^2-1)\,2y_1\,y_2 + 2xy_1^2 = m^2\,2yy_1$$

$$\Rightarrow \quad (x^2-1)\,y_2 + xy_1 - m^2y = 0.$$

Example 9. *If* $y = x^2 e^x$, *show that*

$$\frac{d^ny}{dx^n} = \frac{n\,(n-1)}{2}\frac{d^2y}{dx^2} - n\,(n-2)\frac{dy}{dx} + \frac{1}{2}(n-1)\,(n-2)y.$$

Solution : By Leibnitz's Theorem, we have

$$\frac{d^ny}{dx^n} = D^n\,(x^2 e^x)$$

$$= x^2 D^{n-1} e^x + {}^nC_1\,D^{n-2} e^x\,Dx^2 + {}^nC_2\,D^{n-3} e^x\,D^2x^2$$

$$[\because D^3x^2 = 0]$$

$$= x^2e^x + ne^x\,2x + \frac{n\,(n-1)}{2\cdot 1}\,e^x\,2\cdot 1$$

$$= e^x[x^2 + 2nx + n(n-1)] \qquad \text{... (1)}$$

Here, $\quad y = x^2 e^x,$

therefore $\quad \dfrac{dy}{dx} = x^2 e^x + e^x\, 2x$

and $\quad \dfrac{d^2y}{dx^2} = x^2 e^x + 2xe^x + e^x \cdot 2 + e^x \cdot 2x$

$$= e^x[x^2 + 4x + 2]$$

$\therefore \quad \dfrac{n(n-1)}{2}\dfrac{d^2y}{dx^2} - n(n-2)\dfrac{dy}{dx} + \dfrac{1}{2}(n-1)(n-2)$

$$\frac{1}{2}n(n-1)\cdot[x^2e^x + 4xe^x + 2e^x] - n\,(n-2)[x^2e^x + 2xe^x] + \frac{1}{2}(n-1)(n-2)\,x^2e^x$$

$$= \left[\frac{1}{2}n(n-1) - n(n-2) + \frac{1}{2}(n-1)(n-2)\right].\,x^2e^x + [2n(n-1) - 2n(n-2)]\cdot xe^x + n(n-1)\,e^x$$

$$= \left[\frac{1}{2}n\,\{(n-1)-(n-2)\} - \frac{1}{2}(n-2)\,\{n-(n-1)\}\right]x^2e^x + 2n\,[(n-1)-(n-2)]\,xe^x + n(n-1)\,e^x$$

$$= \left[\frac{1}{2}n - \frac{1}{2}(n-2)\right]x^2e^x + 2n\cdot xe^x + n(n-1)\,e^x$$

$$= x^2e^x + 2n\,xe^x + n(n-1)\,e^x$$

$$= e^x[x^2 + 2nx + n(n-1)] \qquad \text{... (2)}$$

From (1) and (2), we have

$$\frac{d^ny}{dx^n} = \frac{1}{2}n(n-1)\frac{d^2y}{dx^2} - n(n-2)\frac{dy}{dx} + \frac{1}{2}(n-1)(n-2)\,y.$$

Example 10. *If $x + y = 1$, then prove that*

$$\frac{d^n}{dx^n}(x^n y^n) = n!\,[y^n - ({}^nC_1)^2 y^{n-1}\cdot x + ({}^nC_2)^2 y^{n-2} x^2 + \dots + (-1)^n x^n]$$

Solution : $\quad x + y = 1$

$\Rightarrow \quad y = 1 - x$

$\therefore \quad y_1 = -1 \qquad \text{... (1)}$

Now, let $x^n = u$ and $y_n = v$, then

$$u_n = n\,!, u_{n-1} = n\,!\,x, u_{n-2} = \frac{n\,!}{2} x^2, \ldots$$

and
$$v_1 = n\,y^{n-1} \cdot y_1 = -ny^{n-1} \quad \text{[Using (1)]}$$
$$v_1 = -{}^nC_1\,y^{n-1}$$
$$v_2 = -n\,(n-1)\,y^{n-2}\,y_1 = {}^nC_2\,2\,!\,y^{n-2} \quad \text{[Using (1)]}$$

Now, $\dfrac{d^n}{dx^n}(x^n y^n) = D^n\,(u \cdot v)$

$$= u_n v + {}^nC_1 u_{n-1}\,v_1 + {}^nC_2\,u_{n-2}\,v_2 + \ldots + {}^nC_n\,v_n$$
$$= n\,! \cdot y^n + {}^nC_1\,n\,!\,x\,(-{}^nC_1\,y^{n-1})$$
$$+ {}^nC_2\left(\frac{n\,!}{2\,!}x^2\right){}^nC_2\,2\,! \cdot y^{n-2} + \ldots + x^n\,(-1)^n\,n\,!$$
$$= n\,!\,[y^n - ({}^nC_1)^2\,y^{n-1} + ({}^nC_2)^2\,y^{n-2}\,x^2 + \ldots + (-1)^n\,x^n].$$

EXERCISE 1 (C)

1. Find the fourth differential coefficients of the following functions using Leibnitz's Theorem —

(i) $x^2 e^{ax}$

(ii) $x^3 \log x$

(iii) $x^2 \sin 3x$

(iv) $xe^{ax} \sin bx$

(v) $\dfrac{\log x}{x + a}$.

Find the nth differential coefficients of the following functions using Leibnitz's Theorem—

2. $x^2 \sin x$.

3. $x^3 \cos x$.

4. $x^2 \log x$.

5. $e^x \log x$.

6. $x^2 \tan^{-1} x$.

7. $x^2 e^{3x} \cos 4x$.

8. $x^2\,(ax + b)^m$.

9. $x^2 e^{ax}$.

10. $\sin x \log\,(ax + b)$.

11. $\dfrac{x^n}{1 + x}$.

12. Differentiate the following equation n times —

$$(1 - x^2)\frac{d^2y}{dx^2} - x\frac{dy}{dx} + a^2y = 0.$$

13. If $y = x^2 e^x$, show that

$$\frac{d^3y}{dx^3} = 28\frac{d^2y}{dx^2} - 48\frac{dy}{dx} + 21\,y.$$

14. If $y = \sin^{-1} x$, prove that

$(1 + x^2)\, y_2 - xy_1 = 0.$

Also prove that

$(1 - x^2)\, y_{n+2} - (2n + 1)\, xy_{n+1} - n^2 y_n = 0.$

15. If $y = \dfrac{\sin^{-1} x}{\sqrt{1 - x^2}}$, prove that

$(1 - x^2)\, y_{n+1} - (2n + 1)\, xy_n - n^2 y_{n-1} = 0.$

16. If $y = e^{\tan^{-1} x}$, show that

$(1 + x^2)\, y_2 + (2x - 1)\, y_1 = 0,$

and $(1 + x^2)\, y_{n+2} - \{2(n + 1)\, x - 1\}\, y_{n+1} + n(n + 1)\, y_n = 0.$

17. If $x = \cosh\left(\dfrac{1}{m}\log y\right)$, prove that

$$(x^2 - 1)\, y_2 + xy_1 - m^2 y = 0,$$

and $(x^2 - 1)\, y_{n+2} + (2n + 1)\, xy_{n+1} + (n^2 - m^2)\, y_n = 0.$

18. If $y = (x^2 - 1)^n$, prove that

$(x^2 - 1)\, y_{n+2} + 2xy_{n+1} - n\,(n + 1)\, y_n = 0.$

Hence, if $P_n = \dfrac{d^n}{dx^n}(x^2 - 1)^n$, show that

$$\frac{d}{dx}\left\{(1 - x^2)\frac{dP_n}{dx}\right\} + n\,(n + 1)\, P_n = 0.$$

ANSWERS

1. (i) $a^2 e^{ax}(a^2x^2 + 8ax + 12)$ (ii) $\dfrac{6}{x}$

(iii) $3^3\,(3^3 - 4)\sin 3x - 6^3\, x \cdot \cos 3x$

(iv) $(a^2 + b^2)^2\, xe^{ax} \sin\left\{bx + 4\tan^{-1}\dfrac{b}{a}\right\}$

$+ 4\,(a^2 + b^2)^{3/2}\, e^{ax} \sin\left\{bx + 3\tan^{-1}\dfrac{b}{a}\right\}$

(v) $4\,!\,(x+a)^{-5}\log x - \dfrac{2(25x^3+23ax^2+13a^2x+3a^3)}{x^4\,(x+a)^4}$.

2. $x^2 \sin\left(x+\dfrac{n\pi}{2}\right) + 2nx\cos\left[x+(n-1)\dfrac{\pi}{2}\right]$
$+\,n\,(n-1)\cos\left[x+(n-2)\dfrac{\pi}{2}\right]$.

3. $x^3\cos\left(x+\dfrac{n\pi}{2}\right)+3nx^2\cos\left[x+(n-1)\dfrac{\pi}{2}\right]$
$+\,3n\,(n-1)\,x\cos\left[x+(n-2)\dfrac{\pi}{2}\right]$
$+\,n\,(n-1)\,(n-2)\cos\left[x+(n-3)\dfrac{\pi}{2}\right]$.

4. $\dfrac{(-1)^{n-1}\,n\,!}{x^{n-2}}\left[\dfrac{1}{n}-\dfrac{2}{n-1}+\dfrac{1}{n-2}\right]$.

5. $e^x\left[\log x+\dfrac{n}{x}-\dfrac{n\,(n-1)}{2\,!\,x^2}+\,....+\dfrac{(-1)^{n-1}\,(n-1)\,!}{x^n}\right]$

6. $\{(-1)^{n-1}\,(n-1)\,!\,\sin^n\theta\,\sin n\theta\}\,x^2+n\,\{(-1)^{n-2}\,(n-2)\,!\,\sin^{n-1}\theta$
$\sin\,(n-1)\,\theta\}\,2x+\dfrac{n\,(n-1)}{2}\,\{(-1)^{n-3}\,(n-3)\,!\,\sin^{n-2}\theta\,\sin\,(n-2)\,\theta\}\,2,$

where, $\tan\theta=\dfrac{1}{x}$.

7. $5^n e^{3x}\cos\left(4x+n\tan^{-1}\dfrac{4}{3}\right)x^2$
$+\,{}^nC_1\cdot 5^{n-1}\,e^{2x}\cos\left\{4x+(n-1)\tan^{-1}\dfrac{4}{3}\right\}\cdot 2x$
$+\,{}^nC_2\cdot 5^{n-2}\,e^{2x}\cos\left[4x+(n-2)\tan^{-1}\tfrac{4}{3}\right]\cdot 2.$

8. $m\,(m-1)\,(m-2)\,...\,(m-n+3)\,a^{n-2}\,(ax+b)^{m-n}\,x\,[(m-n+2)$
$(m-n+1)\,a^2x^2+2n\,(m-n+2)\,ax\,(ax+b)+n\,(n+1)\,(ax+b)^2]$.

9. $a^{n-2}\,e^{ax}\,\{a^2x^2+2nax+n\,(n-1)\}$.

10. $\dfrac{(-1)^{n-1}\,(n-1)\,!\,a^n}{(ax+b)^n}\cdot\sin x+{}^nC_1\,\dfrac{(-1)^{n-2}\,(n-2)\,!\,a^{n-1}}{(ax+b)^{n-1}}\cos x$
$+\,{}^nC_2\,\dfrac{(-1)^{n-3}\,(n-3)\,!\,a^{n-2}}{(ax+b)^{n-2}}\,(-\sin x)+\,....+\log\,(ax+b)\sin\left(x+\dfrac{n\pi}{2}\right)$.

11. $\dfrac{n\,!}{(x+1)^n}$.

12. $(1-x^2)\,y_{n+2}-(2n+1)\,xy_{n+1}-(n^2-a^2)\,y_n=0$.

1.8. n^{th} Differential Coefficient for Special Values of x

Sometimes we are required to find out the value of nth derivative of y for $x = 0$, *i.e.*, we are required to find out $(y_n)_0$. This may be done even when it is not feasible to find the general value of y_n in a compact form. The following illustrative examples will make the method clear.

ILLUSTRATIVE EXAMPLES

Example 1. *If* $y = e^{a \sin^{-1} x}$, *then show that*

$$(1 - x^2)\, y_{n+2} - (2n+1)\, xy_{n+1} - (n^2 + a^2)\, y_n = 0$$

and find the value of y_n when $x = 0$.

Solution :

$$y = e^{a \sin^{-1} x} \qquad \text{... (1)}$$

$$y_1 = e^{a \sin^{-1} x} \cdot \frac{a}{\sqrt{1-x^2}}$$

$$\Rightarrow \qquad y_1 = \frac{ay}{\sqrt{1-x^2}} \qquad \text{... (2)}$$

$$\Rightarrow \qquad \sqrt{1-x^2}\, y_1 = ay$$

Differentiate it w.r.t. x,

$$(1 - x^2)\, 2y_1 y_2 - 2xy_1^2 - 2a^2 yy_1 = 0$$

$$(1 - x^2)\, y_2 - xy_1 - a^2 y = 0 \qquad \text{... (3)}$$

Differentiate it n times w.r.t. x by Leibnitz's Theorem,

$$(1 - x^2)\, y_{n+2} + ny_{n+1}(-2x) + \frac{n(n-1)}{1 \cdot 2}\, y_n(-2)$$

$$- (xy_{n+1} + ny_n) - a^2 y_n = 0$$

$$\Rightarrow \qquad (1 - x^2)\, y_{n+2} - (2n+1)\, xy_{n+1} - (n^2 + a^2)\, y_n = 0 \qquad \text{... (4)}$$

Putting $x = 0$, in (1), (2), (3) and (4), we get

$$(y)_0 = e^0 = 1$$

$$(y_1)_0 = a\,(y)_0 = a$$

$$(y_2)_0 = a^2\,(y)_0 = a^2$$

....

....

$$(y_{n+2})_0 = (n^2 + a^2)\,(y_n)_0 \qquad \text{... (5)}$$

CASE I. *When n is odd.*

Putting 1, 3, 5, ... $(n-2)$ in (5),

$$(y_3)_0 = (1^2 + a^2)\,(y_1)_0 = (1 + a^2)\,a$$
$$(y_5)_0 = (3^2 + a^2)\,(y_3)_0 = (3^2 + a^2)\,(1 + a^2)\,a$$
$$(y_7)_0 = (5^2 + a^2)\,(y_5)_0$$
$$= (5^2 + a^2)\,(3^2 + a^2)\,(1 + a^2)\,a$$
$$\ldots\ \ldots\ \ldots\ \ldots\ \ldots\ \ldots\ \ldots\ \ldots\ \ldots\ \ldots$$
$$\ldots\ \ldots\ \ldots\ \ldots\ \ldots\ \ldots\ \ldots\ \ldots\ \ldots\ \ldots$$
$$(y_n)_0 = [(n-2)^2 + a^2]\,(y_{n-2})_0$$

$$\Rightarrow \quad (y_n)_0 = [(n-2)^2 + a^2]\,[(n-4)^2 + a^2] \ldots [5^2 + a^2]\,[3^2 + a^2]\,[1 + a^2]\,a.$$

CASE II. *When n is even.*

Putting $n = 2, 4, 6, \ldots\ (n-2)$ in (5),

$$(y_4)_0 = (2^2 + a^2)\,(y_2)_0 = (2^2 + a^2)\,a^2$$
$$(y_6)_0 = (4^2 + a^2)\,(y_4)_0 = (4^2 + a^2)\,(2^2 + a^2)\,a^2$$
$$(y_8)_0 = (6^2 + a^2)\,(y_6)_0 = (6^2 + a^2)\,(4^2 + a^2)\,(2^2 + a^2)\,a^2$$
$$\ldots\ \ldots\ \ldots\ \ldots\ \ldots\ \ldots\ \ldots\ \ldots\ \ldots\ \ldots\ \ldots\ \ldots\ \ldots$$
$$\ldots\ \ldots\ \ldots\ \ldots\ \ldots\ \ldots\ \ldots\ \ldots\ \ldots\ \ldots\ \ldots\ \ldots\ \ldots$$
$$(y_n)_0 = [(n-2)^2 + a^2]\,(y_{n-2})_0$$

$$\Rightarrow \quad (y_n)_0 = [(n-2)^2 + a^2]\,[(n-4)^2 + a^2] \ldots [6^2 + a^2]\,[4^2 + a^2]\,[2^2 + a^2]\,a^2.$$

Example 2. *If* $y = \sin(m \sin^{-1} x)$*, then prove that*

(i) $(1 - x^2)\,y_2 - xy_1 + m^2 y = 0$

(ii) $(1 - x^2)\,y_{n+2} - (2n + 1)\,xy_{n+1} - (n^2 - m^2)\,y_n = 0$

and, then find the value of $(y_n)_0$.

Solution : $$y = \sin(m \sin^{-1} x) \qquad \ldots (1)$$

$$\therefore \quad y_1 = \cos(m \sin^{-1} x)\,\frac{m}{\sqrt{1 - x^2}} \qquad \ldots (2)$$

$$\Rightarrow \quad y_1\sqrt{1 - x^2} = m \cos(m \sin^{-1} x).$$

Squaring both sides, we get

$$y_1^2 (1 - x^2) = m^2 \cos^2 (m \sin^{-1} x)$$

$$\Rightarrow \quad y_1^2 (1 - x^2) = m^2 [1 - \sin^2 (m \sin^{-1} x)]$$

$$\Rightarrow \quad y_1^2 (1 - x^2) = m^2 [1 - y^2].$$

Differentiating again w.r.t. x, we get

$$2y_1 y_2 (1 - x^2) - 2xy_1^2 = -2m^2 yy_1 \quad \text{... (3)}$$

Cancelling $2y$ throughout,

$$(1 - x^2) y_2 - xy_1 + m^2 y = 0.$$

Differentiating this equation n times by Leibnitz's Theorem, we get

$$[(1 - x^2) y_{n+2} + {}^nC_1 \cdot y_{n+1} \cdot (-2x) + {}^nC_2 \cdot y_n \cdot (-2)]$$
$$- [y_{n+1} \cdot x + {}^nC_1 \cdot y_n \cdot 1] + m^2 y_n = 0$$

which on simplification gives,

$$(1 - x^2) y_{n+2} - (2n + 1) xy_{n+1} - (n^2 - m^2) y_n = 0 \quad \text{... (4)}$$

Put $x = 0$ in (1), (2), (3) and (4), we get

$$(y)_0 = 0, \ (y_1)_0 = m, \ (y_2)_0 = 0 \ (y_{n+2})_0 = (n^2 - m^2) (y_n)_0 \quad \text{... (5)}$$

CASE I. ***When n is odd.***

Putting $n = 1, 3, 5, 7, \ldots (n - 2)$ in (5), we get

$$(y_3)_0 = (1^2 - m^2) (y_1)_0 = (1^2 - m^2) m$$
$$(y_5)_0 = (3^2 - m^2) (y_3)_0 = (3^2 - m^2) (1^2 - m^2) m$$
$$(y_7)_0 = (5^2 - m^2) (y_5)_0 = (5^2 - m^2) (3^2 - m^2) (1^2 - m^2) m$$
$$\ldots = \ldots = \ldots$$
$$\ldots = \ldots = \ldots$$

$$\therefore \quad (y_n)_0 = [(n - 2)^2 - m^2] (y_{n-2})_0$$
$$= [(n - 2)^2 - m^2] [(n - 4)^4 - m^2] (y_{n-4})_0$$
$$= [(n - 2)^2 - m^2] [(n - 4)^2 - m^2] \ldots (3^2 - m^2) (1^2 - m^2) m.$$

CASE II. ***When n is even.***

Putting $n = 2, 4, 6, \ldots, (n - 2)$ in (5), we get

$$(y_4)_0 = (2^2 - m^2) (y_2)_0 = 0$$
$$(y_6)_0 = (4^2 - m^2) (y_4)_0 = 0$$
$$\ldots = \ldots = \ldots$$

$$\therefore \quad (y_n)_0 = 0.$$

Example 3. *If* $y = (\sin^{-1} x)^2$, *then show that*

$$(1 - x^2)\, y_2 - xy_1 - 2 = 0$$

and differentiate it n times more, and find $(y_n)_0$.

Solution :
$$y = (\sin^{-1} x)^2 \quad \text{... (1)}$$

Differentiating it w.r.t. x,

$$y_1 = (2 \sin^{-1} x) \frac{1}{\sqrt{1 - x^2}} \quad \text{... (2)}$$

$$\Rightarrow \quad \sqrt{1 - x^2}\, y_1 = 2 \sin^{-1} x.$$

Squaring it,
$$(1 - x^2)\, y_1^2 = 4 (\sin^{-1} x)^2$$

$$(1 - x^2)\, y_1^2 = 4y \quad \text{[Using (1)]}$$

Differentiating it again,

$$(1 - x^2)\, 2y_1 y_2 - 2xy_1^2 = 4y_1$$

$$\Rightarrow \quad (1 - x^2)\, y_2 - xy_1 - 2 = 0 \quad \text{... (3)}$$

Now, differentiate it n times more by Leibnitz's Theorem,

$$(1 - x^2)\, y_{n+2} + n(-2x)\, y_{n+1} + \frac{n(n-1)}{2}(-2)\, y_n - (xy_{n+1} + ny_n) = 0$$

Put $x = 0$ in (1), (2) and (3), we get

$$(y)_0 = 0$$

$$(y_1)_0 = 1$$

$$(y_2)_0 = 0$$

and
$$(y_{n+2})_0 = n^2 (y_n)_0 \quad \text{... (5)}$$

This is known as a recurrence relation.

CASE I. ***When n is odd.***

Putting $n = 1, 3, 5, 7, \ldots, (n-2)$ in (5), we get

$$(y_3)_0 = 1^2 (y_1)_0 = 1^2 \cdot 1$$

$$(y_5)_0 = 3^2 (y_3)_0 = 3^2 \cdot 1^2 \cdot 1$$

$$(y_7)_0 = 5^2 (y_5)_0 = 5^2 \cdot 3^2 \cdot 1^2 \cdot 1$$

....

....

$$\therefore \quad (y_n)_0 = (n-2)^2 (y_{n-2})_0$$
$$= (n-2)^2 (n-4)^2 \ldots 5^2 \cdot 3^2 \cdot 1^2.$$

CASE II. *When n is even.*

Putting $n = 2, 4, 6, \ldots, (n-2)$ in (5), we get

$$(y_4)_0 = 2^2 (y_2)_0 = 0$$
$$(y_6)_0 = 4^2 (y_4)_0 = 0$$
$$(y_8)_0 = 6^2 (y_6)_0 = 0$$
....

....

$$\therefore \quad (y_n)_0 = (n-2)^2 (y_{n-2})_0 = 0.$$

Example 4. *If $y = (x + \sqrt{1+x^2})^m$, then find $(y_n)_0$.*

Solution :
$$y = (x + \sqrt{1+x^2})^m \quad \ldots (1)$$

Put $x = 0$,

$$(y)_0 = 1 \quad \ldots (2)$$
$$y_1 = m\,(x + \sqrt{1+x^2})^{m-1} \left\{1 + \frac{1}{2} \cdot \frac{2x^2}{\sqrt{1+x^2}}\right\}$$
$$\Rightarrow \quad y_1 = m\,(x + \sqrt{1+x^2})^{m-1} \left\{\frac{x + \sqrt{1+x^2}}{\sqrt{1+x^2}}\right\}$$
$$\Rightarrow \quad y_1 = m\,\frac{(x + \sqrt{1+x^2})^m}{\sqrt{1+x^2}}$$
$$\Rightarrow \quad y_1 = \frac{my}{\sqrt{1+x^2}} \quad \ldots (3)$$

Put $x = 0$,

$$(y_1)_0 = m\,(y_0) = m \quad \ldots (4)$$

From (3),

$$y_1\sqrt{1+x^2} = my$$

Squaring,
$$y_1^2 (1+x^2) = m^2 y^2 \quad \ldots (5)$$

Differentiating both sides of (5) w.r.t. x, we get

$$2y_1 y_2 (1+x^2) + 2xy_1^2 = 2m^2 y\, y_1.$$

Cancelling $2y_1$ throughout,

$$y_2(1+x^2) + xy_1 = m^2 y$$

$$\Rightarrow \qquad y_2(1+x^2) + xy_1 - m^2 y = 0 \qquad \text{... (6)}$$

Put $x = 0$,

$$(y_2)_0 - m^2(y) = 0$$

$$\Rightarrow \qquad (y_2)_0 = m^2(y_0)$$

$$\Rightarrow \qquad (y_2)_0 = m^2 \qquad \text{... (7)}$$

Now, differentiating every term of equation (6) n times, we get

$$[(1+x^2)\, y_{n+2} + {}^nC_1 \cdot 2x \cdot y_{n+1} + {}^nC_2 \cdot 2 \cdot y_n]$$
$$+ [xy_{n+1} + {}^nC_1 \cdot 1 \cdot y_n] - m^2 y_n = 0$$

which on simplification gives

$$(1+x^2)\, y_{n+2} + (2n+1)\, xy_{n+1} + (n^2 - m^2) y_n = 0 \qquad \text{... (8)}$$

Put $x = 0$,

$$(y_{n+2})_0 = (m^2 - n^2)\,(y_n)_0 \qquad \text{... (9)}$$

Replace n by $(n-2)$, we get

$$(y_n)_0 = [m^2 - (n-2)^2]\,(y_{n-2})_0 \qquad \text{... (10)}$$

Replace n by $n-2$ in (10), we get

$$(y_{n-2})_0 = [m^2 - (n-4)^2]\,(y_{n-4})_0 \qquad \text{... (11)}$$

Put the value of $(y_{n-1})_0$ in (10),

$$(y_n)_0 = [m^2 - (n-2)^2]\,[m^2 - (n-4)^2]\,(y_{n-4})_0.$$

Now, there arise two cases.

Case I. ***When n is odd;*** then

$$(y_1)_0 = [m^2 - (n-2)^2]\,[m^2 - (n-4)^2]\,[m^2 - (n-6)^2]\,(m^2 - 3^2)\,(m^2 - 1^2)\,(y_1)_0$$
$$= [m^2 - (n-2)^2]\,[m^2 - (n-4)^2]\,[m^2 - (n-6)^2] \ldots$$
$$(m^2 - 3^2)\,(m^2 - 1^2), m. \qquad \text{[From (4)]}$$

Case II. ***When n is even;*** then

$$(y_n)_0 = [m^2 - (n-2)^2]\,[m^2 - (n-4)^2]\,[m^2 - (n-6)^2] \ldots$$
$$\ldots (m^2 - 4^2)\,(m^2 - 2^2)\,(y_2)_0$$

$$= [m^2 - (n-2)^2]\,[m^2 - (n-4)^2]\,[m^2 - (n-6)^2] \,....$$

$$.... (m^2 - 4^2)(m^2 - 2^2)\, m^2. \qquad \text{[From (7)]}$$

Example 5. *If* $y = [\log_e \{x + \sqrt{1+x^2}\}]^2$, *find* $(y_n)_0$.

Solution :
$$y = \{\log(x + \sqrt{1+x^2})\}^2 \qquad ...(1)$$

$\therefore$
$$y_1 = \frac{2\log(x+\sqrt{1+x^2})}{x+\sqrt{1+x^2}}\left\{1 + \frac{1}{2}\cdot\frac{2x}{\sqrt{1+x^2}}\right\}$$

$$= \frac{2\log(x+\sqrt{1+x^2})}{x+\sqrt{1+x^2}}\left\{\frac{x+\sqrt{1+x^2}}{\sqrt{1+x^2}}\right\}$$

$$= \frac{2\log(x+\sqrt{1+x^2})}{\sqrt{1+x^2}} = \frac{2\sqrt{y}}{\sqrt{1+x^2}}.$$

Squaring
$$y_1^2 = \frac{4y}{1+x^2}.$$

$$\Rightarrow \quad (1+x^2)\, y_1^2 = 4y \qquad ...(2)$$

Differentiate both sides w.r.t. x, we get

$$(1+x^2)\, 2y_1 y_2 + 2xy_1^2 = 4y_1$$

Cancelling $2y_1$ throughout,

$$(1+x^2)\, y_2 + xy_1 = 2 \qquad ...(3)$$

Differentiate this equation n times by Leibnitz's Theorem, we get

$$[(1+x^2)\, y_{n+2} + {}^nC_1 \cdot 2x \cdot y_{n+1} + {}^nC_2\, 2y_n] + [xy_{n+1} + {}^nC_1 \cdot 1 \cdot y_n] = 0$$

which on simplification gives,

$$(1+x^2)\, y_{n+2} + (2n+1)\, xy_{n+1} + n^2 y_n = 0 \qquad ...(4)$$

Put $x = 0$,

$$(y_{n+2})_0 = -\, n^2\, (y_n)_0 \qquad ...(5)$$

When $x = 0$, then (1), (2) and (3) give

$$(y)_0 = 0$$
$$(y_1)_0 = 0$$
$$(y_2)_0 = 2.$$

Now, there arise two cases.

CASE I. ***When n is odd.***

From (5), replacing n by $(n-2)$, we get

$$(y_n)_0 = -\,(n-2)^2\, (y_{n+2})_0 \qquad ...(6)$$

Replace n by $(n-2)$ in (6) we get

$$(y_{n-2})_0 = -(n-4)^2 (y_{n-4})_0$$

$$\therefore \quad (y_n)_0 = (-1)^2 (n-2)^2 (n-4)^2 (y_{n-4})_0.$$

Proceeding in this manner

$$(y_n)_0 = (-1)^{(n-1)/2} (n-2)^2 (n-4)^2 (n-6)^2 \ldots 3^2 \cdot 1^2 \cdot (y_1)_0$$
$$= 0 \text{ as } (y_1)_0 = 0.$$

CASE II. ***When n is even.***

$$(y_n)_0 = (-1)^2 (n-2)^2 (n-4)^2 (y_{n-4})_0 \qquad \text{[as in case I]}$$

Proceeding ahead, we get

$$(y_n)_0 = (-1)^{(n-2)/2} (n-2)^2 (n-4)^2 \ldots 6^2 \cdot 4^2 \cdot 2^2 (y_2)_0$$
$$= (-1)^{(n-2)/2} \cdot [2 \cdot 4 \cdot 6 \ldots (n-2)]^2 \cdot 2$$
$$= (-1)^{(n-2)/2} \cdot \left\{2^{(n-2)/2} \cdot \left(1.2.3 \ldots \frac{n-2}{2}\right)\right\}^2 \cdot 2$$
$$= (-1)^{(n-2)/2}\, 2^{n-1} \left[\frac{(n-2)\,!}{2}\right]^2.$$

EXERCISE 1 (D)

1. If $y = e^{m \cos^{-1} x}$, prove that
$$(1-x^2)\, y_{n+2} - (2n+1)\, x y_{n+1} - (n^2+m^2)\, y_n = 0.$$
Hence evaluate $(y_n)_0$.
2. If $y = (\sin^{-1} x)^2$, find $(y_n)_0$.
3. If $y = \tan^{-1} x$, find $(y_n)_0$.
4. If $y = \log [x + \sqrt{x^2 + a^2}]$, prove that $(a^2 + x^2)\, y_2 + x y_1 = 0$.
Differentiate this equation n times and prove that
$$\lim_{x \to 0} \frac{y_{n+2}}{y_n} = -\frac{n^2}{a^2}.$$
5. Prove that the value of nth differential coefficient of $\dfrac{x^3}{x^2-1}$ for $x = 0$, is zero if n is even and $-n\,!$ if n is odd and greater than 1.
6. Prove that the value when $x = 0$ of $D^n (\tan^{-1} x)$ is 0, $(n-1)\,!$ or $-n\,(n-1)\,!$ according as n is of the form $2p$, $4p+1$ or $4p+3$ respectively.

7. Find the nth derivative of $\dfrac{1}{1+x+x^2+x^3}$ and show that for $x=0$, it is zero when n is of the form $4p+2$ or $4p+3$ and $n\,!$ or $-\,n\,!$ according as n is of the form $4p$ or $4p+1$.

ANSWERS

1. $-m\,(1^2+m^2)\,(2^2+m^2)\,.\,\{(n-2)^2\}\,e^{m\pi/2}$, when n is odd.
   $m^2\,(2^2+m^2)\,(4^2+m^2)\,....\,\{(n-2)^2+m^2\}\,e^{m\pi/2}$, when n is even.
2. 0, when n is odd; $(n-2)^2\,(n-4)^2\,....\,4^2.2^2.2$, when n is even.
3. 0, when n is even; $(-1)^{(n-1)/2}\,(n-1)\,!$, when n is odd.

7. $y_n = \dfrac{(-1)^n\,n\,!}{2}\,(x+1)^{-n-1} - \sin^{n+1}\phi\,\cos\,(n+1)\,\phi$

$$+\sin^{n+1}\phi\,\sin\,(n+1)\,\phi,\ \text{when}\ \tan\phi = \frac{1}{x}.$$

OBJECTIVE EXERCISE

The four possible answers of the questions are given, one of these is correct. Select the correct answer.

1. If $f(x)=xe^x$, then value of $f^n(x)$ is —

   (a) $(x+n)\,e^x$ *(b)* xe^x

   (c) e^x *(d)* $(x+1)\,e^x$

2. If $y = \tan x$, then find the value of $f''(0)$ —

   (a) 1 *(b)* 0

   (c) -1 *(d)* 2

3. nth derivative of $\cos ax$ is —

   (a) $\cos\left(ax+\dfrac{n\pi}{2}\right)$ *(b)* $a^n\cos\left(ax+\dfrac{n\pi}{2}\right)$

   (c) $\sin\left(ax+\dfrac{n\pi}{2}\right)$ *(d)* $a^n\sin\left(ax+\dfrac{n\pi}{2}\right)$

4. If $y = e^{3x}$, then find the value of y_n —

   (a) 3^n *(b)* e^{3x}

   (c) $3^n\,e^{3x}$ *(d)* e^{3nx}

5. If $y = (\sin^{-1} x)^2$, then the value of $\frac{d^2y}{dx^2} - x\frac{dy}{dx}$ is —

(a) 0 (b) 4

(c) 2 (d) -2

6. If $y = e^{4x}$, then y_n is —

(a) 4^n (b) e^{4x}

(c) $4^n e^{4x}$ (d) $4n\, e^{4x}$

7. $D^n \log (1 + x)$ is equal to —

(a) $\frac{(-1)^n n!}{(1+x)^n}$ (b) $\frac{(-1)^{n-1} (n-1)!}{(1+x)^n}$

(c) $\frac{(-1)^n (n-1)!}{(1+x)^{n-1}}$ (d) $\frac{(-1)^n (n-1)!}{(1+x)^{n+1}}$

8. The value of $D^n (ax + b)^m$ is equal to —

(a) $\frac{m!}{(m-n)!} a^n (ax+b)^{m-n}$ (b) $m!\, a^n (ax+b)^m$

(c) $\frac{1}{(m-n)!} (ax+b)^{m-n}$ (d) $m!\, (ax+b)^{m-n}$

9. The value of $D^n \cos (ax + b)$ is equal to —

(a) $a^n \cos\left(ax + b + \frac{n\pi}{2}\right)$ (b) $a^n \sin (ax + b)$

(c) $a^n \cos (ax + b + n\pi)$ (d) $a^n \sin\left(ax + b + \frac{n\pi}{2}\right)$

10. If $y = x^5$, then y_5 is —

(a) 5 ! (b) 20

(c) 4 ! (d) 15

11. The value of $D^n \sin (ax + b)$ is —

(a) $a^n \sin (ax + b)$ (b) $a^n \sin\left(ax + b + \frac{\pi}{2}\right)$

(c) $a^n \sin\left(ax + b + \frac{n\pi}{2}\right)$ (d) $b^n \sin\left(ax + b + \frac{n\pi}{2}\right)$

12. The value of $D^n (ax+b)^{-1}$ is —

(a) $n!\, a^n (ax+b)^{-n-1}$ (b) $\dfrac{(-1)^{n-1}(n-1)!}{(ax+b)^n}$

(c) $\dfrac{(-1)^n n!\, a^n}{(ax+b)^{n+1}}$ (d) $(n-1)!\, a^{n-1}(ax+b)^{-n}$

13. If $y = \sin(5-3x)$, then y_n is —

(a) $5^n \sin\left(5-3x+\dfrac{n\pi}{2}\right)$ (b) $3^n \sin\left(5-3x+\dfrac{n\pi}{2}\right)$

(c) $(-3)^n \sin\left(5-3x+\dfrac{n\pi}{2}\right)$ (d) None of these.

14. If $y = e^{-x}\sin x$, then $\dfrac{d^4y}{dx^4}$ is —

(a) $4y$ (b) $-4y$

(c) $\pm y$ (d) $2y$.

ANSWERS

1. (a)	2. (b)	3. (b)	4. (c)	5. (c)	6. (c)
7. (b)	8. (a)	9. (a)	10. (a)	11. (c)	12. (c)
13. (c)	14. (b)				

2
Maclaurin and Taylor Series Expansion

2.1. Infinite Series

We have seen that the ordinary processes of addition, subtraction, multiplication, division, rearrangement of terms, raising to a given power, taking limits, differentiation, etc., though applicable to the sum of a finite number of terms, may break down for infinite series. The expansions in the form of infinite series obtained by the methods given below are therefore to be regarded as formal expansions, which may not be true in exceptional cases.

2.2. Maclaurin's Theorem

Statement : ***If $f(x)$ is a continuous function of x, which can be expanded in ascending powers of x and each term of the expansion can be differentiated any number of times, then***

$$f(x) = f(0) + xf'(0) + \frac{x^2}{2!}f''(0) + \frac{x^2}{3!}f'''(0) + \ldots + \frac{x^n}{n!}f^n(0) + \ldots \text{ ad. inf.}$$

Proof : Let $f(x) = A_0 + A_1x + A_2x^2 + A_3x^3 + A_4x^4 + \ldots$ ... (1)

where, $A_0, A_1, A_2, A_3, A_4, \ldots$ are all constants.

Differentiating (1) successively w.r.t. x,

$$f'(x) = A_1 + 2A_2x + 3A_3x^2 + 4A_4x^3 + \ldots \quad \text{... (2)}$$

$$f''(x) = 2A_2 + 6A_3x + 12A_4x^2 + \ldots \quad \text{... (3)}$$

$$f'''(x) = 6A_3 + 24A_4x + \ldots \quad \text{... (4)}$$

$$f^{iv}(x) = 24A_4 + \ldots \quad \text{... (5)}$$

... ...

... ...

Putting $x = 0$ in (1), (2), (3), (4), (5), ...

$$f(0) = A_0 \Rightarrow A_0 = f(0)$$

$$f'(0) = A_1 \Rightarrow A_1 = f'(0)$$

$$f''(0) = 2A_2 \Rightarrow A_2 = \frac{f''(0)}{2!}$$

$$f'''(0) = 6A_3 \Rightarrow A_3 = \frac{f'''(0)}{3!}$$

$$f^{iv}(0) = 24A_4 \Rightarrow A_4 = \frac{f^{iv}(0)}{4!}$$

....

....

Now, $$f(x) = A_0 + A_1 x + A_2 x^2 + A_3 x^3 + A_4 x^4 + ... + A_{n-1} x^{n-1} + A_n x^n + A_{n+1} x^{n+1} + ...$$

Differentiating this equation n times w.r.t. x, we have

$$f^n(x) = A_n n! + A_{n+1} \cdot (n+1)n(n-1) .... 3 \cdot 2 \cdot x + ...$$

Put $x = 0$ $f^n(0) = A_n n!$

$$\therefore \quad A_n = \frac{f^n(0)}{n!}.$$

Substituting the values of $A_0, A_1, A_2, A_3, A_4, ..., A_n$, in (1), we have

$$f(x) = f(0) + xf'(0) + \frac{x^2}{2!}f''(0) + \frac{x^3}{3!}f'''(0) + \frac{x^4}{4!}f^{iv}(0) + ... + \frac{x^n}{n!}f^n(0) + ...$$

which is Maclaurin's Infinite Series.

Note 1. If $y = f(x)$, then another form in which Maclaurin's Infinite Series can be written as,

$$y = y_0 + x(y_1)_0 + \frac{x^2}{2!}(y_2)_0 + \frac{x^3}{3!}(y_3)_0 + ... + \frac{x^n}{n!}(y_n)_0 + ...$$

Note 2. Maclaurin's Theorem fails to expand a function $f(x)$ if any one of the values $f(0), f'(0), f''(0), ...$, is infinite or does not exist as in case of $\log x$, $\cot x$.

ILLUSTRATIVE EXAMPLES

Example 1. *Expand sin x by Maclaurin's theorem.*

Solution : Let $f(x) = \sin x$ $f(0) = 0$

then $f'(x) = \cos x$ $f'(0) = 1$

$f''(x) = -\sin x$ $f''(0) = 0$

$$f'''(x) = -\cos x \qquad f'''(0) = -1$$
$$f^{iv}(x) = \sin x \qquad f^{iv}(0) = 0$$
$$f^{v}(x) = \cos x \qquad f^{v}(0) = 1$$

... ...

... ...

$$f^n(x) = \sin\left(x + \frac{n\pi}{2}\right), f^n(0) = \sin\frac{n\pi}{2}$$

Now there arise two cases:

CASE I. ***When n is even,*** say $n = 2p$, then

$$f^n(0) = \sin\left(\frac{2p\pi}{2}\right)$$
$$= \sin p\pi = 0.$$

CASE II. ***When n is odd,*** say $n = 2p + 1$, then

$$f^n(0) = \sin\left[(2p+1)\frac{\pi}{2}\right]$$
$$= \sin\left[\frac{\pi}{2} + p\pi\right]$$
$$= \cos p\pi = (-1)^p.$$

Hence,
$$\sin x = 0 + x \cdot 1 + \frac{x^2}{2!}0 + \frac{x^3}{3!}(-1) + \frac{x^4}{4!}0 + \frac{x^5}{5!}\cdot 1 + ... + \frac{x^{2p+1}}{(2p+1)!}(-1)^p + ...$$
$$= x - \frac{x^3}{3!} + \frac{x^5}{5!} - .... + (-1)^p\frac{x^{2p+1}}{(2p+1)!} + ...$$

Example 2. *Expand* $\tan^{-1} x$.

Solution : Let $y = \tan^{-1} x \quad \therefore \quad y_0 = \tan^{-1} 0 = 0$

$$y_1 = \frac{1}{1+x^2} \quad \therefore \quad (y_1)_0 = 1$$

$$\Rightarrow \quad y_1 = (1+x^2)^{-1}$$
$$= 1 - x^2 + x^4 - x^6 + ... \qquad \text{[By Binomial Theorem]}$$

$$y_2 = -2x + 4x^3 - 6x^5 + \ldots.$$

$$\therefore \quad (y_2)_0 = 0$$

$$y_3 = -2 + 12x^2 - 30x^4 + \ldots.$$

$$\therefore \quad (y_3)_0 = -2$$

$$y_4 = 24x - 120x^3 + \ldots.$$

$$\therefore \quad (y_4)_0 = 0$$

$$y_5 = 24 - 360x^2 + \ldots.$$

$$\therefore \quad (y_5)_0 = 24$$

$$\therefore \quad \tan^{-1} x = (y)_0 + x(y_1)_0 + \frac{x^2}{2!}(y_2)_0 + \frac{x^3}{3!}(y_3)_0 + \ldots$$

$$= x - \frac{2x^3}{3!} + \frac{24x^5}{5!} - \ldots.$$

$$= x - \frac{x^3}{3} + \frac{x^5}{5} - \ldots.$$

Example 3. *Expand* $\sin^{-1} x$ *in powers of* x.

Solution : Let $\quad y = \sin^{-1} x \quad \therefore \quad (y)_0 = \sin^{-1} 0 = 0$

$$y_1 = \frac{1}{\sqrt{1-x^2}} \quad \therefore \quad (y_1)_0 = 1$$

Now, $\quad y_1 = (1-x^2)^{-1/2}$

$$= 1 + \tfrac{1}{2}x^2 + \frac{\frac{1}{2}\left(\frac{1}{2}+1\right)}{2!}x^4 + \frac{\frac{1}{2}\left(\frac{1}{2}+1\right)\left(\frac{1}{2}+2\right)}{3!}x^6 + \ldots.$$

[By Binomial Theorem]

$$\Rightarrow \quad y_1 = 1 + \frac{1}{2}x^2 + \frac{3}{8}x^4 + \frac{5}{16}x^6 + \ldots.$$

$$\Rightarrow \quad y_2 = x + \frac{3}{2}x^3 + \frac{15}{8}x^5 + \ldots.$$

$$\Rightarrow \quad (y_2)_0 = 0$$

$$y_3 = 1 + \frac{9}{2}x^2 + \frac{75}{8}x^4 + \ldots.$$

$$\Rightarrow \quad (y_3)_0 = 1$$

$$y_4 = 9x + \frac{75}{2}x^3 + \ldots$$

$$\therefore \quad (y_4)_0 = 0$$

$$y_5 = 9 + \frac{225}{2}x^2 + \ldots$$

$$\therefore \quad (y_5)_0 = 9$$

$$y_6 = 225x + \ldots$$

$$\therefore \quad (y_6)_0 = 0$$

$$\therefore \quad (y_7)_0 = 225 y_7 = 225 + \text{terms containing powers of } x$$

$$\therefore \quad \sin^{-1} x = y_0 + x(y_1)_0 + \frac{x^2}{2!}(y_2)_0 + \ldots$$

$$= x + \frac{x^3}{3!} + \frac{9x^5}{5!} + \frac{225x^7}{7!} + \ldots.$$

$$= x + \frac{1}{2}\cdot\frac{x^3}{3} + \frac{1\cdot 3}{2\cdot 4}\cdot\frac{x^5}{5} + \frac{1\cdot 3\cdot 5}{2\cdot 4\cdot 6}\cdot\frac{x^7}{7} + \ldots.$$

Example 4. *Prove that*

$$e^x \cos x = 1 + x - \frac{2x^2}{3!} - \frac{2^2 x^4}{4!} - \frac{2^3 x^5}{5!} + \ldots$$

Solution: Let $\quad y = e^x \cos x$

$$\therefore \quad (y)_0 = 1$$

$$y_n = (1^2 + 1^2)^{n/2}\, e^x \cos(x + n \tan^{-1} \tfrac{1}{2})$$

$$\Rightarrow \quad y_n = 2^{n/2} e^x \cos\left(x + \frac{n\pi}{4}\right)$$

Put $x = 0$,

$$(y_1)_0 = 2^{n/2} \cos\frac{n\pi}{4}$$

Put $n = 1, 2, 3, 4, \ldots$

$$(y_1)_0 = 2^{1/2} \cos\frac{\pi}{4} = 2^{1/2}\cdot\frac{1}{\sqrt{2}} = 1$$

$$(y_2)_0 = 2^{2/2} \cos\frac{\pi}{2} = 0$$

$$(y_3)_0 = 2^{3/2} \cos\frac{3\pi}{4} = 2\sqrt{2}\left(-\frac{1}{\sqrt{2}}\right) = -2$$

$$(y_4)_0 = 2^{4/2} \cos(\pi) = -4$$

$$(y_5)_0 = 2^{5/2} \cos\left(\frac{5\pi}{4}\right) = 2^{5/2}\left(-\frac{1}{\sqrt{2}}\right) = -2^2 = -4$$

....

....

$$\therefore \quad e^x \cos x = (y)_0 + x(y_1)_0 + \frac{x^2}{2!}(y_2)_0 + \frac{x^3}{3!}(y_3)_0 + \frac{x^4}{4!}(y_4)_0 + \frac{x^5}{5!}(y_5)_0 + \ldots$$

$$= 1 + x \cdot 1 + \frac{x^2}{2!} \cdot 0 + \frac{x^3}{3!} \cdot (-2) + \frac{x^4}{4!} \cdot (-4) + \frac{x^5}{5!}(-4) + \ldots$$

$$= 1 + x - \frac{2x^3}{3!} - \frac{2^2x^4}{4!} - \frac{2^2x^5}{5!} + \ldots$$

Example 5. *Expand* $\log(1+x)$ *by Maclaurin's Theorem.*

Solution : Let

$$y = \log(1+x) \qquad \therefore \quad (y)_0 = 0$$

$$y_1 = \frac{1}{1+x} \qquad \therefore \quad (y_1)_0 = 1$$

$$y_2 = -\frac{1}{(1+x)^2} \qquad \therefore \quad (y_2)_0 = -1$$

$$y_3 = \frac{2}{(1+x)^3} \qquad \therefore \quad (y_3)_0 = 2$$

$$y_4 = -\frac{6}{(1+x)^4} \qquad \therefore \quad (y_4)_0 = -6$$

....

....

$$y_n = \frac{(-1)^{n-1}(n-1)!}{(1+x)^n} \quad \therefore \quad (y_n)_0 = (-1)^{n-1}(n-1)!$$

Hence, by Maclaurin's theorem,

$$\log(1+x) = (y)_0 + x(y_1)_0 + \frac{x^2}{2!}(y_2)_0 + \frac{x^3}{3!}(y_3)_0 + \frac{x^4}{4!}(y_4)_0 + \ldots$$

$$\ldots + \frac{x^n}{n!}(y_n)_0 + \ldots$$

$$= x - \frac{x^2}{2} + \frac{x^3}{3} - \frac{x^4}{4} + \ldots + \frac{(-1)^{n-1}(n-1)!\,x^n}{n!} + \ldots$$

$$= x - \frac{x^2}{2} + \frac{x^3}{3} - \frac{x^4}{4} + \ldots + \frac{(-1)^{n-1}x^n}{n} + \ldots$$

Example 6. *Expand* $e^{x\cos x}$ *by Maclaurin's Theorem.*

Solution : Let $y = e^{x\cos x}$ ... (1)

Then,
$$y_1 = e^{x\cos x}(1 \cdot \cos x - x \sin x) = y(\cos x - x\sin x) \quad \text{... (2)}$$
$$y_2 = y_1(\cos x - x\sin x) + y(-\sin x - \sec x - x\cos x)$$
$$= y_1(\cos x - x\sin x) - y(2\sin x + x\cos x)$$
$$y_3 = y_2(\cos x - x\sin x) + y_1(-\sin x - 1\cdot\sin x - x\cos x)$$
$$- y_1(2\sin x + x\cos x) - y(2\cos x + 1.\cos x - x\sin x)$$
$$= y_2(\cos x - x\sin x) - 2y_1(2\sin x + x\cos x)$$
$$- y(3\cos x - x\sin x).$$
$$y_4 = y_3(\cos x - x\sin x) + y_2(-2\sin x - x\cos x)$$
$$- 2y_2(2\sin x + x\cos x) - 2y_1(3\cos x - x\sin x)$$
$$- y_1(3\cos x - x\sin x) - y(-4\sin x - x\cos x)$$
$$= y_3(\cos x - x\sin x) - 3y_2(2\sin x + x\cos x)$$
$$- 3y_1(3\cos x - x\sin x) + y(4\sin x + x\cos x)$$
$$y_5 = y_4(\cos x - x\sin x) - 4y_3(2\sin x + x\cos x)$$
$$- 6y_2(3\cos x - x\sin x) + 4y_1(4\sin x - x\cos x)$$
$$+ y(5\cos x - x\sin x)$$

Putting $x = 0$ and using $\cos 0 = 1$ and $\sin 0 = 0$, we get
$$(y)_0 = e^0 = 1,\ (y_1)_0 \cdot 1 = 1, (y_2)_0 = (y_1)_0 \cdot 1 = 1$$
$$(y_3)_0 = (y_2)_0 \cdot 1 - (y)_0 \cdot 3 = 1 - 3 = -2$$
$$(y_4)_0 = (y_3)_0 - 3(y_1)_0 \cdot 3 = -2 - 9 = -11$$
$$(y_5)_0 = (y_4)_0 \cdot 1 - 6(y_2)_0 \cdot 3 + (y_0) \cdot 5 = -11 - 18 + 5 = -24$$

By Maclaurin's theorem,
$$e^{x\cos x} = (y)_0 + x(y_1)_0 + \frac{x^2}{2!}(y_2)_0 + \frac{x^3}{3!}(y_3)_0 + \frac{x^4}{4!}(y_4)_0 + \ldots.$$
$$= 1 + x + \frac{x^2}{2} - \frac{x^3}{3} - \frac{11x^4}{24} - \frac{x^5}{5} - \ldots..$$

Example 7. *Expand* $\dfrac{e^x}{\cos x}$ *by Maclaurin's Theorem.*

Solution : Let $y = \dfrac{e^x}{\cos x} = e^x \sec x \quad \therefore \quad (y)_0 = 1$

$$y_1 = e^x \sec x + e^x \sec x \tan x$$

$$\Rightarrow \quad y_1 = y + y \tan x$$

$$\therefore \quad (y_1)_0 = 1 + 1 \cdot 0 = 1$$

$$y_2 = y_1 + y_1 \tan x + y \sec^2 x$$

$$\therefore \quad (y_2)_0 = 1 + 0 + 1 = 2$$

$$y_3 = y_2 + y_2 \tan x + y_1 \sec^2 x + y_1 \sec^2 x + y \cdot 2 \sec x \sec x \tan x$$

$$= y_2 + y_2 \tan x + 2y_1 \sec^2 x + 2y \sec^2 x \tan x$$

$$\therefore \quad (y_2)_0 = 2 + 0 + 2 \times 1 \times 1 + 0 = 4$$

... ...

... ...

$$\therefore \quad \frac{e^x}{\cos x} = y_0 + x(y_1)_0 + \frac{x^2}{2!}(y_2)_0 + \frac{x^3}{3!}(y_3)_0 + \ldots.$$

$$= 1 + x + \frac{x^2}{2!} \cdot 2 + \frac{x^3}{3!} \cdot 4 + \ldots$$

$$= 1 + x + x^2 + \frac{2}{3}x^3 + \ldots$$

Example 8. *Obtain by Maclaurin's Theorem the first five terms in the expansion of* log (1 + sin *x*).

Solution : Let $y = \log(1 + \sin x)$

$$\therefore \quad (y)_0 = \log 1 = 0$$

$$y_1 = \frac{\cos x}{1 + \sin x} = \frac{\sin\left(\frac{\pi}{2} - x\right)}{1 + \cos\left(\frac{\pi}{2} - x\right)}$$

$$= \frac{2 \sin\left(\frac{\pi}{4} - \frac{x}{2}\right) \cos\left(\frac{\pi}{4} - \frac{x}{2}\right)}{2 \cos^2\left(\frac{\pi}{4} - \frac{x}{2}\right)} = \tan\left(\frac{\pi}{4} - \frac{x}{2}\right)$$

$$\therefore \quad (y_1)_0 = \tan\frac{\pi}{4} = 1$$

$$y_2 = -\frac{1}{2} \sec^2\left(\frac{\pi}{4} - \frac{x}{2}\right)$$

$$= -\frac{1}{2}\left[1 + \tan^2\left(\frac{\pi}{4} - \frac{x}{2}\right)\right] = -\frac{1}{2}(1 + y_1^2)$$

$$\therefore \quad (y_2)_0 = -\frac{1}{2}\cdot(1+1^2) = -1$$

$$y_3 = -\frac{1}{2}\cdot 2y_1 y_2 = -y_1 y_2$$

$$\therefore \quad (y_3)_0 = -(y_1)_0 (y_2)_0 = -1\cdot(-1) = 1$$

$$(y_4) = -[y_2\cdot y_2 + y_1\cdot y_2] = -[y_2^{\ 2} + y_1 y_2]$$

$$\therefore \quad (y_4)_0 = -[(y_2)_0^2 + (y_1)_0 (y_3)_0] = -[1+1\cdot 1] = -2$$

$$(y_5) = -[2y_2 y_3 + y_2 y_3 + y_1 y_4]$$

$$\therefore \quad (y_5)_0 = -[2\cdot(-1)(1) + (-1)(1) + (1)(-2)]$$

$$= -[-2-1-2] = 5$$

....

....

$$\therefore \quad \log(1+\sin x) = (y)_0 + x(y_1)_0 + \frac{x^2}{2!}(y_2)_0 + \frac{x^3}{3!}(y_3)_0 + \frac{x^4}{4!}(y_4)_0 + \frac{x^5}{5!}(y_5)_0 + \ldots$$

$$= 0 + x\cdot 1 + \frac{x^2}{2!}\cdot(-1) + \frac{x^3}{3!}\cdot 1 + \frac{x^4}{4!}\cdot(-2) + \frac{x^5}{5!}\cdot 5 + \ldots$$

$$= x - \frac{x^2}{2!} + \frac{x^3}{3!} - \frac{2x^4}{4!} + \frac{5x^5}{5!} + \ldots$$

Example 9. *Prove that* $\log \sec x = \frac{x^2}{2} + \frac{x^4}{12} + \frac{x^6}{45} + \ldots$

Solution : Let $y = \log \sec x.$

Then,
$$y_1 = \frac{1}{\sec x} \times \sec x \tan x = \tan x$$

$$y_2 = \sec^2 x$$

$$y_3 = 2\sec^2 x \tan x = 2\, y_2 y_1$$

$$y_4 = 2(y_2^{\ 2} + y_3 y_1)$$

$$y_5 = 2(2y_4 y_1 + y_3 y_2 + 2y_2 y_3) = 2(y_4 y_1 + 3y_2 y_3)$$

$$y_6 = 2(y_5 y_1 + y_4 y_2 + 3y_2 y_4 + 3y_3^{\ 2})$$

$$= 2\,(y_1 y_5 + 4y_2 y_4 + 3y_3{}^2)$$

....

....

Putting $x = 0$, we get

$$(y)_0 = \log \sec 0 = \log 1 = 0$$

$$(y_1)_0 = \tan 0 = 0$$

$$(y_2)_0 = \sec^2 0 = 1$$

$$(y_3)_0 = 2\,(y_2)_0(y_1)_0 = 0$$

$$(y_4)_0 = 2\,[(y_3)_0\,(y_1)_0 + (y_2)_0{}^2] = 2\,[0 + 1] = 2$$

$$(y_5)_0 = 2\,[(y_4)_0\,(y_1)_0 + 3(y_2)_0\,(y_1)_0] = 2\,[0 + 0] = 0$$

$$(y_6)_0 = 2\,[(y_1)_0\,(y_5)_0 + 4(y_2)_0\,(y_4)_0 + 3\,(y_3)_0{}^2]$$

$$= 2\,[0 + 4 \cdot 1 \cdot 2 + 0] = 16$$

....

....

Using Maclaurin's Theorem, we get

$$y = (y)_0 + x\,(y_1)_0 + \frac{x^2}{2\,!}(y_2)_0 + \frac{x^3}{3\,!}(y_3)_0$$

$$+ \frac{x^4}{4\,!}(y_4)_0 + \frac{x^5}{5\,!}(y_5)_0 + \frac{x^6}{6\,!}(y_6)_0 + \ldots.$$

$$\log \sec x = 0 + x \cdot 0 + \frac{x^2}{2\,!} \cdot 1 + \frac{x^3}{3\,!} \cdot 0 + \frac{x^4}{4\,!} \cdot 2 + \frac{x^5}{5\,!} \cdot 0 + \frac{x^6}{6\,!} \cdot 16 + \ldots.$$

$$\Rightarrow \qquad \log \sec x = \frac{x^2}{2} + \frac{x^4}{12} + \frac{x^6}{45} + \ldots.$$

Example 10. *Expand* $\log(1 + e^x)$ *upto the term containing* x^4 *by Maclaurin's Theorem.*

Solution: Let

$$y = \log(1 + e^x)$$

$$(y)_0 = \log(1 + 1) = \log 2$$

$$y_1 = \frac{e^x}{1 + e^x} = \frac{1 + e^x - 1}{1 + e^x} = 1 - \frac{1}{1 + e^x}$$

$$(y_1)_0 = 1 - \frac{1}{1+1} = 1 - \frac{1}{2} = \frac{1}{2}$$

$$y_2 = \frac{e^x}{(1+e^x)^2} = \frac{e^x}{1+e^x} \cdot \frac{1}{1+e^x} = y_1(1-y_1)$$

$$\therefore \quad (y_2)_0 = \frac{1}{2}\left(1-\frac{1}{2}\right) = \frac{1}{4}$$

$$y_3 = y_2(1-y_1) + y_1(-y_2) = y_2 - 2y_1 y_2$$

$$\therefore \quad (y_3)_0 = \frac{1}{4} - 2 \cdot \frac{1}{2} \cdot \frac{1}{4} = 0$$

$$y_4 = y_3 - 2[y_2 \cdot y_2 + y_1 \cdot y_3] = y_3 - 2y_2^{\,2} - 2y_1 y_3$$

$$\therefore \quad (y_4)_0 = 0 - 2 \cdot \left(\frac{1}{4}\right)^2 - 2 \cdot \frac{1}{2} \cdot 0 = -\frac{1}{8}$$

....

....

$$\therefore \quad \log(1+e^x) = y_0 + x(y_1)_0 + \frac{x^2}{2!}(y_2)_0 + \frac{x^3}{3!}(y_3)_0 + \frac{x^4}{4!}(y_4)_0 + \ldots.$$

$$= \log 2 + x \cdot \frac{1}{2} + \frac{x^2}{2!} \cdot \frac{1}{4} + \frac{x^3}{3!} \cdot 0 + \frac{x^4}{4!}\left(-\frac{1}{8}\right) + \ldots.$$

$$= \log 2 + \frac{x}{2} + \frac{x^2}{8} - \frac{x^4}{192} + \ldots.$$

Example 11. *Expand* $e^{a \sin^{-1} x}$ *by Maclaurin's Theorem and find the general term. Hence show that*

$$e^\theta = 1 + \sin\theta + \frac{1}{2!}\sin^2\theta + \frac{2}{3!}\sin^3\theta + \ldots$$

Solution : Let $\quad y = e^{a \sin^{-1} x}$

$$(y)_0 = e^0 = 1$$

Differentiating, $\quad y_1 = e^{a \sin^{-1} x} \cdot \dfrac{a}{\sqrt{1-x^2}}$

$$\Rightarrow \quad y_1 = \frac{ay}{\sqrt{1-x^2}}$$

$$\Rightarrow \quad y_1\sqrt{1-x^2} = ay \qquad \therefore \quad (y_1)_0 = a$$

$$\Rightarrow \quad y_1^2(1-x^2) = a^2 y^2$$

Differentiating, we get, $2y_1 y_2 (1-x^2) - 2xy_1^2 = 2a^2 yy_1$. Cancelling $2y_1$ throughout, we get

$$y_2(1-x^2) - xy_1 = a^2 y \quad \therefore \quad (y_2)_0 = a^2$$

Differentiating n times by Leibnitz's Theorem, we have

$$[y_{n+2}(1-x^2) + {}^nC_1 \cdot y_{n+1} \cdot (-2x) + {}^nC_2 \cdot y_2(-2)] - [y_{n+1} \cdot x + {}^nC_1 \cdot y_n \cdot 1] = a^2 y_n$$

which on simplification gives,

$$(1-x^2)\, y_{n+2} - (2n+1)\, xy_{n+1} - (n^2+a^2)\, y_n = 0$$

Put $x = 0$,

$$(y_{n+2})_0 = (n^2+a^2)\,(y_n)_0$$

Put $n = 1, 2, 3, \ldots,$

$$(y_3)_0 = (1^2+a^2)\,(y_1)_0 = (1^2+a^2)\,a$$

$$(y_4)_0 = (2^2+a^2)\,(y_2)_0 = (2^2+a^2)\,a^2$$

$$(y_5)_0 = (3^2+a^2)\,(y_3)_0 = (3^2+a^2)\,(1^2+a^2)\,a$$

... ...

... ...

$$\therefore \quad 2^{a\sin^{-1}x} = (y)_0 + x(y_1)_0 + \frac{x^2}{2!}(y_2)_0 + \frac{x^3}{3!}(y_3)_0 + \frac{x^4}{4!}(y_4)_0 + \frac{x^5}{5!}(y_5)_0 + \ldots$$

$$= 1 + x \cdot a + \frac{x^2}{2!} \cdot a^2 + \frac{x^3}{3!} \cdot (1^2+a^2)\,a + \frac{x^4}{4!} \cdot (2^2+a^2)\,a^2 + \frac{x^5}{5!}(3^2+a^2)\,(1^2+a^2) + \ldots$$

Now, to find the general term,

$$(y_n)_0 = [(n-2)^2+a^2]\,[(n-4)^2+a^2] \ldots (1^2+a^2)\,(y_1)_0, \quad \text{when } n \text{ is odd}$$

$$= [n-2)^2+a^2]\,[(n-4)^2+a^2] \ldots (1^2+a^2) \cdot a$$

and
$$(y_n)_0 = [(n-2)^2+a^2]\,[(n-4)^2+a^2] \cdot (2^2+a^2)\,(y_2)_0, \quad \text{when } n \text{ is even}$$

$$= [(n-2)^2+a^2]\,[(n-4)^2+a^2] \ldots (2^2+a^2) \cdot a^2$$

Hence, general term $= \dfrac{x^n}{n!}(y_n)_0$, where $(y_n)_0$ is given by the above expressions.

Deduction : Put $a = 1$ and $x = \sin\theta$, we have

$$e^{\theta} = 1 + \sin\theta + \frac{1}{2!}\sin^2\theta + \frac{2}{3!}\sin^3\theta + \frac{5}{4!}\sin^4\theta + \ldots$$

Example 12. *Expand* $(\sin^{-1} x)^2$ *in the powers of x by Maclaurin's Theorem. Hence deduce that*

(1) $$\frac{\sin^{-1} x}{\sqrt{1-x^2}} = x + \frac{2^2}{2!}x^3 + \frac{2^2 \cdot 4^2}{5!}x^5 + \frac{2^2 \cdot 4^2 \cdot 6^2}{7!}x^7 + \ldots$$

(2) $$\left(\frac{\theta}{\sin\theta}\right)^2 = 1 + \frac{2^2}{3 \cdot 4}\cdot\sin^2\theta + \frac{2^2 \cdot 4^2}{3 \cdot 4 \cdot 5 \cdot 6}\cdot\sin^4\theta + \ldots.$$

Solution : Let

$$y = (\sin^{-1}x)^2$$

$$y_1 = \frac{2\sin^{-1} x}{\sqrt{1-x^2}} \qquad \ldots (1)$$

$$\sqrt{1-x^2}\, y_1 = 2\sin^{-1} x$$

$$(1-x^2)\, y_1^2 = 4(\sin^{-1} x)^2$$

$$(1-x^2)\, y_1^2 = 4y$$

Differentiate it again w.r.t. x,

$$(1 - x^2)\, 2y_1 y_2 - 2xy_1^2 = 4y_1$$

$$(1 - x^2)\, y_2 - xy_1 - 2 = 0 \qquad \ldots (2)$$

Now, differentiate it n times more by Leibnitz Theorem,

$$(1-x^2)\, y_{n+2} + n(-2x)\, y_{n+1} + \frac{n(n-1)}{2}(-2)\, y_n - (xy_{n+1} + ny_n \cdot 1) = 0$$

$$(1-x^2)\, y_{n+2} - 2nxy_{n+1} - n(n-1)\, y_n - xy_{n+1} - ny_n = 0$$

$$\Rightarrow \qquad (1-x^2)\, y_{n+2} - (2n+1)\, xy_{n+1} - n^2 y_n = 0$$

Putting $x = 0$ in the above result, we get

$$(y_{n+2})_0 = n^2 (y_n)_0 \qquad \ldots (3)$$

Again $(y)_0 = (\sin^{-1} 0)^2 = 0$, $(y_1)_0 = 0$ and $(y_2)_0 = 2$

Putting $x = 1, 3, 5, \ldots$ in (3), we get

$$(y_3)_0 = (y_5)_0 = (y_7)_0 \ldots. = (y_1)_0 = 0$$

Putting $n = 2, 4, 6, \ldots.$ in (3), we get

$$(y_4)_0 = 2^2 \cdot (y_2)_0 = 2^2 \cdot 2$$

$$(y_6)_0 = 4^2 \cdot (y_4)_0 = 4^2 \cdot 2^2 \cdot 2$$

$$\ldots \ldots \ldots \ldots \ldots \ldots$$

$$\ldots \ldots \ldots \ldots \ldots \ldots$$

Using these in Maclaurin's Theorem,

$$(\sin^{-1} x)^2 = (y)_0 + x(y_1)_0 + \frac{x^2}{2!}(y_2)_0 + \frac{x^3}{3!}(y_3)_0 + \ldots.$$

$$\Rightarrow \quad (\sin^{-1} x)^2 = \frac{x^2}{2!} \cdot 2 + \frac{x^4}{4!} \cdot 2^2 \cdot 2 + \frac{x^6}{6!} \cdot 4^2 \cdot 2^2 \cdot 2 + \ldots.$$

Deductions : We have proved that

$$(\sin^{-1} x)^2 = \frac{x^2}{2!} \cdot 2 + \frac{x^4}{4!} \cdot 2^2 \cdot 2 + \frac{x^6}{6!} \cdot 4^2 \cdot 2^2 \cdot 2 + \ldots.$$

$$\frac{1}{2}(\sin^{-1} x)^2 = \frac{1}{2!} \cdot x^2 + \frac{2^2}{4!} \cdot x^4 + \frac{2^2 \cdot 4^2}{6!} x^6 + \ldots. \qquad \ldots (4)$$

(1) Differentiating it w.r.t. x, we get

$$\frac{1}{2} \cdot 2 \cdot \frac{\sin^{-1} x}{\sqrt{1-x^2}} = \frac{1}{2!} \cdot 2x + \frac{2^2}{4!} \cdot 4x^3 + \frac{2^2 \cdot 4^2}{6!} \cdot 6x^5 + \ldots.$$

$$\Rightarrow \quad \frac{\sin^{-1} x}{\sqrt{1-x^2}} = x + \frac{2^2}{3!} x^3 + \frac{2^2 \cdot 4^2}{5!} x^5 + \frac{2^2 \cdot 4^2 \cdot 6^2}{7!} x^7 + \ldots.$$

(2) Putting $\sin^{-1} x = 0 \Rightarrow x = \sin \theta$ in the result (iv), we get

$$\frac{1}{2} \cdot \theta^2 = \frac{1}{2!} \sin^2 \theta + \frac{2^2}{4!} \sin^4 \theta + \frac{2^2 \cdot 4^2}{6!} \sin^6 \theta + \ldots.$$

$$\Rightarrow \quad \theta^2 = \sin^2 \theta + \frac{2^2}{3 \cdot 4} \sin^4 \theta + \frac{2^2 \cdot 4^2}{3 \cdot 4 \cdot 5 \cdot 6} \sin^6 \theta + \ldots..$$

$$\left(\frac{\theta}{\sin \theta}\right)^2 = 1 + \frac{2^2}{3 \cdot 4} \sin^2 \theta + \frac{2^2 \cdot 4^2}{3 \cdot 4 \cdot 5 \cdot 6} \sin^4 \theta + \ldots.$$

EXERCISE 2 (A)

Expand the following functions by Maclaurin's Theorem—

1. e^x
2. a^x
3. $\cos x$
4. $\sec x$
5. $\tan x$
6. $(1+x)^m$
7. $e^{\sin x}$
8. $\sinh x$
9. $\cosh x$

10. $\log \sqrt{\dfrac{1+x}{1-x}}$ **11.** $\log (\sec x + \tan x)$.

12. $e^x \log (1+x)$

13. Prove that

$$e^x \sin x = x + x^2 + \frac{2}{3\,!} x^3 - \frac{2^2}{5\,!} x^5 - .... + \sin \left(\frac{n\pi}{4} \right) \frac{2^{n/2}}{n\,!} x^n + ...$$

14. Prove that

$$\cos x \sinh x = x - \frac{2}{3\,!} x^3 - \frac{4}{5\,!} x^5 + ...$$

ANSWERS

1. $1 + x + \dfrac{x^2}{2\,!} + \dfrac{x^3}{3\,!} + ... + \dfrac{x^n}{n\,!} + ...$

2. $1 + x \log a + \dfrac{(x \log a)^2}{2\,!} + \dfrac{(x \log a)^3}{3\,!} + ... + \dfrac{(x \log a)^n}{n\,!} + ...$

3. $1 - \dfrac{x^3}{2\,!} + \dfrac{x^4}{4\,!} - \dfrac{x^6}{6\,!} + ... + (-1)^p \dfrac{x^{2p}}{2p\,!} + ...$

4. $1 + \dfrac{x^2}{2\,!} + \dfrac{5x^4}{4\,!} + 61 \dfrac{x^6}{6\,!} + ...$

5. $x + \dfrac{x^3}{3} + \dfrac{2}{15} x^5$

6. $1 + mx + \dfrac{m(m-1)}{2\,!} x^2 + \dfrac{m(m-1)(m-2)}{3\,!} x^3 + ... + x^m$

7. $1 + x + \dfrac{x^2}{2} - \dfrac{x^4}{8} - \dfrac{x^5}{15} + ...$

8. $x + \dfrac{x^3}{3\,!} + \dfrac{x^5}{5\,!} + ...$

9. $1 + \dfrac{x^2}{2\,!} + \dfrac{x^4}{4\,!} + ...$

10. $x + \dfrac{x^3}{3} + \dfrac{x^5}{5} + ...$

11. $x + \dfrac{x^3}{6} + \dfrac{x^5}{24} + ...$

12. $x + \dfrac{x^2}{2} + \dfrac{x^3}{3} + \dfrac{3x^5}{40} + ...$

TAYLOR'S THEOREMS

2.3. Taylor's Theorem (Infinite Form)

STATEMENT : ***Let $f(x+h)$ be a function of h (x being independent of h) which can be expanded as a convergent series in powers of h, and let the expansion be differentiable any number of times with respect to h, then***

$$f(x+h) = f(x) + hf'(x) + \frac{h^2}{2!}f''(x) + \frac{h^3}{3!}f'''(x) + ... + \frac{h^n}{n!}f^n(x) + ...$$

PROOF : Let,

$$f(x+h) = A_0 + A_1 h + A_2 h^2 + A_3 h^3 + A_4 h^4 + ... \qquad ...(1)$$

where, $A_0, A_1, A_2, A_3, A_4, ...$ are functions of x alone which are to be determined.

Now, since

$$\frac{d}{dh}f(x+h) = \frac{d}{dt}f(t) \cdot \frac{dt}{dh}, \text{ where, } t = x+h$$

$$= f'(t)\frac{d}{dh}(x+h)$$

$$= f(t)\left(\frac{dx}{dh} + \frac{dh}{dh}\right)$$

$$= f'(t)(0+1)$$

$$= f'(t)$$

$$= f'(x+h).$$

Hence, differentiating (1) successively w.r.t. h, we get

$$f'(x+h) = A_1 + 2A_2 h + 3A_3 h^2 + 4A_4 h^3 + ... \qquad ...(1a)$$

$$f''(x+h) = 2 \cdot 1 \cdot A_2 + 3 \cdot 2 \cdot A_3 h + 4 \cdot 3 \cdot A_4 h^2 + ... \qquad ...(1b)$$

$$f'''(x+h) = 3 \cdot 2 \cdot 1 \cdot A_4 + 4 \cdot 3 \cdot 2 \cdot A_4 h + ... \qquad ...(1c)$$

.... etc.

Putting $h = 0$ in (1), (1a), (1b), (1c) .... etc, we have

$f(x) = A_0 \quad \therefore \quad A_0 = f(x)$

$f'(x) = A_1 \quad \therefore \quad A_1 = \frac{f'(x)}{1!}$

$f''(x) = 2!\,A_2 \quad \therefore \quad A_2 = \frac{f''(x)}{2!}$

$f'''(x) = 3!\,A_3 \quad \therefore \quad A_3 = \frac{f'''(x)}{3!}$

.... etc. .... etc.

Substituting these values, (1) becomes

$$f(x+h) = f(x) + hf'(x) + \frac{h^2}{2!}f''(x) + \frac{h^3}{3!}f'''(x) + ... \qquad ...(2)$$

It proves the Taylor's Theorem.

COROLLARY 1. Replacing x by h and h by x in the Taylor's Theorem (2), we have

$$f(x+h) = f(h) + xf'(h) + \frac{x^2}{2!}f''(h) + \frac{x^3}{3!}f'''(h) + \dots \qquad \dots (3)$$

COROLLARY 2. Put $x = a$ in Taylor's theorem (2), then

$$f(a+h) = f(a) + hf'(a) + \frac{h^2}{2!}f''(a) + \frac{h^3}{3!}f'''(a) + \dots. \qquad \dots (4)$$

COROLLARY 3. Put $a = 0$ and $h = x$ in (4), then

$$f(x) = f(0) + xf'(0) + \frac{x^2}{2!}f''(0) + \frac{x^3}{3!}f'''(0) + \dots \qquad \dots (5)$$

which is Maclaurin's Theorem.

COROLLARY 4. Put $h = x - a$ in (4), then

$$f(x) = f(a) + (x-a)f'(a) + \frac{(x-a)^2}{2!}f''(a) + \frac{(x-a)^3}{3!}f'''(a) + \dots. + \frac{(x-a)^n}{n!}f^n(a) + \dots.$$

which gives the expansion of $f(x)$ in powers of $(x - a)$.

ILLUSTRATIVE EXAMPLES

Example 1. *Using Taylor's Series, prove that,*

$$\log(x+h) = \log h + \frac{x}{h} - \frac{x^2}{2h^2} + \frac{x^3}{3h^3} + \dots.$$

Solution : By Taylor's Series, we know that

$$f(x+h) = f(h) + xf'(h) + \frac{x^2}{2!}f''(h) \cdot \frac{x^3}{3!}f'''(h) + \dots. \qquad \dots (1)$$

Here, $f(x+h) = \log(x+h)$. ... (2)

Put $x = 0$,

$$f(h) = \log h$$

Differentiate (2) w.r.t. x,

$$f'(x+h) = \frac{1}{x+h} \qquad \dots (3)$$

Put $x = 0$,

$$f'(h) = \frac{1}{h}$$

Differentiate (3) w.r.t. x,

$$f''(x+h) = -\frac{1}{(x+h)^2} \quad \text{... (4)}$$

Put $x = 0$,

$$f''(h) = -\frac{1}{h^2}$$

Differentiate (4) w.r.t. x,

$$f'''(x+h) = \frac{2}{(x+h)^3}$$

Put $x = 0$,

$$f'''(h) = \frac{2}{h^3}, \text{ and so on.}$$

Substituting the values of $f(x+h)$ and $f(h), f'(h), f''(h), f'''(h), \ldots$ in (1), we get

$$\log(x+h) = \log h + \frac{x}{h} - \frac{x^2}{2h^2} + \frac{x^3}{3h^3} - \ldots$$

Example 2. *Prove that*

$$\log_e \sin(x+h) = \log_e \sin x + h \cot x - \frac{h^2}{2} \operatorname{cosec}^2 x + \frac{h^3}{3} \operatorname{cosec}^2 x \cot x + \ldots$$

Solution : We know by Taylor's Theorem,

$$f(x+h) = f(x) + hf'(x) + \frac{h^2}{2!} f''(x) + \frac{h^3}{3!} f'''(x) + \ldots \quad \text{... (1)}$$

Here, $f(x+h) = \log_e \sin(x+h)$

Put $h = 0$,

$$f(x) = \log_e \sin x$$

$$\therefore \quad f'(x) = \frac{\cos x}{\sin x} = -\cot x$$

$$f''(x) = -\operatorname{cosec}^2 x$$

$$f'''(x) = -2\operatorname{cosec} x(-\operatorname{cosec} x \cot x) = 2\operatorname{cosec}^2 x \cot x, \text{ etc.}$$

$\therefore$ From (i),

$$\log_e \sin(x+h) = \log_e \sin x + h\cot x - \frac{h^2}{2}\operatorname{cosec}^2 x + \frac{h^3}{3}\operatorname{cosec}^2 x \cot x + \ldots.$$

Example 3. *Expand* $\sin x$ *in powers of* $x - \frac{\pi}{2}$.

Solution : Let $f(x) = \sin x$, then $f\left(\frac{\pi}{2}\right) = 1$

$$f'(x) = \cos x, \qquad f'\left(\frac{\pi}{2}\right) = 0$$

$$f''(x) = -\sin x, \qquad f''\left(\frac{\pi}{2}\right) = -1$$

$$f'''(x) = -\cos x, \qquad f'''\left(\frac{\pi}{2}\right) = 0$$

$$f^{iv}(x) = \sin x, \qquad f^{iv}\left(\frac{\pi}{2}\right) = 1 \text{ and so on.}$$

Now, by Taylor's Theorem,

$$f(x) = f(a) + (x-a)f'(a) + \frac{(x-a)^2}{2!}f''(a) + \frac{(x-a)^3}{3!}f'''(a) + \frac{(x-a)^4}{4!}f^{iv}(a) + \ldots.$$

Here, $f(x) = \sin x$ and $a = \frac{\pi}{2}$.

$$\therefore \sin x = f\left(\frac{\pi}{2}\right) + \left(x-\frac{\pi}{2}\right)f'\left(\frac{\pi}{2}\right) + \frac{\left(x-\frac{\pi}{2}\right)^2}{2!} f''\left(\frac{\pi}{2}\right) + \frac{\left(x-\frac{\pi}{2}\right)^3}{3!} f'''\left(\frac{\pi}{2}\right) + \frac{\left(x-\frac{\pi}{2}\right)^4}{4!} f^{iv}\left(\frac{\pi}{2}\right) + \ldots.$$

$$= 1 - \frac{\left(x-\frac{\pi}{2}\right)^2}{2!} + \frac{\left(x-\frac{\pi}{2}\right)^4}{4!} - \ldots$$

Example 4. *Expand* $\tan^{-1}x$ *in the powers of* $\left(x-\frac{\pi}{4}\right)$ *by Taylor's Theorem.*

Solution : Let

$$f(x) = \tan^{-1}x, \text{ then } f'\left(\frac{\pi}{4}\right) = \tan^{-1}\left(\frac{\pi}{4}\right)$$

$$f'(x) = \frac{1}{1+x^2}, \quad f'\left(\frac{\pi}{4}\right) = \frac{1}{1+\left(\frac{\pi}{4}\right)^2} = \frac{1}{1+\frac{\pi^2}{16}}$$

$$f''(x) = \frac{-2x}{(1+x^2)^2}, \quad f''\left(\frac{\pi}{4}\right) = \frac{-2\cdot\frac{\pi}{4}}{\left\{1+\left(\frac{\pi}{4}\right)^2\right\}^2} = \frac{-\frac{\pi}{2}}{\left(1+\frac{\pi^2}{16}\right)^2}$$

and so on.

Now, by Taylor's series,

$$f(x) = f(a)+(x-a)f'(a)+\frac{(x-a)^2}{2!}f''(a)+ \ldots.$$

Here, $f(x) = \tan^{-1}x$ and $a=\frac{\pi}{4}$.

$$\therefore \quad \tan^{-1}x = f\left(\frac{\pi}{4}\right)+\left(x-\frac{\pi}{4}\right)f'\left(\frac{\pi}{4}\right)+\frac{\left(x-\frac{\pi}{4}\right)^2}{2!}f''\left(\frac{\pi}{4}\right)+ \ldots.$$

$$= \tan^{-1}\frac{\pi}{4}+\left(x-\frac{\pi}{4}\right)\cdot\frac{1}{1+\frac{\pi^2}{16}}$$

$$+\frac{\left(x-\frac{\pi}{4}\right)^2}{2!}\cdot\left[-\frac{\pi}{2\left(1+\frac{\pi}{16}\right)^2}\right]+ \ldots.$$

$$= \tan^{-1}\frac{\pi}{4}+\frac{x-\frac{\pi}{4}}{1+\frac{\pi^2}{16}}-\frac{\pi\left(x-\frac{\pi}{4}\right)^2}{4\left(1+\frac{\pi^2}{16}\right)^2}+ \ldots.$$

Example 5. *Expand* $2x^3 + 7x^2 + x - 1$ *in powers of* $(x - 2)$ *by Taylor's Theorem.*

Solution : Let

$$f(x) = 2x^3 + 7x^2 + x - 1, \text{ then } f(2) = 2(2)^3 + 7(2)^2 + 2 - 1 = 45$$

$$f'(x) = 6x^2 + 14x + 1, \qquad f'(2) = 6(2)^2 + 14(2) + 1 = 53$$

$$f''(x) = 12x + 14, \qquad f''(2) = 12(2) + 14 = 38$$

$$f'''(x) = 12, \qquad f'''(2) = 12$$

Since next higher derivatives of $f(x)$ vanish, hence by Taylor's Theorem,

$$f(x) = f(a) + (x - a) f'(a) + \frac{(x-a)^2}{2!} f''(a) + \frac{(x-a)^3}{3!} f'''(a) + \ldots$$

Here, $f(x) = 2x^3 + 7x^2 + x - 1$, and $a = 2$.

$$\therefore 2x^3 + 7x^2 + x - 1 = 45 + (x-2)\cdot 53 + \frac{(x-2)^2}{2!}\cdot 38 + \frac{(x-2)^3}{2!}\cdot 12$$

$$= 45 + (x-2)\cdot 53 + (x-2)^2 \cdot 19 + (x-2)^3 \cdot 2$$

$$= 45 + 53(x-2) + 19(x-2)^2 + 2(x-2)^3.$$

Example 6. *Use Taylor's Theorem to prove that*

$$\tan^{-1}(x + h) = h \sin z \cdot \frac{\sin z}{1} - (h \sin z)^2 \frac{\sin^2 z}{2} + (h \sin z)^3 \frac{\sin^3 z}{3} - \ldots,$$

where, $z = \cot^{-1} x$.

Solution : Let

$$f(x + h) = \tan^{-1}(x + h)$$

$$\therefore \quad f(x) = \tan^{-1} x \qquad \text{[Putting } h = 0\text{]}$$

Now $$f'(x) = \frac{1}{1 + x^2} = \frac{1}{1 + \cot^2 z} = \frac{1}{\text{cosec}^2 z}$$

Since $z = \cot^{-1} x$, therefore

$$\frac{dz}{dx} = -\frac{1}{1 + x^2} = -\frac{1}{1 + \cot^2 z}$$

$$= -\frac{1}{\text{cosec}^2 z} = -\sin^2 z$$

Now, $$f''(x) = \frac{d}{dx} f'(x)$$

$$= \frac{d}{dx} \sin^2 z$$

$$= 2 \sin z \cos z \frac{dz}{dx}$$

$$= -2 \sin z \cos z \sin^2 z \quad \left[\because \frac{dz}{dx} = -\sin^2 z \right]$$

$$= -\sin 2z \sin^2 z$$

and $$f''' = \frac{d}{dx} f''(x)$$

$$= \frac{d}{dx} [-\sin 2z \sin^2 z]$$

$$= -\frac{d}{dz} [\sin 2z \sin^2 z] \frac{dz}{dx}$$

$$= -[2 \cos 2z \sin^2 z + \sin 2z \, 2 \sin z \cos z] (-\sin^2 z)$$

$$= 2 \sin^2 z [\cos 2z \sin z + \sin 2z \cos z]$$

$$= 2 \sin^3 z \sin (z + 2z)$$

$$= 2 \sin^3 z \sin 3z.$$

Now, by Taylor's Theorem

$$f(x+h) = f(x) + h f'(x) + \frac{h^2}{2!} f''(x) + \frac{h^3}{3!} f'''(x) + \ldots$$

$$\therefore \quad \tan^{-1}(x+h) = \tan^{-1} x + h \sin^2 z + \frac{h^2}{2!} (-2 \sin 2z \sin^2 z)$$
$$+ \frac{h^3}{3!} 2 \sin^3 z \sin 3z + \ldots$$

$$\therefore \quad \tan^{-1}(x+h) = \tan^{-1} x + h \sin z . \frac{\sin z}{1} - (h \sin z)^2 \frac{\sin 2z}{2}$$
$$+ \frac{(h \sin z)^3}{3} \sin 3z - \ldots$$

EXERCISE 2 (B)

1. Prove that

$$\log \cos (x+h) = \log \cos x - h \tan x - \frac{h^2}{2} \sec^2 x - \frac{h^3}{3} \sec^2 x \tan x + \ldots$$

2. Prove that

$$\frac{1}{x+h} = \frac{1}{x}\left[1 - \frac{h}{x} + \frac{h^2}{x^2} - \frac{h^3}{x^3} + \ldots\right].$$

3. Prove that

$$\sin^{-1}(x+h) = \sin^{-1} h + \frac{x}{(1-h^2)^{1/2}} + \frac{x}{2\,!}\frac{h}{(1-h^2)^{3/2}} + \frac{x^2}{3\,!}\cdot\frac{1+2h^2}{(1-h^2)^{5/2}} + \ldots$$

4. Prove that

$$\sin^{-1}(x+h) = \sin^{-1} x + \frac{h}{\sqrt{1-x^2}} + \frac{x}{(1-x^2)^{3/2}}\cdot\frac{h^2}{2\,!} + \ldots$$

5. Prove that

$$\tan^{-1}(x+h) = \tan^{-1} x + \frac{h}{1+x^2} - \frac{xh^2}{(1+x^2)^2} + \ldots$$

6. Prove that

$$\tan(x+h) = \tan x + h\sec^2 x + h^2 \sec^2 x \tan x + \frac{h^2}{3}\sec^2 x\,(1 + 3\tan^2 x) + \ldots$$

7. Prove that

$$\sec^{-1}(x+h) = \sec^{-1} x + \frac{h}{x\sqrt{x^2-1}} - \frac{h^2}{2\,!}\cdot\frac{2x^2-1}{x^2(x^2-1)^{3/2}} + \ldots$$

8. Prove that

$$e^{x+h} = e^x + he^x + \frac{h^2}{2\,!}e^x + \frac{h^3}{3\,!}e^x + \ldots$$

9. Expand $\sin\left(\frac{\pi}{4} + \theta\right)$ in powers of θ.

10. Expand $\log \sin x$ in powers of $(x-2)$.

11. Expand e^x in ascending powers of $(x-1)$.

12. Expand $2 + x^2 - 3x^5 + 7x^6$ in powers of $(x-1)$.

13. Expand $\tan x$ in powers of $\left(x - \frac{\pi}{4}\right)$.

14. Prove that

$$f(mx) = f(x) + (m-1)\,x f'(x) + \frac{(m-1)^2}{2!}x^2 f''(x) + \frac{(m-1)^3}{3!}x^3 f'''(x) + \ldots$$

15. Prove that

$$f\left(\frac{x^2}{1+x}\right) = f(x) - \frac{x}{1+x}f'(x) + \frac{x^2}{(1+x)^2}\frac{f''(x)}{2!} - \ldots.$$

16. State and prove Taylor's Theorem.

ANSWERS

9. $\frac{1}{\sqrt{2}}\left(1+\theta-\frac{\theta^2}{2!}-\frac{\theta^3}{3!}+\frac{\theta^4}{4!}+\frac{\theta^5}{5!}+\ldots.\right)$

10. $\log\ \sin 2 + (x-2)\cot 2 - \frac{(x-2)^2}{2!}\operatorname{cosec}^2 2 + \ldots.$

11. $e\left[1+(x-1)+\frac{(x-1)^2}{2!}+\frac{(x-1)^3}{3!}+\ldots\right]$

12. $7 + 29(x-1) + 76(x-1)^2 + 110(x-1)^3 + 90(x-1)^4 + 39(x-1)^5 + 7(x-1)^6$

13. $\tan^{-1}\frac{\pi}{4} + \frac{2\left(x-\frac{\pi}{4}\right)}{1+\frac{\pi^2}{16}} + \frac{\left(x-\frac{\pi}{4}\right)^2}{2!} + \left\{\frac{-\pi/2}{\left(1+\frac{\pi^2}{16}\right)^2}\right\} + \ldots$

3
Rolle and Mean Value Theorem

Rolle's Theorem

3.1. Rolle's Theorem

STATEMENT : *If $f(x)$ is a function of x such that*

(i) *it is continuous in the closed interval $a \leq x \leq b$, i.e., the interval $[a, b]$,*

(ii) *it is differentiable in the open interval $a < x < b$, i.e., the interval (a, b), and*

(iii) $f(a) = f(b)$

GEOMETRICAL PROOF:

Let AB be the graph of the function $y = f(x)$ such that the points A and B of the graph correspond to the numbers a and b of the interval $[a, b]$.

(i) As $f(x)$ is continuous in the interval $a \leq x \leq b$, hence its graph is a continuous curve between A and B.

(ii) As $f(x)$ is differentiable in the interval $a < x < b$, hence the graph of $f(x)$ has a unique tangent at every point between A and B.

(iii) As $f(a) = f(b)$, hence AL and BM, the ordinates of A and B are equal.

Now, the curve may be of the following forms :

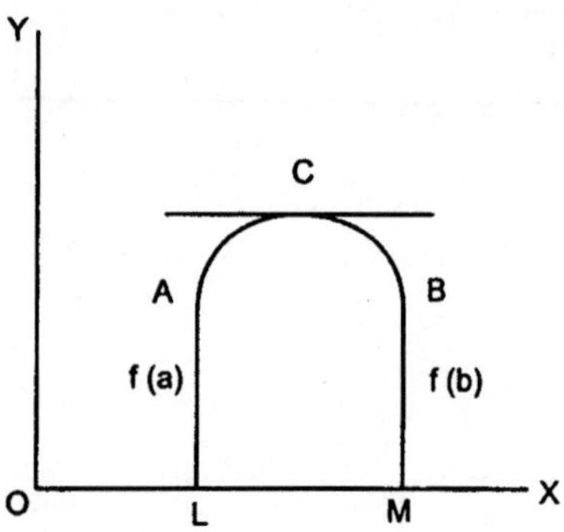

Fig. 3.1

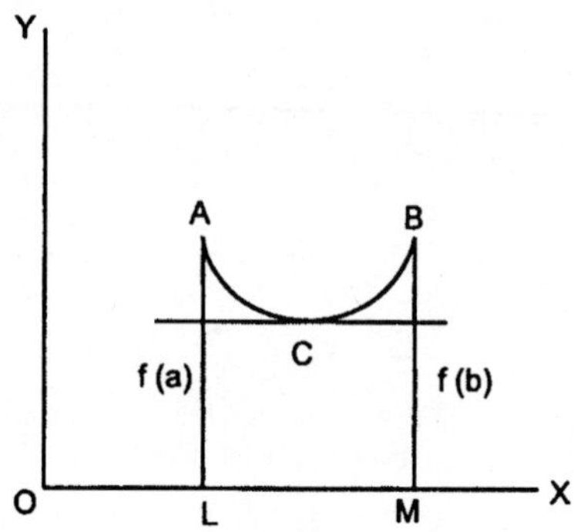

Fig. 3.2

(1) **The curve is a straight line :** Then since $f(a) = f(b)$, the straight line will be parallel to X-axis, *i.e.*, $f(x)$ is constant throughout (a, b), therefore $f'(x) = 0$ for every point of the interval (a, b), hence the theorem is easily proved.

(2) **The curve decreases first and increases afterwards (or increases first and decreases afterwards) :** Since $f(a) = f(b)$ [given], then there will be at least one point $C\ \{x = c,\ a < c < b\}$ where the curve ceases to decrease and begins to increase (or where the curve ceases to increase

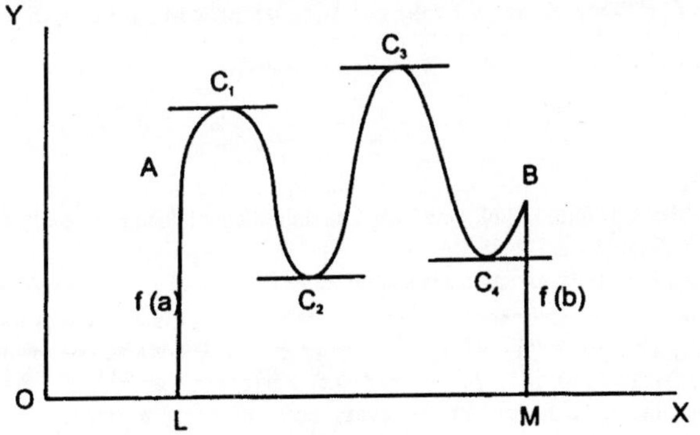

Fig. 3.3

and begins to decrease). At this point, the tangent to the curve is clearly parallel to X-axis. Hence, at the point C, where $x = c\ (a < c < b), f'(x) = 0$, *i.e.*, $f'(c) = 0$.

This proves the theorem.

Note : From Fig. 3.3, it is evident that there may exist more than one point of this type.

ANALYTICAL PROOF:

Since $f(x)$ is continuous in $[a, b]$ and $f(a) = f(b)$, therefore there may be three possibilities :

(i) $f(x)$ is constant in the given interval, $\therefore f'(x) = 0$ for all x in the given interval. Hence, the theorem is easily proved.

(ii) $f(x)$ increases as x takes values greater than a. Then since $f(a) = f(b)$. It will have to cease to increase, and begin to decrease at some point $x = c$ of (a, b). At this point $f(x)$ will have a maximum value. Therefore,

$$f(c+h) - f(c) \text{ and } f(c-h) - f(c)$$

are both negative, where h is small and positive.

Hence, $\dfrac{f(c+h)-f(c)}{h}<0$ and $\dfrac{f(c-h)-f(c)}{-h}>0.$

Now, if we proceed to take the limits as $h \to 0$, we find that the right hand and left hand derivatives of $f(x)$ at $x=c$ are negative and positive respectively. This shows that $f'(c)$ does not exist. This contradicts our hypothesis that $f(x)$ is differentiable in (a, b). This anomaly can be removed only when each of the above limits is zero, *i.e.*,

$$f'(c) = 0 \text{ for } a<c<b.$$

(iii) Proceeding similarly as in case (ii), we can prove the theorem also when $f(x)$ decreases as x takes values greater than a.

Thus the theorem is completely established.

Remember:

Notes

1. Rolle's theorem fails to hold good for a function which does not satisfy even one of the three conditions stated above.
2. Every algebraic polynomial in x is a continuous function of x for every value of x.
3. $\sin x$, $\cos x$, e^x are continuous for all values of x and $\operatorname{Log} x$ is continuous for all $x>0$.
4. If f and g are both continuous in $[a, b]$, then $f \pm g$ and fg are also continuous in $[a, b]$ and f/g is also continuous in $[a, b]$ provided $g(x) \neq 0$ for any $x \in [a, b]$.
5. If a function is differentiable for every point of a given interval, then it must be continuous in that interval, *i.e.*, differentiability $\Rightarrow$ continuity.

3.2. Alternative Form of Rolle's Theorem

STATEMENT : ***If a function $f(x)$ is such that***

(i) ***it is continuous in the interval $a \le x \le a+h$,***

(ii) ***it is differentiable in the interval $a<x<a+h$, and***

(iii) $f(a)=f(a+h)$

then there exists at least one number θ such that

$$f'(a+\theta h)=0, \text{ where } 0<\theta<1$$

(the number c which lies between a and $a+h$ must be greater than a by a fraction of h and hence may be written as

$$c = a + \theta h, \text{ where } 0 < \theta < 1$$

ILLUSTRATIVE EXAMPLES

Example 1. *Examine Rolle's theorem for the function $f(x) = |x|$ in the interval* $[-1, 1]$.

Solution : $\because f(x) = |x| = \begin{cases} x, \text{ when, } 0 \le x \le 1 \\ -x, \text{ when, } -1 \le x \le 0 \end{cases}$

$x = 0$ is a point in the interval $[-1, 1]$. Let us test the differentiability of the function at this point.

$$\text{Right hand derivative} = \lim_{x \to 0+} \frac{f(x) - f(0)}{x - 0}$$

$$= \lim_{x \to 0+} \frac{x - 0}{x}$$

$$= \lim_{x \to 0+} (1) = 1.$$

$$\text{Left hand derivative} = \lim_{x \to 0+} \frac{f(x) - f(0)}{x - 0}$$

$$= \lim_{x \to 0-} \left(\frac{-x - 0}{x}\right)$$

$$= \lim_{x \to 0-} (-1) = -1.$$

$\because$ Right hand derivative $\neq$ Left hand derivative,

$\therefore$ The function is not differentiable at $x = 0$.

$\Rightarrow$ The function is not differentiable in the interval $[-1, 1]$.

Therefore Rolle's theorem cannot be applied for the function $f(x) = |x|$ in the interval $[-1, 1]$.

Example 2. *Verify Rolle's theorem for the function*

$$f(x) = x^2 \text{ in } [-1, 1].$$

Solution : Here, $f(x) = x^2,\ a = -1,\ b = 1.$

(i) $f(a) = (-1)^2 = 1$

and $f(b) = 1^2 = 1$

$\therefore$ $f(a) = f(b)$

(ii) Since every polynomial in x is a continuous function of x for every finite value of x, hence, $f(x)$ is continuous in the closed interval $[-1, 1]$.

(iii) $f'(x) = 2x$ which exists for every value of x in the open interval $(-1, 1)$.

Hence, $f(x)$ satisfies all the three conditions of Rolle's theorem.

Hence, there must exist atleast one number c between -1 and 1 such that $f'(c) = 0 \Rightarrow 2c = 0 \Rightarrow c = 0 \in (-1, 1)$.

This is a point in the interval $(-1, 1)$ and hence the theorem is verified.

Example 3. *Verify Rolle's theorem for the function* $f(x) = x^2 - 6x + 8$ *in the interval* $[2, 4]$.

Solution : Here, $a = 2, b = 4$.

(i) $$f(x) = x^2 - 6x + 8$$

$\therefore f(x)$ is a polynomial.

Since every polynomial of x is a continuous function of x for every finite value of x, hence, $f(x)$ is continuous in the closed interval [2, 4].

(ii) $f'(x) = 2x - 6$ which exists in the open interval (2, 4).

(iii)
$$f(2) = 4 - 12 + 8 = 0$$
$$f(4) = 16 - 24 + 8 = 0$$
$$\therefore \quad f(2) = 0 = f(4)$$

$\therefore f(x)$ satisfies all the three conditions of Rolle's theorem.

$\therefore$ There must exist atleast one number c between 2 and 4 such that $f(c) = 0$.

Now, $$f'(x) = 2x - 6$$
$$\therefore \quad f'(c) = 0 \text{ gives } 2c - 6 = 0$$
$$\therefore \quad c = 3$$

This is a point in the open interval (2, 4) and, therefore, the theorem is verified.

Example 4. *Does Rolle's theorem apply to the function*

$$f(x) = 1 - (x-3)^{2/3}\ ?$$

Solution : $$f(x) = 1 - (x-3)^{2/3}$$

If $f(x) = 0$, then

$$1 - (x-3)^{2/3} = 0$$
$$\Rightarrow \quad (x-3)^{2/3} = 1$$
$$\Rightarrow \quad (x-3)^2 = 1^3 = 1$$
$$\Rightarrow \quad x - 3 = \pm 1$$
$$\Rightarrow \quad x = 3 \pm 1$$
$$\therefore \quad x = 2, 4$$

Hence, $f(x) = 0$ when $x = 2$ and $x = 4$.

Now, $$f'(x) = -\frac{2}{3}(x-3)^{-1/3} = -\frac{2}{3(x-3)^{1/3}}.$$

It is clear that $f'(x)$ does not exist at $x = 3$ which is a point in the interval $2 < x < 4$.

Hence, Rolle's theorem does not apply to the given function.

Example 5. *Verify Rolle's theorem for the function*

$$f(x) = x^2 - 6x^2 + 11x - 6.$$

Solution : $f(x) = x^3 - 6x^2 + 11x - 6$

Put $f(x) = 0$, we have

$$x^3 - 6x^2 + 11x - 6 = 0$$

$$\Rightarrow \quad x^2(x-1) - 5x(x-1) + 6(x-1) = 0$$

$$\Rightarrow \quad (x-1)(x^2 - 5x + 6) = 0$$

$$\Rightarrow \quad (x-1)(x-2)(x-3) = 0$$

$$\Rightarrow \quad x = 1, 2, 3.$$

Hence, $f(x) = 0$ when $x = 1, 2, 3$, *i.e.*, $f(1) = f(2) = f(3) = 0$.

Let us consider the interval $[1, 3]$.

(i) Since $f(x)$ is a polynomial in x, hence it is continuous in the closed interval $[1, 3]$.

(ii) $f'(x) = 3x^2 - 12x + 11$ which exists in the open interval $(1, 3)$.

(iii) Now, $f'(x) = 0$

$$\Rightarrow \quad 3x^2 - 12x + 11 = 0$$

$$\therefore \quad x = \frac{12 \pm \sqrt{144 - 132}}{6}$$

$$= \frac{12 \pm 2\sqrt{3}}{6}$$

$$= 2 \pm \frac{1}{\sqrt{3}} = 2 \pm \frac{1}{1.732}$$

$$= 2 \pm \cdot 58 = 1.42, 2.58 \text{ (appx.)}$$

Both these points lie in the interval (1, 3), and, therefore the theorem is verified.

Example 6. *Verify Rolle's theorem for* $(x-a)^m (x-b)^n$ *in the interval* $[a, b]$, *m, n being positive integers.*

Solution : Here, $f(x) = (x-a)^m (x-b)^n$.

As m and n are +ve integers, $(x-a)^m$ and $(x-b)^n$ are polynomials, on expansion by Binomial Theorem and consequently $f(x)$ is a polynomial of degree $(m+n)$.

(i) Since every polynomial of x is a continuous function of x for every value of x,

$\therefore$ $f(x)$ is continuous in the closed interval $[a, b]$.

(ii) $$f'(x) = m(x-a)^{m-1}(x-b)+n(x-a)^m(x-b)^{n-1}$$
$$= (x-a)^{m-1}(x-b)^{n-1}[m(x-b)+n(x-a)]$$

which exists in the open interval (a, b).

Note : Exists $\Rightarrow$when it is not

(1) imaginary, (2) infinite, (3) indeterminate.

(iii) $f(a)=0=f(b)$

$\therefore f(x)$ satisfies all the three conditions of Rolle's theorem.

$\therefore$ There must exist atleast one number c between a and b such that $f'(c)=0$.

$\therefore f'(x)=(x-a)^{m-1}(x-b)^{n-1}[m(x-b)+n(x-a)]$

$\therefore f'(c)=0$ gives

$$(c-a)^{m-1}(c-b)^{n-1}[m(c-b)+n(c-a)]=0$$

$$\Rightarrow \quad m(c-b)+n(c-a) = 0 \qquad [\because c \neq a, c \neq b \text{ as } a<c<b]$$

$$\Rightarrow \quad (m+n)c = mb+na$$

$$\Rightarrow \quad c = \frac{mb+na}{m+n},$$

which is a point within the interval (a, b) dividing the internal in the ratio $m : n$ internally.

Hence, the theorem is verified.

Example 7. *Verify Rolle's theorem for* $f(x) = x(x+3)e^{-x/2}$ *in* $[-3, 0]$.

Solution : Here, $f(x)=x(x+3)e^{-x/2}=(x^2+3x)e^{-x/2}$

Since (x^2+3x) and $e^{-x/2}$ are continuous for all x.

$\therefore$ their product $= x(x+3)e^{-x/2} = f(x)$ is continuous for all x

$\Rightarrow$ $= f(x)$ is continuous in $[-3, 0]$.

Also,
$$f'(x) = (2x+3)e^{-x/2}+(x^2+3x)e^{-x/2}\left(-\frac{1}{2}\right)$$
$$= e^{-x/2}\left[2x+3-\frac{x^2+3x}{2}\right]$$
$$= -\frac{1}{2}e^{-x/2}(x^2-x-6)$$

which exists in $(-3, 0)$ and $f(-3) = 0 = f(0)$.

$\therefore$ $f(x)$ satisfies all the three conditions of Rolle's theorem.

$\therefore$ There must exist at least one number c in $(-3, 0)$ such that

$$f'(c) = 0$$

i.e.,
$$-\frac{1}{2}(c^2 - c - 6)\, e^{-c/2} = 0$$

$$\Rightarrow \quad c^2 - c - 6 = 0 \quad [\because e^x \neq 0 \text{ for any finite value of } x]$$

$$\Rightarrow \quad (c-3)(c+2) = 0$$

$$\Rightarrow \quad c = 3, -2$$

But $c = 3 \notin (-3, 0)$ while $c = -2 \in (-3, 0)$

Hence the verification.

Example 8. *Verify Rolle's theorem for the function*

$$f(x) = \log \frac{x^2 + ab}{(a+b)\,x} \text{ in } [a, b].$$

Solution : We have,

$$f(x) = \log \frac{x^2 + ab}{(a+b)\,x}$$

$$\therefore \quad f(a) = \log \frac{a^2 + ab}{(a+b)\,a} = \log 1 = 0$$

$$f(b) = \log \frac{b^2 + ab}{(a+b)\,b} = \log 1 = 0$$

$$\therefore \quad f(a) = f(b) = 0$$

To test the differentiability of $f(x)$, we have

$$Rf'(x) = \lim_{h \to 0} \frac{f(x+h) - f(x)}{h}$$

$$= \lim_{h \to 0} \frac{1}{h}\left[\log\left\{\frac{(x+h)^2 + ab}{(a+b)(x+h)}\right\} - \log\left\{\frac{x^2 + ab}{(a+b)\,x}\right\}\right]$$

$$= \lim_{h \to 0} \frac{1}{h}\left[\log\left\{\frac{(x^2 + 2hx + h^2 + ab)(a+b)\,x}{(a+b)(x+h)(x^2 + ab)}\right\}\right]$$

$$= \lim_{h \to 0} \frac{1}{h}\left[\log\left\{\frac{x + 2hx + h^2 + ab}{x^2 + ab} \cdot \frac{x}{x+h}\right\}\right] \quad \text{[Note]}$$

$$= \lim_{h \to 0} \frac{1}{h}\left[\log\left\{1+\frac{2hx+h^2}{x^2+ab}\right\} - \log\left\{1+\frac{h}{x}\right\}\right] \quad \text{[Note]}$$

$$= \lim_{h \to 0} \frac{1}{h}\left[\frac{2hx}{x^2+ab} - \frac{h}{x}\right] \quad [\because h \to 0]$$

$$= \frac{2x}{x^2+ab} - \frac{1}{x}$$

Also, $$Lf'(x) = \lim_{h \to 0}\left[\frac{f(x-h)-f(x)}{-h}\right]$$

$$= \lim_{h \to 0} \frac{1}{(-h)}\left[\frac{-2hx}{x^2+ab} - \left(\frac{-h}{x}\right)\right]$$

[Replacing h by $-h$]

$$= \frac{2x}{x^2+ab} - \frac{1}{x}.$$

Thus, $Rf'(x) = Lf'(x)$.

Hence, the function $f(x)$ is differentiable in (a, b).

It is therefore continuous also in $[a, b]$. We thus find that all the conditions of Rolle's theorem are satisfied by $f(x)$ in (a, b).

Therefore, $f'(x) = 0$ for atleast one point in (a, b)

Now, $f'(x) = 0$ gives

$$\frac{2x}{x^2+ab} - \frac{1}{x} = 0$$

$$\Rightarrow \quad 2x^2 = x^2 + ab$$

$$\Rightarrow \quad x = \pm\sqrt{ab}$$

Value $x = \sqrt{ab}$ clearly it lies in between a and b as $\sqrt{ab}$ is the Geometric Mean of a and b.

Hence, the Rolle's theorem is completely verified.

Example 9. *If $f(x), \phi(x), \Psi(x)$ have the derivatives when $a \le x \le b$, show that there is a value ξ of x lying between a and b such that*

$$\begin{vmatrix} f(a) & \phi(a) & \Psi(a) \\ f(b) & \phi(b) & \Psi(b) \\ f'(\xi) & \phi'(\xi) & \Psi'(\xi) \end{vmatrix} = 0.$$

Solution : Consider the function $F(x)$, where

$$F(x) = \begin{vmatrix} f(a) & \phi(a) & \Psi(a) \\ f(b) & \phi(b) & \Psi(b) \\ f(x) & \phi(x) & \Psi(x) \end{vmatrix}$$

Putting $x = a$ in the above determinant, we find that first and third rows become identical, therefore $F(a) = 0$.

Similarly, by putting $x = b$, the second and third rows become identical, therefore $F(b) = 0$.

Therefore, we have

(i) $F(a) = F(b)$.

(ii) $F(x)$ is continuous in $[a, b]$, since $f(x), \phi(x), \Psi(x)$ have derivatives when $a \le x \le b$ and derivability $\Rightarrow$ continuity.

(iii) $F(x)$ is differentiable in (a, b).

Hence, all the conditions of Rolle's theorem are satisfied.

Hence, there exists a point $x = \xi$ in the interval (a, b) such that

$$F'(\xi) = 0$$

i.e.,
$$\begin{vmatrix} f(a) & \phi(a) & \Psi(a) \\ f(b) & \phi(b) & \Psi(b) \\ f'(\xi) & \phi'(\xi) & \Psi'(\xi) \end{vmatrix} = 0.$$

EXERCISE 3 (A)

Verify Rolle's Theorem for the following functions :

1. $8x - x^2$ in $[0, 8]$

2. $x^2 - 4x$ in $[-2, 2]$

3. $x^2 - 6x + 8$ in $[2, 4]$

4. $x^3 - 12x$ in $[0, 2\sqrt{3}]$

5. $x^2(1 - x)^2$ in $[0, 1]$

6. $2x^2 - 7x + 10$ in $[2, 5]$

7. $2x^3 + x^2 - 4x - 2$

8. $(x - a)^2 (x - b)^4$ in $[a, b]$

9. $\log \left(\dfrac{x^2 + 12}{7x} \right)$ in $[3, 4]$

10. $\sin x$ in $[-\pi, \pi]$

11. $e^x \sin x$ in $[0, \pi]$

12. $\dfrac{\sin x}{e^x}$ in $[0, \pi]$

13. $e^x (\sin x - \cos x)$ in $\left[\dfrac{\pi}{4}, \dfrac{5\pi}{4}\right]$

14. $\sqrt{4 - x^2}$ in $[-2, 2]$

15. $\sin 2x$ in $\left[0, \dfrac{\pi}{2}\right]$

16. Can Rolle's Theorem be applied to :

(i) $f(x) = \tan x$ in $[0, \pi]$,

(ii) $f(x) = \sec x$ in $[0, 2\pi]$?

17. Does Rolle's Theorem apply to the function $f(x) = \dfrac{x^2 - 3x}{x - 1}$?

18. Discuss the applicability of Rolle's theorem in the interval [0, 2] to the function $f(x) = 2+(x-1)^{2/3}$.

19. Let $f(x) = 1-(x-1)^{2/3}$, $0 \le x \le 2$. Explain why Rolle's theorem is not applicable to this function, *i.e.*, there is no value c of x for which $f'(c) = 0, 0 < c < 2$.

20. Show that under suitable conditions there exists atleast one number c, where $a < c < b$, such that

$$\begin{vmatrix} f(a) & f(b) \\ \phi(a) & \phi(b) \end{vmatrix} = (b-a) \begin{vmatrix} f(a) & f'(c) \\ \phi(a) & \phi(c) \end{vmatrix}$$

[**Hint :** Take the auxiliary function

$$\Psi(x) = \begin{vmatrix} f(a) & f(x) \\ \phi(a) & \phi(x) \end{vmatrix} \frac{x-a}{b-a} \begin{vmatrix} f(a) & f(b) \\ \phi(a) & \phi(b) \end{vmatrix}$$

which satisfies all the conditions of Rolle's theorem in the interval $[a, b]$.]

21. State and prove the Rolle's theorem.

22. Write the Rolle's theorem and explain the geometrical meaning.

ANSWERS

16. (i) No, (ii) No. **17.** No.

MEAN VALUE THEOREM

3.3. First Mean Value Theorem (Lagrange's Mean Value Theorem)

STATEMENT : ***If f(x) is a function of x such that :***

(i) ***it is continuous in the closed interval*** **[*a*, *b*],**

(ii) ***it is differentiable in the open interval*** **(*a*, *b*),**

then there exists one value c of x in the open interval **(*a*, *b*)** ***such that***

$$\frac{f(b)-f(a)}{b-a} = f'(c).$$

Proof : Consider the function

$$F(x) = f(x) + Ax, \quad \text{...(1)}$$

where A is a constant to be determined such that

$$F(a) = F(b)$$

Now,
$$F(a) = f(a)+Aa$$
$$F(b) = f(b)+Ab$$

Since, $F(a) = F(b)$

$\therefore$ $f(a) + Aa = f(b) + Ab$

$\Rightarrow$ $f(b) - f(a) = -A(b-a)$

$$-A = \frac{f(b)-f(a)}{b-a} \qquad \text{... (2)}$$

Now, it is given that $f(x)$ is continuous in $a \le x \le b$ and differentiable in $a < x < b$.

Also, A being a constant, Ax is also continuous in $a \le x \le b$ and differentiable in $a < x < b$.

$$F(x) = [f(x) + Ax] \text{ is :}$$

(1) continuous in the interval $a \le x \le b$.

(2) differentiable in the interval $a < x < b$.

Also, (3) $F(a) = F(b)$.

$\therefore$ $F(x)$ satisfies all the three conditions of Rolle's theorem.

$\therefore$ There must exist atleast one value c of x in the open interval (a, b) such that $F'(c) = 0$.

Now, $F'(x) = f'(x) + A$

$\therefore$ $F'(c) = 0$ gives

$$f'(c) + A = 0$$

$\Rightarrow$ $-A = f'(c)$ ... (3)

From (2) and (3), $\dfrac{f(b)-f(a)}{b-a} = f'(c)$.

3.4 Alternative Form of Lagrange's Mean Value Theorem

STATEMENT : ***If a function $f(x)$ is such that***

(i) it is continuous in the closed interval $[a, a+h]$,

(ii) it is differentiable in the open interval $(a, a+h)$.

then there exists atleast one number θ such that

$$f(a+h) = f(a) + hf'(a+\theta h), \text{ where } 0 < \theta < 1.$$

PROOF : $a + h = b$, then by art. 3.3 we have

$$\frac{f(b)-f(a)}{b-a} = f'(c) \qquad \text{... (4)}$$

$\because$ $a + h = b$

$\therefore$ $b - a = h$, the length of the interval.

The number c which lies between a and $a+h$ must be greater than a by a fraction of h and may be written as $c=a+\theta h$, where θ is some positive fraction lying between 0 and 1, *i.e.*, $0<\theta<1$.

Hence, (4) becomes

$$\frac{f(a+h)-f(a)}{h} = f'(a+\theta h)$$

$$\Rightarrow \qquad f(a+h) = f(a)+hf'(a+\theta h), \text{ where } 0<\theta<1.$$

3.5. Geometrical Interpretation of Lagrange's Mean Value Theorem

Let A and B be two points on the graph of the function $y=f(x)$ corresponding to $x = a$ and $x = b$ respectively.

$\therefore$ The co-ordinate of the points A and B are $[a, f(a)]$ and $[b, f(b)]$ respectively.

$$\therefore \text{ Slope of chord } AB = \frac{\text{Difference of ordinates}}{\text{Difference of abscissae}}$$

$$= \frac{f(b)-f(a)}{b-a}.$$

Now,

(1) ***Since $f(x)$ is continuous in the interval $a \le x \le b$.*** **(Fig. 3.4).**
$\therefore$ Its graph is a continuous curve from A to B.

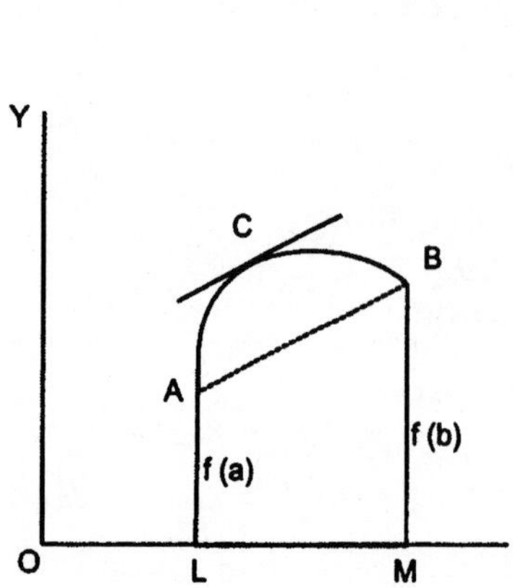

Fig. 3.4

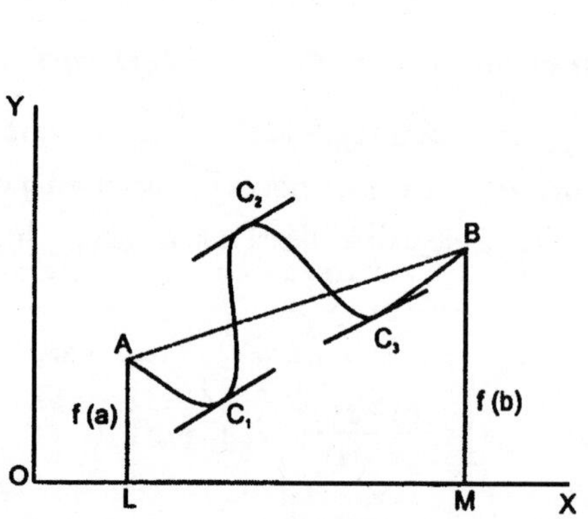

Fig. 3.5

(2) ***Since $f(x)$ is derivable in the interval $a < x < b$.* (Fig. 3.5).**

$\therefore$ It possesses a unique tangent at every point between A and B.

It is evident from the figures that there exists at least one point C between A and B, the tangent at which it is parallel to the chord AB.

If c be the abscissa of this point, then slope of tangent there at is $f'(c)$.

Hence, $$\frac{f(b)-f(a)}{b-a} = f'(c) \qquad [\because\ m_1 = m_2 \text{ for parallelism}]$$

(3) ***The result of mean value theorem is***

$$\frac{f(b)-f(a)}{b-a} = f'(c)$$

$$\Rightarrow \qquad f(b)-f(a) = (b-a)f'(c).$$

This equation gives the magnitude of the increment $f(b)-f(a)$. That is why the above form of mean value theorem is frequently called the formula of finite increments.

(4) ***If the derivative of a function is everywhere zero in the open interval (a, b), then $f(x)$ is constant in (a, b).***

PROOF : $f'(x) = 0$ in (a, b), [given].

Let ξ be any point of the interval (a, b).

$\therefore$ By Lagrange's mean value theorem,

$$f(\xi)-f(a) = (\xi-a)f'(\xi) = 0$$

$$\Rightarrow \qquad f(\xi) = f(a)$$

Since ξ is any point of (a, b), hence $f(x)$ is constant in (a, b).

Hence, the result.

(5) ***If the derivative of two functions $f(x)$ and $g(x)$ are equal everywhere in the open interval (a, b), then $f(x)$ and $g(x)$ differ by a constant in (a, b).***

PROOF : $\qquad f'(x) = g'(x)$ for all x in (a, b), [Given].

Let us define a new function $F(x)$ as

$$F(x) = f(x) - g(x) \text{ for all } x \text{ in } (a, b)$$

Then, $$F'(x) = f'(x) - g'(x) \text{ in } (a, b)$$
$$= 0 \text{ in } (a, b) \text{ for all } x$$

Hence, by art. 3.5,

$$F(x) = \text{constant (say } h) \text{ in } (a, b)$$

so that, $\qquad f(x) - g(x) = h$

Hence, the result.

This result is known as the ***Constant Difference Theorem.***

3.6. Second Mean Value Theorem (Cauchy's Mean Value Theorem)

STATEMENT : *If $f(x)$ and $\phi(x)$ are two functions of x such that*

(i) both are continuous in the closed interval [a, b],

(ii) both are differentiable in the open interval (a, b),

(iii) $\phi'(x) \neq 0$ for any value of x in the open interval (a, b),

then there exists one value c of x in the open interval (a, b) such that

$$\frac{f(b) - f(a)}{\phi(b) - \phi(a)} = \frac{f'(c)}{\phi'(c)}.$$

PROOF : Consider the function

$$F(x) = f(x) + A\,\phi(x), \qquad \text{...(1)}$$

where the constant A is to be determined such that

$$F(a) = F(b).$$

Now, $F(a) = f(a) + A\,\phi(a)$

$F(b) = f(b) + A\,\phi(b)$

Since, $F(a) = F(b)$

$\therefore$ $f(a) + A\,\phi(a) = f(b) + A\,\phi(b)$

$\Rightarrow$ $f(b) - f(a) = -A[\phi(b) - \phi(a)]$

$$\Rightarrow \qquad -A = \frac{f(b) - f(a)}{\phi(b) - \phi(a)}, \qquad \text{...(2)}$$

where $\phi(b) - \phi(a) \neq 0$.

[If $\phi(b) - \phi(a) = 0, \phi(a) = \phi(b)$

$\therefore$ $\phi(x)$ satisfies all the three conditions of Rolle's theorem.

$\therefore$ $\phi'(x) = 0$ for at least one value of x in the interval $a < x < b$ which is contrary to the given condition that $\phi'(x) \neq 0$ for any value of x in the interval $a < x < b$].

Since $f(x)$ and $\phi(x)$ both are continuous in the interval $a \le x \le b$ and differentiable in the interval $a < x < b$.

Also, since A is a constant,

$\therefore$ $F(x) = [f(x) + A\,\phi(x)]$ is :

(1) continuous in the interval $a \le x \le b$.

(2) differentiable in the interval $a < x < b$.

Also, (3) $F(a) = F(b)$.

$\therefore$ $F(x)$ satisfies all the three conditions of Rolle's theorem.

$\therefore$ There exists at least one value c of x in the interval $a < x < b$ such that

$$F'(c) = 0$$

Now, $$F'(x) = f'(x) + A\phi'(x)$$

$\therefore$ $F'(c) = 0$ gives

$$f'(c) + A\phi'(c) = 0$$

$\therefore$ $$-A = \frac{f'(c)}{\phi'(c)} \quad \text{... (3)}$$

From (2) and (3),

$$\frac{f(b) - f(a)}{\phi(b) - \phi(a)} = \frac{f'(c)}{\phi'(c)}$$

3.7. Deduction of Lagrange's Mean Value Theorem from Cauchy's Mean Value Theorem

If $$\phi(x) = x, \text{ then}$$

$$\phi(b) = b,$$

$$\phi(a) = a$$

and $$\phi'(x) = 1 \text{ for all } x.$$

$\therefore$ The result of Cauchy's mean value theorem, *viz.*,

$$\frac{f(b) - f(a)}{\phi(b) - \phi(a)} = \frac{f'(c)}{\phi'(c)}$$

reduces to $$\frac{f(b) - f(a)}{b - a} = \frac{f'(c)}{1} = f'(c)$$

which is Lagrange's mean value theorem.

3.8. Alternative Form of Cauchy's Mean Value Theorem

STATEMENT : ***If two functions $f(x)$ and $\phi(x)$ are such that***

(i) both are continuous in the closed interval $[a, a + h]$,

(ii) both are differentiable in the open interval $(a, a + h)$,

(iii) $\phi'(x) \neq 0$ for any value of x in the open interval $(a, a + h)$, then there exists at least one number θ such that

$$\frac{f(a+h) - f(a)}{\phi(a+h) - \phi(a)} = \frac{f'(a+\theta h)}{\phi'(a+\theta h)}, \text{ where } 0 < \theta < 1.$$

3.9. Monotonic Increasing or Decreasing Functions

DEFINITION : *If x_1 and x_2 be any two numbers in the interval (a, b) such that $x_2 > x_1$, then $f(x)$ is said to be a monotonic increasing function if $f(x_2) > f(x_1)$ and a monotonic decreasing function if $f(x_2) < f(x_1)$.*

3.10. Theorems

STATEMENT 1. ***If $f(x)$ is continuous in the interval $a \leq x \leq b$ and $f'(x)$ is positive for every value of x in the interval $a < x < b$, then $f(x)$ is monotonic increasing function of x in the interval $a \leq x \leq b$.***

PROOF : Let x_1 and x_2 be any two values of x between a and b such that $x_2 > x_1$. Then by mean value theorem,

$$f(x_2) - f(x_1) = (x_2 - x_1)f'(c).$$

where c lies between x_1 and x_2.

$\because \quad x_2 > x_1$

$\therefore$ $x_2 - x_1$ is + ve. Also, since $f'(c)$ is +ve

$\therefore \quad f(x_2) - f(x_1) > 0$

$\Rightarrow \quad f(x_2) > f(x_1).$

Hence, $f(x)$ is a monotonic increasing function in the interval (a, b).

STATEMENT 2. ***If $f(x)$ is continuous in the interval $a \leq x \leq b$ and $f'(x)$ is negative for every value of x in the interval $a < x < b$, then $f(x)$ is a monotonic decreasing function of x in the interval $a \leq x \leq b$.***

PROOF : Let x_1 and x_2 be any two values of x between a and b such that $x_2 > x_1$. Then by mean value theorem

$$f(x_2) - f(x_1) = (x_2 - x_1)f'(c),$$

where c lies between x_1 and x_2.

$\because \quad x_2 > x_1$

$\therefore x_2 - x_1$ is + ve. Also, $f'(c)$ is –ve

$\therefore f(x_2) - f(x_1)$ is negative

$\therefore \quad f(x_2) - f(x_1) < 0$

$\Rightarrow \quad f(x_2) < f(x_1).$

Hence, $f(x)$ is a monotonic decreasing function in the interval $[a, b]$.

3.11. General Mean Value Theorem or Taylor's Theorem in Finite Form

STATEMENT : ***If $f(x)$ is continuous in the closed interval $[a, b]$ and the first n derivatives $f'(x), f''(x), \ldots, f^n(x)$ exist for all values of x in the open interval (a, b), then***

$$f(b) = f(a) + (b-a)f'(a) + \frac{(b-a)^2}{2!}f''(a) + \ldots.$$

$$\dots + \frac{(b-a)^{n-1}}{(n-1)!} f^{n-1}(a) + \frac{(b-a)^n}{n!} f^n [a + \theta(b-a)]$$

where $0 < \theta < 1.$

PROOF : Consider the function $F(x)$ given by

$$F(x) = f(b) - f(x) - (b-x) f'(x) - \frac{(b-x)^2}{2!} f''(x) - \dots - \frac{(b-x)^r}{r!} f^r(x)$$

$$- \dots - \frac{(b-x)^{n-1}}{(n-1)!} f^{n-1}(x) - \frac{(b-x)^n}{n!} \cdot A \qquad \dots (1)$$

where A is a constant given by $F(a) = 0$, *i.e.*,

$$f(b) - f(a) - (b-a) f'(a) - \frac{(b-a)^2}{2!} f''(a)$$

$$- \dots - \frac{(b-a)^{n-1}}{(n-1)!} f^{n-1}(a) - \frac{(b-a)^n}{n!} \cdot A = 0 \qquad \dots (2)$$

Now, $F(x)$ satisfies all the conditions of Rolle's theorem because :

(1) $F(x)$ is continuous in the closed interval $[a, b]$ because $F(x)$ is algebraic sum of $(n+1)$ continuous functions of x

(2) $F(x)$ is differentiable in the open interval (a, b) as $f'(x), f''(x), \dots, f^n(x)$ exist in the interval (a, b).

(3) $F(a) = F(b) = 0$.

Hence, $F'(c) = 0$, where $a < c < b$. [By Rolle's theorem]

But, $F'(x) = -f'(x) + f'(x) - (b-x) f''(x) + (b-x) f'''(x)$

$$- \dots - \frac{(b-x)^{n-1}}{(n-1)!} f^n(x) + \frac{(b-x)^{n-1}}{(n-1)!} \cdot A$$

$$= \frac{(b-x)^{n-1}}{(n-1)!} \{A - f^n(x)\}$$

$\therefore$ $F'(c) = 0$ gives

$$A = f^n(c)$$

$$= f^n \{a + \theta(b-a)\}, \text{ where } 0 < \theta < 1 \qquad \dots (3)$$

Hence, from (2),

$$f(b) = f(a) + (b-a) f'(a) + \frac{(b-a)^3}{2!} f''(a) + \dots + \frac{(b-a)^{n-1}}{(n-1)!} f^{n-1}(a)$$

$$+\frac{(b-a)^n}{n!}f^n\{a+\theta(b-a)\} \text{ where } 0<\theta<1 \quad \text{.... (4)}$$

COROLLARY : Putting $b-a=h$, *i.e.*, $b=a+h$, this becomes

$$f(a+h)=f(a)+hf'(a)+\frac{h^2}{2!}f''(a)+\ldots+\frac{h^{n-1}}{(n-1)!}f^{n-1}(a)$$

$$+\frac{h^n}{n!}f^n(a+\theta h), \quad \text{where } 0<\theta<1. \quad \text{.... (5)}$$

Deductions:

(i) Writing x for b in (4), we get

$$f(x)=f(a)+(x-a)f'(a)+\frac{(x-a)^2}{2!}f''(a)+\ldots.$$

$$\ldots.+\frac{(x-a)^{n-1}}{(n-1)!}f^{n-1}(a)+\frac{(x-a)^n}{n!}f^n\{a+\theta(x-a)\} \quad \text{.... (6)}$$

(ii) Putting x in place of a in (5), we get

$$f(x+h)=f(x)+hf'(x)+\frac{h^2}{2!}f''(x)+\ldots.+\frac{h^{n-1}}{(n-1)!}f^{n-1}(x)$$

$$+\frac{h^n}{n!}f^n(x+\theta h) \quad \text{.... (7)}$$

(iii) Putting $a=0$ *in* (6), we get

$$f(x)=f(0)+xf'(0)+\frac{x^2}{2!}f''(0)+\ldots.+\frac{x^{n-1}}{(n-1)!}f^{n-1}(0)$$

$$+\frac{x^n}{n!}f^n(\theta x) \quad \text{.... (8)}$$

which is Maclaurin's theorem in finite form.

3.12. Taylor's Series

From equation (6) above, we have

$$f(x)=f(a)+(x-a)f'(a)+\frac{(x-a)^2}{2!}f''(a)+\ldots.$$

$$\ldots.+\frac{(x-a)^{n-1}}{(n-1)!}f^{n-1}(a)+\frac{(n-a)^n}{n!}f^n\{a+\theta(x-a)\}$$

$$=S_n(x)+R_n(x),$$

where

$$S_n(x)=f(a)+(x-a)f'(a)+\frac{(x-a)^2}{2!}f''(a)+\ldots.+\frac{(x-a)^{n-1}}{(n-1)!}f^{n-1}(a)$$

and $R_n(x)=\dfrac{(x-a)^n}{n!}f^n\{a+\theta(x-a)\}$.

Let us suppose that $f(x)$ possesses derivatives of all orders and let $R_n(x) \to 0$ as $h \to \infty$, then

$$f(x) = f(a) + (x-a)f'(a) + \frac{(x-a)}{2!}f''(a) + \ldots$$

$$\ldots + \frac{(x-a)^n}{n!} f^n(a) + \ldots \qquad \ldots (9)$$

3.13. Remainder in Taylor's Theorem

(1) Lagrange's form of the remainder : Consider equation (7) obtained above

$$f(x+h) = f(x) + hf'(x) + \frac{h^2}{2!}f''(x) + \ldots + \frac{h^{n-1}}{(n-1)!}f^{n-1}(x)$$

$$+ \frac{h^n}{n!}f^n(x+\theta h).$$

Here, the terms $R_n(h) = \frac{h^n}{n!}f^n(x+\theta h),\ 0 < \theta < 1$ is called ***Lagrange's form of the remainder*** after n terms of the Taylor's series.

(2) Schlomilch and Roche's form of the remainder : Let us consider the function $\phi(x)$, where

$$\phi(x) = f(b) - f(x) - (b-x)f'(x) - \frac{(b-x)^2}{2!}f''(x) - \ldots$$

$$\ldots - \frac{(b-x)^{n-1}}{(n-1)!}f^{n-1}(x) - \frac{(b-x)^m}{(b-a)^m}F^n(a).$$

Since $\phi(x)$ satisfies all the three conditions of Rolle's theorem, hence there must exist at least one value ξ of x in (a, b) such that

$$\phi'(\xi) = 0.$$

Now, $$\phi'(x) = -f'(x) + f'(x) - (b-x)f''(x) + (b-x)f''(x) - \ldots$$

$$\ldots - \frac{(b-x)^{n-1}}{(n-1)!}f^n(x) + \frac{m(b-x)^{m-1}}{(b-a)^m}F_n(a)$$

$$= -\frac{(b-x)^{n-1}}{(n-1)!}f^n(x) + \frac{m(b-x)^{m-1}}{(b-a)^m}F_n(a)$$

$\therefore$ $\phi'(\xi) = 0$ gives

$$-\frac{(b-\xi)^{n-1}}{(n-1)!}f^n(\xi) + \frac{m(b-\xi)^{m-1}}{(b-a)^m}F_n(a) = 0$$

$$\Rightarrow \quad F_n(a) = \frac{(b-a)^m(b-\xi)^{n-m}}{m(n-1)!}f^n(\xi), \text{ where } a < \xi < b.$$

Now, put $b - a = h$ and $\xi = a + \theta h,\ 0 < \theta < 1$

$$\therefore \quad b - \xi = b - a - \theta h$$

$$= h - \theta h = h(1 - \theta)$$

$$\therefore \quad F_n(a) = \frac{h^n (1-\theta)^{n-m}}{m(n-1)!} f^n(a + \theta h), \text{ where } 0 < \theta < 1.$$

Hence,
$$f(b) = f(a) + (b-a) f(a) + \ldots + \frac{(b-a)^{n-1}}{(n-1)!} f^{n-1}(a)$$
$$+ \frac{(b-a)^n (1-\theta)^{n-m}}{m(n-1)!} f^n(a+\theta h) \qquad [\text{Since } \phi(a) = 0]$$

Now, replace b by $a + h$, we obtain

$$f(a+h) = f(a) + hf'(a) + \ldots + \frac{h^{n-1}}{(n-1)!} f^{n-1}(a)$$
$$+ \frac{h^n (1-\theta)^{n-m}}{m(n-1)!} f^n(a+\theta h).$$

Here the term $R_n(h) = \dfrac{h^n (1-\theta)^{n-m}}{m(n-1)!} f^n(a+\theta h)$,

where $0 < \theta < 1$, and m is a positive integer, is known as ***Schlomilch and Roche's form of the remainder.***

It is to be noted here that if we put $m = n$, we get ***Lagrange's form of the remainder.***

(3) **Cauchy's form of the remainder :** Put $m = 1$ in Schlomilch and Roche's form of the remainder, we get

$$R_n(h) = \frac{h^n (1-\theta)^{n-1}}{(n-1)!} f^n(a+\theta h), \text{ where } 0 < \theta < 1.$$

Here $R_n(h)$ is called ***Cauchy's form of the remainder.***

ILLUSTRATIVE EXAMPLES

Example 1. *Prove that*

$$\sin ax = ax - \frac{a^3 x^3}{3!} + \frac{a^5 x^5}{5!} - \ldots + \frac{a^{n-1} x^{n-1}}{(n-1)!} \sin\left(\frac{n-1}{2}\pi\right)$$
$$+ \frac{a^n x^n}{n!} \sin\left(a\theta x + \frac{n\pi}{2}\right).$$

Solution : We know that

$$f(x) = f(0) + xf'(0) + \frac{x^2}{2!} f''(0) + \frac{x^3}{3!} f'''(0) + \frac{x^4}{4!} f^{iv}(0) + \frac{x^5}{5!} f^{v}(0) +$$

$$\dots + \frac{x^{n-1}}{(n-1)!} f^{n-1}(0) + \frac{x^n}{n!} f^n(\theta x) \qquad \dots \text{(i)}$$

[Maclaurin's theorem in finite form]

Here, $f(x) = \sin ax$

$\therefore$ $f^n(x) = a^n \sin\left(ax + \frac{n\pi}{2}\right)$

$\therefore$ $f^n(\theta x) = a^n \sin\left(a\theta x + \frac{n\pi}{2}\right)$

$$f^n(0) = a^n \sin\left(\frac{n\pi}{2}\right)$$

$$f(0) = \sin 0 = 0$$

$$f'(0) = a \sin\frac{\pi}{2} = a$$

$$f''(0) = a^2 \sin \pi = 0$$

$$f'''(0) = a^3 \sin\frac{3\pi}{2} = -a^3$$

$$f^{iv}(0) = a^4 \sin 2\pi = 0$$

$$f^{v}(0) = a^5 \sin\frac{5\pi}{2} = a^5$$

....

....

$$f^{(n-1)}(0) = a^{n-1} \sin\left(\frac{n-1}{2}\pi\right).$$

Substituting all these values in (i), we get

$$\sin ax = ax - \frac{a^3x^3}{3!} + \frac{a^5x^5}{5!} - \dots + \frac{a^{n-1}x^{n-1}}{(n-1)!}\sin\left(\frac{n-1}{2}\pi\right) + \frac{a^n x^n}{n!}\sin\left(a\theta x + \frac{n\pi}{2}\right).$$

Example 2. *Prove the mean value theorem for the function*

$$f(x) = 2x^2 - 7x + 10, a = 2 \text{ and } b = 5.$$

Solution : Here, $f(x) = 2x^2 - 7x + 10$.

(i) For each value of x in the interval [2, 5] the value of the function is unique and finite, the function will be continuous in the interval [2, 5].

(ii) $f'(x) = 4x - 7$, which exists for each value of x in [2, 5], therefore, $f(x)$ is differentiable in (2, 5).

(iii) $$f(a) = f(2) = 2 \times 2^2 - 7 \times 2 + 10 = 4$$

and $$f(b) = f(5) = 2 \times 5^2 - 7 \times 5 + 10 = 25$$

Thus, $f(a) \neq f(b)$

By Lagrange's theorem,

$$f'(c) = \frac{f(b) - f(a)}{b - a}$$

$$\Rightarrow \quad 4c - 7 = \frac{25 - 4}{5 - 2}$$

$$\Rightarrow \quad 4c = 7 + 7$$

$$\Rightarrow \quad c = \frac{14}{4} = 3.5 \in (2, 5)$$

Thus, the Lagrange's theorem is satisfied.

Example 3. *Verify Lagrange's mean value theorem for the function*

$$f(x) = \sqrt{x^2 - 4}$$

in the interval [2, 4].

Solution : $$f(x) = \sqrt{x^2 - 4}$$

Here, $a = 2$, $b = 4$.

$\because$ $f(x)$ has a unique and definite value for each x in the closed interval [2, 4].

$\therefore f(x)$ is continuous in [2, 4].

$$f'(x) = \frac{1}{2}(x^2 - 4)^{-1/2} . 2x = \frac{x}{\sqrt{x^2 - 4}}$$

which exists for all x in the open interval (2, 4), *i.e.*, for $2 < x < 4$.

Thus, $f(x)$ is differentiable in the interval (2, 4).

Again, $$f(a) = f(2) = \sqrt{4 - 4} = 0$$

and $$f(b) = f(4) = \sqrt{16 - 4} = 2\sqrt{3}$$

Thus, $$f(a) \neq f(b).$$

$\therefore$ By Lagrange's mean value theorem, we have

$$f'(c) = \frac{f(b) - f(a)}{b - a}$$

$$= \frac{f(4) - f(2)}{4 - 2} = \frac{\sqrt{12} - 0}{2} = \sqrt{3}$$

$$\Rightarrow \qquad \frac{c}{\sqrt{c^2-4}} = \sqrt{3} \qquad\qquad \Rightarrow c = \sqrt{3}\sqrt{(c^2-4)}.$$

Squaring both the sides

$$c^2 = 3(c^2-4)$$

$$\Rightarrow \qquad 2c^2 = 12$$

$$\Rightarrow \qquad c^2 = 6$$

$$\Rightarrow \qquad c = \pm\sqrt{6}$$

Discarding the value $c = -\sqrt{6}$ which does not lie in (2, 4).

$\therefore \qquad c = \sqrt{6},$ a value which lies in (2, 4).

Hence, the theorem is verified.

Example 4. *Does* $f(x) = x+\frac{1}{x}$ *in* $\left[\frac{1}{2}, 3\right]$ *satisfy the conditions of the mean value theorem ? If yes, find c in (a, b) to satisfy*

$$f'(c) = \frac{f(b)-f(a)}{b-a} \left(a = \frac{1}{2}, b=3\right).$$

Solution : $\qquad f(x) = x + \frac{1}{x}$

Here, $a = \frac{1}{2},\ b = 3.$

$\because f(x)$ has a unique and definite value for each x in the closed interval $\left[\frac{1}{2}, 3\right]$,

$\therefore f(x)$ is continuous in $\left[\frac{1}{2}, 3\right]$.

$$f'(x) = 1-\frac{1}{x^2}$$

which exists for all x in the open interval $\left(\frac{1}{2}, 3\right)$, *i.e.*, for $\frac{1}{2} < x < 3$.

$\therefore f(x)$ is derivable in $\left(\frac{1}{2}, 3\right)$.

Hence, $f(x)$ satisfies the conditions of Lagrange's mean value theorem.

$\therefore$ By Lagrange's mean value theorem, we have

$$f'(c) = \frac{f(b)-f(a)}{b-a}$$

$$= \frac{\left(3+\frac{1}{2}\right)-\left(\frac{1}{2}+2\right)}{3-\frac{1}{2}} = \frac{1}{3}$$

$$\therefore \qquad 1-\frac{1}{c^2}=\frac{1}{3}$$

$$\Rightarrow \qquad \frac{1}{c^2}=1-\frac{1}{3}=\frac{2}{3}$$

$$\Rightarrow \qquad c^2=\frac{3}{2}$$

$$\Rightarrow \qquad c=\pm\sqrt{\tfrac{3}{2}}=\pm\ 1.22 \text{ (approx.)}$$

Discarding the value $c = -1.22$ which does not lie in $\left(\frac{1}{2}, 3\right)$.

$\therefore c = 1.22$ which is a value lying in $\left(\frac{1}{2}, 3\right)$.

Example 5. *If* $f(x) = (x-1)(x-3)(x-5)$, $a=0, b=4$, *find the value of c such that* $f'(c)$ *has the same value as the slope of the chord joining the points for which* $x=0$ and $x=4$.

Solution : We have $f(x) = (x-1)(x-3)(x-5)$

$$\therefore \qquad f(a)=f(0)=-15$$

$$f(b)=f(4)=-3$$

$$\therefore \qquad \frac{f(b)-f(a)}{b-a}=3.$$

Now,

$$f'(x)=(x-3)(x-5)+(x-1)(x-5)+(x-1)(x-3)$$
$$=3x^2-18x+23.$$

$$\therefore \qquad f'(c)=3c^2-18c+23.$$

By mean value theorem, we have

$$\frac{f(b)-f(a)}{b-a}=f(c),\ a<c<b$$

$$\Rightarrow \qquad 3c^2-18c+23=3$$

$$\Rightarrow \qquad 3c^2-18c+20=0$$

$$\Rightarrow \qquad c=\frac{9\pm\sqrt{21}}{3}$$

Since, $a < c < b$, *i.e.*, $0 < c < 4$, therefore the required value of c is $\dfrac{9-\sqrt{21}}{3}$.

Example 6. *Prove that for* $x > 0$ *:*

(a) $\dfrac{x}{1+x} < \log(1+x) < x$,

(b) $x-\dfrac{x^2}{2} < \log(1+x) < x-\dfrac{x^2}{2(1+x)}$.

Solution :

(a) Let
$$f(x) = \log(1+x) - \frac{x}{1+x}$$

$\therefore$
$$f'(x) = \frac{1}{1+x} - \frac{(1+x)\cdot 1 - x\cdot 1}{(1+x)^2}$$
$$= \frac{1}{1+x} - \frac{1}{(1+x)^2}$$
$$= \frac{1+x-1}{(1+x)^2}$$
$$= \frac{x}{(1+x)^2} \text{ which is} > 0 \text{ for } x > 0.$$

$\therefore f(x)$ is a monotonic increasing function for $x > 0$,

i.e., $f(x) > f(0)$ for $x > 0$

$$\Rightarrow \log(1+x) - \frac{x}{1+x} > 0$$

$$\Rightarrow \quad \frac{x}{1+x} < \log(1+x). \qquad \text{.... (i)}$$

Let $\phi(x) = x - \log(1+x)$

$\therefore$
$$\phi'(x) = 1 - \frac{1}{1+x}$$
$$= \frac{x}{1+x}, \text{ which is} > 0 \text{ for } x > 0.$$

$\therefore \quad \phi(x) > \phi(0)$, for $x > 0$.

i.e., $x - \log(1+x) > 0$

$$\Rightarrow \quad \log(1+x) < x \qquad \text{.... (ii)}$$

Combining (i) and (ii),

$$\frac{x}{1+x} < \log(1+x) < x, \text{ for } x > 0.$$

(b) Let
$$f(x) = \log(1+x) - \left(x - \frac{x^2}{2}\right)$$

$\therefore$
$$f'(x) = \frac{1}{1+x} - 1 + x$$
$$= \frac{x^2}{1+x}, \text{ which is} > 0 \text{ for } x > 0.$$

$\therefore$ $f(x)$ is a monotonic increasing function for $x > 0$.

$\therefore$ $$f(x) > f(0) \text{ for } x > 0,$$

i.e., $$\log (1+x) - \left(x - \frac{x^2}{2}\right) > 0$$

$\Rightarrow$ $$x - \frac{x^2}{2} < \log(1+x) \qquad \text{.... (i)}$$

Let $$\phi(x) = x - \frac{x^2}{2(1+x)} - \log(1+x)$$

$\therefore$ $$\phi'(x) = 1 - \frac{1}{2}\frac{(1+x).2x - x^2.1}{(1+x)^2} - \frac{1}{1+x}$$

$$= \frac{2(1+x)^2 - (2x+x^2) - 2(1+x)}{2(1+x)^2}$$

$$= \frac{x^2}{2(1+x)^2} \text{ which is} > 0 \text{ for } x > 0.$$

$\therefore$ $\phi(x)$ is a monotonic increasing function for $x > 0$.

$\therefore$ $$\phi(x) > \phi(0) \text{ for } x > 0,$$

i.e., $$x - \frac{x^2}{2(1+x)} - \log(1+x) > 0$$

$\Rightarrow$ $$\log(1+x) < x - \frac{x^2}{2(1+x)} \qquad \text{.... (ii)}$$

Combining (i) and (ii),

$$x - \frac{x^2}{2} < \log(1+x) < x - \frac{x^2}{2(1+x)}.$$

Example 7. *Find c of Cauchy's mean value theorem for the following pair of functions in [a, b] :*

$$f(x) = e^x, \quad \phi(x) = e^{-x}.$$

Solution : $$f(x) = e^x, \quad \phi(x) = e^{-x}$$

$\therefore$ $$f'(x) = e^x, \quad -\phi'(x) = -e^{-x}$$

Both $f(x)$ and $\phi(x)$ are continuous in $[a, b]$ and derivable in (a, b).

$\therefore$ By Cauchy's mean value theorem, we have

$$\frac{f(b) - f(a)}{\phi(b) - \phi(c)} = \frac{f'(c)}{\phi'(c)}$$

$$\Rightarrow \qquad \frac{e^b - e^a}{e^{-b} - e^{-a}} = \frac{e^c}{-e^{-c}}$$

$$\Rightarrow \qquad \frac{e^b - e^a}{\frac{1}{e^b} - \frac{1}{e^a}} = -e^{+2c}$$

$$\Rightarrow \qquad (e^b - e^a).\frac{e^a . e^b}{e^a - e^b} = -e^{2c}$$

$$\Rightarrow \qquad -e^{a+b} = -e^{2c}$$

$$\Rightarrow \qquad a + b = 2c$$

$$\Rightarrow \qquad c = \frac{a+b}{2}.$$

Example 8. *Deduce from Cauchy's mean value theorem*

$$f(b) - f(a) = \xi f'(\xi) \log\left(\frac{b}{a}\right)$$

where $f(x)$ is continuous and differentiable in $[a, b]$ and $a < \xi < b$.

Solution : Cauchy's mean value theorem is

$$\frac{f(b) - f(a)}{\phi(b) - \phi(a)} = \frac{f'(\xi)}{\phi'(\xi)} \text{ where } a < \xi < b \qquad \text{.... (i)}$$

Now, let $\qquad \phi(x) = \log x$ so that $\phi'(x) = \frac{1}{x}$.

$\therefore \qquad \Psi(b) - \phi(a) = \log b - \log a = \log\left(\frac{b}{a}\right)$ and $\phi'(\xi) = \frac{1}{\xi}$

$\therefore$ (i) gives,

$$\frac{f(b) - f(a)}{\log\left(\frac{b}{a}\right)} = \frac{f'(\xi)}{\frac{1}{\xi}}$$

$$\Rightarrow f(b) - f(a) = \xi f'(\xi) \log\left(\frac{b}{a}\right).$$

Example 9. *Prove that if $f''(x)$ is continuous,*

$$\lim_{h \to 0} \frac{f(x+2h) - 2f(x+h) + f(x)}{h^2} = f''(x).$$

Solution : Since $f''(x)$ is continuous.

$\therefore f'(x)$ and $f(x)$ are both continuous.

Applying Taylor's theorem for the interval $(x, x+2h)$, we have

$$f(x+2h) = f(x) + 2hf'(x) + \frac{(2h)^2}{2!} f''(x + 2\theta_1 h) \qquad \text{.... (i)}$$

where $0 < \theta_1 < 1$.

Again applying Taylor's theorem for the interval $(x, x+h)$, we have

$$f(x+h) = f(x) + hf'(x) + \frac{h^2}{2!} f''(x + \theta_2 h) \qquad \text{.... (ii)}$$

where $0 < \theta_2 < 1$.

Therefore, $\lim_{h \to 0} \dfrac{f(x+2h) - 2f(x+h) + f(x)}{h^2}$

$$= \lim_{h \to 0} \frac{2h^2 f''(x + 2\theta_1 h) - h^2 f''(x + \theta_2 h)}{h^2}$$

$$= \lim_{h \to 0} [2f''(x + 2\theta_1 h) - f''(x + \theta_2 h)]$$

[From (i) and (ii)]

$$= 2f''(x) - f'(x)$$

$$= f''(x) \text{ as } f''(x) \text{ is continuous.}$$

Example 10. *Prove that*

$$\frac{F(x+h) + F(x-h) - 2F(x)}{h^2} = F''(x + \theta h),$$

where θ lies between -1 *and* 1.

Solution : By Lagrange's form of the remainder of Taylor's theorem (Art 3·13), we get

$$F(x+h) = F(x) + hF'(x) + \frac{1}{2} h^2 F''(\theta_1 h + x), \quad 0 < \theta_1 < 1$$

and $$F(x-h) = F(x) - hF'(x) + \frac{1}{2} h^2 F''(\theta_2 h + x), \quad -1 < \theta_2 < 0.$$

Adding these

$$F(x+h) + F(x-h) = 2F(x) + \frac{1}{2} h^2 [F''(x + \theta_1 h) + F''(x + \theta_2 h)]$$

$$\frac{F(x+h) + F(x-h) - 2F(x)}{h^2} = \frac{F''(x + \theta_1 h) + F''(x + \theta_2 h)}{2} \qquad \text{.... (i)}$$

If $-1 < \theta < 1$, then suppose $A < F''(x + \theta_1 h) < B$.

$$\therefore \qquad A < F''(x + \theta_1 h) < B$$

and $$A < F''(x + \theta_2 h) < B$$

Adding these inequalities

$$2A = F''(x + \theta_1 h) + F''(x + \theta_1 h) < 2B$$

$$\Rightarrow \qquad A < \frac{1}{2} F''(x + \theta_1 \theta) + F'(x + \theta_2 h) < B.$$

Thus, in between – 1 and 1 the value of this can be found such that R.H.S. of (i) will be equal to $F''(x + \theta h)$.

Therefore, $\dfrac{F(x+h) + F(x-h) - 2F(x)}{h^2} = F''(x + \theta h), \; -1 < \theta < 1.$

EXERCISE 3(B)

Find c of the Lagrange's mean value theorem, if :

1. $f(x) = x(x-1)(x-2), a = 0, b = \frac{1}{2}.$
2. $f(x) = \log x, a = 1, b = e.$
3. $f(x) = e^x, a = 0, b = 1.$
4. $f(x) = x^2 - 3x - 1, a = -\frac{11}{7}, b = \frac{13}{7}.$
5. $f(x) = x^3 - 5x^2 - 3x, a = 1, b = 3.$
6. $f(x) = x^2 + 3x + 2, a = 1, b = 2.$
7. For the function $f(x) = \dfrac{2x-1}{3x-4}$; $a = 1, b = 2$; prove that there is no number c in the open interval (a, b) which satisfies the mean value theorem. Determine the conditions of the theorem which fail to hold.
8. $F(x) = x \sin \frac{1}{x}$ in $(-1, 1)$, $F(0) = 0$. Do the conditions of the mean value theorem of differential calculus are satisfied in this case ? Does the theorem hold for $F(x)$ in $(-1, 1)$?
9. A function $f(x)$ is continuous in the closed interval [0, 1] and differentiable in the open interval (0, 1). Prove that
$$f'(x) = f(1) - f(0), \text{ where } 0 < x_1 < 1.$$
10. By mean value theorem, prove that
$$\frac{x}{1+x^2} < \tan^{-1} x < x, x > 0.$$
11. Find c of Cauchy's mean value theorem for the following pair of functions in (a, b) :

(i) $f(x) = \sqrt{x}$, $\phi(x) = \dfrac{1}{\sqrt{x}}$, (ii) $f(x) = \sin x$, $\phi(x) = \cos x$, (iii) $f(x) = \dfrac{1}{x^2}$, $\phi(x) = \dfrac{1}{x}$.

12. With the help of the mean value theorem, show that if $x > 0$,

$$\log_{10}(x+1) = \frac{x \log_{10} e}{1+\theta x}, \text{ where } 0 < \theta < 1.$$

13. Show that

$$\sin x = x - \frac{x^3}{3!} + \frac{x^5}{5!} - \dots + (-1)^{n-1}\frac{x^{2n-1}}{(2n-1)!} + (-1)^n \frac{x^{2n}}{2n!}\sin(\theta x)$$

for every value of x.

14. Show that, for $x > -1$:

$$\log(1+x) = x - \frac{x^2}{2} + \frac{x^3}{3} - \dots + (-1)^{n-2}\frac{x^{n-1}}{n-1} + (-1)^{n-1}\frac{x^n}{n(1+\theta x)^n}.$$

15. Show that

$$e^x = 1 + x + \frac{x^2}{2!} + \dots + \frac{x^{n-1}}{(n-1)!} + \frac{x^n}{n!}e^{\theta x}.$$

16. If $f(x) = f(0) + xf'(0) + \frac{x^2}{2!}f''(\theta x)$, then find the value of θ as $x \to 1$, $f(x)$ being $(1-x)^{3/2}$.

17. If $f(x+h) = f(x) + hf'(x) + \frac{h^2}{2!}f''(x+\theta h)$, then find the value of θ as $x \to a, f(x)$ being $(x-a)^{5/2}$.

18. Prove that

$$e^{ax}\cos bx = 1 + ax + \frac{a^2-b^2}{2!}x^2 \frac{a(a^2-3b^2)}{3!}x^3 + \dots$$

and remainder $= \dfrac{(a^2+b^2)^{n/2}}{n!}x^n e^{a\theta x}\cos\left\{b\theta x + n\tan^{-1}\dfrac{b}{a}\right\}.$

19. State and prove the first mean value theorem of Lagrange.

20. State and prove Cauchy' mean value theorem (second mean value theorem).

ANSWERS

1. $1 \pm \sqrt{\frac{1}{2}}$	**2.** $e - 1$
3. $\log(e-1)$	**4.** $\frac{1}{7}$
5. $\frac{7}{3}$	**6.** $\frac{3}{2}$

7. $f(x)$ is not continuous in $[1, 2]$ and $f(x)$ is not derivable in $(1, 2)$.

8. Not satisfied. Theorem does not hold.

11. (i) $\sqrt{ab}$, (ii) $\frac{a+b}{2}$, (iii) $\frac{2ab}{a+b}$ 16. $\frac{9}{25}$

17. $\frac{64}{225}$.

OBJECTIVE EXERCISE

Select the correct answer :

1. The Maclaurin's theorem for $\cos x$ is :

   (a) $1+\frac{x^2}{2!}+\frac{x^4}{x!}+\frac{x^6}{6!}+\ldots$ (b) $1-\frac{x^2}{2!}+\frac{x^4}{4!}-\frac{x^6}{6!}+\ldots$

   (c) $1-\frac{x^2}{2}+\frac{x^4}{4}-\frac{x^6}{6}+\ldots$ (d) $1+\frac{x^2}{2}+\frac{x^4}{4}+\frac{x^6}{6}+\ldots$

2. The value of the series $x-\frac{x^3}{3}+\frac{x^5}{5}-\ldots.$ is :

   (a) $\sin x$ (b) $\tan x$

   (c) $\sin^{-1}x$ (d) $\tan^{-1}x$.

3. In which of the following interval the condition of Rolle's theorem is true :

   (a) $[a, b]$ (b) (a, b)

   (c) $(a, b]$ (d) None.

4. If all conditions of Rolle's theorem are satisfied then in the given interval there will be :

   (a) one value of c where derivative of function is zero.

   (b) two values of c where the given function is zero.

   (c) many points like c where derivative of the function will be zero.

   (d) None.

5. If $f(x) = e^x$ the value of c of Lagrange's theorem will be :

   (a) $e-1$ (b) $\log(e-1)$

   (c) $e(e-1)$ (d) $\frac{1}{e-1}$.

ANSWERS

1. (*b*)	**2.** (*d*)
3. (*a*)	**4.** (*c*)
5. (*b*)	

4 Indeterminate Forms

4.1. Indeterminate Forms

Let $f(x) = \dfrac{\phi(x)}{\Psi(x)}$

Then from the theory of limits, we know that $\lim_{x \to a} f(x) = \dfrac{\lim_{x \to a} \phi(x)}{\lim_{x \to a} \Psi(x)}$

Now if $\lim_{x \to a} \phi(x) = 0$ and $\lim_{x \to a} \Psi(x) = 0$, then $\dfrac{\lim_{x \to a} \phi(x)}{\lim_{x \to a} \Psi(x)}$

takes the form $\dfrac{0}{0}$ which is meaningless. But it does not mean that

$$\frac{\lim_{x \to a} \phi(x)}{\lim_{x \to a} \Psi(x)}$$

is meaningless or that it does not exist. The reality is that the method adopted by us to evaluate the limit is not suitable here. So we shall discuss an appropriate method of evaluating the limits in such and similar other cases.

The form $\dfrac{0}{0}$ *is called an indeterminate form. The other indeterminate forms* are*

$$\frac{\infty}{\infty}, \quad \infty - \infty, \quad 0 \times \infty, \quad 0^0, \quad \infty^0 \text{ and } 1^{\infty}.$$

---

* It should be very clearly borne in mind that 0×0 and $\infty \times \infty$ are not indeterminate forms as their values are 0 and ∞ respectively.

4.2. De L' Hospital's Rule (The form $\frac{0}{0}$)

Let $\phi(x)$ and $\Psi(x)$ be two functions of x capable of being expanded by Taylor's Theorem in the neighbourhood of $x = a$ and let $\phi(a) = \Psi(a) = 0$, then

$\lim_{x \to a} \frac{\phi(x)}{\Psi(x)} = \lim_{x \to a} \frac{\phi'(x)}{\Psi'(x)}$ provided the later limit exists, whether finite or infinite.

Proof. We know that

$$\phi(x) = \phi\,[a + (x-a)]$$

$$= \phi(a) + (x-a)\,\phi' + \frac{(x-a)^2}{\lfloor 2}\,\phi''(a)$$

$$+ \ldots\ldots + R_1 \quad \text{where}$$

$$R_1 = \frac{(x-a)^n}{\lfloor n}\,\phi^n\,[a + \phi_1(x-a)],\ 0 < \phi_1 < 1$$

expanding by Taylor's theorem

and $$\psi(x) = \psi\,[a + (x-a)]$$

$$\psi(a) + (x-a)\,\psi'(a) + \frac{(x-a)^2}{\lfloor 2}\,\psi''(a) + \ldots\ldots + R_2 \quad \text{where}$$

$$R_2 = \frac{(x-a)^n}{\lfloor n}\,\psi^n\,[a + \theta_2\,(x-a)],\ 0 < \theta_2 < 1$$

expanding by Taylor's theorem

$$\therefore \lim_{x \to a} \frac{\phi(x)}{\psi(x)}$$

$$= \lim_{x \to a} \frac{\phi(a) + (x-a)\,\phi'(a) + \frac{(x-a)^2}{\lfloor 2}\,\phi''(a) + \ldots\ldots + R_1}{\Psi(a) + (x-a)\,\Psi'(a) + \frac{(x-a)^2}{\lfloor 2}\,\Psi''(a) + \ldots\ldots + R_2}$$

$$= \lim_{x \to a} \frac{(x-a)\,\phi'(a) + \frac{(x-a)^2}{\lfloor 2}\,\phi''(a) + \ldots\ldots + R_1}{(x-a)\,\Psi'(a) + \frac{(x-a)^2}{\lfloor 2}\,\Psi''(a) + \ldots\ldots + R_2}$$

$$\because \phi(a) = \Psi(a) = 0$$

Dividing the numerator and denominator by $(x-a)$

$$= \lim_{x \to a} \frac{\phi'(a) + \frac{(x-a)}{\lfloor 2} \phi''(a) + \ldots\ldots}{\Psi'(a) + \frac{(x-a)}{\lfloor 2} \Psi''(a) + \ldots\ldots}$$

$$= \frac{\phi'(a)}{\Psi'(a)}$$

$$= \lim_{x \to a} \frac{\phi'(x)}{\Psi'(x)}$$

Notes:

1. If $\phi(a) = \phi'(a) = \phi''(a) = \ldots\ldots = \phi^{(n-1)}(a) = 0$
   and $\Psi(a) = \Psi'(a) = \Psi'(a) = \ldots\ldots = \Psi^{(n-1)}(a) = 0$
   but $\phi^n(a)$ and $\Psi^n(a)$ are not both zero, then
   $$\lim_{x \to a} \frac{\phi(x)}{\Psi(x)} = \lim_{x \to a} \frac{\phi^n(x)}{\Psi^n(x)}$$
2. The above proposition is true even if $x \to \infty$ instead of a *i.e.*, if $\lim_{x \to \infty} \phi(x) = 0$ and $\lim_{x \to a} \Psi(x) = 0$, then
   $$\lim_{x \to \infty} \frac{\phi(x)}{\Psi(x)} = \lim_{x \to \infty} \frac{\phi'(x)}{\Psi'(x)}$$
   Put $x = \frac{1}{z}$ so that when $x \to \infty$
   $z \to 0$. Then we get $\lim_{x \to \infty} \frac{\phi(x)}{\Psi(x)} = \lim_{z \to 0} \frac{\phi\left(\frac{1}{z}\right)}{\Psi\left(\frac{1}{z}\right)}$
   $$= \lim_{z \to 0} \frac{\phi'\left(\frac{1}{z}\right) \cdot -\left(\frac{1}{z^2}\right)}{\Psi'\left(\frac{1}{z}\right) \cdot \left(\frac{-1}{z^2}\right)}$$
   by the above proposition
   $$= \lim_{z \to 0} \frac{\phi'\left(\frac{1}{z}\right)}{\Psi'\left(\frac{1}{z}\right)}$$
   $$= \lim_{x \to \infty} \frac{\phi'(x)}{\Psi'(x)}$$
3. **Rule.** If $\frac{\phi(x)}{\Psi(x)}$ takes the form $\frac{0}{0}$ when $x \to a$, differentiate the numerator and denominator simultaneously but separately [i.e., the students are not supposed to differentiate $\frac{\phi(x)}{\Psi(x)}$ as a function] till the occurrence of the indeterminate form disappears and then evaluate the limit.
   [For this the students are required to check at each step whether the indeterminate form has disappeared or not].

4. We are not finding out here the value of $\frac{0}{0}$. We are simply finding out the limit of the combination of the functions which has assumed the form $\frac{0}{0}$, when the limits of the functions are taken separately. A similar proposition holds for other indeterminate forms also.

Alternate Proof:

We have

$$\lim_{x \to a} f(x) = \frac{\lim_{x \to a} \phi(x)}{\lim_{x \to a} \Psi(x)}$$

Put $x = a + h$ so that as $x \to a, \quad h \to 0$

$$\therefore \frac{\lim_{x \to a} \phi(x)}{\lim_{x \to a} \Psi(x)}$$

$$= \frac{\lim_{h \to 0} \phi(a+h)}{\lim_{h \to 0} \Psi(a+h)}$$

$$= \frac{\lim_{h \to 0} \phi(a+h) - \phi(a)}{\lim_{h \to 0} \Psi(a+h) - \Psi(a)}$$

$$\because \phi(a) = \Psi(a) = 0$$

$$= \frac{\lim_{h \to 0} \frac{\phi(a+h) - \phi(a)}{h}}{\lim_{h \to 0} \frac{\Psi(a+h) - \Psi(a)}{h}}$$

$$= \frac{\phi'(a)}{\Psi'(a)}$$

$$= \frac{\lim_{x \to a} \phi'(x)}{\lim_{x \to a} \Psi'(x)}$$

4.3. Algebraic Methods

The expansions of the functions involved are used sometimes to shorten the work of evaluating the given limits which assume the indeterminate forms.

ILLUSTRATIVE EXAMPLES

Example 1. *Evaluate*

$$\lim_{x \to 0} \frac{(1+x)^n - 1}{x}$$

Solution :

$$\lim_{x \to 0} \frac{(1+x)^n - 1}{x} \qquad \left[\text{Form } \frac{0}{0}\right]$$

$$= \lim_{x \to 0} \frac{n(1+x)^{n-1}}{1}$$

Differentiating the numerator and denominator w.r.t. x

$$= n$$

Aliter:

$$\lim_{x \to 0} \frac{(1+x)^n - 1}{x}$$

$$= \lim_{x \to 0} \frac{\{1 + nx + n(n-1)\frac{x^2}{\lfloor 2} + \dots\} - 1}{x}$$

$$= \lim_{x \to 0} n + n(n-1)\frac{x}{\lfloor 2} + \dots$$

$$= n$$

Example 2. *Evaluate*

$$\lim_{x \to 0} \frac{xe^x - \log(1+x)}{x^2}$$

Solution :

$$\lim_{x \to 0} \frac{xe^x - \log(1+x)}{x^2} \qquad \left[\text{Form } \frac{0}{0}\right]$$

Differentiating the numerator and denomiantor w.r.t. x

$$= \lim_{x \to 0} \frac{xe^x + e^x - \frac{1}{1+x}}{2x} \qquad \left[\text{Form } \frac{0}{0}\right]$$

Again differentiating the numerator and denominator

$$= \lim_{x \to 0} \frac{xe^x + e^x + e^x + \frac{1}{(1+x)^2}}{2}$$

$$= \frac{3}{2}$$

Aliter: $$\lim_{x \to 0} \frac{xe^x - \log(1+x)}{x^2}$$

$$= \lim_{x \to 0} x\left[1 + x + \frac{x^2}{\lfloor 2} + \ldots\right] - \frac{[x - \frac{x^2}{2} + \frac{x^3}{3} - \ldots]}{x^2}$$

$$= \lim_{x \to 0} \frac{\frac{3}{2}x^2 + \frac{1}{6}x^3 + \ldots}{x^2}$$

$$= \lim_{x \to 0} \left[\frac{3}{2} + \frac{1}{6}x + \text{terms containing higher powers of } x\right]$$

$$= \frac{3}{2}$$

Example 3. *Find* $\lim_{x \to 0} \frac{\log(1-x^2)}{\log \cos x}$

$$\lim_{x \to 0} \frac{\log(1-x^2)}{\log \cos x} \qquad \left[\text{Form } \frac{0}{0}\right]$$

$$= \lim_{x \to 0} \frac{\frac{-2x}{1-x^2}}{-\frac{\sin x}{\cos x}}$$

$$= \lim_{x \to 0} \frac{2x}{(1-x^2)\tan x} \qquad \left[\text{Form } \frac{0}{0}\right]$$

$$= \lim_{x \to 0} \frac{2x}{(1-x^2)\sec^2 x - 2x \tan x}$$

$$= 2$$

Aliter: We know that

$$\log \cos x = \log\left[1 - \frac{x^2}{2} + \ldots\right]$$

$$= -\left(\frac{x^2}{2}\right) - \frac{1}{2}\left(\frac{x^2}{2}\right)^2 - \ldots$$

$$\therefore \lim_{x \to 0} \frac{\log(1-x^2)}{\log \cos x}$$

$$= \lim_{x \to 0} \frac{-x^2 - \frac{x^4}{2} - \frac{x^6}{3} - \dots}{-\frac{x^2}{2} - \frac{1}{2}\left(\frac{x^2}{2}\right)^2 - \dots}$$

$$= \lim_{x \to 0} \frac{1 + \frac{1}{2}x^2 + \frac{1}{3}x^4 + \dots}{\frac{1}{2} + \frac{1}{2}\frac{x^2}{4} + \dots}$$

$$= 2$$

Example 4. *Find* $\lim_{x \to 0} \frac{x - \sin x}{\tan^3 x}$

Solution :

$$\lim_{x \to 0} \frac{x - \sin x}{\tan^3 x} \qquad \left[\text{Form } \frac{0}{0}\right]$$

$$= \lim_{x \to 0} \frac{1 - \cos x}{3\tan^2 x \sec^2 x} \qquad \left[\text{Form } \frac{0}{0}\right]$$

$$= \lim_{x \to 0} \frac{\sin x}{6 \tan x \sec^4 x + 6 \tan^3 x \sec^2 x} \qquad \left[\text{Form } \frac{0}{0}\right]$$

$$= \lim_{x \to 0} \frac{\cos x}{6 \sec^6 x + 24 \tan^2 x \sec^4 x + 18 \tan^2 x \sec^4 x + 12 \tan^4 x \sec^2 x}$$

$$= \frac{1}{6 \cdot 1 + 24 \cdot 0.1 + 18 \cdot 0.1 + 12 \cdot 0.1}$$

$$= \frac{1}{6}$$

Aliter:

$$\lim_{x \to 0} \frac{x - \sin x}{\tan^3 x}$$

$$= \lim_{x \to 0} \frac{x - \left[x - \frac{x^3}{\lfloor 3} + \frac{x^5}{\lfloor 5} - \dots\right]}{\left[x + \frac{1}{3}x^3 + \frac{2}{15}x^5 + \dots\right]^3}$$

$$= \lim_{x \to 0} \frac{\frac{x^3}{6} - \frac{x^5}{120} + \dots}{x^3 + \text{terms containing higher powers of } x}$$

$$= \frac{1}{6}$$

Example 5. *Evaluate*

$$\lim_{x \to 0} \frac{\sin 2x + 2\sin^2 x - 2\sin x}{\cos x - \cos^2 x}$$

Solution :

$$\lim_{x \to 0} \frac{\sin 2x + 2\sin^2 x - 2\sin x}{\cos x - \cos^2 x} \qquad \left[\text{Form } \frac{0}{0}\right]$$

$$= \lim_{x \to 0} \frac{2\cos 2x + 2\sin 2x - 2\cos x}{-\sin x + \sin 2x} \qquad \left[\text{Form } \frac{0}{0}\right]$$

$$= \lim_{x \to 0} \frac{-4\sin 2x + 4\cos 2x + 2\sin x}{-\cos x + 2\cos 2x}$$

$$= \frac{-4.0 + 4.1 + 2.0}{-1 + 2}$$

$$= 4$$

Example 6. *Evaluate*

$$\lim_{x \to 0} \frac{e^x - e^{x \cos x}}{x - \sin x}$$

Solution : We have,

$$e^x - e^{x \cos x} = e^x - e^x \left(1 - \frac{x^2}{\lfloor 2} + \frac{x^4}{\lfloor 2} - \dots\right)$$

$$= e^x \left[1 - e^{\left(-\frac{x^3}{\lfloor 2} + \frac{x^5}{\lfloor 4} - - - -\right)}\right]$$

$$= e^x \left[1 - \left\{1 + \left(\frac{-x^3}{\lfloor 2} + \frac{x^5}{\lfloor 4} - \dots\right) + \dots\right\}\right]$$

$$= e^x \left[\frac{x^3}{\lfloor 2} - \frac{x^5}{\lfloor 4} + \dots\right]$$

and

$$x - \sin x = x - \left(x - \frac{x^3}{\lfloor 3} + \frac{x^5}{\lfloor 5} - \dots\right)$$

$$= \frac{x^3}{\lfloor 3} - \frac{x^5}{\lfloor 5} + \dots$$

$$\therefore \qquad \lim_{x \to 0} \frac{e^x - e^{x \cos x}}{x - \sin x}$$

$$= \lim_{x \to 0} \frac{e^x \left[\dfrac{x^3}{\lfloor 2} - \dfrac{x^5}{\lfloor 4} + \ldots \right]}{\dfrac{x^3}{\lfloor 3} - \dfrac{x^5}{\lfloor 5} + \ldots}$$

$$= \lim_{x \to 0} \frac{e^x \left[\dfrac{1}{\lfloor 2} - \dfrac{x^2}{\lfloor 4} + \ldots \right]}{\left[\dfrac{1}{\lfloor 3} - \dfrac{x^2}{\lfloor 5} + \ldots \right]} \quad \text{Dividing by } x^3$$

$$= \frac{\dfrac{1}{\lfloor 2}}{\dfrac{1}{\lfloor 3}} = 3$$

Example 7. *Evaluate*

$$\lim_{x \to 0} \frac{e^x - e^{-x} - 2 \log (1 + x)}{x \sin x}$$

Solution :

$$\lim_{x \to 0} \frac{e^x - e^{-x} - 2 \log (1 + x)}{x \sin x} \qquad \left[\text{Form } \frac{0}{0}\right]$$

$$= \lim_{x \to 0} \frac{\left(1 + x + \dfrac{x^2}{\lfloor 2} + \dfrac{x^3}{\lfloor 3} + \dfrac{x^4}{\lfloor 4} + \ldots\right) - \left(1 - x + \dfrac{x^2}{\lfloor 2} - \dfrac{x^3}{\lfloor 3} + \dfrac{x^4}{\lfloor 4} - \ldots\right) - 2\left(x - \dfrac{x^2}{2} + \dfrac{x^3}{3} - \dfrac{x^4}{4} + \ldots\right)}{x\left(x - \dfrac{x^3}{\lfloor 3} + \dfrac{x^5}{\lfloor 5} - \ldots\right)}$$

$$= \lim_{x \to 0} \frac{x^2 + 2x^3 \left(\dfrac{1}{\lfloor 3} - \dfrac{1}{3}\right) + \ldots}{x^2 - \dfrac{x^4}{\lfloor 3} + \dfrac{x^6}{\lfloor 5} - \ldots}$$

$$= \lim_{x \to 0} \frac{1 + 2x \left(\dfrac{1}{\lfloor 3} - \dfrac{1}{3}\right) + \ldots}{1 - \dfrac{x^2}{\lfloor 3} + \dfrac{x^4}{\lfloor 5} - \ldots} \quad \text{Dividing by } x^2$$

$$= 1$$

Example 8. *Evaluate*

$$\lim_{x \to b} \frac{x^b - b^x}{x^x - b^b}$$

Solution :

$$\lim_{x \to b} \frac{x^b - b^x}{x^x - b^b} \qquad \left[\text{Form } \frac{0}{0}\right]$$

$$= \lim_{x \to b} \frac{b\, x^{b-1} - b^x \log b}{x^x (\log x + 1) - 0}$$

$$= \frac{b \,.\, b^{b-1} - b^b \log b}{b^b (\log b + 1)}$$

$$= \frac{b^b (1 - \log b)}{b^b (1 + \log b)}$$

$$= \frac{1 - \log b}{1 + \log b}$$

Example 9. *If* $\lim_{x \to 0} \frac{\sin 2x + a \sin x}{x^3}$ *be finite, find the value of 'a' and the limit.*

Solution :

$$\lim_{x \to 0} \frac{\sin 2x + a \sin x}{x^3} \qquad \left[\text{Form } \frac{0}{0}\right]$$

$$= \lim_{x \to 0} \frac{\left\{2x - \frac{(2x)^3}{\lfloor 3} + \frac{(2x)^5}{\lfloor 5} - \ldots\right\} + a\left\{x - \frac{x^3}{\lfloor 3} + \frac{x^5}{\lfloor 5} - \ldots\right\}}{x^3}$$

$$= \lim_{x \to 0} \frac{(2 + a)\, x + x^3 \left[-\frac{8}{\lfloor 3} - \frac{a}{\lfloor 3}\right] + \ldots}{x^3}$$

In order that the limit is finite, we must have $2 + a = 0$

$\Rightarrow \quad a = -2$

and then

$$\text{limit} = -\frac{8}{\lfloor 3} - \frac{a}{\lfloor 3}$$

$$= -\frac{8}{\lfloor 3} + \frac{2}{\lfloor 3}$$

$$= -\frac{6}{\lfloor 3}$$

$$= -1$$

Example 10. *Find the values of a, b and c so that*

$$\lim_{x \to 0} \frac{ae^x - b\cos x + c\,e^{-x}}{x \sin x} = 2$$

Solution : $\lim_{x \to 0} \frac{ae^x - b\cos x + c\,e^{-x}}{x \sin x}$

$$= \lim_{x \to 0} \frac{a\left[1 + x + \frac{x^2}{\lfloor 2} + \frac{x^3}{\lfloor 3} + \frac{x^4}{\lfloor 4} + \ldots\right] - b\left[1 - \frac{x^2}{\lfloor 2} + \frac{x^4}{\lfloor 4} - \ldots\right] + c\left[1 - x + \frac{x^2}{\lfloor 2} - \frac{x^3}{\lfloor 3} + \frac{x^4}{\lfloor 4} - \ldots\right]}{x\left[x - \frac{x^3}{\lfloor 3} + \frac{x^5}{\lfloor 5} - \ldots\right]}$$

$$= \lim_{x \to 0} \frac{(a - b + c) + x\left(a + \frac{b}{2} - c\right) + x^2\left(\frac{a}{\lfloor 2} + \frac{b}{\lfloor 2} + \frac{c}{\lfloor 2}\right) + \ldots}{x^2 - \frac{x^4}{\lfloor 3} + \frac{x^6}{\lfloor 5} - \ldots}$$

Since the limit is finite (2),

$$\therefore a - b + c = 0$$

$$a + \frac{b}{2} - c = 0$$

and $\frac{a}{\lfloor 2} + \frac{b}{\lfloor 2} + \frac{c}{\lfloor 2} = 2$

$\Rightarrow$ $a - b + c = 0$ (1)

$2a + b - 2c = 0$ (2)

$a + b + c = 4$ (3)

Solving (1), (2) and (3) for a, b, c, we get

$a = 1$

$b = 2$

$c = 1$

EXERCISE 4 (A)

Evaluate the following limits :

1. $\lim_{x \to 1} \frac{\log_e x}{x - 1}$

2. $\lim_{x \to 0} \frac{e^x - 1}{x}$

3. $\lim_{x \to 0} \frac{a^x - 1}{x}$

4. $\lim_{x \to 0} \frac{1 - \cos x}{3x^2}$

5. $\lim_{x \to 0} \frac{a^x - 1}{b^x - 1}$

6. $\lim_{x \to 0} \frac{a^x - b^x}{x}$

7. $\lim_{x \to 0} \frac{x - \sin x}{x^3}$

8. $\lim_{x \to 0} \frac{x - \tan x}{x^3}$

9. $\lim_{x \to 0} \frac{\sin ax}{\sin bx}$

10. $\lim_{x \to 0} \frac{\log (1 + k x^2)}{1 - \cos x}$

11. $\lim_{x \to 0} \frac{e^x + \log \left(\frac{1 - x}{e}\right)}{\tan x - x}$

12. $\lim_{x \to 0} \frac{a^x - 1 - x \log a}{x^2}$

13. $\lim_{x \to 0} \frac{1 - \cos x}{x \log (1 + x)}$

14. $\lim_{x \to 0} \frac{\cosh x - \cos x}{x \sin x}$

15. $\lim_{x \to 0} \frac{1 - \sqrt{1 - x^2}}{x^2}$

16. $\lim_{x \to 1} \left(\frac{1}{\log_e x} - \frac{x}{\log_e x}\right)$

17. $\lim_{x \to 0} \frac{x^{\frac{1}{2}} \tan x}{(e^x - 1)^{3/2}}$

18. $\lim_{x \to \frac{1}{2}} \frac{\cos^2 \pi x}{e^{2x} - 2ex}$

19. $\lim_{x \to 0} \frac{\sin x \sin^{-1} x}{x^2}$

20. $\lim_{x \to 0} \frac{\sin x \sin^{-1} x - x^2}{x^6}$

21. $\lim_{x \to 0} \frac{5 \sin x - 7 \sin 2x + 3 \sin 3x}{\tan x - x}$

22. $\lim_{x \to a} \frac{a^x - x^a}{x^x - a^a}$

23. $\lim_{x \to 0} \frac{\sin x - x + \frac{x^3}{6}}{x^5}$

24. $\lim_{x \to \frac{\pi}{2}} \frac{\cos x}{x - 1}$

25. $\lim_{x \to 1} \frac{x^x - x}{1 - x + \log x}$

26. $\lim_{x \to 0} \frac{\tan x - \sin x}{x^3}$

27. $\lim_{x \to 0} \frac{3^x - 2^x}{\sqrt{x}}$

28. $\lim_{x \to 0} \dfrac{1 + \sin x + \log(1 - x) - \cos x}{x \tan^2 x}$

29. $\lim_{x \to \frac{\pi}{2}} \dfrac{\left(\frac{\pi}{2} - x\right)^2 \sin x}{\cos^2 x}$

30. $\lim_{x \to 0} \dfrac{\sinh x - x}{\sin x - x \cos x}$

31. $\lim_{x \to 1} \dfrac{x\sqrt{3x - 2x^4} - x^{6/5}}{1 - x^{2/3}}$

32. $\lim_{x \to 1} \dfrac{2 \sin \pi x + \pi (x^2 - 1)^2}{(x^2 - 1)^2}$

33. Find the values of a and b in order that

$\lim_{x \to 0} \dfrac{x(1 + a \cos x) - b \sin x}{x^3}$ may be equal to 1.

34. Find the values of a and b in order that

$\lim_{x \to 0} \dfrac{x(1 - a \cos x) + b \sin x}{x^3}$ may be equal to $\dfrac{1}{3}$.

35. Find the values of a, b and c so that

$$\lim_{x \to 0} \frac{ae^x - b \cos x + c e^{-x}}{x \sin x} = 2$$

36. Find the values of a, b and c so that

$\lim_{x \to 0} \dfrac{a \sin x - bx + cx^2 + x^3}{2x^2 \log_e(1 + x) - 2x^3 + x^4}$ be finite and determine the value also.

ANSWERS

1. 1	**2.** 1
3. log a	**4.** $\frac{1}{6}$
5. $\frac{\log a}{\log b}$	**6.** $\log\left(\frac{a}{b}\right)$
7. $\frac{1}{6}$	**8.** $-\frac{1}{3}$

9. $\frac{a}{b}$	10. 2k
11. $-\frac{1}{2}$	12. $\frac{(\log_e a)^2}{2}$
13. $\frac{1}{2}$	14. 1
15. $\frac{1}{2}$	16. -1
17. 1	18. $\frac{\pi^2}{2e}$
19. 1	20. $\frac{1}{18}$
21. -15	22. $\frac{\log a - 1}{\log a + 1}$
23. $\frac{1}{120}$	24. 1
25. -2	26. $\frac{1}{2}$
27. 0	28. $-\frac{1}{2}$
29. 1	30. $\frac{1}{2}$
31. $\frac{81}{20}$	32. π
33. $a = -\frac{5}{2}, b = -\frac{3}{2}$	34. $a = \frac{1}{2}, b = -\frac{1}{2}$
35. $a = 1$ $b = 1$ $c = 1$	36. $a = 6$ $b = 6$ $c = 0$. limit $= \frac{3}{40}$

4.4. The form $\frac{\infty}{\infty}$

If $\lim_{x \to a} \phi(x) = \infty$ and $\lim_{x \to a} \Psi(x) = \infty$, then to show that

$$\lim_{x \to a} \frac{\phi(x)}{\Psi(x)} = \lim_{x \to a} \frac{\phi'(x)}{\Psi'(x)}$$

PROOF : We have

$$\lim_{x \to a} \frac{\phi(x)}{\Psi(x)}$$

$$= \lim_{x \to a} \frac{\dfrac{1}{\Psi(x)}}{\dfrac{1}{\phi(x)}} \qquad \left[\text{Form } \frac{0}{0}\right]$$

$$= \lim_{x \to a} \frac{-\dfrac{\Psi'(x)}{\{\Psi(x)\}^2}}{-\dfrac{\phi'(x)}{\{\phi(x)\}^2}} \qquad \text{Differentiating the numerator and denominator}$$

$$= \lim_{x \to a} \left[\frac{\phi(x)}{\Psi(x)}\right]^2 \frac{\Psi'(x)}{\phi'(x)} \qquad (1)$$

Let $\lim_{x \to a} \dfrac{\phi(x)}{\Psi(x)} = \lambda$ (2)

Then there arise three cases.

CASE I $\lambda \neq 0, \lambda \neq \infty$

Then from (1), we have

$$\lambda = \lambda^2 \lim_{x \to a} \frac{\Psi'(x)}{\phi'(x)}$$

Since λ is neither zero nor infinite, so dividing both sides of the above equation by λ^2, we get

$$\lambda^{-1} = \lim_{x \to a} \frac{\Psi'(x)}{\phi'(x)}$$

$$\Rightarrow \lim_{x \to a} \frac{\phi'(x)}{\Psi'(x)} = \lambda$$

CASE II $\lambda = 0$

Adding 1 to each side of equation (2), we get

$$\lambda + 1 = \lim_{x \to a} \frac{\phi(x)}{\Psi(x)} + 1$$

$$= \lim_{x \to a} \frac{\phi(x)}{\Psi(x)} + \lim_{x \to a} \frac{\Psi(x)}{\Psi(x)}$$

$$= \lim_{x \to a} \frac{\phi(x) + \Psi(x)}{\Psi(x)}$$

$$= \lim_{x \to a} \frac{\phi'(x) + \Psi'(x)}{\Psi'(x)} \qquad \text{by Case I}$$

$$= \lim_{x \to a} \frac{\phi'(x)}{\Psi'(x)} + 1$$

Hence $\lambda = \lim_{x \to a} \frac{\phi'(x)}{\Psi'(x)}$

Case III $\quad \lambda = \infty$

In this case, we have

$$\frac{1}{\lambda} = \frac{1}{\lim_{x \to a} \frac{\phi(x)}{\Psi(x)}}$$

$$= \lim_{x \to a} \frac{\Psi(x)}{\phi(x)}$$

$$= \lim_{x \to a} \frac{\Psi'(x)}{\phi'(x)} \qquad \text{by Case II}$$

$$\Rightarrow \lambda = \lim_{x \to a} \frac{\phi'(x)}{\Psi'(x)}$$

Thus in every case in which

$$\lim_{x \to a} \phi(x) = \infty$$

and $\lim_{x \to a} \Psi(x) = \infty$,

we get

$$\lim_{x \to a} \frac{\phi(x)}{\Psi(x)} = \lim_{x \to a} \frac{\phi'(x)}{\Psi'(x)}$$

Notes

1. By putting $x = \frac{1}{z}$ we can prove that the above proposition is also true even if $x \to \infty$. [The treatment is similar as in case of article 4.2]
2. We have seen that the methods of evaluating the limits of the quotients of the two functions are similar in each of the two cases when they assume the form $\frac{0}{0}$ or $\frac{\infty}{\infty}$. Though

the quotients of the two functions can immediately be put into a form which would change the indeterminate form $\frac{\infty}{\infty}$ to $\frac{0}{0}$ and vice-versa, however before evaluating the limits we should choose the convenient form so that the limit may be obtained more quickly. Suppose if we have the form $\frac{1}{x}$ in the numerator or the denominator, and the limit as x approaches to zero is to be found, then the process of differentiation would not terminate, since it would involve $x^{-1}, x^{-2}, \ldots$, etc., which would all tend to ∞ as $x \to 0$. Hence we should change to the form $\frac{0}{0}$ at a suitable stage.

4.5. Compound Indeterminate Forms

If a function is the product of several factors, the limit of each of which correspond to a simple indeterminate form for the same value of x, then the limit of such a function will be equal to the product of the limits of the factors provided the product is not in itself an indeterminate form. A similar rule is applicable in the cases of a quotient, sum, difference, or power.

4.6. Some Important Algebraic Expansions

(i) $(x + a)^n = x^n + {}^nC_1\, x^{n-1} a + {}^nC_2\, x^{n-2} a^2 + \ldots + {}^nC_n\, a^n$

(ii) $(1+x)^{-1} = 1 - x + x^2 - x^3 + x^4 - \ldots \infty$

(iii) $(1-x)^{-1} = 1 + x + x^2 + x^3 + x^4 + \ldots \infty$

(iv) $(1+x)^{-2} = 1 - 2x + 3x^2 - 4x^3 + 5x^4 - \ldots \infty$

(v) $(1-x)^{-2} = 1 + 2x + 3x^2 + 4x^3 + 5x^4 + \ldots \infty$

(vi) $(1 + x)^{-3} = 1 - 3x + 6x^2 - 10x^3 + \ldots \infty$

(vii) $(1 - x)^{-3} = 1 + 3x + 6x^2 + 10x^3 + \ldots \infty$

(viii) $e^x = 1 + \frac{x}{\lfloor 1} + \frac{x^2}{\lfloor 2} + \frac{x^3}{\lfloor 3} + \ldots \infty$

(ix) $e^{-x} = 1 - \frac{x}{\lfloor 1} + \frac{x^2}{\lfloor 2} - \frac{x^3}{\lfloor 3} + \ldots \infty$

(x) $\sin x = x - \frac{x^3}{\lfloor 3} + \frac{x^5}{\lfloor 5} - \ldots \infty$

(xi) $\cos x = 1 - \frac{x^2}{\lfloor 2} + \frac{x^4}{\lfloor 4} - \frac{x^6}{\lfloor 6} - \cdots \infty$

(xii) $\sinh x = x + \frac{x^3}{\lfloor 3} + \frac{x^5}{\lfloor 5} + \ldots \infty$

(xiii) $\cosh x = 1 + \frac{x^2}{\lfloor 2} + \frac{x^4}{\lfloor 4} + \ldots. \infty$

(xiv) $\tan x = x + \frac{1}{3} x^3 + \frac{2}{15} x^5 + \ldots. \infty$

(xv) $\tan^{-1} x = x - \frac{1}{3} x^3 + \frac{x^5}{5} - \frac{x^7}{7} + \ldots. \infty$

(xvi) $\log (1 + x) = x - \frac{x^2}{2} + \frac{x^3}{3} - \frac{x^4}{4} + \ldots. \infty$

(xvii) $\log (1 - x) = -x - \frac{x^2}{2} - \frac{x^3}{3} - \frac{x^4}{4} - \ldots. \infty$

(xviii) $\sin^{-1} x = x + 1^2 . \frac{x^3}{\lfloor 3} + 3^2 . 1^2 . \frac{x^5}{\lfloor 5} + 5^2 . 3^2 . 1^2 . \frac{x^7}{\lfloor 7} + \ldots. \infty$

4.7. Use of Standard Limits

Sometimes it is desirable to use the standard limits. It shortens the work of evaluating the limits. Some standard limits are being given below for the sake of convenience.

(i) $\lim_{x \to 0} \frac{\sin x}{x} = 1$

(ii) $\lim_{x \to 0} \cos x = 1$

(iii) $\lim_{x \to 0} \frac{\tan x}{x} = 1$

(iv) $\lim_{x \to 0} (1 + x)^{\frac{1}{x}} = e$

(v) $\lim_{x \to \infty} \left(1 + \frac{1}{x}\right)^x = e$

ILLUSTRATIVE EXAMPLES

Example 1. *Evaluate the limit*

$$\lim_{x \to \frac{\pi}{2}} \frac{\tan 5x}{\tan x}$$

Solution :

$$\lim_{x \to \frac{\pi}{2}} \frac{\tan 5x}{\tan x} \qquad \left[\text{Form } \frac{\infty}{\infty}\right]$$

[**Note :** If we differentiate the numerator and denominator, we would have the form $\frac{\infty}{\infty}$. Hence it is advisable here to change it to the form $\frac{0}{0}$].

$$= \lim_{x \to \frac{\pi}{2}} \frac{\cot x}{\cot 5x} \qquad \left[\text{Form } \frac{0}{0}\right]$$

Dividing the numerator and denominator by tan x tan 5x

$$= \lim_{x \to \frac{\pi}{2}} \frac{-\operatorname{cosec}^2 x}{-5 \operatorname{cosec}^2 5x}$$

Differentiating the numerator and denominator

$$= \frac{1}{5}$$

Example 2. *Evaluate*

$$\lim_{x \to 0} \frac{\log x}{\cot x}$$

Solution :

$$\lim_{x \to 0} \frac{\log x}{\cot x} \qquad \left[\text{Form } \frac{\infty}{\infty}\right]$$

$$= \lim_{x \to 0} \frac{\frac{1}{x}}{-\operatorname{cosec}^2 x}$$

Differentiating the numerator and denominator

$$= \lim_{x \to 0} -\frac{\sin^2 x}{x}$$

$$= \lim_{x \to 0} (-\sin x) \left(\frac{\sin x}{x}\right)$$

$$= \lim_{x \to 0} (-\sin x) \left(\lim_{x \to 0} \frac{\sin x}{x}\right)$$

$$= (0)(1)$$

$$= 0$$

Example 3. *Evaluate*

$$\lim_{x \to 0} \frac{\log \sin 2x}{\log \sin x}$$

Solution :

$$\lim_{x \to 0} \frac{\log \sin 2x}{\log \sin x} \qquad \left[\text{Form } \frac{\infty}{\infty}\right]$$

$$= \lim_{x \to 0} \frac{\dfrac{\cos 2x}{\sin 2x}}{\dfrac{\cos x}{\sin x}}$$

Differentiating the numerator and denominator

$$= \lim_{x \to 0} \frac{2 \sin x \cos 2x}{\cos x \sin 2x}$$

$$= \lim_{x \to 0} \frac{2 \sin x \cos 2x}{\cos x \,.\, 2 \sin x \cos x}$$

$$= \lim_{x \to 0} \frac{\cos 2x}{\cos^2 x}$$

$$= \lim_{x \to 0} \left(\frac{\cos^2 x - \sin^2 x}{\cos^2 x}\right)$$

$$= \lim_{x \to 0} \left(1 - \tan^2 x\right)$$

$$= 1$$

Example 4. *Prove that*

$$\lim_{x \to 0} \log_{\tan^2 x} (\tan^2 2x) = 1$$

Solution : $\lim_{x \to 0} \log_{\tan^2 x} (\tan^2 2x)$

$$= \lim_{x \to 0} \log_e \tan^2 2x \,.\, \log_{\tan^2 x} e \qquad \because \log_a m = \log_b m \times \log_a b$$

$$= \lim_{x \to 0} \frac{\log_e \tan^2 2x}{\log_e \tan^2 x} \qquad \because \log_b a = \frac{1}{\log_a b}$$

$$= \lim_{x \to 0} \frac{2 \log \tan 2x}{2 \log \tan x}$$

$$= \lim_{x \to 0} \frac{\log \tan 2x}{\log \tan x} \qquad \left[\text{Form } \frac{\infty}{\infty}\right]$$

$$= \lim_{x \to 0} \frac{\dfrac{\sec^2 2x}{\tan 2x} . 2}{\dfrac{\sec^2 x}{\tan x}}$$

$$= \lim_{x \to 0} \frac{2 \sec^2 2x . \tan x}{\tan 2x . \sec^2 x}$$

$$= \lim_{x \to 0} \frac{2}{\cos^2 2x} . \frac{\sin x}{\cos x} . \frac{\cos^2 x}{1} . \frac{\cos 2x}{\sin 2x}$$

$$= \lim_{x \to 0} \frac{2 \cos x \quad \sin x}{\cos 2x \quad \sin 2x}$$

$$= \lim_{x \to 0} \frac{2 \sin x \cos x}{\cos 2x . 2 \sin x \cos x}$$

$$= \lim_{x \to 0} \frac{1}{\cos 2x}$$

$$= 1$$

Example 5. *Evaluate*

$$\lim_{x \to 0} \left(\frac{1}{x^2} - \cot^2 x \right)$$

Solution :

$$\lim_{x \to 0} \left(\frac{1}{x^2} - \cot^2 x \right) \qquad [\text{Form } \infty - \infty]$$

$$= \lim_{x \to 0} \left(\frac{1}{x^2} - \frac{\cos^2 x}{\sin^2 x} \right)$$

$$= \lim_{x \to 0} \left(\frac{\sin^2 x - x^2 \cos^2 x}{x^2 \sin^2 x} \right) \qquad \left[\text{Form } \frac{0}{0}\right]$$

$$= \lim_{x \to 0} \frac{\left(x - \dfrac{x^3}{\lfloor 3} + \dfrac{x^5}{\lfloor 5} - \dots \right)^2 - x^2 \left(1 - \dfrac{x^2}{\lfloor 2} + \dots \right)^2}{x^2 \left(x - \dfrac{x^3}{\lfloor 3} + \dots \right)^2}$$

$$= \lim_{x \to 0} \frac{\frac{2}{3} x^4 + \text{terms containing higher powers of } x}{x^4 + \text{terms containing higher powers of } x}$$

$$= \frac{2}{3}$$

Example 6. *Evaluate*

$$\lim_{x \to 1} \sec \frac{\pi x}{2} \log x$$

Solution :

$$\lim_{x \to 1} \sec \frac{\pi x}{2} \log x \qquad [\text{Form } \infty \times 0]$$

$$= \lim_{x \to 1} \frac{\log x}{\cos \frac{\pi x}{2}} \qquad \left[\text{Form } \frac{0}{0}\right]$$

$$= \lim_{x \to 1} \frac{\frac{1}{x}}{\sin \frac{\pi x}{2} \cdot \frac{\pi}{2}}$$

$$= \frac{2}{\pi}$$

Example 7. *Evaluate*

$$\lim_{x \to 0} x^x$$

Solution :

$$\text{Let } P = \lim_{x \to 0} x^x \qquad [\text{Form } 0^\circ]$$

$$\therefore \log P = \lim_{x \to 0} x^{(\log x)} \qquad [\text{Form } 0 \times \infty]$$

$$= \lim_{x \to 0} \frac{\log x}{\frac{1}{x}} \qquad \left[\text{Form } \frac{\infty}{\infty}\right]$$

$$= \lim_{x \to 0} \frac{\frac{1}{x}}{-\frac{1}{x^2}}$$

$$= \lim_{x \to 0} (-x)$$

$$= 0$$

$$\Rightarrow P = e^\circ = 1$$

Example 8. *Evaluate*

$$\lim_{x \to 0} (\cos x)^{\frac{1}{x}}$$

Solution :

$$\text{Let} \qquad P = \lim_{x \to 0} (\cos x)^{\frac{1}{x}} \qquad [\text{Form } 1^{\infty}]$$

$$\therefore \qquad \log P = \lim_{x \to 0} \frac{1}{x} \log \cos x \qquad [\text{Form } \infty \times 0]$$

$$= \lim_{x \to 0} \frac{\log \cos x}{x} \qquad \left[\text{Form } \frac{0}{0}\right]$$

$$= \lim_{x \to 0} \frac{-\tan x}{1}$$

$$= 0$$

$$\Rightarrow \quad P = e^{0} = 1$$

Example 9. *Evaluate*

$$\lim_{x \to 0} \left(\frac{\tan x}{x}\right)^{\frac{1}{x^3}}$$

Solution :

$$\lim_{x \to 0} \left(\frac{\tan x}{x}\right)^{\frac{1}{x^3}}$$

$$= \lim_{x \to 0} \left(\frac{x + \frac{1}{3}x^3 + \ldots}{x}\right)^{\frac{1}{x^3}}$$

$$= \lim_{x \to 0} \left(1 + \frac{x^2}{3} + \ldots\right)^{\frac{1}{x^3}}$$

$$= \lim_{x \to 0} \left(1 + \frac{x^2}{3}\right)^{\frac{1}{x^3}}$$

$$= \lim_{x \to 0} \left[\left(1 + \frac{x^2}{3}\right)^{\frac{3}{x^2}}\right]^{\frac{1}{3x}}$$

$$= \lim_{\substack{x \to 0 \\ y \to 0}} \left[(1 + y)^{\frac{1}{y}}\right]^{\frac{1}{3x}}$$

where $y = \dfrac{x^2}{3}$

$$= \lim_{x \to 0} e^{\frac{1}{3x}}$$
$$= e^{\infty}$$
$$= \infty$$

Example 10. *Evaluate*

$$\lim_{x \to 0} \left(\frac{1}{x}\right)^{\tan x}$$

Solution : Let $P = \lim_{x \to 0} \left(\frac{1}{x}\right)^{\tan x}$ [Form ∞°]

$\therefore \quad \log P = \lim_{x \to 0} \tan x \log \left(\frac{1}{x}\right)$ [Form $0 \times \infty$]

$$= \lim_{x \to 0} \frac{\log\left(\frac{1}{x}\right)}{\cot x} \quad \left[\text{Form } \frac{\infty}{\infty}\right]$$

$$= \lim_{x \to 0} \frac{-\log x}{\cot x} \quad \left[\text{Form } \frac{\infty}{\infty}\right]$$

$$= \lim_{x \to 0} \frac{-\frac{1}{x}}{-\ cosec^2\ x} \quad \text{[By De L' Hospital's rule]}$$

$$= \lim_{x \to 0} \frac{\sin^2 x}{x} \quad \left[\text{Form } \frac{0}{0}\right]$$

$$= \lim_{x \to 0} \frac{\sin x}{x} \cdot \lim_{x \to 0} (\sin x)$$
$$= 1 . 0 = 0$$

$\Rightarrow \quad P = e^{\circ} = 1$

Example 11. *Evaluate*

$$\lim_{x \to 0} \left[\frac{2\ (\cosh x - 1)}{x^2}\right]^{\frac{1}{x^2}}$$

Solution :

$$\text{Let } P = \lim_{x \to 0} \left[\frac{2\ (\cosh x - 1)}{x^2}\right]^{\frac{1}{x^2}}$$

$$\therefore \log P = \lim_{x \to 0} \frac{1}{x^2} \log \left[\frac{2(\cosh x - 1)}{x^2}\right]$$

$$= \lim_{x \to 0} \frac{1}{x^2} \log \frac{2}{x^2} \left\{1 + \frac{x^2}{\lfloor 2} + \frac{x^4}{\lfloor 4} + ---- - 1\right\}$$

$$= \lim_{x \to 0} \frac{1}{x^2} \log \left(1 + \frac{x^2}{12} + ----\right)$$

$$= \lim_{x \to 0} \frac{1}{x^2} \left[\left(\frac{x^2}{12} + ---\right) - \frac{1}{2}\left(\frac{x^2}{12} + ---\right)^2\right] + ---$$

$$= \frac{1}{12}$$

$$= P = e^{1/12}$$

Example 12. *Evaluate*

(i) $\lim_{x \to 0} \dfrac{(1 + x)^{1/x} - e}{x}$

(ii) $\lim_{x \to 0} \dfrac{(1 + x)^{\frac{1}{x}} - e + \dfrac{ex}{2}}{x^2}$

Solution :

(i) Let $y = (1 + x)^{\frac{1}{x}}$

Then $\log y = \dfrac{1}{x} \log (1 + x)$

$$= \frac{1}{x}\left(x - \frac{x^2}{2} + \frac{x^3}{3} - -----\right)$$

$$= 1 - \frac{x}{2} + \frac{x^2}{3} - ----$$

$$\Rightarrow \quad y = e^{1 - \frac{x}{2} + \frac{x^2}{3} - ----}$$

$$= e \cdot e^{\frac{-x}{2} + \frac{x^2}{3} - ----}$$

$$= e\left[1 + \left(-\frac{x}{2} + \frac{x^2}{3} ---\right) + \frac{1}{\lfloor 2}\left(-\frac{x}{2} + \frac{x^2}{3} ---\right)^2 + ---\right]$$

$$= e\left[1 - \frac{x}{2} + x^2\left(\frac{1}{3} + \frac{1}{8}\right) + ----\right]$$

$$= e\left[1 - \frac{x}{2} + \frac{11}{24}x^2 + ----\right]$$

$$\therefore \lim_{x \to 0} \frac{(1+x)^{1/x} - e}{x} \qquad \left[\text{Form } \frac{0}{0}\right]$$

$$= \lim_{x \to 0} \frac{e\left(1 - \frac{x}{2} + \frac{11}{24}x^2 + ----\right) - e}{x}$$

$$= -\frac{e}{2}$$

(ii) $$\lim_{x \to 0} \frac{(1+x)^{1/x} - e + \frac{ex}{2}}{x^2}$$

$$= \lim_{x \to 0} \frac{e\left(1 - \frac{x}{2} + \frac{11}{24}x^2 - ----\right) - e + \frac{ex}{2}}{x^2}$$

$$= \frac{11}{24}e$$

EXERCISE 4.2 (B)

Evaluate each of the following limit:

1. $\lim_{x \to \infty} \frac{\log x}{x}$

2. $\lim_{x \to \infty} \frac{\log e^x}{a^x}$; $a > 1$

3. $\lim_{x \to 0} \frac{\cot 2x}{\cot x}$

4. $\lim_{x \to 0} \frac{\log \tan x}{x}$

5. $\lim_{x \to 0} x \log x$

6. $\lim_{x \to 0} x^2 \log x$

7. $\lim_{x \to 0} \sin x \,.\, \log x$

8. $\lim_{x \to 0} x \log_e \sin x$

9. $\lim_{x \to 0} \log_x \sin x$

10. $\lim_{x \to \frac{\pi}{2}} \frac{\log\left(x - \frac{\pi}{2}\right)}{\tan x}$

11. $\lim_{x \to 0} \frac{\log \sin ax}{\log \sin bx}$ $(a, b > 0)$

12. $\lim_{x \to 0} \frac{\log x^2}{\cot x^2}$

13. $\lim_{x \to 0} \log_{\sin x} \sin 2x$

14. $\lim_{x \to \infty} x^n e^{-x}$

15. $\lim_{x \to \infty} \dfrac{\log x}{x^m}$; $m > 0$

16. $\lim_{x \to 0} \dfrac{\operatorname{cosec} x}{\log x}$

17. $\lim_{x \to a} \dfrac{\log (x - a)}{\log (e^x - e^a)}$

18. $\lim_{x \to \infty} \dfrac{3x + 4}{\sqrt{2x^2 + 5}}$

19. $\lim_{x \to \frac{\pi}{2}} \dfrac{\tan x}{\tan 3x}$

20. $\lim_{x \to \infty} \dfrac{x^3 - 8x^2 + 2x + 1}{x^4 - x^2 + 2x - 3}$

21. $\lim_{x \to \infty} x \tan \dfrac{1}{x}$

22. $\lim_{x \to \infty} \left(a^{1/x} - 1\right) x$

23. $\lim_{x \to 1} (1 - x) \tan \dfrac{\pi x}{2}$

24. $\lim_{x \to \infty} 2^x \sin \dfrac{a}{2x}$

25. $\lim_{x \to \frac{\pi}{2}} (\sec x - \tan x)$

26. $\lim_{x \to 0} (\operatorname{cosec} x - \cot x)$

27. $\lim_{x \to 0} \left(\dfrac{1}{x^2} - \dfrac{1}{\sin^2 x}\right)$

28. $\lim_{x \to 0} \left(\dfrac{1}{x} - \cot x\right)$

29. $\lim_{x \to 0} \left[\dfrac{1}{x} - \dfrac{1}{x^2} \log (1 + x)\right]$

30. $\lim_{x \to 0} \dfrac{\cot x - \dfrac{1}{x}}{x}$

31. $\lim_{x \to \frac{\pi}{2}} (\cos x)^{\cos x}$

32. $\lim_{x \to \frac{\pi}{2}} (\sin x)^{\tan x}$

33. $\lim_{x \to \infty} \left(\dfrac{\pi}{2} - \tan^{-1} x\right)^{\frac{1}{x}}$

34. $\lim_{x \to \infty} \left(\dfrac{1}{x}\right)^{2 \sin x}$

35. $\lim_{x \to 0} \left(\dfrac{\sin x}{x}\right)^{\frac{1}{x^2}}$

36. $\lim_{x \to 0} \left(\dfrac{\sinh x}{x}\right)^{\frac{1}{x^2}}$

37. $\lim_{x \to 0} \dfrac{c\left[e^{\frac{1}{x - a}} - 1\right]}{e^{\frac{1}{x - a}} + 1}$

38. $\lim_{x \to a} \left(2 - \dfrac{x}{a}\right)^{\tan \frac{\pi x}{2a}}$

39. $\lim_{x \to \infty} \left(a_0 x^m + a_1 x^{m-1} + \dots + a_m\right)^{\frac{1}{x}}$

40. $\lim_{x \to 0} \frac{e^x - 1}{x^4 \sin x} \left(\frac{3 \sin x - \sin 3x}{\cos x - \cos 3x} \right)^4$

ANSWERS

1. 0	**2.** 0
3. $\frac{1}{2}$	**4.** 1
5. 0	**6.** 0
7. 0	**8.** 0
9. 1	**10.** 0
11. 1	**12.** 0
13. 1	**14.** 0
15. 0	**16.** $-\infty$
17. 1	**18.** $\frac{3}{\sqrt{2}}$
19. 3	**20.** 0
21. 1	**22.** $\log a$
23. $\frac{2}{\pi}$	**24.** a
25. 0	**26.** 0
27. $-\frac{1}{3}$	**28.** 0
29. $\frac{1}{2}$	**30.** $-\frac{1}{3}$
31. 1	**32.** 1
33. 1	**34.** 1
35. $e^{-1/6}$	**36.** $e^{1/6}$
37. c	**38.** $e^{2/x}$
39. 1	**40.** 1

5
Tangents and Normals

5.1. Definition of Tangent

Let P be a given point on the curve $y = f(x)$. Let $Q(x + \delta x, y + \delta y)$ by any other point on the curve in the neighbourhood of P. Then the limiting position of the chord PQ when the point Q, travelling along the curve approaches P, is called the tangent to the curve at the point P.

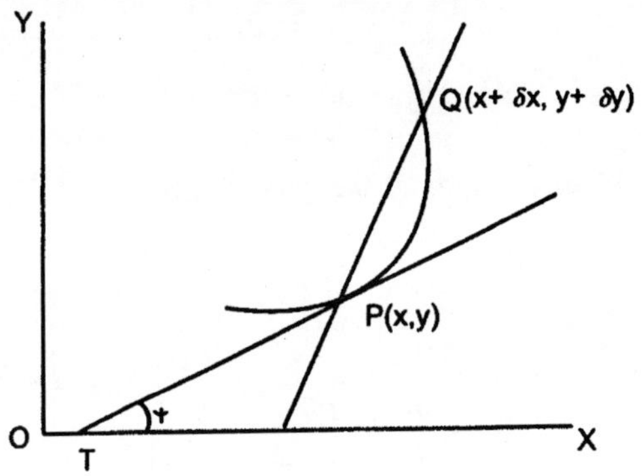

Fig. 5.1

5.2. Equation of the Tangent

To find the equation of the tangent at any point $P(x, y)$ of the curve $y = f(x)$.

The curve is $y = f(x)$. Let in the neighbourhood P there is point $Q\ (x + \delta x, y + \delta y)$.

Equation of the chord PQ is,

$$Y - y = \frac{(y + \delta y) - y}{(x + \delta x) - x}\ (X - x)$$

$$\Rightarrow \qquad Y - y = \frac{\delta y}{\delta x}\ (X - x),$$

where X, Y are taken as current coordinates.

Now as $Q \to P$; then $\delta x \to 0, \frac{\delta y}{\delta x} \to \frac{dy}{dx}$ and the chord PQ becomes the tangent to the curve at P.

Therefore the equation of the tangent at the point $P(x, y)$ of the curve $y = f(x)$ is,

$$Y - y = \frac{dy}{dx}(X - x). \quad \text{...(1)}$$

COROLLARY 1. If the equation of the curve is given by

$$f(x, y) = 0,$$

then we know that

$$\frac{dy}{dx} = -\frac{\partial f/\partial x}{\partial f/\partial y}$$

Hence, the equation of the tangent becomes

$$Y - y = -\frac{\partial f/\partial x}{\partial f/\partial y}(X - x)$$

$$\Rightarrow \quad (Y - y)\frac{\partial f}{\partial y} + (X - x)\frac{\partial f}{\partial x} = 0 \quad \text{....(2)}$$

COROLLARY 2. If the equation of the curve is given in the parametric form say

$$x = f(t)$$
$$y = \phi(t),$$

then we have,

$$\frac{dy}{dx} = \frac{dy/dt}{dx/dt} = \frac{\phi'(t)}{f'(t)}$$

Hence, the equation of the tangent is,

$$Y - \phi(t) = \frac{\phi'(t)}{f'(t)}\{X - f(t)\} \quad \text{....(3)}$$

Note : This is known as the equation of the tangent at the point 't' of the curve.

5.3. Geometrical Meaning of *dy/dx*

The equation of the tangent is,

$$Y - y = \frac{dy}{dx}(X - x)$$

$$\Rightarrow \quad Y = \frac{dy}{dx} \cdot X + \left(y - x\frac{dy}{dx}\right).$$

Also, the equation of a straight line which makes an angle Ψ with OX and cuts an intercept c on OY is given by,

$$Y = mX + c,$$

where $m = \tan \Psi$,

If we compare these two equations, we get

$$\frac{dy}{dx} = m = \tan \Psi = \text{slope of the tangent at } (x, y).$$

Hence, "*the differential coefficient* $\frac{dy}{dx}$ *at the point* (x, y) *of a curve* $y = f(x)$ *is equal to the trigonometrical tangent of the angle which the tangent to the curve at* (x, y), *makes with the positive direction of the axis of* x."

COROLLARY 1. If the tangent is parallel to X-axis, then $\Psi = 0$.

$$\therefore \quad \frac{dy}{dx} = \tan \Psi = \tan 0 = 0.$$

COROLLARY 2. If the tangent is perpendicular to X-axis then $\phi = 90°$.

$$\therefore \quad \frac{dy}{dx} = \tan \Psi = \tan 90° = \infty$$

or
$$\frac{dx}{dy} = \frac{1}{dy/dx} = \frac{1}{\infty} = 0.$$

5.4. Definition of Normal

The normal to a curve at any point P is a straight line which passes through the point of contact P and is at right angle to the tangent at P.

5.5. Equation of Normal

Tangent at point $P(x, y)$ is,

$$Y - y = \frac{dy}{dx}(X - x)$$

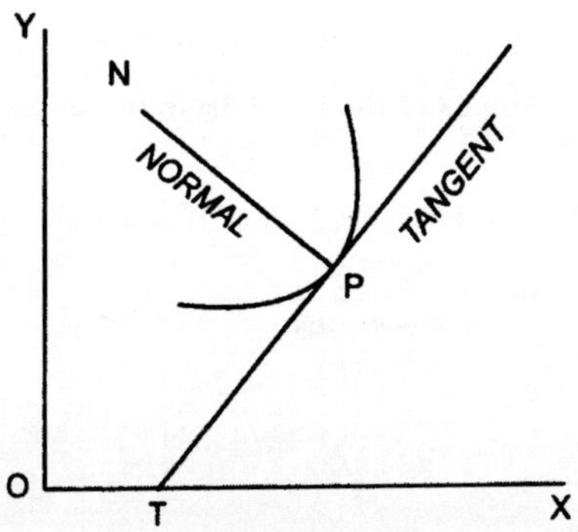

Fig. 5.2

If m is the gradient of the normal, then

$$m \frac{dy}{dx} = -1,$$

i.e.,
$$m = -1 \frac{1}{(dy/dx)}$$

$\therefore$ Equation of the normal at the point $P(x, y)$ is,

$$Y - y = -\frac{1}{(dy/dx)} (X - x)$$

$$\Rightarrow \qquad (Y - y) \frac{dy}{dx} + (X - x) = 0.$$

COROLLARY : If the equation of the curve is given in the parametric form, say

$$x = f(t)$$
$$y = \phi(t).$$

then we have

$$\frac{dy}{dx} = \frac{\phi'(t)}{f'(t)}$$

Hence, the equation of the normal is,

$$Y - \phi(t) = -\frac{1}{\frac{\phi'(t)}{f'(t)}} \{X - f(t)\}$$

$$\Rightarrow \qquad Y - \phi(t) = -\frac{f'(t)}{\phi'(t)} \{X - f(t)\}.$$

Note : This is the equation of the normal at point 't' of the curve.

5.6. Angle of the Intersection of Two Curves

The angle of intersection of two curves is the angle between the tangents to the two curves at their point of intersection.

Let us consider two curves C_1 and C_2 intersecting at the point P.

Let m_1 and m_2 be the slopes of the tangents at the point P to the two curves respectively.

Let θ be the angle between the tangents at P. Then,

$$\tan\theta = \frac{m_1 \sim m_2}{1 + m_1 m_2}$$

$$\Rightarrow \qquad \theta = \tan^{-1} \left\{\frac{m_1 \sim m_2}{1 + m_1 m_2}\right\}.$$

Notes :

1. The other angle of intersection is $\pi - \theta$.

2. The curves C_1 and C_2 are said to cut orthogonally, if

$$\theta = \frac{\pi}{2}$$

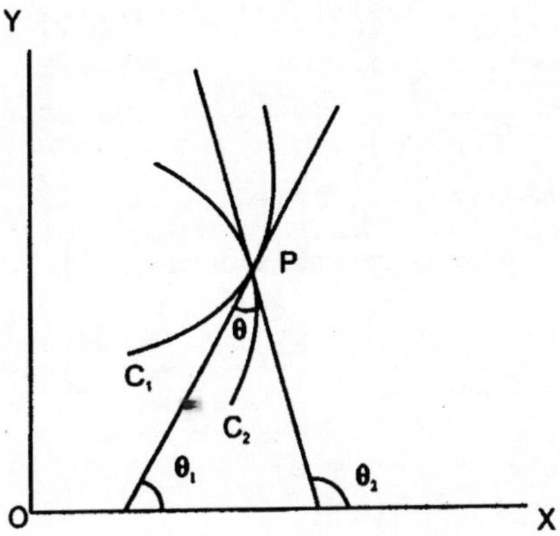

Fig. 5.3

i.e., $$\tan\theta = \tan\frac{\pi}{2} = \infty$$

i.e., $$1 + m_1 m_2 = 0$$

i.e., $$m_1 m_2 = -1.$$

3. The curve C_1 and C_2 will touch each other if

$$m_1 = m_2.$$

ILLUSTRATIVE EXAMPLES

Example 1. *Find the slope of the tangent at the point* (1, 1) *of the curve* $y = x^2$.

Solution : $$y = x^2$$

$\Rightarrow$ $$\frac{dy}{dx} = 2x$$

At point (1, 1),

$$\frac{dy}{dx} = 2 \times 1 = 2$$

$\Rightarrow$ Required slope = 2

Example 2. *Find the equations of the tangent and the normal at the point* (1, 1) *of the curve,*

$$2y = 3 - x^2.$$

Solution : Equation to the curve be

$$2y = 3 - x^2$$

$$2\frac{dy}{dx} = -2x \quad \text{[On differentiating w.r.t. } x\text{]}$$

$$\Rightarrow \quad \frac{dy}{dx} = -x$$

$\Rightarrow$ at the point (1, 1), $\quad \dfrac{dy}{dx} = -1.$

Hence, the equation of the tangent at the point (1, 1) is,

$$y - 1 = \left(\frac{dy}{dx}\right)_{(1,\,1)} (x - 1)$$

$$\Rightarrow \quad y - 1 = (-1)(x - 1)$$

$$\Rightarrow \quad y + x = 2$$

Again, slope of normal at (1, 1) = 1.

Hence, the equation of the normal at the point (1, 1) is,

$$y - 1 = 1\,(x - 1)$$

$$\Rightarrow \quad y - x = 0.$$

Example 3. *Find the equation of normal to the curve* $9x^2 - 4y^2 = 108$ *at the point* (4, 3).

Solution :

$$9x^2 - 4y^2 = 108$$

$$\therefore \quad 18x - 8y\frac{dy}{dx} = 0 \quad \text{[On differentiating w.r.t. } x\text{]}$$

$$\therefore \quad 9x - 4y\frac{dy}{dx} = 0$$

$$\therefore \quad \frac{dy}{dx} = \frac{9x}{4y}$$

At the point (4, 3), $\quad \dfrac{dy}{dx} = \dfrac{9 \times 4}{4 \times 3} = 3.$

$\therefore$ Slope of the normal at the point (4, 3) is $-\frac{1}{3}$.

Hence, the equation of the normal at the point (4, 3) is,

$$y - 3 = -\tfrac{1}{3}(x - 4)$$

$$\Rightarrow \quad 3y - 9 = -x + 4$$

$$\Rightarrow \quad 3y + x = 13.$$

Example 4. *Find the*

(i) *equation to tangent.*

(ii) *equation to normal.*

at the point where curve $y(x-2)(x-3)-x+7=0$ *meets the X-axis.*

Solution : Equation to curve is,

$$y(x-2)(x-3)-x+7=0 \quad \text{.... (i)}$$

This curve will meet the X-axis where $y = 0$.

Putting $y = 0$ in (i), we get the curve meets at the point (7, 0).

Differentiate (i) with respect to x,

$$y.[(x-2).1+(x-3).1]+(x-2)(x-3)\frac{dy}{dx}-1=0$$

$$\frac{dy}{dx} = \frac{1-y(2x-5)}{(x-2)(x-3)}$$

The value of $\frac{dy}{dx}$ at (7, 0),

$$\frac{dy}{dx} = \frac{1-0}{5\times 4} = \frac{1}{20}$$

Equation to tangent at (7, 0),

$$y-0 = \frac{1}{20}(x-7)$$

$$20y-x+7 = 0$$

$$20y = x-7.$$

Equation to normal at (7, 0) whose gradient = – 20 is,

$$y-0 = -20(x-7)$$

$$y+20x = 140.$$

Example 5. *The equation of the tangent at the point* (2, 3) *of the curve* $y^2 = ax^3 + b$ *is* $y = 4x - 5$. *Find the value of a and b.*

Solution :

$$y^2 = ax^3 + b$$

$$\therefore \quad 2y\frac{dy}{dx} = 3ax^2 \quad \text{[On differentiating w.r.t. } x\text{]}$$

$$\therefore \quad \frac{dy}{dx} = \frac{3ax^2}{2y}$$

$\therefore$ at the point (2, 3), $\frac{dy}{dx} = \frac{3a(4)}{2(3)} = 2a.$

Hence, the equation of the tangent at the point (2, 3) is,

$$y-3 = 2a(x-4)$$

$$\Rightarrow \quad y = 2ax + 3 - 4a. \quad \text{.... (i)}$$

But the equation of the tangent at the point (2, 3) is given to be,

$$y = 4x - 5. \qquad \text{.... (ii)}$$

Hence, (i) and (ii) will represent the same line. Therefore, comparing the coefficients of (i) and (ii), we get

$$2a = 4 \quad \therefore \quad a = 2$$

and also

$$3 - 4a = -5 \quad \therefore \quad a = 2.$$

Again, the point (2, 3) lies on the curve,

$$y^2 = ax^3 + b$$

$$\therefore \quad 3^2 = a.2^3 + b$$

$$\Rightarrow \quad 9 = 8a + b$$

$$\Rightarrow \quad 9 = 8 \times 2 + b \qquad \text{[Putting } a = 2\text{]}$$

$$\Rightarrow \quad 9 = 16 + b$$

$$\Rightarrow \quad b = -7.$$

Hence, the required values of a and b are 2 and -7 respectively.

Example 6. *Find the equation of the tangent at the point (x, y) to the curve*

$$\frac{x^m}{a^m} + \frac{y^m}{b^m} = 1.$$

Solution : The equation of the curve is,

$$\frac{x^m}{a^m} + \frac{y^m}{b^m} = 1 \qquad \text{... (1)}$$

Differentiating it w.r.t. x, we get

$$\frac{mx^{m-1}}{a^m} + \frac{my^{m-1}}{b^m} \frac{dy}{dx} = 0$$

$$\therefore \quad \frac{dy}{dx} = -\frac{x^{m-1}}{a^m} \cdot \frac{b^m}{y^{m-1}}$$

Hence, the equation of the tangent at (x, y) is,

$$Y - y = -\frac{x^{m-1}}{a^m} \cdot \frac{b^m}{y^{m-1}} (X - x)$$

$$\Rightarrow \quad \frac{y^{m-1}}{b^m}(Y - y) = -\frac{x^{m-1}}{a^m}(X - x)$$

$$\Rightarrow \quad Y\frac{y^{m-1}}{b^m} + X\frac{x^{m-1}}{a^m} = \frac{x^m}{a^m} + \frac{y^m}{b^m} = 1 \qquad \text{[From (i)]}$$

$$\Rightarrow \quad \frac{Xx^{m-1}}{a^m} + \frac{Yy^{m-1}}{b^m} = 1$$

which is the required equation of the tangent at the point (x, y).

Example 7. *Find the equations of the tangent and normal at the point $(at^2, 2at)$ on the parabola $y^2 = 4ax$.*

Solution : We have

$$x = at^2, \qquad y = 2at$$

Differentiating $y^2 = 4ax$ w.r.t. x

$$2y\frac{dy}{dx} = 4a \Rightarrow \frac{dy}{dx} = \frac{2a}{y}$$

$$\frac{dy}{dx} \text{ at } (at^2, 2at) = \frac{2a}{2at} = \frac{1}{t}.$$

Hence, the equation of the tangent at the point $(at^2, 2at)$ on the parabola $y^2 = 4ax$ is,

$$y - 2at = \frac{1}{t}(x - at^2)$$

$$\Rightarrow \qquad ty - 2at^2 = x - at^2$$

$$\Rightarrow \qquad ty = x + at^2.$$

Example 8. *Find the point on the curve $x^2 + y^2 + 2x - 3 = 0$ so that tangent at that point is parallel to X-axis.*

Solution : Let (x_1, y_1) be the point of curve

$$x^2 + y^2 + 2x - 3 = 0 \qquad \text{....(i)}$$

Therefore,

$$x_1^2 + y_1^2 + 2x_1 - 3 = 0 \qquad \text{...(ii)}$$

Differentiating (i) w.r.t. x,

$$2x + 2y\frac{dy}{dx} - 2 = 0 \Rightarrow \frac{dy}{dx} = \frac{-(x+1)}{y}$$

$\therefore$ The value of $\frac{dy}{dx}$ at $(x_1, y_1) = \frac{-(1+x_1)}{y_1}$

The tangent will be parallel to X-axis if $\frac{dy}{dx} = 0$, therefore

$$\frac{-x_1 - 1}{y_1} = 0 \Rightarrow x_1 = -1$$

Putting this in (ii),

$$1 + y_1^2 - 2 - 3 = 0 \Rightarrow y_1 = \pm 2$$

The required point will be $(-1, \pm 2)$.

Example 9. *Find the equation to the tangent on the curve $xy + 4 = 0$ which makes the angle 45° with X-axis. Find the point of contact also.*

Solution : The given curve is,

$$xy + 4 = 0 \quad \text{.... (i)}$$

Let the required point be (x_1, y_1), therefore

$$x_1 y_1 + 4 = 0 \quad \text{.... (ii)}$$

Differentiating (i) w.r.t. x

$$x\frac{dy}{dx} + y = 0 \Rightarrow \frac{dy}{dx} = \frac{-y}{x}$$

The value of $\frac{dy}{dx}$ at $(x_1, y_1) = \frac{-y_1}{x_1}$

But $\frac{dy}{dx} = \tan 45° = 1$

Therefore $\frac{-y_1}{x_1} = 1 \Rightarrow y_1 = -x_1.$

Putting in (ii),

$$x_1^2 = 4 \Rightarrow x_1 = \pm 2 \text{ and } y_1 = -x_1 = \mp 2.$$

The points will be (2, –2) and (–2, 2).

The equation to tangent at (2, –2) is,

$$y + 2 = 1\,(x - 2) \Rightarrow x - y = 4.$$

Equation to tangent at (– 2, 2) is,

$$y - 2 = 1\,(x + 2) \Rightarrow y - x = 4.$$

Example 10. *Prove that the straight line* $\frac{x}{a} + \frac{y}{b} = 1$, *touches the curve* $y = be^{-x/a}$ *at the point where the curve cuts the Y-axis.*

Solution : The curve is,

$$y = be^{-x/a} \quad \text{.... (i)}$$

It cuts Y-axis at the point for which $x = 0$.

$\therefore$ Putting $x = 0$ in (i), we get $y = b$

Hence, the point of intersection on Y-axis is $(0, b)$.

Now, differentiating (i), we get

$$\frac{dy}{dx} = be^{-x/a}\left(-\frac{1}{a}\right) - \frac{b}{a}e^{-x/a}$$

At the point $(0, b)$, $\frac{dy}{dx} = -\frac{b}{a}$

Hence, the equation of the tangent to (i) at the point (a, b) is,

$$y - b = -\frac{b}{a}(x - 0)$$

$$\Rightarrow \qquad \frac{y}{b} - 1 = -\frac{x}{a}$$

$$\Rightarrow \qquad \frac{x}{a} + \frac{y}{b} = 1.$$

Example 11. *Find the equation to the tangent and normal at the point* $(a \cos \phi, b \sin \phi)$ *to the ellipse*

$$\frac{x^2}{a^2} + \frac{y^2}{b^2} = 1.$$

Solution : Equation of the curve in parametric form is,

$$\left.\begin{aligned} x &= a \cos \phi \\ y &= b \sin \phi \end{aligned}\right\}$$

$$\therefore \qquad \frac{dx}{d\phi} = -a \sin \phi$$

$$\frac{dy}{d\phi} = b \cos \phi$$

$$\therefore \qquad \frac{dy}{dx} = \frac{dy/d\phi}{dx/d\phi} = -\frac{b \cos \phi}{a \sin \phi}$$

= slope of the tangent at the point 'ϕ'

Hence, the equation of the tangent at the point ϕ is,

$$y - b \sin \phi = -\frac{b \cos \phi}{a \sin \phi}(x - a \cos \phi)$$

$$\Rightarrow \qquad ay \sin \phi - ab \sin^2 \phi = -bx \cos \phi + ab \cos^2 \phi$$

$$\Rightarrow \qquad bx \cos \phi + ay \sin \phi = ab (\sin^2 \phi + \cos^2 \phi) = ab$$

Dividing by ab

$$\frac{x}{a} \cos \phi + \frac{y}{b} \sin \phi = 1$$

The slope of the normal at the point ϕ is $\dfrac{a \sin \phi}{b \cos \phi}$.

Hence, the equation of the normal at the point ϕ is,

$$y - b \sin \phi = \frac{a \sin \phi}{b \cos \phi}(x - a \cos \phi)$$

$$\Rightarrow \qquad \frac{y}{\sin \phi} - b = \frac{a}{b}\left(\frac{x}{\cos \phi} - a\right)$$

$$\Rightarrow \qquad \frac{by}{\sin \phi} - b^2 = \frac{ax}{\cos \phi} - a^2$$

$$\Rightarrow \qquad \frac{ax}{\cos \phi} - \frac{by}{\sin \phi} = a^2 - b^2.$$

Example 12. *Show that curve* $\left(\frac{x}{a}\right)^n + \left(\frac{y}{b}\right)^n = 2$ *touches the line* $\frac{x}{a} + \frac{y}{b} = 2$ *at the point* (a, b), *whatever may be the value of* n.

Solution : The equation to curve

$$\left(\frac{x}{a}\right)^n + \left(\frac{y}{b}\right)^n = 2$$

$$\Rightarrow \qquad \frac{x^n}{a^n} + \frac{y^n}{b^n} = 2$$

Differentiate it w.r.t. x,

$$\frac{nx^{n-1}}{a^n} - \frac{ny^{n-1}}{b^n}\frac{dy}{dx} = 0 \Rightarrow \frac{dy}{dx} = \frac{-b^n x^{n-1}}{a^n y^{n-1}}$$

$$\frac{dy}{dx} \text{ at } (a, b) = \frac{-b^n a^{n-1}}{a^n b^{n-1}} = \frac{-b}{a}$$

Equation to the tangent at (a, b) will be.

$$y - b = \frac{-b}{a}(x - a) \Rightarrow ay + bx = 2ab \Rightarrow \frac{x}{a} + \frac{y}{b} = 2$$

which is free from n. Hence this line will meet the curve for every value of n.

Example 13. *If the curve* $\sqrt{x} + \sqrt{y} = \sqrt{a}$ *touches the X-axis at P and Q then show that OP + OQ = a.*

Solution : The curve is $\sqrt{x} + \sqrt{y} = \sqrt{a}$ .... (i)

Let (x_1, y_1), be on the curve, then

$$\sqrt{x_1} + \sqrt{y_1} = \sqrt{a} \qquad \text{.... (ii)}$$

Differentiating (i) w.r.t. x,

$$\frac{1}{2\sqrt{x}} + \frac{1}{2\sqrt{y}}\frac{dy}{dx} = 0 \Rightarrow \frac{dy}{dx} = \frac{-\sqrt{y}}{\sqrt{x}}$$

$$\left(\frac{dy}{dx}\right)_{\text{at } (x_1, y_1)} = -\sqrt{\frac{y_1}{x_1}}$$

Equation to tangent at (x_1, y_1) is,

$$y - y_1 = -\sqrt{\frac{y_1}{x_1}}(x - x_1). \qquad \text{.... (iii)}$$

This tangent meets the X-axis at P, then $y = 0$

$$0 - y_1 = -\sqrt{\frac{y_1}{x_1}}(OP - x_1)$$

$$OP - x_1 = \sqrt{x_1 y_1} \Rightarrow OP = x_1 + \sqrt{x_1 y_1}$$

$$OP = \sqrt{x_1}\,(\sqrt{x_1} + \sqrt{y_1}) = \sqrt{ax_1} \qquad \text{[Using (ii)]}$$

Again, the tangent (iii) will meet Y-axis at Q if $x = 0$,

$$y = OQ$$

$$OQ - y_1 = -\sqrt{\frac{y_1}{x_1}}\,(0 - x_1)$$

$$OQ = y_1 + \sqrt{x_1 y_1}$$

$$= \sqrt{y_1}\,(\sqrt{x_1} + \sqrt{y_1}) = \sqrt{y_1 a} \qquad \text{[Using (ii)]}$$

$$OP + OQ = \sqrt{x_1 a} + \sqrt{y_1 a}$$

$$= \sqrt{a}\,(\sqrt{x_1} + \sqrt{y_1}) = \sqrt{a}\,\sqrt{a} = a.$$

Example 14. *If p and q are the intercepts of any tangent line to the curve* $\left(\frac{x}{a}\right)^{1/2} + \left(\frac{y}{b}\right)^{1/2} = 1$, *then prove that*

$$\frac{p}{a} + \frac{q}{b} = 1.$$

Solution :

$$\left(\frac{x}{a}\right)^{1/2} + \left(\frac{y}{b}\right)^{1/2} = 1 \qquad \text{.... (i)}$$

Differentiating with respect to x, we get

$$\frac{1}{2}\left(\frac{x}{a}\right)^{-1/2}\frac{1}{a} + \frac{1}{2}\left(\frac{y}{b}\right)^{-1/2}\frac{1}{b}\frac{dy}{dx} = 0$$

$$\Rightarrow \quad \left(\frac{x}{a}\right)^{1/2}\frac{1}{a} + \left(\frac{y}{b}\right)^{1/2}\frac{1}{b}\frac{dy}{dx} = 0$$

$$\Rightarrow \quad \frac{1}{\sqrt{ax}} + \frac{1}{\sqrt{by}}\frac{dy}{dx} = 0$$

$$\Rightarrow \quad \frac{\sqrt{by}}{\sqrt{ax}} + \frac{dy}{dx} = 0$$

$$\Rightarrow \quad \frac{dy}{dx} = -\frac{\sqrt{by}}{\sqrt{ax}}.$$

Hence, the equation of any tangent line to the given curve is,

$$Y - y = \frac{dy}{dx}(X - x)$$

$$\Rightarrow \quad Y - y = -\sqrt{\frac{by}{ax}}\,(X - x)$$

$$\Rightarrow \quad \sqrt{ax}\,Y - y\sqrt{ax} = -\sqrt{by}\,X + x\sqrt{by}$$

$$\Rightarrow \quad \sqrt{bx}\,X + \sqrt{ax}\,Y = y\sqrt{ax} + x\sqrt{by}$$

$$\Rightarrow \qquad \frac{X}{\dfrac{y\sqrt{ax}+x\sqrt{by}}{\sqrt{by}}} + \frac{Y}{\dfrac{y\sqrt{ax}+x\sqrt{by}}{\sqrt{ax}}} = 1$$

$$\therefore \qquad p = \frac{y\sqrt{ax}+x\sqrt{by}}{\sqrt{by}}$$

and
$$q = \frac{y\sqrt{ax}+x\sqrt{by}}{\sqrt{ax}}$$

Now,
$$\frac{p}{a}+\frac{q}{b} = \frac{y\sqrt{ax}+x\sqrt{by}}{a\sqrt{by}} + \frac{y\sqrt{ax}+x\sqrt{by}}{b\sqrt{ax}}$$

$$= \frac{\sqrt{bx}\{y\sqrt{ax}+x\sqrt{by}\}+\sqrt{ay}\{y\sqrt{ax}+x\sqrt{by}\}}{ab\sqrt{xy}}$$

$$= \frac{(y\sqrt{ax}+x\sqrt{by})(\sqrt{bx}+\sqrt{ay})}{ab\sqrt{xy}}$$

$$= \frac{\sqrt{x}\sqrt{y}\,(\sqrt{ay}+\sqrt{bx})(\sqrt{bx}+\sqrt{ay})}{ab\sqrt{xy}}$$

$$= \frac{(\sqrt{ay}+\sqrt{bx})^2}{ab} \qquad \text{.... (ii)}$$

Now, from (i),

$$\sqrt{bx}+\sqrt{ay} = \sqrt{ab}$$

$\therefore$ From (ii),

$$\frac{p}{a}+\frac{q}{b} = \frac{(\sqrt{ab})^2}{ab}$$

$$\Rightarrow \qquad \frac{p}{a}+\frac{q}{b} = \frac{ab}{ab} = 1.$$

Example 15. *If the normal to the curve* $x^{2/3}+y^{2/3}=a^{2/3}$ *makes an angle* ϕ *with the axis of x, show that its equation is,*

$$y\cos\phi - x\sin\phi = a\cos 2\phi.$$

Solution : The equation to the curve is,

$$x^{2/3}+y^{2/3} = a^{2/3} \qquad \text{.... (i)}$$

Differentiating w.r.t. x, we get

$$\frac{2}{3}x^{-1/3}+\frac{2}{3}y^{-1/3}\frac{dy}{dx} = 0$$

$$\Rightarrow \qquad \frac{dy}{dx} = -\frac{y^{1/3}}{x^{1/3}}$$

$$\therefore \qquad \text{The gradient of the normal} = \frac{x^{1/3}}{y^{1/3}} = \tan\phi \text{ (given)}$$

$$\therefore \qquad x^{1/3} = y^{1/3}\tan\phi$$

$$\Rightarrow \qquad x = y\tan^3\phi$$

$$\therefore \text{ From (i),} \quad y^{2/3}\tan^2\phi + y^{2/3} = a^{2/3}$$

$$\Rightarrow \quad y^{2/3}\sec^2\phi = a^{2/3}$$

$$\Rightarrow \quad y = a\cos^3\phi$$

$$\therefore \quad x = y\tan^3\phi$$

$$= a\cos^3\phi\cdot\frac{\sin^3\phi}{\cos^3\phi} = a\sin^3\phi$$

Hence, the equation of the normal at the point $(a\sin^3\phi, a\cos^3\phi)$ is,

$$y - a\cos^3\phi = \tan\phi\,(x - a\sin^3\phi)$$

$$\Rightarrow \quad y - a\cos^3\phi = \frac{\sin\phi}{\cos\phi}(x - a\sin^3\phi)$$

$$\Rightarrow \quad y\cos\phi - a\cos^4\phi = x\sin\phi - a\sin^4\phi$$

$$\Rightarrow \quad y\cos\phi - x\sin\phi = a(\cos^4\phi - \sin^4\phi)$$

$$= a(\cos^2\phi + \sin^2\phi)(\cos^2\phi - \sin^2\phi)$$

$$= a\,.\,1\,.\cos 2\phi$$

$$= a\cos 2\phi.$$

Example 16. *Find all those points on the curve*

$$y^2 = 4a\left\{x + a\sin\left(\frac{x}{a}\right)\right\},$$

where the tangent lines are parallel to X-axis and show that all those points lie on a parabola.

Solution :

$$y^2 = 4a\left\{x + a\sin\left(\frac{x}{a}\right)\right\} \quad (i)$$

Differentiating both sides w.r.t. x, we get

$$2y\frac{dy}{dx} = 4a\left\{1 + a\cos\left(\frac{x}{a}\right)\cdot\frac{1}{a}\right\}$$

$$= 4a\left(1 + \cos\frac{x}{a}\right)$$

$$\therefore \quad \frac{dy}{dx} = \frac{4a\left(1 + \cos\frac{x}{a}\right)}{2y} = \frac{2a}{y}\left(1 + \cos\frac{x}{a}\right)$$

$\because$ The tangent line is parallel to X-axis,

$$\therefore \quad \frac{dy}{dx} = 0$$

$$\therefore \quad \frac{2a}{y}\left(1 + \cos\frac{x}{a}\right) = 0$$

$$\Rightarrow \quad 1 + \cos\frac{x}{a} = 0$$

$$\Rightarrow \quad \cos\frac{x}{a} = -1 = \cos\pi$$

$$\therefore \quad \frac{x}{a} = 2n\pi \pm \pi \qquad \left[\begin{array}{l} \because \cos\theta = \cos\alpha \\ \Rightarrow \quad \theta = 2n\pi \pm \alpha \end{array}\right]$$

$$\therefore \quad x = (2n+1)\,a\pi,$$

where n is 0 or a +ve integer or a −ve integer.

Now, putting the value of x in (i), we get

$$y^2 = 4a\,[(2n+1)\,a\pi + a\sin(2n+1)\pi] = 4a^2\,(2n+1)\,\pi$$

$$y = \pm 2a\,\sqrt{(2n+1)\,\pi}$$

The required points are : $\{(2n \pm 1)\,a\pi, \pm 2a\sqrt{(2n+1)\,\pi}\}$.

Let the tangent at this point is parallel to X-axis, then

$$\left(\frac{dy}{dx}\right)_{(h,\,k)} = 0 \Rightarrow \cos\frac{h}{a} = -1$$

$$\sin\left(\frac{h}{a}\right) = \sqrt{1-\cos^2\frac{h}{a}} = \sqrt{1-1} = 0.$$

Point (h, k) lies on (i), therefore,

$$k^2 = 4a\left[h + a\sin\frac{h}{a}\right] = 4a\,[h + a \times 0] = 4\,ah$$

The locus of (h, k) will be $y^2 = 4ax$, which is a parabola

Example 17. *If the line* $x\cos\alpha + y\sin\alpha = p$ *touches the curve* $\frac{x^2}{a^2} + \frac{y^2}{b^2} = 1$, *then show that* $p^2 = a^2\cos^2\alpha + b^2\sin^2\alpha$.

Solution : Equation to tangent at the point $(a\cos\phi, b\sin\phi)$ of ellipse

$$\frac{x^2}{a^2} + \frac{y^2}{b^2} = 1$$

will be

$$\frac{x}{a}\cos\Phi + \frac{y}{b}\sin\Phi = 1 \qquad \text{.... (i)}$$

If the given line $\quad x\cos\alpha + y\sin\alpha = p \qquad$ .... (ii)

touches the ellipse, then equation (i) and (ii) will represent the same line. Therefore coefficients of x, y and constants of both are in the same ratio.

$$\frac{\frac{\cos\phi}{a}}{\cos\alpha} = \frac{\frac{\sin\phi}{b}}{\sin\alpha} = \frac{1}{p}$$

$$\cos\phi = \frac{a\cos\alpha}{p} \text{ and } \sin\phi = \frac{b\sin\alpha}{p}.$$

Squaring and adding these,

$$\frac{a^2\cos^2\alpha}{p^2} + \frac{b^2\sin^2\alpha}{p^2} = \cos^2\phi + \sin^2\phi = 1$$

$$\therefore \qquad p^2 = a^2\cos^2\alpha + b^2\sin^2\alpha.$$

Example 18. *If* $x\cos\alpha + y\sin\alpha = p$ *touches the curve*

$$\left(\frac{x}{a}\right)^{n/(n-1)} + \left(\frac{y}{b}\right)^{n/(n-1)} = 1$$

prove that $\qquad (a\cos\alpha)^n + (b\sin\alpha)^n = p^n.$

Solution : The equation of the given curve is,

$$\left(\frac{x}{a}\right)^{n/(n-1)} + \left(\frac{y}{b}\right)^{n/(n-1)} = 1 \qquad \text{.... (i)}$$

Differentiating w.r.t. x, we get

$$\frac{1}{a^{n/(n-1)}}\cdot\frac{n}{n-1}x^{\left(\frac{n}{n-1}-1\right)} + \frac{1}{b^{n/(n-1)}}\cdot\frac{n}{n-1}y^{\left(\frac{n}{n-1}-1\right)}\frac{dy}{dx} = 0$$

$$\Rightarrow \frac{1}{a^{n/(n-1)}}\cdot x^{1/(n-1)} + \frac{1}{b^{n/(n-1)}}\cdot y^{1/(n-1)}\frac{dy}{dx} = 0$$

$$\Rightarrow \qquad \frac{dy}{dx} = -\frac{x^{1/(n-1)}\cdot b^{n/(n-1)}}{y^{1/(n-1)}\cdot a^{n/(n-1)}}$$

Hence, the equation of the tangent at (x, y) is,

$$Y - y = -\frac{x^{1/(n-1)}}{y^{1/(n-1)}}\cdot\frac{b^{n/(n-1)}}{a^{n/(n-1)}}\cdot(X - x)$$

$$\Rightarrow \frac{X x^{1/(n-1)}}{a^{n/(n-1)}} + \frac{Y y^{1/(n-1)}}{b^{n/(n-1)}} = \left(\frac{x}{a}\right)^{n/(n-1)} + \left(\frac{y}{b}\right)^{n/(n-1)} = 1 \qquad \text{.... (ii)}$$

[By virtue of (i)]

Also, $x\cos\alpha + y\sin\alpha = p$ ... (iii)

is a tangent to the curve.

Therefore comparing (ii) and (iii), we get

$$\frac{\cos\alpha}{\dfrac{x^{1/(n-1)}}{a^{n/(n-1)}}} = \frac{\sin\alpha}{\dfrac{y^{1/(n-1)}}{b^{n/(n-1)}}} = \frac{p}{1}$$

This gives $\qquad a\cos\alpha = p\left(\frac{x}{a}\right)^{1/(n-1)}$

and $\qquad b\sin\alpha = p\left(\frac{y}{b}\right)^{1/(n-1)}$

Raising both relations to power n and adding, we have

$$(a\cos\alpha)^n + (b\sin\alpha)^n = p^n\left[\left(\frac{x}{a}\right)^{n/(n-1)} + \left(\frac{y}{b}\right)^{n/(n-1)}\right]$$

$$\Rightarrow\ (a\cos\alpha)^n + (b\sin\alpha)^n = p^n \qquad \text{.... (iv)}$$

[By virtue of (i)]

which is the required condition.

Example 19. *Find the angle of intersection of the parabolas $2y^2 = ax$ and $x^2 = 4ay$.*

Solution :

$$2y^2 = ax \qquad \text{.... (i)}$$

$$x^2 = 4ay \qquad \text{.... (ii)}$$

From (i),

$$a = \frac{2y^2}{x}$$

Put this value of a in (ii),

$$x^2 = 4\cdot\frac{2y^2}{x}\cdot y$$

$$\Rightarrow\quad x^2 = \frac{8y^3}{x}$$

$$\Rightarrow\quad x^3 = 8y^3$$

$$\Rightarrow\quad x^3 - 8y^3 = 0$$

$$\Rightarrow\quad (x-2y)(x^2+4y^2+2xy) = 0$$

$$\Rightarrow\quad x - 2y = 0 \text{ and}$$

$$x^2 + 2xy + 4y^2 = 0.$$

CASE I. When $x - 2y = 0$, then

$$x = 2y$$

$\therefore$ From (i),

$$2y^2 = ax$$

$$\Rightarrow\quad 2y^2 = a.2y$$

$$\Rightarrow\quad 2y^2 = 2ay$$

$$\Rightarrow\quad y^2 - ay = 0$$

$$\Rightarrow\quad y(y-a) = 0$$

$$\Rightarrow\quad y = 0,\ a \qquad \therefore x = 0,\ 2a$$

$\therefore$ The points of intersection are $(0, 0)$ and $(2a, a)$.

CASE II. When $x^2 + 2xy + 4y^2 = 0$

$$\therefore \qquad x = \frac{-2y \pm \sqrt{4y^2 - 16y^2}}{2}$$

$$= \frac{-2y \pm i\,2\sqrt{3}\,y}{2}$$

$$= (-1 \pm i\sqrt{3})\,y,$$

which gives imaginary points of intersection.

Now, from (i), $$4y\frac{dy}{dx} = a$$

$$\therefore \qquad \frac{dy}{dx} = \frac{a}{4y} \text{ (say } m_1) \qquad \text{.... (iii)}$$

and from (ii), $$2x = 4a\frac{dy}{dx}$$

$$\therefore \qquad \frac{dy}{dx} = \frac{x}{2a} \text{ (say } m_2) \qquad \text{.... (iv)}$$

Angle of Intersection θ at the point (2*a*, *a*) :

For (i), $\frac{dy}{dx}$ at $(2a, a) = \frac{a}{4a} = \frac{1}{4} = m_1$

For (ii), $\frac{dy}{dx}$ at $(2a, a) = \frac{2a}{2a} = 1 = m_2$

$$\therefore \qquad \tan\theta = \frac{m_1 - m_2}{1 + m_1 m_2}$$

$$= \frac{1 - \frac{1}{4}}{1 + 1 \cdot \frac{1}{4}}$$

$$= \frac{\frac{3}{4}}{\frac{5}{4}} = \frac{3}{5}$$

$$\therefore \qquad \theta = \tan^{-1}\left(\frac{3}{5}\right).$$

Angle of Intersection θ at the point (0, 0) :

$$\therefore \qquad m_1 = \frac{a}{4y} = \frac{a}{4 \times 0} = \infty = \tan 90^\circ$$

$$m_2 = \frac{x}{2a} = \frac{0}{2a} = 0 = \tan 0^\circ$$

Thus, the tangent to curve (i) at (0, 0) is perpendicular to *X*-axis.

The tangent to curve (ii) at (0, 0) is *X*-axis itself.

Therefore, the angle between these tangents will be 90°.

Example 20. *Prove that the two curves* $ax^2 + by^2 = 1$ and $a'x^2 + b'y^2 = 1$ *will intersect orthogonally, if*

$$\frac{1}{a} - \frac{1}{b} = \frac{1}{a'} - \frac{1}{b'}.$$

Hence show that the curves

$$\frac{x^2}{a^2+\lambda_1} + \frac{y^2}{b^2+\lambda_1} = 1 \text{ and } \frac{x^2}{a^2+\lambda_2} + \frac{y^2}{b^2+\lambda_2} = 1$$

cut orthogonally.

Solution : Let (x_1, y_1) be the point of intersection of the two curves, then

$$ax_1^{\,2} + by_1^{\,2} = 1 \qquad \text{.... (i)}$$

and

$$a'x_1^{\,2} + b'y_1^{\,2} = 1 \qquad \text{.... (ii)}$$

From (i) and (ii),

$$(a - a')\,x_1^{\,2} + (b - b')\,y_1^{\,2} = 0$$

$$\Rightarrow \quad \frac{x_1^{\,2}}{y_1^{\,2}} = -\frac{b-b'}{a-a'} \qquad \text{.... (iii)}$$

For the first curve,

$$\frac{dy}{dx} = -\frac{ax}{by}$$

$$\therefore \quad m_1 = \left(\frac{dy}{dx}\right)_{(x_1, y_1)} = -\frac{ax_1}{by_1}$$

For the second curve,

$$\frac{dy}{dx} = -\frac{a'x}{b'y}$$

$$\therefore \quad m_2 = \left(\frac{dy}{dx}\right)_{(x_1, y_1)} = -\frac{a'x_1}{b'y_1}$$

The two curves will cut orthogonally if,

$$m_1 m_2 = -1$$

i.e.,

$$\left(-\frac{ax_1}{by_1}\right) \times \left(-\frac{a'x_1}{b'y_1}\right) = -1$$

$$\Rightarrow \quad aa'x_1^{\,2} + bb'y_1^{\,2} = 0$$

$$\Rightarrow \quad \frac{x_1^{\,2}}{y_1^{\,2}} = -\frac{bb'}{aa'} \qquad \text{.... (iv)}$$

Hence, from (iii) and (iv),

$$-\frac{b-b'}{a-a'} = -\frac{bb'}{aa'}$$

$$\Rightarrow \quad \frac{b-b'}{bb'} = \frac{a-a'}{aa'}$$

$$\Rightarrow \qquad \frac{1}{b'} - \frac{1}{b} = \frac{1}{a'} - \frac{1}{a}$$

$$\Rightarrow \qquad \frac{1}{a} - \frac{1}{b} = \frac{1}{a'} - \frac{1}{b'}$$

which proves the first part.

Again for the given two curves, we have

$$(a^2 + \lambda_1) - (b^2 + \lambda_1) = (a^2 + \lambda_2) - (b^2 + \lambda_2)$$

which is true.

Hence these curves cut orthogonally.

Example 21. *If $lx + my = 1$ is normal to the parabola $y^2 = 4ax$ then show that $al^3 + 2alm^2 = m^2$.*

Solution : Differentiating the equation

$$y^2 = 4ax \qquad .... \text{(i)}$$

with respect to x,

$$2y\frac{dy}{dx} = 4a \Rightarrow \frac{dy}{dx} = \frac{2a}{y}$$

Let (x_1, y_1) be any point on the parabola (i), then

$$y_1^2 = 4ax_1 \qquad .... \text{(ii)}$$

The value of $\frac{dy}{dx}$ at $(x_1, y_1) = \frac{2a}{y_1}$.

The value of gradient of normal at $(x_1, y_1) = \frac{-y_1}{2a}$

The equation to normal at (x_1, y_1) is,

$$y - y_1 = -\frac{y_1}{2a}(x - x_1)$$

$$\Rightarrow \qquad 2ay - 2ay_1 = -y_1x + x_1y_1$$

$$\Rightarrow \qquad xy_1 + 2ay = y_1(x_1 + 2a) \qquad .... \text{(iii)}$$

If the line $lx + my = 1$ is perpendicular to normal then equations (iii) and (ii) will be equation to the same line, then

$$\frac{y_1}{l} = \frac{2a}{m} = \frac{(x_1 + 2a)\,y_1}{1}$$

$$\Rightarrow \qquad y_1 = \frac{2al}{m} \text{ and } x_1 = \frac{1}{l} - 2a.$$

Putting the values of x_1 and y_1 in (ii), we get

$$\left(\frac{2al}{m}\right)^2 = 4a\left(\frac{1}{l} - 2a\right)$$

$$\Rightarrow \qquad a^2 l^2 = m^2 a\left(\frac{1}{l} - 2a\right)$$

$$\Rightarrow \qquad al^3 = m^2 - 2alm^2$$

$$\Rightarrow \qquad al^3 + 2alm^2 = m^2$$

EXERCISE 5 (A)

1. Find the equation of the tangent and normal at the points (x, y) on each of the following curves :

(i) $x^2 + y^2 = a^2$.

(ii) $\frac{x^2}{a^2} + \frac{y^2}{b^2} = 1$.

(iii) $\frac{x^2}{a^2} - \frac{y^2}{b^2} = 1$.

(iv) $y^2 = 4ax$.

(v) $xy = c^2$.

(vi) $y = c \cosh\left(\frac{x}{c}\right)$.

(vii) $y = a \log \sec\left(\frac{x}{a}\right)$.

(viii) $e^y = \sin x$.

(ix) $x^3 + y^3 = 3axy$.

(x) $(x^2 + y^2)^2 = a^2 (x^2 - y^2)$.

2. Find the equation of the tangent and normal at the point 't' on each of the following curves :

(i) $x = a \sin^3 t$, $y = b \cos^3 t$.

(ii) $x = a (1 + \sin t)$, $y = a (1 - \cos t)$.

(iii) $x = a (2 \cos t - \cos 2t)$, $y = a (2 \sin t - \sin 2t)$.

(iv) $x = t^2 - a$, $y = t^3 - b$.

(v) $x = a \cosh t$, $y = b \sinh t$.

(vi) $x = a \sec \theta$, $y = b \tan \theta$.

3. Find the points on the curve $x^2 + y^2 - 2x - 3 = 0$ where the tangent lines are parallel to the axis of X.

4. On the curve $3b^2 y = x^3 - 3ax^2$, find the points at which the tangents are parallel to the axis of X.

5. Find the points on the curve $y = x^4 - 6x^3 + 13x^2 - 10x + 5$ where the tangent is parallel to $y = 2x$. Also prove that two of these points have the same tangents.

6. At what points of the curve $y = \sin x$, is the tangent parallel to the X-axis ?

7. Find the coordinates of the points on the curve $y = x^2 + 3x + 4$, the tangents at which pass through the origin.

8. At what point on the curve $y = 3x^2 - 4$ is the tangent perpendicular to the line $3y + x = 2$?

9. Find the equations of the normals to the curve $2x^2 - y^2 = 14$ parallel to $x + 3y = 4$.

10. Find the equation to the tangent to $x^3 = ay^2$ at $(4am^2, 8am^3)$ and also the point in which the tangent cuts the curve again. Show that if $9m^2 = 2$, the tangent is also a normal to the curve.

11. Tangents are drawn from the origin to the curve $y = \sin x$. Prove that their points of contact lie on $x^2 - y^2 = x^2 y^2$.

12. Show that the tangents to the Folium of Descrates $x^3 + y^3 = 3axy$ at the points where it meets the parabola $y^2 = ax$ are parallel to the axis of y.

13. Show that the line $3x + 4y = 5$ touches the curve $48xy = 25$.

14. If line $x \cos \alpha + y \sin \alpha = p$ touches the curve $xy = a^2$, then prove that $p^2 = 4a^2 \cos \alpha \sin \alpha$.

15. Prove that the condition that $\cos \alpha + y \sin \alpha = p$ should touch

$$a^m y^n = a^{m+n}$$

is

$$p^{m+n} m^m n^n = (m+n)^{m+n} a^{m+n} \cos^m \alpha \sin^n \alpha.$$

16. In the curve $x^m y^n = a^{m+n}$, prove that the portion of the tangent intercepted between the axes is divided at its points of contact into segments which are in a constant ratio.

17. If $Ax + By = C$ is a normal to the curve $a^{n-1} y = x^n$ then prove that

$$n^n A^n C^{n-1} = a^{n-1} B (nA^2 + B^2)^{n-1}.$$

18. (a) Show that the curves $x^3 - 3xy^2 + 2 = 0$ and $3x^2 y - y^2 = 2$ cut orthogonally.

(b) Find the angle of intersection of the curves :

(i) $y^2 = 4ax, x^2 = 4by$.

(ii) $x^2 - y^2 = a^2, x^2 + y^2 = a^2\sqrt{2}$.

(iii) $2y^2 = x^3, y^2 = 32x$.

(iv) $xy = a^2, x^2 + y^2 = 2a^2$.

(v) $y = 4 - x^2, y = x^2$.

ANSWERS

1. (i) $xX + yY = a^2,\ xY = yX.$

(ii) $\dfrac{xX}{a^2} + \dfrac{yY}{b^2} = 1,\ \dfrac{X-x}{b^2x} = \dfrac{Y-y}{a^2y}.$

(iii) $\dfrac{xX}{a^2} + \dfrac{yY}{b^2} = 1,\ \dfrac{X-x}{b^2x} + \dfrac{Y-y}{a^2y} = 0.$

(iv) $yY = 2a(X+x),\ Y-y = -\dfrac{y}{2a}(X-x).$

(v) $\dfrac{X}{x} + \dfrac{Y}{y} = 2,\ \dfrac{Y-y}{x} = \dfrac{X-x}{y}.$

(vi) $Y - y = \left\{\sinh\left(\dfrac{x}{c}\right)\right\}(X-x),$

$X - x + (Y-y)\sinh\left(\dfrac{x}{c}\right) = 0.$

(vii) $Y - y = \tan\left(\dfrac{x}{a}\right)(X-x),$

$\tan\left(\dfrac{x}{a}\right)(Y-y) + (X-x) = 0.$

(viii) $Y - y = \cot x\,(X-x),\ \dfrac{Y-y}{e^y} + \dfrac{X-x}{\cos x} = 0.$

(ix) $X(x^2 - ay) + Y(y^2 - ax) = axy,\ \dfrac{Y-y}{y^2-ax} = \dfrac{X-x}{x^2-ay}.$

(x) $\{2y(x^2+y^2) + a^2y\}X + \{2x(x^2+y^2) - a^2x\}Y = a^2(x^2-y^2),$
$(2x^2+2y^2+a^2)yX - (2x^2+2y^2-a^2)xY = 2a^2xy.$

2. (i) $\dfrac{X}{a\sin t} + \dfrac{Y}{b\cos t} = 1,$

$aX\sin t - bY\cos t = a\sin^4 t - b\cos^4 t.$

(ii) $X\sin\dfrac{t}{2} - Y\cos\dfrac{t}{2} = at\sin\dfrac{t}{2},$

$X\cos\dfrac{t}{2} + Y\sin\dfrac{t}{2} = at\cos\dfrac{t}{2} + 2a\sin\dfrac{t}{2}.$

(iii) $Y - a(2\sin t - \sin 2t) = \tan\left(\dfrac{3t}{2}\right)\{X - a(2\cos t - \cos 2t)\}$

$\{Y - a(2\sin t - \sin 2t)\}\tan\left(\dfrac{3t}{2}\right) + [X - a(2\cos t - \cos 2t)] = 0$

(iv) $2Y = 3tX - t^3 + 3at - 2b,$
$2x + 3ty = 3t^4 + 2t^2 - 3bt - 2a.$

(v) $Xb \cosh t - Ya \sinh t = ab,$

$Yb \cosh t + Xa \sinh t = \frac{a^2+b^2}{4} \sinh 2t.$

(vi) $\frac{X}{a} \sec\theta - \frac{Y}{b} \tan\theta = 1,\ \frac{aX}{\sec\theta} + \frac{bY}{\tan\theta} = a^2 + b^2.$

3. $(1, 2)$ and $(1, -2)$.

4. $(0, 0)$ and $\left(2a, -\frac{4a^3}{3b^2}\right)$.

5. $(1, 3), (2, 5)$ and $\left(\frac{3}{2}, \frac{65}{16}\right)$.

6. $\left[(2n+1)\frac{\pi}{2}, (-1)^n\right]; n \in I.$

7. $(2, 14), (-2, 2)$.

8. $\left(\frac{1}{2}, -\frac{13}{4}\right)$.

9. $x + 3y - 9 = 0,\ x + 3y + 9 = 0.$

10. $y = 3mx - 4am^3;\ (am^2, -am^3).$

18. (b) (i) $\tan^{-1}\left\{\frac{3a^{1/2}b^{1/2}}{2(a^{2/3}+b^{2/3})}\right\} \frac{\pi}{2}.$ (ii) $\pm\frac{\pi}{4}.$ (iii) $\frac{\pi}{2}, \tan^{-1}\frac{1}{2}.$

(iv) 0, the curves touch other. (v) $\tan^{-1}\left(\frac{4\sqrt{2}}{7}\right).$

GEOMETRICAL RESULTS

5.7. Cartesian Subtangent and Subnormal

Let $y = f(x)$ be the equation of the curve. Let $P(x, y)$ be any point on it. Let PT and PN be the tangent and normal at point P; T and N being the points

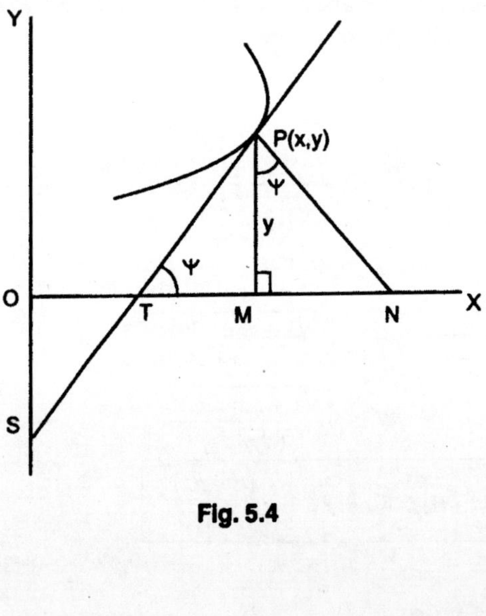

Fig. 5.4

where these cut the axis of X. Draw PM perpendicular from P on the axis of X.

Then length TM is called the *subtangent* and length MN is called the *subnormal*. Let Ψ be the angle which the tangent at P makes with the axis of X.

Then, $$\tan \Psi = \frac{dy}{dx}.$$

Thus, $\angle PTM = \Psi$ and $\angle MPN = \Psi$.

From $\Delta\, PTM$,

$$\cot \Psi = \frac{TM}{y}$$

$$\Rightarrow \quad TM = y \cot \Psi$$

$$\Rightarrow \quad TM = \frac{y}{\tan \Psi} = \frac{y}{(dy/dx)}$$

$$\therefore \quad \boxed{\textbf{Subtangent} = \frac{y}{(dy/dx)}}$$

From $\Delta\, PMN$,

$$\tan \Psi = \frac{MN}{y}$$

$$\Rightarrow \quad MN = y \tan \Psi = y \frac{dy}{dx}$$

$$\therefore \quad \boxed{\textbf{Subnormal} = y \frac{dy}{dx}}$$

5.8. Length of the Tangent and Normal

Refer to the fig. of art. 5.7 above. The lengths TP and PN are called the lengths of the tangent and normal respectively.

From ΔPTM,

$$\operatorname{cosec} \Psi = \frac{PT}{y}$$

$$\Rightarrow \quad PT = y \operatorname{cosec} \Psi$$

$$\Rightarrow \quad PT = y \sqrt{1 + \cot^2 \Psi}$$

$$\Rightarrow \quad PT = y \sqrt{1 + \frac{1}{\tan^2 \Psi}}$$

$$\Rightarrow \quad PT = y \frac{\sqrt{1 + \tan^2 \Psi}}{\tan \Psi}$$

$$\Rightarrow \quad PT = y \frac{\sqrt{1 + (dy/dx)^2}}{(dy/dx)}$$

$$\therefore \quad \boxed{\textbf{Length of tangent} = y \frac{\sqrt{1 + (dy/dx)^2}}{(dy/dx)}}$$

From ΔPMN,

$$\sec\Psi = \frac{PN}{y}$$

$$\Rightarrow \quad PN = y\sec\Psi$$

$$\Rightarrow \quad PN = y\sqrt{1+\tan^2\Psi}$$

$$\Rightarrow \quad PN = y\sqrt{1+(dy/dx)^2}$$

$$\therefore \quad \boxed{\textbf{Length of normal} = y\sqrt{1+(dy/dx)^2}}$$

COROLLARY : It can be easily shown from the above results that in any curve,

$$\frac{\text{Subnormal}}{\text{Subtangent}} = \frac{(\text{length of the normal})^2}{(\text{length of the tangent})^2}$$

5.9. Intercepts

Equation to the tangent at $P(x, y)$ is,

$$Y - y = \frac{dy}{dx}(X - x)$$

It meets the X-axis at the point, where

$$y = 0.$$

i.e., where
$$0 - y = \frac{dy}{dx}(X - x)$$

$$\Rightarrow \quad X = x - \frac{y}{(dy/dx)}.$$

Hence, the length of the intercept, that the tangent cuts off from the axis of X, *i.e.*,

$$OT = x - \frac{y}{dy/dx}.$$

Again, the tangent meets the Y-axis where $X = 0$

i.e., where
$$Y - y = \frac{dy}{dx}(0 - x)$$

i.e., where
$$Y = y - x\frac{dy}{dx}$$

Hence, the length of the intercept OS, that the tangent cuts off from the axis of Y, *i.e.*,

$$OS = y - x\frac{dy}{dx}.$$

ILLUSTRATIVE EXAMPLES

Example 1. *Find the lengths of the tangent, normal, subtangent and subnormal at the point (a, a) on the curve $ay^2 = x^3$.*

Solution : Equation to the given curve is,

$$ay^2 = x^3 \quad \text{.... (i)}$$

Differentiating the equation w.r.t. x, we get

$$2ay\frac{dy}{dx} = 3x^2$$

$$\Rightarrow \quad \frac{dy}{dx} = \frac{3x^2}{2ay}$$

$$\Rightarrow \quad \tan\Psi = \frac{3x^2}{2ay}$$

$$\Rightarrow \quad \tan\Psi \text{ at the point } (a, a) = \frac{3a^2}{2a.a} = \frac{3}{2}$$

$$\therefore \quad \sin\Psi = \frac{3}{\sqrt{13}} \text{ and } \cos\Psi = \frac{2}{\sqrt{13}}$$

Now,

$$\text{Length of tangent} = y \operatorname{cosec}\Psi = \frac{a\sqrt{13}}{3}$$

$$\text{Length of normal} = y\sec\Psi = \frac{a\sqrt{13}}{2}$$

$$\text{Subtangent} = y\cot\Psi = \frac{a.2}{3} = \frac{2a}{3}$$

$$\text{Subnormal} = y\tan\Psi = a\cdot\frac{3}{2} = \frac{3a}{2}.$$

Example 2. *In the tractrix* $x = a\left(\cos t + \frac{1}{2}\log\tan^2\frac{t}{2}\right)$, $y = a\sin t$, *show that the portion of the tangent intercepted between the curve and axis of X is of constant length.*

Solution :

$$x = a\left(\cos t + \tfrac{1}{2}\log\tan^2\frac{t}{2}\right).$$

Since

$$x = a\left(\cos t + \log\tan\frac{t}{2}\right)$$

$$\therefore \quad \frac{dx}{dt} = a\left[-\sin t + \frac{1}{\tan t/2}\cdot\sec^2\frac{t}{2}\cdot\frac{1}{2}\right]$$

$$= a\left[-\sin t + \frac{1}{2\sin\frac{t}{2}\cos\frac{t}{2}}\right]$$

$$= a\left[-\sin t + \frac{1}{\sin t}\right] = a\left[\frac{1-\sin^2 t}{\sin t}\right]$$

$$= \frac{a\cos^2 t}{\sin t}$$

$$y = a\sin t$$

$$\therefore \quad \frac{dy}{dt} = a\cos t$$

$$\therefore \quad \frac{dy}{dt} = \frac{dy/dt}{dx/dt} = \frac{a\cos t}{\dfrac{a\cos^2 t}{\sin t}} = \tan t$$

$$\therefore \quad \text{Length of the tangent} = \frac{y\sqrt{1+(dy/dx)^2}}{(dy/dx)}$$

$$= \frac{a\sin t\sqrt{1+\tan^2 t}}{\tan t}$$

$$= \frac{a\sin t\sec t}{\tan t}$$

$= a$, which is constant.

Example 3. *In the catenary* $y = c\cosh\left(\frac{x}{c}\right)$, *prove that the length of the portion of the normal intercepted between the curve and the axis of X (i.e., length of the normal) is* y^2/c.

Solution : Equation of the curve is,

$$y = c\cosh\left(\frac{x}{c}\right) \quad \text{.... (i)}$$

Differentiating w.r.t. x, we get

$$\frac{dy}{dx} = c\sinh\left(\frac{x}{c}\right)\cdot\frac{1}{c} = \sinh\left(\frac{x}{c}\right)$$

$$\therefore \quad \text{Length of normal} = y\sqrt{1+(dy/dx)^2}$$

$$= y\sqrt{1+\sinh^2(x/c)}$$

$$= y\cosh\left(\frac{x}{c}\right)$$

$$= y\cdot\frac{y}{c} \quad \text{[From (i)]}$$

$$= \frac{y^2}{c}.$$

Example 4. *Prove that the sum of the intercepts of the coordinate axes of any tangent to* $\sqrt{x} + \sqrt{y} = \sqrt{a}$ *is constant.*

Solution : Equation of the curve is,

$$\sqrt{x} + \sqrt{y} = \sqrt{a} \qquad \text{.... (i)}$$

Differentiating both the sides w.r.t. x, we get

$$\frac{1}{2}x^{-1/2} + \frac{1}{2}y^{-1/2}\frac{dy}{dx} = 0$$

$$\therefore \qquad \frac{dy}{dx} = -\frac{\sqrt{y}}{\sqrt{x}}$$

Length of the intercept of the tangent on X-axis

$$= x - \frac{y}{(dy/dx)}$$

$$= x - \frac{y}{(-\sqrt{y}/\sqrt{x})}$$

$$= x + \sqrt{x}\sqrt{y}$$

$$= \sqrt{x}(\sqrt{x} + \sqrt{y})$$

$$= \sqrt{x}\sqrt{a} \qquad \text{[From (i)]}$$

Length of the intercept of the tangent on Y-axis

$$= y - x\frac{dy}{dx}$$

$$= y - x\left(-\frac{\sqrt{y}}{\sqrt{x}}\right)$$

$$= y + \sqrt{x}\sqrt{y}$$

$$= \sqrt{y}(\sqrt{y} + \sqrt{x})$$

$$= \sqrt{y}\sqrt{a} \qquad \text{[From (i)]}$$

$\therefore$ Sum of these intercepts

$$= \sqrt{x}\sqrt{a} + \sqrt{y}\sqrt{a}$$

$$= \sqrt{a}(\sqrt{x} + \sqrt{y})$$

$$= \sqrt{a}\sqrt{a}$$

$$= a, \text{ which is a constant.}$$

Example 5. *Show that the subtangent and subnormal of the curve*

$$y^n = a^{n-1}x \textit{ are } nx \text{ and } y^2/nx.$$

Solution : Equation of the curve is,

$$y^n = a^{n-1}x \qquad \text{.... (i)}$$

Differentiating both sides the given equation w.r.t. x, we get

$$ny^{n-1}\frac{dy}{dx} = a^{n-1}$$

$$\therefore \quad \frac{dy}{dx} = \frac{a^{n-1}}{ny^{n-1}} = \frac{ya^{n-1}}{ny^{n}} = \frac{ya^{n-1}}{na^{n-1}x} \quad \text{[From (i)]}$$

$$\frac{dy}{dx} = \frac{y}{nx}$$

$$\therefore \quad \text{Subtangent} = \frac{y}{dy/dx} = \frac{y}{y/nx} = nx$$

and
$$\text{Subnormal} = y\frac{dy}{dx} = y\cdot\frac{y}{nx} = \frac{y^2}{nx}.$$

Example 6. *Show that in the curve $x^{m+n} = a^{m-n}y^{2n}$, the mth power of the subtangent varies as the nth power of the subnormal.*

Solution : The equation of the curve is,

$$x^{m+n} = a^{m-n}y^{2n} \quad \text{.... (i)}$$

Take log on both sides,

$$(m+n)\log x = (m-n)\log a + 2n\log y$$

Differentiating both sides w.r.t. x, we get,

$$\frac{m+n}{x} = \frac{2n}{y}\frac{dy}{dx}$$

$$\therefore \quad \frac{dy}{dx} = \frac{y}{x}\cdot\frac{m+n}{2n}$$

$$\text{Subtangent} = \frac{y}{\left(\frac{dy}{dx}\right)} = \frac{y}{\frac{y}{x}\cdot\left(\frac{m+n}{2n}\right)}$$

$$= \frac{2nx}{m+n}$$

$$\therefore \quad (\text{Subtangent})^m = \left(\frac{2nx}{m+n}\right)^m \quad \text{.... (i)}$$

$$\text{Subnormal} = y\frac{dy}{dx} = y\cdot\frac{y}{x}\cdot\frac{m+n}{2n}$$

$$= \frac{y^2}{x}\cdot\frac{m+n}{2n}$$

$$\therefore \quad (\text{Subnormal})^n = \left\{\frac{y^2(m+n)}{2nx}\right\}^n$$

$$= \left(\frac{m+n}{2n}\right)^n\frac{y^{2n}}{x^n}$$

$$= \left(\frac{m+n}{2n}\right)^n \frac{x^{m+n}}{a^{m-n} \cdot x^n} \qquad \text{[From (i)]}$$

$$= \left(\frac{m+n}{2n}\right)^n \frac{x^m}{a^{m-n}}$$

$$\therefore \qquad \frac{(\text{Subtangent})^m}{(\text{Subnormal})^n} = \frac{\left(\dfrac{2nx}{m+n}\right)}{\dfrac{(m+n)^n x^m}{(2n)^n a^{m-n}}}$$

$$= \frac{(2n)^{m-n} a^{m-n}}{(m+n)^{m+n}} = \text{constant}$$

$$\therefore \qquad (\text{Subtangent})^m \propto (\text{Subnormal})^n.$$

Example 7. *Prove that in the ellipse* $\frac{x^2}{a^2} + \frac{y^2}{b^2} = 1$, *the length of the normal varies inversely as the perpendicular from the origin on the tangent.*

Solution : Equation of the ellipse is,

$$\frac{x^2}{a^2} + \frac{y^2}{b^2} = 1 \qquad \text{.... (i)}$$

Differentiating w.r.t. x, we get

$$\frac{2x}{a^2} + \frac{2y}{b^2}\frac{dy}{dx} = 0$$

$$\therefore \qquad \frac{dy}{dx} = -\frac{b^2 x}{a^2 y}$$

Equation of tangent at (x, y) is,

$$Y - y = -\frac{b^2 x}{a^2 y}(X - x)$$

$$\Rightarrow \qquad \frac{yY}{b^2} - \frac{y^2}{b^2} = -\frac{xX}{a^2} + \frac{x^2}{a^2}$$

$$\Rightarrow \qquad \frac{xX}{a^2} + \frac{yY}{b^2} = \frac{x^2}{a^2} + \frac{y^2}{b^2} = 1 \qquad \text{[From (i)]}$$

Length of perpendicular from (0, 0) on the tangent at (x, y)

$$= \frac{1}{\sqrt{\dfrac{x^2}{a^4} + \dfrac{y^2}{b^4}}} = \frac{a^2 b^2}{\sqrt{b^4 x^2 + a^4 y^2}}$$

$$= p \text{ (say)} \qquad \text{.... (ii)}$$

$$\text{Length of normal} = y\sqrt{1+(dy/dx)^2}$$
$$= y\sqrt{1+(b^4x^2/a^4y^2)}$$
$$= y\frac{\sqrt{a^4y^2+b^4x^2}}{a^2y}$$
$$= \frac{\sqrt{a^4y^2+b^4x^2}}{a^2b^2}\cdot b^2$$
$$= \frac{b^2}{p} \propto \frac{1}{p} \qquad \text{[From (ii)]}$$

Thus, the length of the normal is inversely proportional to the length of the tangent drawn from the origin.

Example 8. *If* x_1, y_1 *be the parts of the axes of x and y intercepted by the tangents at any point (x, y) on the curve*

$$\left(\frac{x}{a}\right)^{2/3} + \left(\frac{y}{b}\right)^{2/3} = 1$$

show that

$$\frac{x_1{}^2}{a^2} + \frac{y_1{}^2}{b_1{}^2} = 1.$$

Solution : The equation of the curve is,

$$\left(\frac{x}{a}\right)^{2/3} + \left(\frac{y}{b}\right)^{2/3} = 1 \qquad \text{.... (i)}$$

Differentiating (i) w.r.t. x, we get

$$\frac{2}{3}\left(\frac{x}{a}\right)^{2/3-1}\cdot\frac{1}{a} + \frac{2}{3}\left(\frac{y}{b}\right)^{2/3-1}\cdot\frac{1}{b}\frac{dy}{dx} = 0$$

$$\Rightarrow \quad \left(\frac{x}{a}\right)^{-1/3}\frac{1}{a} + \left(\frac{y}{b}\right)^{-1/3}\frac{1}{b}\frac{dy}{dx} = 0$$

$$\Rightarrow \quad \frac{dy}{dx} = -\frac{b}{a}\frac{(y/b)^{1/3}}{(x/a)^{1/3}} = -\frac{y^{1/3}}{x^{1/3}}\frac{b^{2/3}}{a^{2/3}}.$$

Here, the equation of the tangent at any point (x, y) is,

$$Y - y = -\frac{y^{1/3}\,b^{2/3}}{x^{1/3}\,a^{2/3}}(X - x)$$

$$\Rightarrow \quad \frac{Y}{y^{1/3}\,b^{2/3}} - \frac{y^{2/3}}{b^{2/3}} = -\frac{X}{x^{1/3}\,a^{2/3}} + \frac{x^{2/3}}{a^{2/3}}$$

$$\Rightarrow \quad \frac{X}{x^{1/3}\,a^{2/3}} + \frac{Y}{y^{1/3}\,b^{2/3}} = \frac{x^{2/3}}{a^{2/3}} + \frac{y^{2/3}}{b^{2/3}} = 1 \qquad \text{[From (i)]}$$

This equation is of the form $\frac{x}{\alpha} + \frac{y}{\beta} = 1$, hence the intercepts made by the tangent on the axes are $x^{1/3} a^{2/3}$ and $y^{1/3} a^{2/3}$

$$\therefore \qquad x_1 = x^{1/3} a^{2/3}, \; y_1 = y^{1/3} a^{2/3}$$

$$\frac{x_1^{\;2}}{a^2} + \frac{y_1^{\;2}}{b^2} = \frac{x^{2/3} a^{4/3}}{a^2} + \frac{y^{2/3} b^{4/3}}{b^2}$$

$$= \frac{x^{2/3}}{a^{2/3}} + \frac{y^{2/3}}{b^{2/3}} = 1, \qquad \text{[Using (i)]}$$

Therefore, $$\frac{x_1^{\;2}}{a^2} + \frac{y_1^{\;2}}{b^2} = 1.$$

EXERCISE 5 (B)

1. Show that the length of the sub-tangent is constant for the curve $y = a^x$.
2. Show that in the curve $y = be^{-a/x}$, the subtangent varies as the square of the abscissa.
3. Show that in the exponential curve $y = be^{x/a}$
   (i) the subtangent at any point is of constant length,
   (ii) the subnormal varies as the square of the ordinate.
4. Show that the subtangent at any point of the curve $x^m y^n = a^{m+n}$ varies as the abscissa of the point.
5. Show that in case of the curve $\beta y^2 = (x + a)^3$, the square of the subtangent varies as the subnormal.
6. Find the length of subnormal of the curve
$$y = \tfrac{1}{2} a \left(e^{x/a} + e^{-x/a}\right).$$
7. Show that the subnormal at any point of a parabola is of constant length and the subtangent varies as the abscissa of the point of contact.
8. Prove that the subnormal at any point of the curve
$$y^2 x^3 = a^2 (x^2 - a^2)$$
varies inversely as the cube of its abscissa.
9. Show that in the curve $y^2 = ax^3$, the square of the subtangent varies as subnormal.
10. Show that for any point of the curve $y = a \log (x^2 - y^2)$, the sum of the tangent and sub-tangent varies as the product of the coordinates of the point of contact.

11. Find the abscissa of the point on the curve $ay^2 = x^3$, the normal at which cuts equal intercepts from the coordinate axes.

12. What should be the value of n in the equation of the curve $y = a^{1-n} x^n$ in order that the subnormal may be of constant length ?

13. Show that the length of tangent for the curve

$$x = \sqrt{a^2 - y^2} + \frac{a}{2} \log \frac{a - \sqrt{a^2 - y^2}}{a + \sqrt{a^2 - y^2}} \text{ is } a^2.$$

14. Show that the length of the normal at any point P of the rectangular hyperbola $x^2 - y^2 = a^2$ is equal to the distance of P from the origin.

15. Prove that in the curve $x^{2/3} + y^{2/3} = a^{2/3}$, the intercepts, made by the tangent at any point, on the coordinate axes are $a^{2/3} x^{2/3}$ and $a^{2/3} y^{2/3}$ respectively. Hence verify that the length of tangent intercepted by the axes is constant.

16. Show that at any point of the hyperbola $xy = c^2$, the subtangent varies as the abscissa and subnormal varies as the cube of the ordinate of the point of contact. Also prove that product of the intercepts of the tangent on the coordinate axes is constant.

17. Prove that for the catenary $y = c \cosh (x/c)$, the perpendicular dropped from the foot of the ordinate upon the tangent is of constant length.

18. Find the subtangent, tangent, subnormal, normal, and the intercepts by the tangent on the axes at the point t of the cycloid

$$x = a(t + \sin t),\ y = a(1 - \cos t).$$

19. Show that the normal at any point of the curve

$$x = a(\cos t + t \sin t),\ y = a(\sin t - t \cos t)$$

is at a constant distance from the origin.

20. (i) Find the lengths of tangent, normal, subtangent and subnormal to $x = a \cos^3 \theta, y = a \sin^3 \theta$ at the point θ.

(ii) Find the locus of the mid-point of the portion of the tangent intercepted between the axes and the curve

$$y = a \cos^3 \theta,\ y = a \sin^3 \theta.$$

ANSWERS

6. $\frac{1}{2} a \sinh (2x/a)$ **11.** $4a/9$ **12.** $\frac{1}{2}$

18. $a \sin t,\ 2a \sin^2 \frac{t}{2},\ 2a \sin \frac{t}{2} \tan \frac{t}{2},\ 2a \sin \frac{t}{2},\ at,\ -at \tan \frac{t}{2}.$

20. (i) $a\sin^2\theta, a\sin^2\theta\tan\theta, a\sin^2\theta\cos\theta, a\sin^3\theta\tan\theta$.

(ii) $x^2+y^2=a^2/4$.

5.10. Polar Coordinates

Let O be a fixed point called the ***Pole*** and OX be a fixed straight line called the ***Initial line***.

The position of a point P in the plane is known if we know $OP = r$ and the angle $POX = \theta$.

Here r is called the radius vector and θ the vectorial angle. Also (r, θ) are called the polar coordinates of P.

Angle θ is regarded as positive when measured in anti-clockwise direction and negative when measured in clockwise direction. r is regarded as positive when measured along the line bounding the vectorial angle and negative when measured in the opposite direction.

A little consideration will show that the point (r, θ) in the plane can also be represented by

$$(r, \theta \pm 2\pi), \ (r, \theta \pm 4\pi), \ \ldots\ldots, \ (r, \theta \pm 2n\pi)$$

or
$$(-r, \theta + \pi), \ (-r, \theta + 3\pi), \ \ldots\ldots, \ (-r, \theta + \overline{2n-1}\,\pi),$$

where $n \in I$.

Hence, unlike cartesian coordinates, the polar coordinates of any point P are not unique.

If (x, y) be the cartesian coordinates of P referred to the pole O as origin, the initial line OX as X-axis and a perpendicular line through O as Y-axis, then,

$$x = OM = r\cos\theta \qquad \ldots\text{(i)}$$

and
$$y = MP = r\sin\theta \qquad \ldots\text{(ii)}$$

Squaring and adding (i) and (ii),

$$x^2 + y^2 = r^2$$

$$\Rightarrow \qquad r = \sqrt{x^2+y^2} \qquad \ldots\text{(iii)}$$

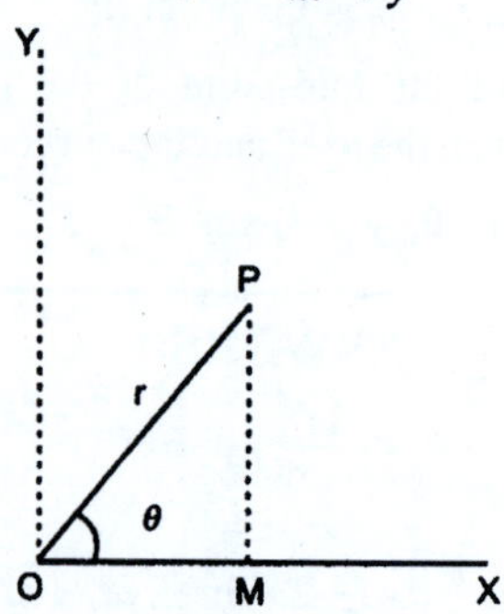

Fig. 5.5

Dividing (ii) by (i),

$$\tan\theta = \frac{y}{x}$$

$$\Rightarrow \qquad \theta = \tan^{-1}(y/x) \qquad \text{.... (iv)}$$

With the help of the relations (i) and (ii), we can transform cartesian coordinates into polar coordinates and with the help of relations (iii) and (iv), we can convert polar coordinates into cartesian coordinates.

5.11. Angle between Radius Vector and Tangent

Let $P(r, \theta)$ and $Q(r+\delta r, \theta+\delta\theta)$ be two neighbouring points on the curve $r = f(\theta)$. Let PT be the tangent to the curve at P. OP is the radius vector. The angle between the radius vector OP and the tangent PT is OPT. This is denoted by ϕ. We are required to determine ϕ. Draw QM perpendicular to OP (produce if necessary).

From ΔPQM, we have

$$QM = OQ \sin\delta\theta$$
$$= (r+\delta r)\sin\delta\theta$$

and

$$OM = OQ\cos\delta\theta$$
$$= (r+\delta r)\cos\delta\theta$$

$$\therefore \quad PM = OM - OP$$
$$= (r+\delta r)\cos\delta\theta - r$$

$$\therefore \tan\angle QPM = \frac{QM}{PM}$$
$$= \frac{(r+\delta r)\sin\delta\theta}{(r+\delta r)\cos\delta\theta - r}.$$

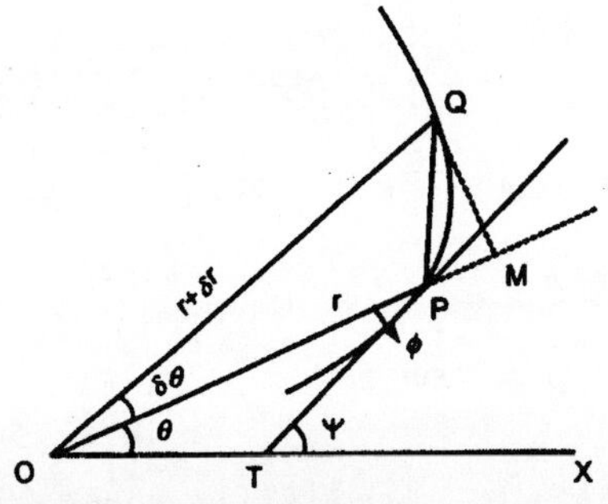

Fig. 5.6

When $Q \to P$; $\delta\theta \to 0$, the chord PQ tends to become the tangent PT at P and hence the angle $QPM \to \phi$.

$$\therefore \quad \tan \phi = \lim_{\delta\theta \to 0} \tan \angle QPM$$

$$= \lim_{\delta\theta \to 0} \frac{(r+\delta r)\sin \delta\theta}{(r+\delta r)\cos \delta\theta - r}$$

$$= \lim_{\delta\theta \to 0} \frac{(r+\delta r)\ \delta\theta}{(r+\delta r)\ 1 - r} \qquad \left[\begin{array}{r} \because \text{ When } \delta\theta \to 0; \\ \sin \delta\theta \to \delta\theta \\ \text{and } \cos \delta\theta \to 1 \end{array}\right]$$

$$= \lim_{\delta\theta \to 0} \frac{r\delta\theta + \delta r\,\delta\theta}{\delta r}$$

$$= \lim_{\delta\theta \to 0} \frac{r\delta\theta}{\delta r}$$ [Neglecting $\delta r \delta\theta$, it being a very small quantity of second order]

$$= \frac{r\,d\theta}{dr}$$

Therefore $$\tan \phi = \frac{rd\theta}{dr}.$$

Note : The relation between θ, ϕ and Ψ is $\Psi = \theta + \phi$.

5.12. Three Important Relations

If p denotes the length of the perpendicular from pole on the tangent and r, θ, ϕ have their usual meanings, then we have following three important relations :

(1) $p = r \sin \phi$.

(2) $$\frac{1}{p^2} = \frac{1}{r^2} + \frac{1}{r^4}\left(\frac{dr}{d\theta}\right)^2.$$

(3) $$\frac{1}{p^2} = x^2 + \left(\frac{dr}{d\theta}\right)^2, \text{ where } x = \frac{1}{r}.$$

PROOF:

(1) From right angled triangle OPM, we have

$$\sin \phi = \frac{p}{r}$$

$$\therefore \quad p = r \sin \phi.$$

(2) We have,

$$p = r \sin \phi$$

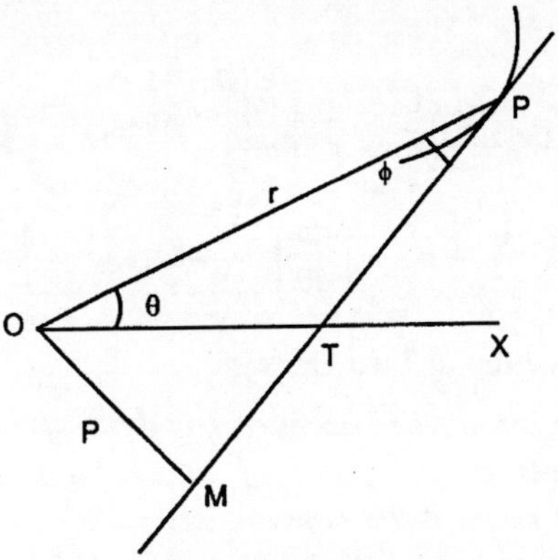

Fig. 5.7

$$\Rightarrow \quad p^2 = r^2 \sin^2 \phi$$

$$\Rightarrow \quad \frac{1}{p^2} = \frac{1}{r^2} \operatorname{cosec}^2$$

$$\Rightarrow \quad \frac{1}{p^2} = \frac{1}{r^2} [1 + \cot^2 \phi]$$

$$\Rightarrow \quad \frac{1}{p^2} = \frac{1}{r^2} \left[1 + \frac{1}{\tan^2 \phi}\right]$$

$$\Rightarrow \quad \frac{1}{p^2} = \frac{1}{r^2} \left[1 + \frac{1}{\left(r\frac{d\theta}{dr}\right)^2}\right] \qquad \left[\because \tan \phi = r\frac{d\theta}{dr}\right]$$

$$\Rightarrow \quad \frac{1}{p^2} = \frac{1}{r^2} + \frac{1}{r^4}\left(\frac{dr}{d\theta}\right)^2.$$

(3) We have, $u = \dfrac{1}{r}$

$$\therefore \quad \frac{du}{d\theta} = -\frac{1}{r^2}\frac{dr}{d\theta}$$

$$\therefore \quad \left(\frac{du}{d\theta}\right)^2 = \frac{1}{r^4}\left(\frac{dr}{d\theta}\right)^2$$

Hence the result (2),

$$\frac{1}{p^2} = \frac{1}{r^2} + \frac{1}{r^4}\left(\frac{dr}{d\theta}\right)^2 \quad \text{gives,}$$

$$\frac{1}{p^2} = u^2 + \left(\frac{du}{d\theta}\right)^2, \quad \text{where } u = \frac{1}{r}.$$

5.13. Angle of Intersection of Two Curves

The angle of intersection of the two curves is defined as the acute angle between their tangents at the point of intersection. At the point of intersection the radius vector is the same.

Let ϕ_1 and ϕ_2 be the angles between the common radius vector OP and the tangents PT_1 and PT_2 to the two curves C_1 and C_2 at their point of intersection P respectively. Then it is clear that the angle between the tangents $= \phi_1 \sim \phi_2$.

$$\therefore \text{ Angle of intersection } = \phi_1 \sim \phi_2 = \tan^{-1} \tan(\phi_1 \sim \phi_2)$$

$$= \tan^{-1}\left(\frac{\tan\phi_1 \sim \tan\phi_2}{1 + \tan\phi_1 \tan\phi_2}\right)$$

Note : The two curves will cut orthogonally if $\tan\phi_1 \tan\phi_2 = -1$.

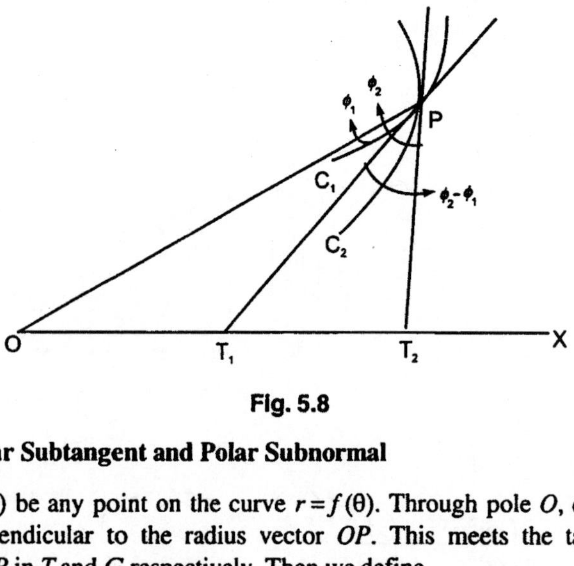

Fig. 5.8

5.14. Polar Subtangent and Polar Subnormal

Let $P(r, \theta)$ be any point on the curve $r = f(\theta)$. Through pole O, draw a line GOT perpendicular to the radius vector OP. This meets the tangent and normal at P in T and G respectively. Then we define,

$$\text{length } OT = \text{polar subtangent}$$

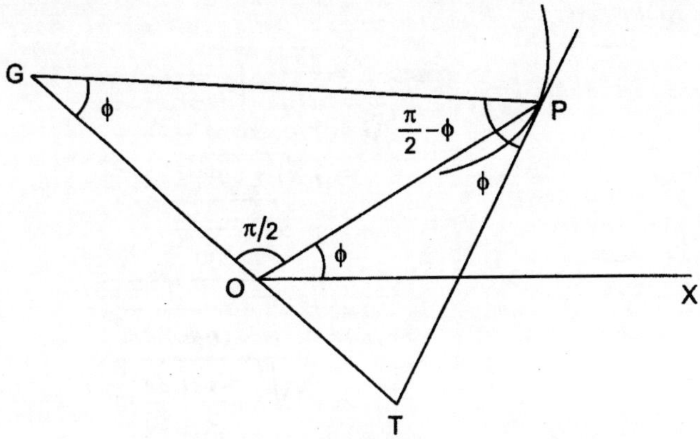

Fig. 5.9

$$\text{length } OG = \text{polar subnormal}$$

Also, $\text{length } PT = \text{polar tangent}$

and $\text{length } PG = \text{polar normal.}$

From ΔOPT,

$$\tan\phi = \frac{OT}{r}$$

$$\therefore \quad OT = r\tan\phi = r\,.\,r\frac{d\theta}{dr} = r^2\frac{d\theta}{dr}$$

$$\therefore \quad \boxed{\textbf{Polar subtangent} = r^2\frac{d\theta}{dr}.}$$

and from ΔOPG,

$$\cot\phi = \frac{OG}{r}$$

$$\therefore \quad OG = r\cot\phi = \frac{r}{\tan\phi} = \frac{r}{r\,(d\theta/dr)} = \frac{dr}{d\theta}$$

$$\therefore \quad \boxed{\textbf{Polar subnormal} = \frac{dr}{d\theta}.}$$

Again, from ΔOPT,

$$\sec\phi = \frac{PT}{r}$$

$$\therefore \quad PT = r\sec\phi$$

$$= r\sqrt{1+\tan^2\phi}$$

$$= r\sqrt{1+r^2\,(d\theta/dr)^2}$$

$$\therefore \quad \boxed{\textbf{Polar tangent} = r\sqrt{1+r^2\,(d\theta/dr)^2}}$$

and from ΔOPG,

$$\text{cosec}\ \phi = \frac{PG}{r}$$

$$\therefore \quad PG = r\ \text{cosec}\ \phi$$

$$= r\sqrt{1+\cot^2\phi}$$

$$= r\sqrt{1+\frac{1}{\tan^2\phi}}$$

$$= r\sqrt{1+\frac{1}{(r\,d\theta/dr)^2}}$$

$$= r\sqrt{1+\frac{1}{r^2}\left(\frac{dr}{d\theta}\right)^2}$$

$$= \sqrt{r^2+(dr/d\theta)^2}$$

$$\therefore \quad \boxed{\textbf{Polar normal} = \sqrt{r^2+(dr/d\theta)^2}}$$

5.15. Pedal Equation

A relation between p and r for a curve is called the ***pedal equation*** *of the curve, where p is the length of the perpendicular on the tangent from the pole and r is the radius vector to that point of the curve.*

5.16. The Pedal Equation of a Curve whose Cartesian Equation is given

Let the cartesian equation of the curve be

$$f(x, y) = 0 \qquad \text{.... (1)}$$

Equation of tangent at any point (x, y) on the curve is,

$$Y - y = \frac{dy}{dx}(X - x)$$

$$\Rightarrow \quad X\frac{dy}{dx} - Y + y - x\frac{dy}{dx} = 0$$

$\therefore$ p = length of the perpendicular from the origin (0, 0) upon the tangent

$$= \frac{y - x\dfrac{dy}{dx}}{\sqrt{(dy/dx)^2+1}} \qquad \text{.... (2)}$$

Also, $$r^2 = x^2 + y^2 \qquad \text{.... (3)}$$

Eliminating x and y from (1), (2) and (3), we obtain a relation between p and r which is the required pedal equation of the curve.

5.17. The Pedal Equation of a Curve whose Polar Equation is given

Method I. Let the polar equation of the curve be

$$r = f(\theta) \quad \text{.... (1)}$$

Also we have, $$\tan \phi = r\frac{d\theta}{dr} \quad \text{.... (2)}$$

and $$p = r \sin \phi \quad \text{.... (3)}$$

Eliminating θ and ϕ from (1), (2) and (3), we get a relation between p and r which is the required pedal equation of the curve.

Method II. Let $$r = f(\theta) \quad \text{.... (1)}$$

be the polar curve.

Also we know that

$$\frac{1}{p^2} = \frac{1}{r^2} + \frac{1}{r^4}\left(\frac{dr}{d\theta}\right)^2 \quad \text{.... (2)}$$

Eliminating θ between (1) and (2), we get a relation between p and r, which is the required pedal equation of the curve.

ILLUSTRATIVE EXAMPLES

Example 1. *Find the angle ϕ in the case of the curve*

$$r^n = a^n \sec(n\theta + \alpha)$$

and prove that this curve is intersected by the curve

$$r^n = b^n \sec(n\theta + \beta)$$

at an angle which is independent of a and b.

Solution : $$r^n = a^n \sec(n\theta + \alpha).$$

Taking log on both sides, we get

$$n \log r = n \log a + \log \sec (n\theta + \alpha).$$

Differentiating with respect to θ, we get

$$\frac{n}{r}\frac{dr}{d\theta} = \frac{\sec(n\theta + \alpha)\cdot\tan(n\theta + \alpha)\,.\,n}{\sec(n\theta + \alpha)}$$

$$\Rightarrow \quad \cot \phi_1 = \tan(n\theta + \alpha)$$

$$\Rightarrow \quad \cot \phi_1 = \cot\left[\frac{\pi}{2} - (n\theta + \alpha)\right]$$

$$\therefore \quad \phi_1 = \frac{\pi}{2} - (n\theta + \alpha).$$

Similarly, for the second curve, we can get,

$$r^n = b^n \sec(n\theta + \beta)$$

$$\phi_2 = \frac{\pi}{2} - (n\theta + \beta).$$

Hence, the angle of intersection $= \phi_1 - \phi_2 = \beta - \alpha$

which is independent of a and b.

Example 2. *Find the angle of intersection of the curves*

$$r = a(1+\cos\theta),\ r = b(1-\cos\theta).$$

Solution : $r = a(1+\cos\theta)$.

Taking log on both sides,

$$\log r = \log a + \log(1+\cos\theta).$$

Differentiating w.r.t. θ, we get

$$\frac{1}{r}\frac{dr}{d\theta} = -\frac{\sin\theta}{1+\cos\theta}$$

$$\Rightarrow \quad \cot\phi_1 = -\frac{2\sin\frac{\theta}{2}\cos\frac{\theta}{2}}{2\cos^2\frac{\theta}{2}} = -\tan\frac{\theta}{2}$$

$$\Rightarrow \quad \cot\phi_1 = \cot\left(\frac{\pi}{2}+\frac{\theta}{2}\right)$$

$$\Rightarrow \quad \phi_1 = \frac{\pi}{2}+\frac{\theta}{2}.$$

Again, $r = b(1-\cos\theta)$.

Taking log on both sides,

$$\log r = \log b + \log(1-\cos\theta).$$

Differentiating w.r.t. θ, we get

$$\frac{1}{r}\frac{dr}{d\theta} = \frac{\sin\theta}{1-\cos\theta}$$

$$\Rightarrow \quad \cot\phi_2 = \frac{2\sin\frac{\theta}{2}\cos\frac{\theta}{2}}{2\sin^2\frac{\theta}{2}} = \cot\frac{\theta}{2}$$

$$\Rightarrow \quad \theta_2 = \frac{\theta}{2}.$$

$\therefore$ Angle of intersection $= \phi_1 - \phi_2$

$$= \left(\frac{\pi}{2}+\frac{\theta}{2}\right) - \frac{\theta}{2} = \frac{\pi}{2}.$$

Hence, the two curves intersect at right angles, *i.e.*, orthogonally.

Example 3. *Find the angle of intersection of the curves*

$$r^2 = 16\sin 2\theta \quad \text{and} \quad r^2\sin 2\theta = 4.$$

Solution : $r^2 = 16\sin 2\theta$.

Taking log on both sides,

$$2 \log r = \log 16 + \log \sin 2\theta.$$

Differentiating w.r.t. θ, we get

$$\frac{2}{r}\frac{dr}{d\theta} = \frac{\cos 2\theta}{\sin 2\theta} \cdot 2$$

$$\Rightarrow \quad \cot \phi_1 = \cot 2\theta$$

$$\Rightarrow \quad \phi_1 = 2\theta.$$

Again, $\quad r^2 \sin 2\theta = 4.$

Taking log on both sides,

$$2 \log r + \log \sin 2\theta = \log 4.$$

Differentiating w.r.t. θ, we get

$$\frac{2}{r}\frac{dr}{dt} + \frac{\cos 2\theta}{\sin 2\theta} \cdot 2 = 0$$

$$\Rightarrow \quad \cot \phi_2 + \cot 2\theta = 0$$

$$\Rightarrow \quad \cot \phi_2 = -\cot 2\theta$$

$$\Rightarrow \quad \cot \phi_2 = \cot (\pi - 2\theta)$$

$$\Rightarrow \quad \phi_2 = \pi - 2\theta$$

$$\therefore \quad \phi_2 - \phi_1 = \pi - 2\theta - 2\theta = \pi - 4\theta.$$

Now, at the point where the two curves intersect, the radius vector is the same. Hence, equating the values of r^2 from the two curves, we get

$$16 \sin 2\theta = \frac{4}{\sin 2\theta}$$

$$\Rightarrow \quad \sin^2 2\theta = \frac{1}{4}$$

$$\Rightarrow \quad \sin 2\theta = \pm\frac{1}{2}$$

$$\Rightarrow \quad 2\theta = \frac{\pi}{6} \text{ or } \frac{7\pi}{6}$$

$$\Rightarrow \quad \theta = \frac{\pi}{12} \text{ or } \frac{7\pi}{12}.$$

Hence, the given curves intersect at the points where

$$\theta = \frac{\pi}{12} \text{ or } \frac{7\pi}{12}.$$

When $\theta = \frac{\pi}{12}$. Then the required angle of intersection

$$= \pi - 4.\frac{\pi}{12}$$

$$= \pi - \frac{\pi}{3} = \frac{2\pi}{3}$$

When $\theta = \frac{7\pi}{12}$. Then the required angle of intersection

$$= \pi - 4 \cdot \frac{7\pi}{12} = \pi - \frac{7\pi}{3}$$

$$= -\frac{4\pi}{3} = \frac{4\pi}{3}. \qquad \text{[Leaving –ve sign]}$$

Example 4. *If ϕ is the angle between the tangent to a curve and the radius vectors drawn from the origin of the coordinates to the point of contact, prove that*

$$\tan \phi = \frac{x\frac{dy}{dx} - y}{x + y\frac{dy}{dx}}.$$

Solution : Let $P(r, \theta)$ be a point on the curve. The line PT is tangent at the point P. The line PT makes the angle Ψ with the initial line OX.

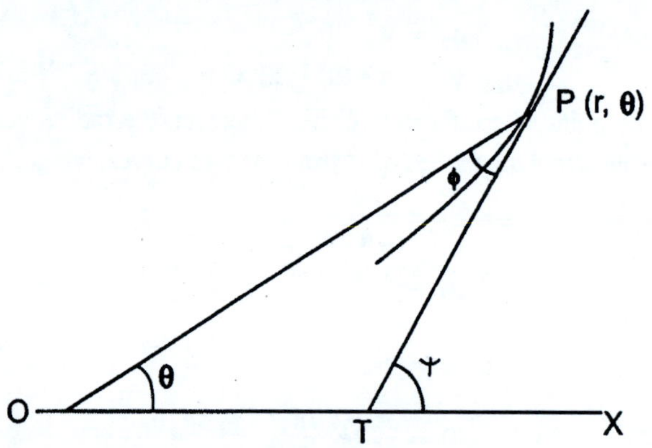

Fig. 5.10

The line OP is radius vector and $\angle OPT = \phi$.

Let the cartesian coordinates of the point P be (x, y), then

$$\frac{y}{x} = \frac{r \sin \theta}{r \cos \theta} = \tan \theta$$

and
$$\frac{dy}{dx} = \tan \Psi.$$

It is clear from the figure that

$$\Psi = \theta + \phi$$
$$\phi = \Psi - \theta$$

$$\tan\phi = \tan(\Psi - \theta) = \frac{\tan\Psi - \tan\theta}{1 + \tan\Psi\tan\theta}$$

$$= \frac{\frac{dy}{dx} - \frac{y}{x}}{1 + \frac{dy}{dx}\cdot\frac{y}{x}} = \frac{x\frac{dy}{dx} - y}{x + y\frac{dy}{dx}}.$$

Thus, $$\tan\phi = \frac{x\frac{dy}{dx} - y}{x + y\frac{dy}{dx}}.$$

Example 5. *Prove that the normal at any point* (r, θ) *of the curve* $r^n = a^n \cos n\theta$ *makes an angle* $(n + 1)\,\theta$ *with the initial line.*

Solution : $$r^n = a^n \cos n\theta.$$

Take log on both sides,

$$n \log r = n \log a + \log \cos n\theta.$$

Differentiating w.r.t. θ, we get

$$\frac{n}{r}\frac{dr}{d\theta} = -\frac{\sin n\theta}{\cos n\theta}\cdot n$$

$$\Rightarrow \quad \cot\phi = -\tan n\theta = \cot\left(\frac{\pi}{2} + n\theta\right)$$

$$\Rightarrow \quad \phi = \frac{\pi}{2} + n\theta.$$

From the figure, it is clear that the angle which the normal PG at P makes with the initial line is $\angle PGT = n\theta + \theta = (n + 1)\,\theta$.

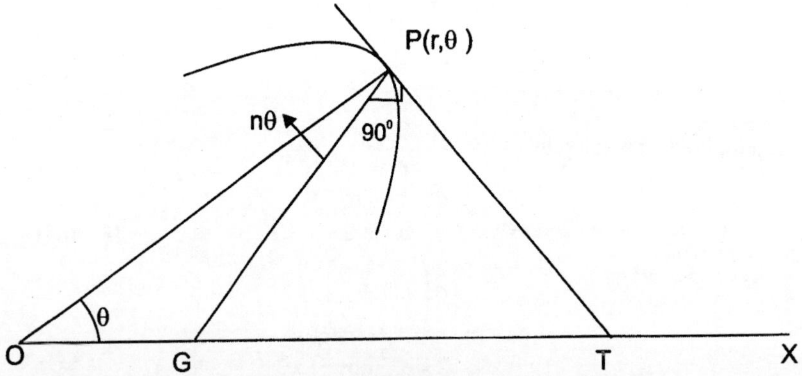

Fig 5.11

Example 6. *For the cardiod* $r = a(1 - \cos\theta)$, *prove that*

(i) $\phi = \frac{\theta}{2}$. (ii) $2ap^2 = r^3$.

(iii) *Polar subtangent* $= 2a \sin^2 \frac{\theta}{2} \tan \frac{\theta}{2}$.

Solution : $$r = a(1 - \cos\theta). \quad \text{.... (i)}$$

Taking log on both sides,

$$\log r = \log a + \log(1 - \cos\theta).$$

Differentiating w.r.t. θ, we get

$$\frac{1}{r}\frac{dr}{d\theta} = \frac{\sin\theta}{1 - \cos\theta}$$

$$\Rightarrow \quad \cot\phi = \frac{2\sin\frac{\theta}{2}\cos\frac{\theta}{2}}{2\sin^2\frac{\theta}{2}} = \cot\frac{\theta}{2}$$

$$\Rightarrow \quad \phi = \frac{\theta}{2}.$$

Hence, the first part is proved.

Now, $$p = r\sin\phi$$

$$\therefore \quad p = r\sin\frac{\theta}{2}$$

$$\Rightarrow \quad p^2 = r^2\sin^2\frac{\theta}{2}$$

$$\Rightarrow \quad 2p^2 = r^2\left(2\sin^2\frac{\theta}{2}\right)$$

$$\Rightarrow \quad 2p^2 = r^2(1 - \cos\theta)$$

$$\Rightarrow \quad 2p^2 = r^2 \cdot \frac{r}{a} \quad \text{[From (i)]}$$

$$\Rightarrow \quad 2ap^2 = r^3.$$

Hence, the second part is proved.

Again, Polar sub-tangent $$= r^2\frac{d\theta}{dr} = \frac{r^2}{(dr/d\theta)}$$

$$= \frac{a^2(1 - \cos\theta)^2}{a\sin\theta} \quad \text{[From (i)]}$$

$$= \frac{a\left(2\sin^2\frac{\theta}{2}\right)^2}{2\sin\frac{\theta}{2}\cos\frac{\theta}{2}}$$

$$= 2a\sin^2\frac{\theta}{2}\tan\frac{\theta}{2}.$$

Example 7. *Find the pedal equation to the curve $r = ae^{\theta \cot \alpha}$.*

Solution :
$$r = ae^{\theta \cot \alpha} \qquad \text{.... (i)}$$

$$\frac{dr}{d\theta} = ae^{\theta \cot \alpha} \cot \alpha - r \cot \alpha \qquad \text{[Using (i)]}$$

We know that

$$\frac{1}{p^2} = \frac{1}{r^2} + \frac{1}{r^4}\left(\frac{dr}{d\theta}\right)^2$$

$$= \frac{1}{r^2} + \frac{1}{r^4} \cdot \cot^2 \alpha$$

$$= \frac{1}{r^2}(1 + \cot^2 \alpha) = \frac{1}{r^2} \operatorname{cosec}^2 \alpha = \frac{1}{r^2 \sin^2 \alpha}$$

$$\therefore \qquad \frac{1}{p^2} = \frac{1}{r^2 \sin^2 \alpha}$$

$$\Rightarrow \qquad p^2 = r^2 \sin^2 \alpha \Rightarrow p = r \sin \alpha.$$

Example 8. *Show that the pedal equation of the spiral $r = a$ sech $n\theta$ is of the form*

$$\frac{1}{p^2} = \frac{A}{r^2} + B.$$

Solution : The equation to the spiral is,

$$r = a \operatorname{sech} n\theta. \qquad \text{.... (i)}$$

Differentiating the given equation w.r.t. θ,

$$\frac{dr}{d\theta} = -a\, n \operatorname{sech} n\theta . \tanh n\theta$$

$$= -nr \tanh n\theta. \qquad \text{[Using (i)]}$$

Now,
$$\frac{1}{p^2} = \frac{1}{r^2} + \frac{1}{r^4}\left(\frac{dr}{d\theta}\right)^2$$

$$= \frac{1}{r^2} + \frac{1}{r^4}[n^2 r^2 \tanh^2 n\theta]$$

$$= \frac{1}{r^2} + \frac{n^2}{r^2} \tanh^2 n\theta$$

$$= \frac{1}{r^2}[1 + n^2 \tanh^2 n\theta]$$

$$= \frac{1}{r^2}[1 + n^2 (1 - \operatorname{sech}^2 n\theta)]$$

$$= \frac{1}{r^2}\,[(1+n^2) - n^2\,\text{sech}^2\, n\theta]$$

$$= \frac{1+n^2}{r^2} - \frac{n^2}{r^2}\cdot\frac{r^2}{a^2}, \qquad \text{[Using (i)]}$$

$$\therefore \qquad \frac{1}{p^2} = \frac{1+n^2}{r^2} - \frac{n^2}{a^2}$$

$$\Rightarrow \qquad \frac{1}{p^2} = \frac{A}{r^2} + B,$$

where $A = 1+n^2$ and $B = -\frac{n^2}{a^2}$.

Example 9. *Find the pedal equation of the curve*

$$\frac{l}{r} = 1 + e\cos\theta.$$

Solution :

$$\frac{l}{r} = 1 + e\cos\theta. \qquad \text{.... (i)}$$

Differentiating w.r.t. θ, we get

$$-\frac{l}{r^2}\frac{dr}{d\theta} = -e\sin\theta$$

$$\therefore \qquad \frac{dr}{d\theta} = \frac{r^2 e\sin\theta}{l}$$

Now,

$$\frac{1}{p^2} = \frac{1}{r^2} + \frac{1}{r^4}\left(\frac{dr}{d\theta}\right)^2$$

$$= \frac{1}{r^2} + \frac{1}{r^4}\cdot\frac{r^4 e^2\sin^2\theta}{l^2}$$

$$= \frac{1}{r^2} + \frac{e^2}{l^2}(1-\cos^2\theta)$$

$$= \frac{1}{r^2} + \frac{e^2}{l^2} - \frac{e^2}{l^3}\cos^2\theta$$

$$= \frac{1}{r^2} + \frac{e^2}{l^2} - \frac{1}{l^2}\left(\frac{l}{r} - 1\right)^2 \qquad \text{[From (i)]}$$

$$= \frac{1}{r^2} + \frac{e^2}{l^2} - \frac{1}{l^2}\left(\frac{l^2}{r^2} + 1 - \frac{2l}{r}\right)$$

$$= \frac{1}{r^2} + \frac{e^2}{l^2} - \frac{1}{r^2} - \frac{1}{l^2} + \frac{2}{lr}$$

$$= \frac{e^2}{l^2} - \frac{1}{l^2} + \frac{2}{lr}$$

$$= \frac{1}{l^2}\left[e^2 - 1 + \frac{2l}{r}\right]$$

which is the required pedal equation.

Example 10. *Show that the pedal equation to ellipse*

$$\frac{x^2}{a^2} + \frac{y^2}{b^2} = 1 \text{ is } \frac{1}{p^2} = \frac{1}{a^2} + \frac{1}{b^2} - \frac{r^2}{a^2 b^2}.$$

Solution : The equation to the ellipse is,

$$\frac{x^2}{a^2} + \frac{y^2}{b^2} = 1. \qquad \text{....(i)}$$

Let (x_1, y_1) be any point on (i), then

$$\frac{x_1^2}{a^2} + \frac{y_1^2}{b^2} = 1. \qquad \text{.... (ii)}$$

Differentiating (i) with respect to x

$$\frac{2x}{a^2} + \frac{2y}{b^2}\frac{dy}{dx} = 0 \Rightarrow \frac{dy}{dx} = \frac{-b^2 x}{a^2 y}.$$

The value of $\frac{dy}{dx}$ at (x_1, y_1) $= \frac{-b^2 x_1}{a^2 y_1}$

Equation to tangent at the point (x_1, y_1) on the ellipse is,

$$y - y_1 = \frac{-b^2 x_1}{a^2 y_1}(x - x_1)$$

$$\Rightarrow \quad \frac{yy_1}{b^2} - \frac{y_1^2}{b^2} = -\frac{xx_1}{a^2} + \frac{x_1^2}{a^2}$$

$$\Rightarrow \quad \frac{y_1 y}{b^2} + \frac{xx_1}{a^2} = \frac{x_1^2}{a^2} + \frac{y_1^2}{b^2}$$

$$\Rightarrow \quad \frac{yy_1}{b^2} + \frac{xx_1}{a^2} = 1. \qquad \text{[Using (ii)]}$$

Now, p = length of perpendicular from (0, 0) on the tangent

$$= \frac{1}{\sqrt{\frac{x_1^2}{a^4} + \frac{y_1^2}{b^4}}}$$

$$\therefore \quad \frac{1}{p^2} = \frac{x_1^2}{a^4} + \frac{y_1^2}{b^4}$$

$$= \frac{1}{a^2}\left[\frac{x_1^2}{a^2} + \frac{y_1^2}{b^2}\right] + \frac{1}{b^2}\left[\frac{x_1^2}{a^2} + \frac{y_1^2}{b^2}\right] - \frac{x_1^2 + y_1{}^2}{a^2 b^2}$$

$$= \frac{1}{a^2} + \frac{1}{b^2} - \frac{r^2}{a^2 b^2},$$

[Using (i) and the relation $x_1^2 + y_1^2 = r^2$]

$$\therefore \quad \frac{1}{p^2} = \frac{1}{a^2} + \frac{1}{b^2} - \frac{r^2}{a^2 b^2}.$$

Example 11. *Find the pedal equation of the astroid*

$$x^{2/3} + y^{2/3} = a^{2/3}.$$

Solution :

$$x^{2/3} + y^{2/3} = a^{2/3}. \quad \text{.... (i)}$$

Differentiating w.r.t. x, we get

$$\frac{2}{3}x^{-1/3} + \frac{2}{3}y^{-1/3}\frac{dy}{dx} = 0$$

$$\therefore \quad \frac{dy}{dx} = -\frac{y^{1/3}}{x^{1/3}}$$

Hence, the equation of the tangent at (x, y) is,

$$Y - y = -\frac{y^{1/3}}{x^{1/3}}(X - x)$$

$$\Rightarrow \quad \frac{Y}{y^{1/3}} - y^{2/3} = -\frac{X}{x^{1/3}} + x^{2/3}$$

$$\Rightarrow \quad \frac{X}{x^{1/3}} + \frac{Y}{y^{1/3}} = x^{2/3} + y^{2/3} = a^{2/3} \text{ [From (i)]} \quad \text{....(ii)}$$

$\therefore$ p = length of perpendicular (0, 0) on (ii)

$$= \frac{a^{2/3}}{\sqrt{\frac{1}{x^{2/3}} + \frac{1}{y^{2/3}}}}$$

$$= \frac{a^{2/3}\, x^{1/3}\, y^{2/3}}{\sqrt{x^{2/3}+y^{2/3}}}$$

$$= \frac{a^{2/3}\, x^{1/3}\, y^{1/3}}{\sqrt{a^{2/3}}} \qquad \text{[From (i)]}$$

$$= a^{1/3}\, x^{1/3}\, y^{1/3}. \qquad \text{.... (iii)}$$

Also,

$$r^2 = x^2 + y^2$$

$$= (x^{2/3})^3 + (y^{2/3})^3$$

$$= (x^{2/3}+y^{2/3})^3 - 3x^{2/3}\, y^{2/3}\, (x^{2/3}+y^{2/3})$$

$$= (a^{2/3})^3 - 3x^{2/3}\, y^{2/3}\, a^{2/3} \qquad \text{[From (i)]}$$

$$= a^2 - 3p^2. \qquad \text{[From (ii)]}$$

Hence, the required pedal equation is,

$$r^2 = a^2 - 3p^2.$$

Example 12. *Prove that*

$$\frac{d\phi}{d\theta} = \frac{u\dfrac{d^2u}{d\theta^2} - \left(\dfrac{du}{d\theta}\right)^2}{u^2 + \left(\dfrac{du}{d\theta}\right)^2}, \text{ where } u = \frac{1}{r}.$$

Solution : We know that

$$\tan\phi = r\frac{d\theta}{dr} = \left(\frac{r}{dr/d\theta}\right)$$

$$= \frac{1/u}{\dfrac{d}{d\theta}\left(\dfrac{1}{u}\right)} \qquad \left[\because r = \frac{1}{u}\right]$$

$$= \frac{1/u}{-\dfrac{1}{u^2}\dfrac{du}{d\theta}} = -\frac{u}{\left(\dfrac{du}{d\theta}\right)}$$

$$\therefore \quad \phi = \tan^{-1}\left[-\frac{u}{(du/d\theta)}\right]$$

$$= -\tan^{-1}\left[\frac{u}{(du/d\theta)}\right]$$

$$\left[\because \tan^{-1}(-\theta) = -\tan^{-1}\theta\right]$$

$$\therefore \quad \frac{d\phi}{d\theta} = -\frac{1}{1+\left\{\dfrac{u}{(du/d\theta)}\right\}^2} \cdot \frac{\dfrac{du}{d\theta}\cdot\dfrac{du}{d\theta} - u\cdot\dfrac{d^2u}{d\theta^2}}{\left(\dfrac{du}{d\theta}\right)^2}$$

$$= \frac{u - \dfrac{d^2u}{d\theta^2} - \left(\dfrac{du}{d\theta}\right)^2}{u^2 + (du/d\theta)^2}$$

Example 13. *Show that the locus of the extremity of the polar subtangent of the curve* $\frac{1}{r} + f(\theta) = 0$ *is,*

$$u = f'\left(\frac{\pi}{2} + \theta\right), \quad \textit{where } u = \frac{1}{r}.$$

Hence show that the locus of the extremity of the polar subtangent of the curve

$$r = \frac{1 + \tan\dfrac{\theta}{2}}{m + n\tan\dfrac{\theta}{2}} \quad \textit{is a cardioid.}$$

Solution : Let $P(r, \theta)$, be any point on the curve. OP is the radius vector. PT is the tangent to the curve at P.

Draw a line OT at right angle to OP which meets the tangent at T. Then T is the extremity of the polar subtangent. Let $T \to (r_1, \theta_1)$.

We want to find out the locus of T.

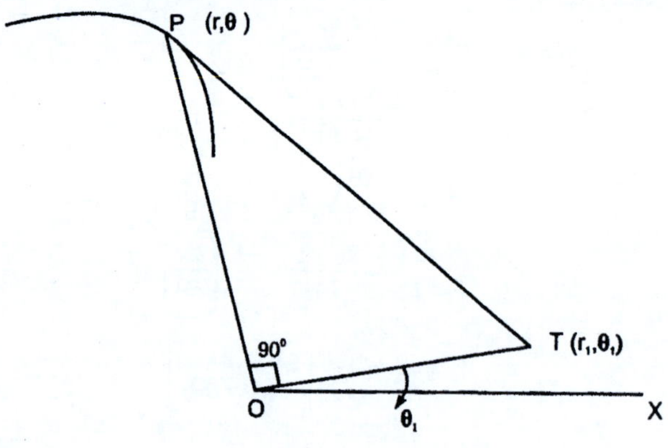

Fig. 5.12

The equation of the curve is,

$$\frac{1}{r} = -f(\theta). \qquad \text{.... (i)}$$

Differentiating w.r.t. θ, we get

$$-\frac{1}{r^2}\frac{dr}{d\theta} = -f'(\theta)$$

$$\Rightarrow \qquad \frac{1}{r^2}\frac{dr}{d\theta} = f'(\theta) \qquad \text{.... (ii)}$$

$\therefore$ Polar subtangent $= OT$

$$= r_1$$

$$= r^2 \frac{d\theta}{dr}$$

$$= \frac{1}{f'(\theta)} \qquad \text{[From (ii)]}$$

$$\Rightarrow \qquad \frac{1}{r_1} - f'(\theta) = 0. \qquad \text{.... (iii)}$$

From figure, it is evident that

$$\theta = \theta_1 + \frac{\pi}{2}.$$

Hence, from (iii), we have,

$$\frac{1}{r_1} - f'\left(\theta_1 + \frac{\pi}{2}\right) = 0$$

$\therefore$ The locus of $T(r_1, \theta_1)$ is,

$$\frac{1}{r} - f'\left(\theta + \frac{\pi}{2}\right) = 0$$

$$\Rightarrow \qquad u - f'\left(\theta + \frac{\pi}{2}\right) = 0 \qquad \left[\because u = \frac{1}{r}\right]$$

$$u = f'\left(\frac{\pi}{2} + \theta\right)$$

which is the required locus.

Now, the given curve is

$$r = \frac{1 + \tan\frac{\theta}{2}}{m + n\tan\frac{\theta}{2}}$$

$$\Rightarrow \qquad \frac{1}{r} = \frac{m + n\tan\frac{\theta}{2}}{1 + \tan\frac{\theta}{2}}$$

Comparing it with (i), we have

$$-f(\theta) = \frac{m + n\tan\frac{\theta}{2}}{1 + \tan\frac{\theta}{2}}$$

Differentiating w.r.t. θ, we get

$$-f(\theta) = \frac{\left(1+\tan\frac{\theta}{2}\right) n \sec^2\frac{\theta}{2}\cdot\frac{1}{2} - \left(m + n\tan\frac{\theta}{2}\right)\sec^2\frac{\theta}{2}\cdot\frac{1}{2}}{\left(1+\tan\frac{\theta}{2}\right)^2}$$

$$= \frac{\frac{1}{2}\sec^2\frac{\theta}{2}\left(n + n\tan\frac{\theta}{2} - m - n\tan\frac{\theta}{2}\right)}{\left(1+\tan\frac{\theta}{2}\right)^2}$$

$$= \frac{n-m}{2}\cdot\frac{1+\tan^2\frac{\theta}{2}}{\left(1+\tan\frac{\theta}{2}\right)^2}$$

$$\therefore -f'\left(\theta+\frac{\pi}{2}\right) = \frac{n-m}{2}\cdot\frac{1+\tan^2\left(\frac{\theta}{2}+\frac{\pi}{4}\right)}{\left[1+\tan\left(\frac{\theta}{2}+\frac{\pi}{4}\right)\right]^2}, \quad \left[\text{Replacing } \theta \text{ by } \left(\theta+\frac{\pi}{2}\right)\right]$$

$$= \frac{n-m}{2}\cdot\frac{1+\left[\dfrac{1+\tan\frac{\theta}{2}}{1-\tan\frac{\theta}{2}}\right]^2}{\left[1+\dfrac{1+\tan\frac{\theta}{2}}{1-\tan\frac{\theta}{2}}\right]^2}$$

$$= \frac{n-m}{2}\cdot\frac{2\left(1+\tan^2\frac{\theta}{2}\right)}{4}$$

$$= \frac{n-m}{4}\cdot\sec^2\frac{\theta}{2}$$

$$\therefore f'\left(\theta+\frac{\pi}{2}\right) = \frac{m-n}{4}\sec^2\frac{\theta}{2}.$$

Hence, the locus of the extremity of the polar subtangent of the given curve is,

$$u = f'\left(\theta + \frac{\pi}{2}\right)$$

$$\Rightarrow \qquad \frac{1}{r} = \frac{m-n}{4}\sec^2\frac{\theta}{2}$$

$$\Rightarrow \qquad r = \frac{4}{m-n}\cdot\frac{1}{\sec^2\frac{\theta}{2}}$$

$$\Rightarrow \qquad r = \frac{2}{m-n}2\cos^2\frac{\theta}{2}$$

$$\Rightarrow \qquad r = \frac{2}{m-n}(1+\cos\theta)$$

which being of the form

$$r = A(1+\cos\theta), \quad \text{where } A = \frac{2}{m-n}$$

represents a cardoid.

EXERCISE 5 (C)

1. Find the value of ϕ for the curve $r^m = a^m(\cos m\theta - \sin m\theta)$ at the point $\theta = 0$.
2. Show that in the equiangular spiral $r =$ exp $(\theta \cot \alpha)$, the tangent is inclined at a constant angle to the medium vector.
3. Find the angle at which the radius vector cuts the curve
$$\frac{l}{r} = -\, e\cos\theta.$$
4. Show that in the curve
$$\theta = \frac{\sqrt{r^2-a^2}}{a} - \cos^{-1}\left(\frac{a}{r}\right), \cos\phi = \frac{a}{r}.$$
5. Prove that the spirals
$$r^n = a^n\cos n\theta \text{ and } r^n = b^n\sin n\theta$$
intersect orthogonally.
6. Show that the angle between the tangent at any point P and the line joining P to the origin is the same at all points of the curve
$$\log(x^2+y^2) = k\tan^{-1}(y/x).$$
[**Hint :** Change the equation of the curve into polar form.]
7. Prove that the tangent at any point (r, θ) on the curve $r^2 = a^2\sin 2\theta$ makes an angle 3θ with the initial line.

8. Find the angle between the radius vector and the tangent in each of the following curves :

(i) $r = a(1 + \sin\theta)$ at $\theta = \frac{\pi}{6}$.

(ii) $r = a\,\text{cosec}^2\frac{\theta}{2}$ at $\theta = \frac{\pi}{2}$.

(iii) $r = a(1 + \cos\theta)$ at $\theta = \frac{\pi}{2}$.

(iv) $r^2 = a^2\cos 2\theta$ at $\theta = \frac{\pi}{6}$.

(v) $r = a\theta$.

9. Find the angle of intersection of the following curves :

(i) $r = \dfrac{a}{1 + \cos\theta}$, $r = \dfrac{b}{1 - \cos\theta}$

(ii) $r = 2\sin\theta$, $r = 2\cos\theta$.

(iii) $r = a(1 + \sin\theta)$, $r = a(1 - \sin\theta)$.

(iv) $r^n = a^n\sec(n\theta + 30°)$, $r^n = b^n\sec(n\theta + 45°)$.

(v) $r = \sin\theta + \cos\theta$, $r = 2\sin\theta$.

(vi) $r = a\sin\theta$, $r = a\cos\theta$.

(vii) $r = a\cos\theta$, $2r = a$.

(viii) $r = a\theta$, $r\theta = a$.

(ix) $r = ae^{\theta}$, $re^{\theta} = b$.

(x) $r = a\sin 2\theta$, $r = a\cos 2\theta$.

10. Show that the circle $r = b$ cuts the curve $r^2 = a^2\cos 2\theta + b^2$ at an angle $\tan^{-1}(a^2/b^2)$.

11. Prove that the curves $r^2 = a^2\cos 2\theta$ and $r = a(1 + \cos\theta)$ intersect at an angle $3\sin^{-1}(3/4)^{1/4}$

12. Show that in the curve $r = a\theta$, the polar subnormal, and in the curve $r\theta = a$, the polar subtangent is constant.

13. Find the polar subtangent for the following curves :

(i) $\frac{l}{r} = 1 + e\cos\theta$. (ii) $r = a(1 + \cos\theta)$. (iii) $r = ae^{\theta\cot\alpha}$

14. Find the polar subnormal for the following curves :

(i) $r = ae^{\theta\cot\alpha}$ (ii) $r = a + b\cos\theta$. (iii) $r^2 = a^2\cos 2\theta$.

15. Show that for the curve

$$\theta = \cos^{-1}\frac{r}{k} - \sqrt{\frac{k^2 - r^2}{r^2}}$$

the length of polar tangent is constant.

16. Find the length of the polar tangent and polar normal of the curve

$$r = a\,(1 + \cos\theta).$$

17. Prove that, in the parabola

$$\frac{2a}{r} = 1 - \cos\theta \quad \text{or } a = r\sin^2\frac{\theta}{2},$$

(i) $\phi = \pi - \frac{\theta}{2}$ (ii) $p = a\,\text{cosec}^2\,\frac{\theta}{2}$

(iii) $p^2 = ar$

(iv) Polar subtangent $= 2a\,\text{cosec}\,\theta$

(v) Polar subnormal $= r\cot\frac{\theta}{2}$

18. Find the pedal equation of the following curves :

(i) $r^m = a^m \cos m\theta$ (Cosine spiral)

(ii) $r^2 = a^2\cos 2\theta$ (Lemniscate)

(iii) $r^2\cos 2\theta = a^2$ (Hyperbola)

(iv) $r^n = a^n \sin n\theta$.

(v) $r = a\theta$ (Archimedian spiral)

(vi) $r\theta = a$.

19. Show that for the curve $2\theta = \cosh^{-1}(r^2/a^2) + \cos^{-1}(a^2/r^2)$, the pedal equation is $r^2 + a^2 = 2p^2$.

20. Find the pedal equation of the following curves :

(i) $y^2 = 4a\,(x + a)$. (ii) $x^2 + y^2 = 2ax$.

(iii) $\frac{x^2}{a^2} - \frac{y^2}{b^2} = 1$. (iv) $y^2 = 4ax$.

21. Show that the pedal equation of the curve $x = a\cos^3\theta, y = a\sin^3\theta$ is

$$r^2 = a^2 - 3p^2.$$

22. Prove that the locus of the extremity of the polar subnormal of the curve $r = f(\theta)$ is

$$r = f'\left(\theta - \frac{\pi}{2}\right).$$

Hence prove that the locus of the extremity of the polar subnormal of the equiangular spiral $r = ae^{m\theta}$ is another equiangular spiral.

ANSWERS

1. $\frac{3\pi}{4}$ **3.** $\tan^{-1}\left(\frac{1 + e\cos\theta}{\sin\theta}\right)$

8. (i) $\frac{\pi}{3}$ (ii) $\frac{3\pi}{4}$ (iii) $\frac{3\pi}{4}$ (iv) $\frac{5\pi}{6}$ (v) $\tan^{-1}\theta$.

9. (i) $\frac{\pi}{2}$ (ii) $\frac{\pi}{2}$ (iii) $\frac{\pi}{2}$ (iv) 15° (v) $\frac{\pi}{4}$ (vi) $\frac{\pi}{2}$

(vii) $\frac{\pi}{3}$ (viii) $\frac{\pi}{2}$ (ix) $\frac{\pi}{2}$ (x) $\tan^{-1}\frac{4}{3}$.

13. (i) $\frac{l}{e\sin\theta}$ (ii) $2a\cos^2\frac{\theta}{2}\cot\frac{\theta}{2}$ (iii) $r\tan\alpha$.

14. (i) $r\cot\alpha$ (ii) $-b\sin\theta$ (iii) $-a\sqrt{\cos 2\theta}\tan 2\theta$.

16. $2a\cos\frac{\theta}{2}\cot\frac{\theta}{2}$; $2a\cos\frac{\theta}{2}$.

18. (i) $pa^m = r^{m+1}$ (ii) $pa^2 = r^3$ (iii) $pr = a^2$ (iv) $pa^n = r^{n+1}$

(v) $p^2 = \frac{r^4}{r^2 + a^2}$ (vi) $\frac{1}{p^2} = \frac{1}{r^2} + \frac{1}{a^2}$

20. (i) $p^2 = ar$ (ii) $r^2 = 2ap$ (iii) $r^2 = a^2 - b^2 + \frac{a^2 b^2}{p^2}$

(iv) $2c^2r^2 = p^4 + 6p^2c^2 + p(p^2 + 4c^2)^{3/2}$.

5.18. Differential Coefficient of the Length of Arc (Cartesian Coordinates)

If A be a fixed point on the curve $y = f(x)$, let $P(x, y)$ be any point on it and the length of arc AP be s, then

(i) $\frac{ds}{dx} = \sqrt{1 + \left(\frac{dy}{dx}\right)^2}$ (ii) $\frac{ds}{dy} = \sqrt{1 + \left(\frac{dx}{dy}\right)^2}$

PROOF :

A is the fixed point on the curve from which the arc lengths are measured. Let P is the point (x, y) so that arc $PA = s$.

Let $Q\,(x + \delta x, y + \delta y)$ be a point in the neighbourhood of P such that

$$\text{arc } AQ = s + \delta s$$

$$\therefore \quad \text{arc } PQ = \delta s$$

Draw $PM \perp OX$ and $QN \perp OX$.

Also draw $PR \perp QN$.

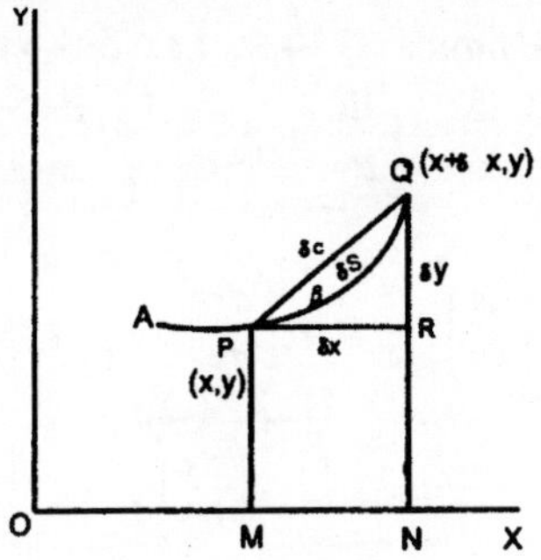

Fig. 5.13

Join PQ.

Let chord $PQ = \delta c$

Let $\angle QPR = \beta$.

Then, from right angled triangle PQR,

$$PQ^2 = PR^2 + RQ^2$$

$$\therefore \quad (\delta c)^2 = (\delta x)^2 + (\delta y)^2 \qquad \text{.... (1)}$$

Dividing both sides by $(\delta x)^2$, we have

$$\left(\frac{\delta c}{\delta x}\right)^2 = 1 + \left(\frac{\delta y}{\delta x}\right)^2$$

i.e.,

$$\left(\frac{\delta c}{\delta s}\right)^2 \cdot \left(\frac{\delta s}{\delta x}\right)^2 = 1 + \left(\frac{\delta y}{\delta x}\right)^2 \qquad \text{.... (2)}$$

Now, proceeding to the limit as $Q \to P$, *i.e.,* $\delta x \to 0$, we have

$$\frac{\delta c}{\delta s} \to 1, \quad \frac{\delta s}{\delta x} \to \frac{ds}{dx} \quad \text{and} \quad \frac{\delta y}{\delta x} \to \frac{dy}{dx}.$$

Hence we have from (2),

$$\left(\frac{ds}{dx}\right)^2 = 1 + \left(\frac{dy}{dx}\right)^2$$

$$\therefore \quad \frac{ds}{dx} = \sqrt{1 + \left(\frac{dy}{dx}\right)^2}$$

Positive sign before the radical is taken on the assumption that s increases with x so that ds/dx is +ve.

Again, dividing both sides of (1) by $(\delta y)^2$, we get

$$\left(\frac{dc}{dy}\right)^2 = \left(\frac{\delta x}{\delta y}\right)^2 + 1$$

i.e.,

$$\left(\frac{\delta c}{\delta s}\right)^2 \left(\frac{\delta s}{\delta y}\right)^2 = \left(\frac{\delta x}{\delta y}\right)^2 + 1 \qquad \text{.... (3)}$$

Now, proceeding to the limit as $Q \to P$, *i.e.*, $\delta y \to 0$, we have

$$\frac{\delta c}{\delta s} \to 1, \quad \frac{\delta s}{\delta y} \to \frac{ds}{dy} \text{ and } \frac{\delta x}{\delta y} \to \frac{dx}{dy}.$$

$\therefore$ (3) gives us that

$$\left(\frac{ds}{dy}\right)^2 = 1 + \left(\frac{dx}{dy}\right)^2$$

$$\therefore \qquad \frac{ds}{dy} = \sqrt{1 + \left(\frac{dx}{dy}\right)^2}.$$

Positive sign before the radical is taken on the assumption that s increases with y so that ds/dy is +ve.

5.19. Differential Coefficient of the Length of Arc (Parametric Coordinates)

If s is the length of an arc of the curve $x = f(t)$, $y = \phi(t)$ measured from a fixed point on it to the point (x, y), then

$$\frac{ds}{dt} = \sqrt{\left(\frac{dx}{dt}\right)^2 + \left(\frac{dy}{dt}\right)^2}.$$

Proof :

We have from article **5.18**,

$$\frac{ds}{dx} = \sqrt{1 + \left(\frac{dy}{dx}\right)^2}.$$

Multiplying both sides by $\frac{dx}{dt}$, we have

$$\frac{ds}{dx} \cdot \frac{dx}{dt} = \sqrt{1 + \left(\frac{dy}{dx}\right)^2} \cdot \frac{dx}{dt}$$

$$\Rightarrow \qquad \frac{ds}{dt} = \sqrt{\left(\frac{dx}{dt}\right)^2 + \left(\frac{dy}{dx}\right)^2 \left(\frac{dx}{dt}\right)^2}$$

$$\Rightarrow \qquad \frac{ds}{dt} = \sqrt{\left(\frac{dx}{dt}\right)^2 + \left(\frac{dy}{dt}\right)^2}.$$

5.20. To show that $\cos\Psi = \frac{dx}{ds}$ and $\sin\Psi = \frac{dy}{ds}$

We have, $\qquad \tan\Psi = \frac{dy}{dx}$

$$\therefore \qquad \cos\Psi = \frac{1}{\sqrt{sec^2\Psi}} = \frac{1}{\sqrt{1 + \tan^2\Psi}}$$

$$= \frac{1}{\sqrt{1+\left(\frac{dy}{dx}\right)^2}} = \frac{1}{\frac{ds}{dx}}$$

$$\cos\Psi = \frac{dx}{ds}$$

Again,
$$\sin\Psi = \frac{1}{\sqrt{\text{cosec}^2\,\Psi}} = \frac{1}{\sqrt{1+\cot^2\Psi}}$$

$$= \frac{1}{\sqrt{1+\frac{1}{\tan^2\Psi}}} = \frac{\tan\Psi}{\sqrt{1+\tan^2\Psi}}$$

$$= \frac{\frac{dy}{dx}}{\sqrt{1+\left(\frac{dy}{dx}\right)^2}} = \frac{\frac{dy}{dx}}{\frac{ds}{dx}}$$

$$\sin\Psi = \frac{dy}{ds}.$$

Note : $\because\ \sin^2\Psi + \cos^2\Psi = 1,\ \therefore\ \left(\frac{dy}{ds}\right)^2 + \left(\frac{dx}{ds}\right)^2 = 1.$

Aid to memory : The hypothetical triangle given here is a good device to remember the values of $\sin\Psi$, $\cos\Psi$ and $\tan\Psi$.

$$\tan\Psi = \frac{dy}{dx}$$

$$\sin\Psi = \frac{dy}{ds}$$

$$\cos\Psi = \frac{dx}{ds}$$

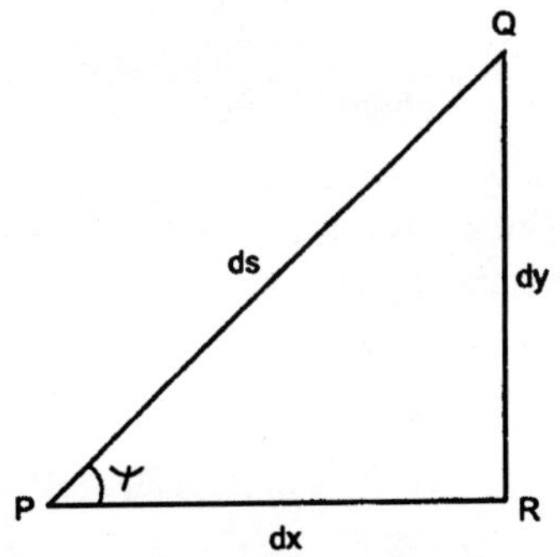

Fig. 5.14

5.21. Differential Coefficient of the Length of Arc (Polar Coordinates)

If s is the length of an arc of the curve $r = f(\theta)$ measured from a fixed point on it to the variable point (r, θ), then

(i) $\frac{ds}{d\theta} = \sqrt{r^2 + \left(\frac{dr}{d\theta}\right)^2}$ (ii) $\frac{ds}{dr} = \sqrt{1 + r^2\left(\frac{d\theta}{dr}\right)^2}$

PROOF :

(i) Let AB be the curve $r = f(\theta)$. Let A be the fixed point on it from which the arc lengths are measured. Let P be the point (r, θ) such that

$$\text{arc } AP = s.$$

Let $Q\ (r + \delta r, \theta + \delta\theta)$ be a point in the neighbourhood of $P\ (r, \theta)$ such that

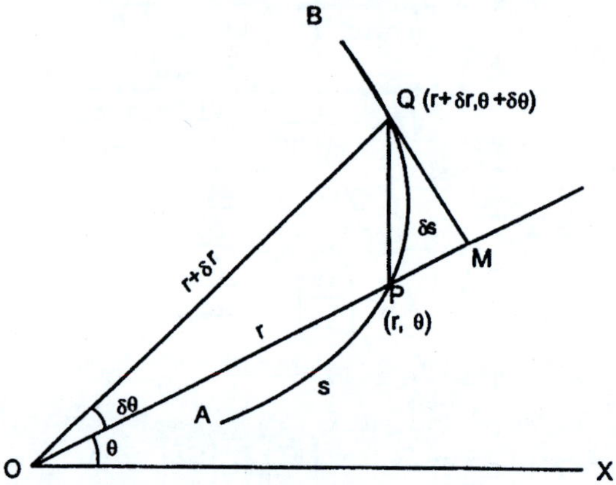

Fig. 5.15

$$\text{arc } AQ = s + \delta s.$$

$\therefore$ $$\text{arc } PQ = \delta s$$

Join PQ. Let chord $PQ = \delta c$.

Join OP and OQ.

From Q draw perpendicular QM on OP (produced if necessary).

Now, from the right angled triangle PQM,

$$PQ^2 = PM^2 + MQ^2$$

$$\Rightarrow \quad (\delta c)^2 = (OM - OP)^2 + MQ^2$$

$$= \{(r + \delta r)\cos\delta\theta - r\}^2 + \{(r + \delta r)\sin\delta\theta\}^2$$

But when $Q \to P$, $\delta\theta \to 0$

$\therefore$ $\cos\delta\theta \to 1$ and $\sin\delta\theta \to \delta\theta$

$\therefore$ In limit $Q \to P$

$$\Rightarrow \quad (\delta c)^2 = \{(r + \delta r).1 - r\}^2 + \{(r + \delta r)\,\delta\theta\}^2$$

$$= (\delta r)^2 + (r\,\delta\theta)^2$$ (Neglecting $\delta r\,\delta\theta$, it being a very small quantity of second order)

Dividing both sides by $(\delta\theta)^2$, we get

$$\left(\frac{\delta c}{\delta\theta}\right)^2 = \left(\frac{\delta r}{\delta\theta}\right)^2 + r^2$$

i.e., $$\left(\frac{\delta c}{\delta s}\cdot\frac{\delta s}{\delta \theta}\right)^2 = \left(\frac{\delta r}{\delta \theta}\right)^2 + r^2$$

In limit $Q \to P,\ \delta\theta \to 0,$

$$\therefore \quad \frac{\delta c}{\delta s} \to 1,\ \frac{\delta s}{\delta \theta} \to \frac{ds}{d\theta} \quad \text{and} \quad \frac{\delta r}{\delta \theta} \to \frac{dr}{d\theta}$$

$$\therefore \quad \left(\frac{ds}{d\theta}\right)^2 = r^2 + \left(\frac{dr}{d\theta}\right)^2$$

$$\therefore \quad \frac{ds}{d\theta} = \sqrt{r^2 + \left(\frac{dr}{d\theta}\right)^2}.$$

(ii) We have,

$$\frac{ds}{d\theta} = \sqrt{r^2 + \left(\frac{dr}{d\theta}\right)^2}$$

Multiplying both sides $d\theta/dr$, we get

$$\frac{ds}{d\theta}\cdot\frac{d\theta}{dr} = \sqrt{r^2 + \left(\frac{dr}{d\theta}\right)^2}\cdot\frac{d\theta}{dr}$$

$$\therefore \quad \frac{ds}{dr} = \sqrt{r^2\left(\frac{d\theta}{dr}\right)^2 + \left(\frac{dr}{d\theta}\right)^2\left(\frac{d\theta}{dr}\right)^2}$$

$$\therefore \quad \frac{ds}{dr} = \sqrt{r^2\left(\frac{d\theta}{dr}\right)^2 + 1}.$$

5.22. With Usual Notations, to prove that

(i) $\sin\phi = r\,\dfrac{d\theta}{ds}$, **(ii)** $\cos\phi = \dfrac{dr}{ds}$.

PROOF :

(i) We know that $$\tan\phi = r\frac{d\theta}{dr}$$

$$\therefore \quad \sin\phi = \frac{1}{\sqrt{\operatorname{cosec}^2\phi}} = \frac{1}{\sqrt{1+\cot^2\phi}}$$

$$= \frac{1}{\sqrt{1+\left(\dfrac{dr}{r\,d\theta}\right)^2}}$$

$$= \frac{r}{\sqrt{r^2+\left(\dfrac{dr}{d\theta}\right)^2}}$$

$$= \frac{r}{ds/d\theta} = \frac{r\,d\theta}{ds}.$$

(ii) We know that $\tan\phi = r\dfrac{d\theta}{dr}$

$$\therefore \quad \cos\phi = \frac{1}{\sqrt{\sec^2\phi}}$$

$$= \frac{1}{\sqrt{1+\tan^2\phi}}$$

$$= \frac{1}{\sqrt{1+r^2\left(\dfrac{d\theta}{dr}\right)^2}}$$

$$= \frac{1}{ds/dr} = \frac{dr}{ds}.$$

Aid to memory : The hypothetical triangle given here is a good device to remember the values of sin ϕ, cos ϕ and tan ϕ.

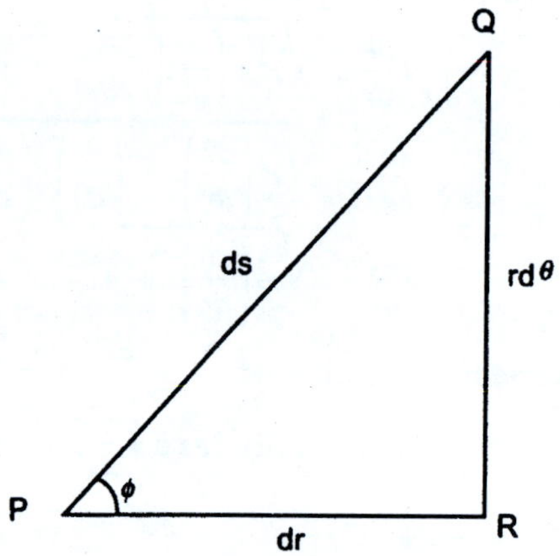

Fig. 5.16

Note : $\because \quad \sin^2\phi + \cos^2\phi = 1$

$$\therefore \quad \left(r\frac{d\theta}{ds}\right)^2 + \left(\frac{dr}{ds}\right)^2 = 1.$$

5.23. To prove that

(i) $\dfrac{ds}{d\theta} = \dfrac{r^2}{p}$;

(ii) $\dfrac{ds}{dr} = \dfrac{r}{\sqrt{r^2-p^2}}$.

PROOF :

(i) We know that,

$$p = r \sin \phi$$

$$\Rightarrow \quad p = r \cdot r \frac{d\theta}{ds} \qquad \left[\because \sin \phi = r \frac{d\theta}{ds}\right]$$

$$\Rightarrow \quad \frac{ds}{d\theta} = \frac{r^2}{p}.$$

(ii) We know that $\cos \phi = \dfrac{dr}{ds}$

$$\Rightarrow \quad \sqrt{1 - \sin^2 \phi} = \frac{dr}{ds}$$

$$\Rightarrow \quad \sqrt{1 - \frac{p^2}{r^2}} = \frac{dr}{ds} \qquad [\because p = r \sin \phi]$$

$$\Rightarrow \quad \frac{\sqrt{r^2 - p^2}}{r} = \frac{dr}{ds}$$

$$\Rightarrow \quad \frac{ds}{dr} = \frac{r}{\sqrt{r^2 - p^2}}.$$

ILLUSTRATIVE EXAMPLES

Example 1. *For the curve* $y = a \log \sec (x/a)$, *prove that*

(i) $\dfrac{ds}{dx} = \sec \dfrac{x}{a}$, (ii) $\dfrac{ds}{dy} = \operatorname{cosec} \dfrac{x}{a}$,

(iii) $x = a\Psi$.

Solution : Equation of the curve is,

$$y = a \log \sec (x/a).$$

Differentiating w.r.t. x, we get

$$\frac{dy}{dx} = a \cdot \frac{1}{\sec \frac{x}{a}} \cdot \sec \frac{x}{a} \tan \frac{x}{a} \cdot \frac{1}{a}$$

$$\Rightarrow \quad \frac{dy}{dx} = \tan \frac{x}{a}$$

$$\Rightarrow \quad \tan \Psi = \tan \frac{x}{a}$$

$$\Rightarrow \qquad \Psi = \frac{x}{a}$$

$$\Rightarrow \qquad x = a\Psi$$

Now,

$$\frac{ds}{dx} = \sqrt{1+\left(\frac{dy}{dx}\right)^2} = \sqrt{1+\tan^2\frac{x}{a}}$$

$$= \sqrt{\sec^2\frac{x}{a}} = \sec\frac{x}{a}$$

$$\frac{ds}{dy} = \sqrt{1+\left(\frac{dx}{dy}\right)^2}$$

$$= \sqrt{1+\frac{1}{(dy/dx)^2}} = \sqrt{1+\frac{1}{\tan^2(x/a)}}$$

$$= \sqrt{1+\cot^2\frac{x}{a}} = \sqrt{\text{cosec}^2\frac{x}{a}}$$

$$= \text{cosec}\,\frac{x}{a}.$$

Example 2. *Prove that for the ellipse* $\frac{x^2}{a^2}+\frac{y^2}{b^2} = 1$, *if* $x = a\cos\phi$,

$$\frac{ds}{d\phi} = a\sqrt{1-e^2\cos^2\phi}.$$

Solution : $\frac{x^2}{a^2}+\frac{y^2}{b^2} = 1.$

Put $x = a\cos\phi$, in the above equation, we get

$$\frac{a^2\cos^2\phi}{a^2}+\frac{y^2}{b^2} = 1.$$

$$\Rightarrow \qquad y^2 = b^2\sin^2\phi$$

$$\Rightarrow \qquad y = b\sin\phi$$

$$\therefore \qquad \frac{ds}{d\phi} = \sqrt{\left(\frac{dx}{d\phi}\right)^2+\left(\frac{dy}{d\phi}\right)^2}$$

$$= \sqrt{(-a\sin\phi)^2+(b\cos\phi)^2}$$

$$= \sqrt{a^2\sin^2\phi+b^2\cos^2\phi}$$

$$= \sqrt{a^2\sin^2\phi+a^2(1-e^2)\cos^2\phi}$$

$$\left[\because b^2 = a^2(1-e^2) \text{ for ellipse}\right]$$

$$= \sqrt{a^2\sin^2\phi+a^2\cos^2\phi-a^2e^2\cos^2\phi}$$

$$= \sqrt{a^2-a^2e^2\cos^2\phi} = a\sqrt{1-e^2\cos^2\phi}.$$

Example 3. *In the curve $r^m = a^m \cos m\theta$, prove that*

(i) $\dfrac{ds}{d\theta} = a \sec^{(m-1)/m}(m\theta)$ (ii) $a^{2m}\dfrac{d^2r}{ds^2} + mr^{2m-1} = 0.$

Solution :

(i) The equation of the curve is,

$$r^m = a^m \cos m\theta. \quad \text{.... (i)}$$

Take log on both sides,

$$m \log r = m \log a + \log \cos m\theta.$$

Differentiate w.r.t. θ, we get

$$\frac{m}{r}\frac{dr}{d\theta} = -\frac{\sin m\theta}{\cos m\theta}. m$$

$$\therefore \quad \frac{dr}{d\theta} = -r \tan m\theta$$

$$\therefore \quad \frac{ds}{d\theta} = \sqrt{r^2 + \left(\frac{dr}{d\theta}\right)^2}$$

$$= \sqrt{r^2 + r^2 \tan^2 m\theta}$$

$$= r \sec m\theta$$

$$= \frac{a(\cos m\theta)^{1/m}}{\cos m\theta} \quad \text{[From (i)]}$$

$$= a \cos^{(1/m)-1}(m\theta)$$

$$= a \cos^{(1-m)/m}(m\theta)$$

$$= \frac{a}{\cos^{(m-1)/m}(m\theta)}$$

$$= a \sec^{(m-1)/m}(m\theta)$$

which proves the first part.

(ii) We have,

$$\frac{ds}{d\theta} = r \sec m\theta$$

$$= \frac{r}{\cos m\theta}$$

$$= \frac{r}{(r^m/a^m)} \quad \text{[From (i)]}$$

$$= a^m r^{1-m}$$

and $$\frac{dr}{d\theta} = -r\tan m\theta$$

$$\therefore \quad \frac{dr}{ds} = \frac{dr}{d\theta}\cdot\frac{d\theta}{ds} = \frac{dr/d\theta}{ds/d\theta}$$

$$= -\frac{r\tan m\theta}{a^m r^{1-m}} = -a^{-m} r^m \tan m\theta$$

$$\therefore \quad \frac{d^2r}{ds^2} = \frac{d}{ds}\left(\frac{dr}{ds}\right)$$

$$= \frac{d}{dr}\left(\frac{dr}{ds}\right)\frac{dr}{ds}$$

$$= \frac{dr}{ds}\cdot\frac{d}{dr}\left(\frac{dr}{ds}\right)$$

$$= -a^{-m} r^m \tan m\theta \frac{d}{dr}\,[-a^{-m} r^m \tan m\theta]$$

$$= a^{-2m} r^m \tan m\theta \left[mr^{m-1}\tan m\theta + mr^m \sec^2 m\theta.\frac{d\theta}{dr}\right]$$

$$= ma^{-2m} r^m \tan m\theta \left[r^{m-1}\tan m\theta + r^m \sec^2 m\theta\left(-\frac{1}{r}\cot m\theta\right)\right]$$

$$= ma^{-2m} r^m \tan m\theta.\,[r^{m-1}\tan m\theta - r^{m-1}\sec^2 m\theta\cot m\theta]$$

$$= ma^{-2m} r^{2m-1}\tan m\theta\,[\tan m\theta - \sec^2 m\theta\cot m\theta]$$

$$= ma^{-2m} r^{2m-1}\,[tan^2 m\theta - \sec^2 m\theta]$$

$$= ma^{-2m} r^{2m-1}\,(-1)$$

$$= -ma^{-2m} r^{2m-1}$$

$$\Rightarrow \quad \frac{d^2r}{ds^2} + ma^{-2m} r^{2m-1} = 0$$

$$\Rightarrow \quad a^{2m}\frac{d^2r}{ds^2} + mr^{2m-1} = 0$$

which proves the (ii) part.

Example 4. *For the cycloid* $x = a\,(1-\cos t)$, $y = a\,(t+\sin t)$ *find* $\frac{ds}{dt}, \frac{ds}{dx}$ and $\frac{ds}{dy}$.

Solution : The given equation is

$$x = a(1-\cos t) \text{ and } y = a(t+\sin t)$$

$$\therefore \quad \frac{dx}{dt} = a\sin t \text{ and } \frac{dy}{dt} = a(1+\cos t)$$

$$\frac{ds}{dt} = \sqrt{\left(\frac{dx}{dt}\right)^2 + \left(\frac{dy}{dt}\right)^2}$$

$$= \sqrt{a^2\sin^2 t + a^2(1+\cos t)^2}$$

$$= \sqrt{a^2\sin^2 t + a^2(1+2\cos t+\cos^2 t)}$$

$$= a\sqrt{\sin^2 t+\cos^2 t+1+2\cos t}$$

$$= a\sqrt{2+2\cos t} = a\sqrt{2+2\left(2\cos^2\frac{t}{2}-1\right)}$$

$$= 2a\cos\frac{t}{2}$$

Now,

$$\frac{ds}{dx} = \frac{ds}{dt}\cdot\frac{dt}{dx} = 2a\cos\frac{t}{2}\times\frac{1}{a\sin t}$$

$$= 2a\cot\frac{t}{2}\times\frac{1}{2a\sin\frac{t}{2}\cos\frac{t}{2}} = \text{cosec}\left(\frac{t}{2}\right).$$

Again,

$$\frac{ds}{dy} = \frac{ds}{dt}\cdot\frac{dt}{dy} = 2a\cos\frac{t}{2}\times\frac{1}{a(1+\cos t)}$$

$$= \frac{2a\cos\frac{t}{2}}{2a\cos^2\frac{t}{2}} = \sec\frac{t}{2}.$$

Example 5. *For the cycloid* $x = a(t+\sin t)$, $y = a(1-\cos t)$ *prove that*

(i) $\Psi = \dfrac{t}{2}$ (ii) $\dfrac{ds}{dt} = 2a\cos\dfrac{t}{2}$

(iii) $\dfrac{ds}{dy} = \sqrt{\dfrac{2s}{y}}$ (iv) $\dfrac{ds}{dx} = \sec\dfrac{t}{2}.$

Solution : Equation of the curve is,

$$x = a(t+\sin t)$$

$$y = a(1-\cos t)$$

$$\therefore \quad \frac{dx}{dt} = a(1+\cos t)$$

$$\frac{dy}{dt} = a\sin t.$$

(i)
$$\tan \Psi = \frac{dy}{dx} = \frac{dy/dt}{dx/dt}$$
$$= \frac{a \sin t}{a(1+\cos t)} = \frac{a \sin t}{1+\cos t}$$
$$= \frac{2 \sin \frac{t}{2} \cos \frac{t}{2}}{2 \cos^2 \frac{t}{2}} = \tan \frac{t}{2}$$

$$\therefore \quad \Psi = \frac{t}{2}.$$

(ii)
$$\frac{ds}{dt} = \sqrt{\left(\frac{dx}{dt}\right)^2 + \left(\frac{dy}{dt}\right)^2}$$
$$= \sqrt{a^2(1+\cos t)^2 + a^2 \sin^2 t}$$
$$= a\sqrt{1+\cos^2 t + 2\cos t + \sin^2 t}$$
$$= a\sqrt{2+2\cos t} = a\sqrt{2(1+\cos t)}$$
$$= a\sqrt{2.2 \cos^2 \frac{t}{2}} = 2a \cos \frac{t}{2}.$$

(iii)
$$\frac{ds}{dy} = \frac{ds}{dt} \cdot \frac{dt}{dy}$$
$$= \frac{ds/dt}{dy/dt}$$
$$= \frac{2a \cos \frac{t}{2}}{a \sin t}$$
$$= \frac{2a \cos \frac{t}{2}}{2a \sin \frac{t}{2} \cos \frac{t}{2}} = \frac{1}{\sin \frac{t}{2}}$$
$$= \frac{1}{\sqrt{\sin^2 \frac{t}{2}}} = \frac{\sqrt{2}}{\sqrt{2 \sin^2 \frac{t}{2}}}$$
$$= \frac{\sqrt{2}}{\sqrt{1-\cos t}}$$
$$= \frac{\sqrt{2}}{\sqrt{y/a}} = \sqrt{\frac{2a}{y}}.$$

(iv) $$\frac{ds}{dx} = \frac{ds}{dt}\cdot\frac{dt}{dx} = \frac{ds/dt}{dx/dt}$$

$$= \frac{2a\cos\frac{t}{2}}{a(1+\cos t)} = \frac{2a\cos\frac{t}{2}}{2a\cos^2\frac{t}{2}}$$

$$= \frac{1}{\cos\frac{t}{2}} = \sec\frac{t}{2}.$$

Example 6. *Prove that*

$$\sin^2\phi\frac{d\phi}{d\theta} + r\frac{d^2r}{ds^2} = 0.$$

Solution : $$\frac{dr}{ds} = \cos\phi$$

$$\therefore \quad \frac{d^2r}{ds^2} = \frac{d}{ds}(\cos\phi) = -\sin\phi\frac{d\phi}{ds}$$

$$\therefore \quad r\frac{d^2r}{ds^2} = -r\sin\phi\frac{d\phi}{ds}$$

$$\therefore \quad r\frac{d^2r}{ds^2} = -r\sin\phi\frac{d\phi}{d\theta}\frac{d\theta}{ds}$$

$$\therefore \quad r\frac{d^2r}{ds^2} = -r\sin\phi\frac{d\phi}{d\theta}\left(r\frac{d\theta}{ds}\right)$$

$$\therefore \quad r\frac{d^2r}{ds^2} = -\sin\phi\frac{d\phi}{d\theta}\sin\phi$$

$$\therefore \quad r^2\frac{d^2r}{ds^2} = -\sin\phi\frac{d\phi}{d\theta}$$

$$\therefore \quad \sin^2\phi\frac{d\phi}{d\theta} + r\frac{d^2r}{ds^2} = 0.$$

Example 7. *For the curve* $r = ae^{\theta\cot\alpha}$, *prove that* $\frac{s}{r}$ = *constant, r being measured from the origin.*

Solution : The equation of the curve is,

$$r = ae^{\theta\cot\alpha}$$

$$\therefore \quad \frac{dr}{d\theta} = ae^{\theta \cot \alpha}.\ \cot \alpha = r \cot \alpha$$

$$\frac{ds}{dr} = \sqrt{1 + r^2\left(\frac{d\theta}{dr}\right)^2}$$

$$= \sqrt{1 + \frac{r^2}{r^2 \cot^2 \alpha}} = \sec\ \alpha$$

$$\therefore \quad ds = \sec \alpha\, dr.$$

Integrating both sides, we get

$$s = r \sec \alpha + c,$$

where c is a constant of integration.

At the origin, $r = 0,\ s = 2$ (given)

$$\therefore \quad 0 = 0 + c$$

$$\therefore \quad c = 0$$

$$\therefore \quad s = r \sec \alpha$$

$$\therefore \quad \frac{s}{r} = \sec \alpha = \text{constant.}$$

Example 8. *For the curve* $\theta = \cos^{-1}\left(\frac{r}{k}\right) - \frac{\sqrt{k^2 - r^2}}{r}$, *prove that*

$$r\frac{ds}{dr} = \textit{constant}.$$

Solution : The equation of the curve is,

$$\theta = \cos^{-1}\left(\frac{r}{k}\right) - \frac{\sqrt{k^2 - r^2}}{r}.$$

Differentiating both sides w.r.t. r,

$$\frac{d\theta}{dr} = \frac{-1}{\sqrt{1 - \frac{r^2}{k^2}}} \times \frac{1}{k} - \frac{r \cdot \frac{(-2r)}{2\sqrt{k^2 - r^2}} - \sqrt{k^2 - r^2}}{r^2}$$

$$= \frac{-1}{\sqrt{k^2 - r^2}} + \frac{r^2 + k^2 - r^2}{r^2\sqrt{k^2 - r^2}}$$

$$= \frac{k^2}{r^2\sqrt{k^2 - r^2}} - \frac{1}{\sqrt{k^2 - r^2}}$$

$$= \frac{k^2 - r^2}{r^2\sqrt{k^2 - r^2}} = \frac{\sqrt{k^2 - r^2}}{r^2}$$

$$\therefore \quad \frac{d\theta}{dr} = \frac{\sqrt{k^2 - r^2}}{r^2}$$

Now,
$$\frac{ds}{dr} = \sqrt{1+\left(r\frac{d\theta}{dr}\right)^2}$$
$$= \sqrt{1+r^2\frac{k^2-r^2}{r^4}} = \sqrt{1+\frac{k^2-r^2}{r^2}}$$
$$= \sqrt{\frac{r^2+k^2-r^2}{r^2}} = \frac{k}{r}.$$
$$r\frac{ds}{dr} = k = \text{constant.}$$

EXERCISE 5 (D)

1. Find ds/dx for the curves :

   (i) $y^2 = 4ax$. (ii) $x^{2/3} + y^{2/3} = a^{2/3}$.

   (iii) $y = a\cosh\left(\frac{x}{t}\right)$. (iv) $y = \log\cos x$.

   (v) $y = a\log\frac{a^2}{a^2-x^2}$.

2. Find ds/dy for the following curve :
$$27a^4 \cdot y^2 = \{4(x^2+y^2)-x^2\}^3.$$

3. Prove that $\frac{ds}{dt} = a\sqrt{1-e^2\cos^2 t}$

   for ellipse $x = a\cos t,\ y = b\sin t$.

4. Find ds/dt for the following curves, t being the parameter :

   (i) $x = a\cos^3 t, y = b\sin^3 t$. (ii) $x = a\sec t,\ y = b\tan t$.

   (iii) $x = 2\sin t,\ y = \cos 2t$. (iv) $x = e^t\sin t,\ y = e^t\cos t$.

   (v) $x = a(\cos t + t\sin t),\ y = a(\sin t - t\cos t)$.

5. Show that in the curve $r^m = a^m\cos m\theta$, $ds/d\theta$ varies inversely as the $(m-1)$th power of r.

6. Find $ds/d\theta$ for the following curves :

   (i) $r^2 = a^2\cos 2\theta$. (ii) $r = ae^{\theta\cot\alpha}$.

   (iii) $r = a(1+\cos\theta)$. (iv) $r^n = a^n\sin n\theta$.

7. Show that for the hyperbolical spiral $r\theta = a$,
$$\frac{ds}{dr} = \frac{\sqrt{r^2+a^2}}{r}.$$

ANSWERS

1. (i) $\sqrt{1+\frac{a}{x}}$ (ii) $\left(\frac{a}{x}\right)^{1/3}$

(iii) $\cosh\left(\frac{x}{c}\right)$ (iv) $\sec x$

(v) $\frac{a^2+x^2}{a^2-x^2}$.

2. $\frac{a^{2/3}(a^{2/3}+2y^{2/3})}{4xy^{1/3}}$.

4. (i) $3\sin t\cos t\sqrt{a^2\cos^2 t+b^2\sin^2 t}$

(ii) $\sec t\sqrt{a^2\tan^2 t+b^2\sec^2 t}$ (iii) $2\cos t\sqrt{1+4\sin^2 t}$

(iv) $\sqrt{2}\,e^t$ *(v)* at.

6. (i) $a\sqrt{\sec 2\theta}$ (ii) $a \operatorname{cosec} \alpha\, e^{\theta\cot\alpha}$

(iii) $2a\cos\frac{\theta}{2}$ (iv) $\frac{a^n}{r^{n-1}}$.

OBJECTIVE EXERCISE

Select the correct answer :

1. The angle of intersection of the curves $2y^2 = x^3$ and $y^2 = 32x$ is :

(a) $\frac{\pi}{3}, \tan^{-1}\left(\frac{1}{4}\right)$ *(b)* $\pi, \tan^{-1}\left(\frac{3}{2}\right)$

(c) $\frac{\pi}{2}, \tan^{-1}\left(\frac{1}{2}\right)$ *(d)* $\frac{2\pi}{3}, \tan^{-1}\left(\frac{1}{2}\right)$.

2. Sub-tangent is given by :

(a) $\frac{y}{dy/dx}$ *(b)* $\frac{dy}{dx}$

(c) $y-x\frac{dy}{dx}$ *(d)* $y+x$.

3. The angle between radius vector and tangent of the curve $r = ce^{\theta\cot\alpha}$ is :

(a) α *(b)* $-\alpha$

(c) 2α *(d)* -2α.

4. The angle of intersection between the cardioids $r = a(1 + \cos\theta)$ and $r = b(1 - \cos\theta)$ is :

(a) $a\theta$ (b) $\frac{\pi}{2} - b\theta$

(c) $\frac{\pi}{2}$ (d) π.

5. Equation to tangent to the curve $y^2(a + x) = (a - x)x^2$ at the origin is :

(a) $x = -a$ (b) $x = 0$

(c) $x = a$ (d) $y^2 = x^2$.

6. The formula for $dr/d\theta$ is :

(a) $\sqrt{r^2 + \left(\frac{dr}{d\theta}\right)^2}$ (b) $\sqrt{1 + r^2\left(\frac{d\theta}{dr}\right)^2}$

(c) $\sqrt{r^2 + \left(\frac{d\theta}{dr}\right)^2}$ (d) $\sqrt{1 + r^2\left(\frac{dr}{d\theta}\right)^2}$.

7. The length of normal is :

(a) $\frac{y}{dy/dx}$ (b) $y\frac{dy}{dx}$

(c) $\frac{y\sqrt{1 + y_1^2}}{y_1}$ (d) $\sqrt{y + \left(\frac{dy}{dx}\right)^2}$.

8. The length of normal between the X-axis and the normal to the curve $y = \cosh\frac{x}{a}$ is :

(a) $\frac{x^2}{a}$ (b) $\frac{x}{a}$

(c) $\frac{y^2}{a}$ (d) $\frac{y}{a}$.

9. If the inclination of normal at any point of any curve is 3 then dy/dx is :

(a) 3 (b) -3 (c) $\frac{1}{3}$ (d) $-\frac{1}{3}$.

ANSWERS

1. (c)	2. (a)	3. (a)	4. (c)	5. (d)	6. (a)
7. (d)	8. (c)	9. (d)			

6

Curvature

6.1. Conception of Curvature

In the adjoining figure, we are given two curves *AB* and *CD*. We observe that *AB* bends more sharply than *CD* at the point *P*. This geometrical

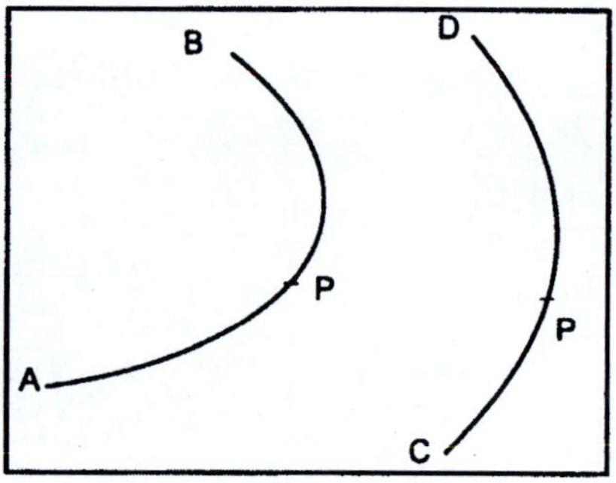

Fig. 6.1

difference between *AB* and *CD* is expressed by saying that the curvature of *AB* is greater than that of *CD*. Again, if a small arc in the neighbourhood of *P* of any curve is regarded as circular, then it is quite clear from this figure that the radius of such a circular arc will be smaller for *AB* as compared to *CD*, *i.e., greater the curvature, lesser the radius and vice-versa.* This forms the basis of curvature. Now, in the next article we shall present a mathematical definition and quantitative estimation of curvature at any point of a curve.

6.2. Definition

Let *P* and *Q* be two neighbouring points on a curve *AB*. Let *N* be the point of intersection of normals at *P* and *Q* to the curve. If *N* tends to a definite position *C*, as *Q* tends to *P*, then the point *C* is called the *centre of curvature* of the curve at the point *P*. The length *CP* is called the *radius of curvature* of

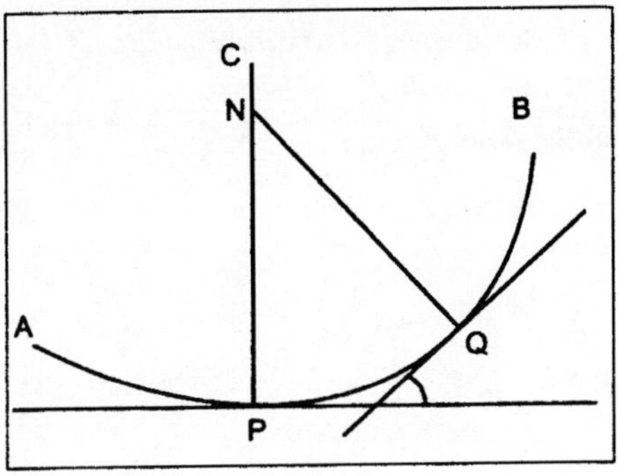

Fig. 6.2

the curve at P and is denoted by the Greek letter ρ. The reciprocal of the distance CP, *i.e.*, $\frac{1}{CP}$, *i.e.*, $\frac{1}{\rho}$ is called the *curvature of the curve* at P.

A circle with C as centre and CP as radius is called the ***circle of curvature*** *of the curve at P. A chord passing through P, of the circle of curvature at P, is called the* ***chord of curvature.***

6.3. Alternative Definition

Let P and Q be two neighbouring points on the curve AB. Let arc AP = s, and arc AQ = s + δs, so that arc PQ = δs, A being a fixed point on the curve from which arc lengths are measured. Draw tangents at P and Q. Let δΨ be the angle between these tangents. Then δΨ is called the ***angle of contingence*** *of the arc PQ. It is also called the* ***total curvature*** *of arc PQ.*

The fraction $\frac{\text{angle of contingence}}{\text{length of arc}}$, *i.e.*, $\frac{\delta\Psi}{\delta s}$ *is called the* ***average curvature*** *of the arc PQ. It is also called* ***average bending*** *of the arc PQ.*

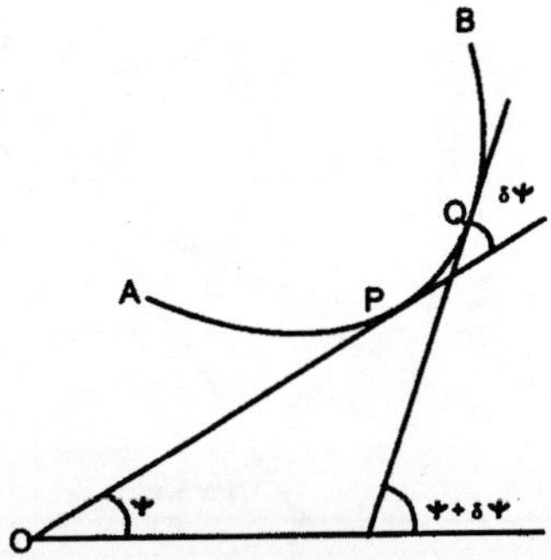

Fig. 6.3

The limiting value of the average curvature when $Q \to P$ is called the ***curvature of the curve*** *at the point P.*

Thus, curvature at the point P

$$= \lim_{Q \to P} \frac{\delta\Psi}{\delta s}$$

$$= \lim_{\delta s \to 0} \frac{\delta\Psi}{\delta s}$$

$$= \frac{d\Psi}{ds}$$

and radius of curvature at the point P

$$= \frac{1}{\text{curvature}} = \frac{ds}{d\Psi} = \rho.$$

Note : A relation between s and Ψ is called the intrinsic equation of the curve. The formula $\rho = \frac{ds}{d\Psi}$ is used to find ρ when the equation of the curve is given in intrinsic form (s, Ψ form).

6.4. Formula for Radius of Curvature in Intrinsic Form

Let P, Q be two neighbouring points on a curve AB. Let arc $AP = s$, arc $AQ = s + \delta s$ so that arc $PQ = \delta s$, A being a fixed point on the curve from which arc lengths are measured. Draw tangents at P and Q. Let the tangents at P and Q make angle Ψ and $\Psi + \delta\Psi$ respectively with OX.

Also let the normals at P and Q intersect in N. Join PQ.

$\therefore \qquad \angle PNQ = \delta\Psi.$

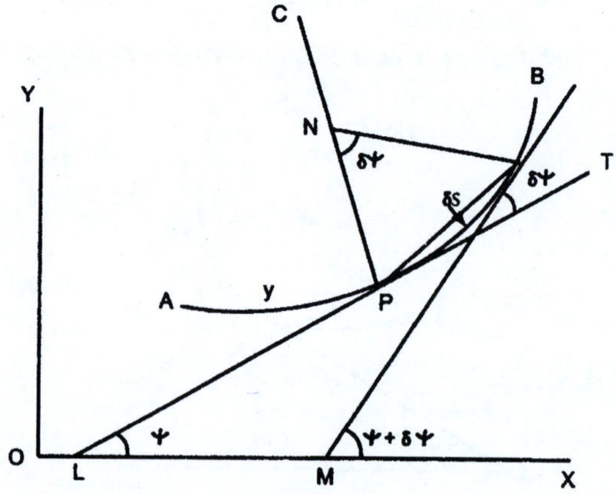

Fig. 6.4

From ΔPQN, we have sine formula of trigonometry.

$$\frac{PN}{\sin PQN} = \frac{\text{chord } PQ}{\sin \delta\Psi}$$

$$\therefore \qquad PN = \frac{\text{chord } PQ}{\sin \delta\Psi} \cdot \sin PQN$$

$$= \frac{\text{chord } PQ}{\delta s} \cdot \frac{\delta s}{\delta\Psi} \cdot \frac{\delta\Psi}{\sin \delta\Psi} \cdot \sin PQN$$

Let ρ be the radius of curvature at P. Then

$$\rho = \lim_{Q \to P} PN$$

$$= \lim_{\delta s \to 0} \frac{\text{chord } PQ}{\delta s} \cdot \frac{\delta s}{\delta\Psi} \cdot \frac{\delta\Psi}{\sin \delta\Psi} \cdot \sin PQN$$

$$= 1 \cdot \frac{ds}{d\Psi} \cdot 1 \cdot \sin \frac{\pi}{2}$$

[$\because$ $\angle PQN \to \angle NPT$ because $PQ \to PT$ when $Q \to P$]

$$\Rightarrow \qquad \rho = \frac{ds}{d\Psi}.$$

6.5. Cartesian Formula for Radius of Curvature

Equation of the curve is,

$$y = f(x).$$

Let Ψ be the angle which the tangent at any point (x, y) makes with the X-axis. Then,

$$\tan\Psi = \frac{dy}{dx}.$$

Differentiating both sides with respect to s, we have

$$\sec^2\Psi \frac{d\Psi}{ds} = \frac{d}{ds}\left(\frac{dy}{dx}\right)$$

$$\Rightarrow \qquad \sec^2\Psi \cdot \frac{1}{\rho} = \frac{d}{dx}\left(\frac{dy}{dx}\right)\frac{dx}{ds}$$

$$\Rightarrow \qquad \sec^2\Psi \frac{1}{\rho} = \frac{d^2y}{dx^2} \cdot \cos\Psi \qquad \left[\because \cos\Psi = \frac{dx}{ds}\right]$$

$$\therefore \qquad \rho = \frac{\sec^3\Psi}{d^2y/dx^2}$$

$$= \frac{(1 + \tan^2\Psi)^{3/2}}{d^2y/dx^2}$$

Hence,

$$\rho = \frac{\left\{1 + \left(\frac{dy}{dx}\right)^2\right\}^{3/2}}{d^2y/dx^2}$$

i.e.,
$$\rho = \frac{(1+y_1^2)^{3/2}}{y_2},$$

where, $y_1 = \frac{dy}{dx}$, $y_2 = \frac{d^2y}{dx^2}$.

Notes :

1. Since ρ is a length, hence if it comes out to be negative, the sign is left to give a positive value of ρ.
2. The above formula does not hold good when the tangent is parallel to X-axis, *i.e.,* $\frac{dy}{dx}$ is infinite. Since the value of ρ depends only on the shape of the curve and not on the choice of axes, therefore, in such case we interchange the axes of x and y and the formula then used is
$$\rho = \frac{\left[1+\left(\frac{dx}{dy}\right)^2\right]^{3/2}}{d^2x/dy^2}.$$
This formula will be found convenient when the tangent is parallel to Y-axis.

6.6. Parametric Formula for Radius of Curvature

Let the equations of the curve in the parametric form be
$$x = f(t), \quad y = \phi(t).$$

Then,
$$\frac{dy}{dx} = \frac{dy/dt}{dx/dt} = \frac{\phi'(t)}{f'(t)} = \frac{y'}{x'}$$

where, x' and y' denote the differential coefficients of x and y with respect to t, and
$$\frac{d^2y}{dx^2} = \frac{d}{dx}\left(\frac{dy}{dx}\right)$$
$$= \frac{d}{dt}\left(\frac{dy}{dx}\right)\frac{dt}{dx}$$
$$= \frac{\frac{d}{dt}\left(\frac{y'}{x'}\right)}{dx/dt}$$
$$= \frac{\frac{x'y''-y'x''}{x'^2}}{x'} = \frac{x'y''-y'x''}{x'^3}$$

Putting the values of $\frac{dy}{dx}$ and $\frac{d^2y}{dx^2}$ in the formula

$$\rho = \frac{\left\{1+\left(\frac{dy}{dx}\right)^2\right\}^{3/2}}{d^2y/dx^2}$$

$$= \frac{\left\{1+\left(\frac{y'}{x'}\right)^2\right\}^{3/2}}{\frac{x'y''-y'x''}{x'^3}}$$

Hence, $$\rho = \frac{(x'^2+y'^2)^{3/2}}{x'y''-y'x''}.$$

6.7. Formula for Radius of Curvature when x and y are given to be Functions of s

We are given $x = f(s),\ y = \phi(s).$

We know that $\cos\Psi = \frac{dx}{ds}.$

Differentiating both sides w.r.t. s, we get

$$-\sin\Psi\cdot\frac{d\Psi}{ds} = \frac{d^2x}{ds^2} \quad \text{.... (1)}$$

$$\Rightarrow \quad -\frac{dy}{ds}\frac{1}{\rho} = \frac{d^2x}{ds^2}$$

$$\Rightarrow \quad \rho = -\left(\frac{dy/ds}{d^2x/ds^2}\right). \quad \text{.... (2)}$$

Also, we know that

$$\sin\Psi = \frac{dy}{ds}.$$

Differentiating both sides w.r.t. s, we get

$$\cos\Psi = \frac{d\Psi}{ds} = \frac{d^2y}{ds^2} \quad \text{.... (3)}$$

$$\Rightarrow \quad \frac{dx}{ds}\frac{1}{\rho} = \frac{d^2y}{ds^2}$$

$$\Rightarrow \quad \rho = \frac{dx/ds}{d^2y/ds^2}. \quad \text{.... (4)}$$

Now, squaring and adding (1) and (3), we get

$$\left(\frac{d\Psi}{ds}\right)^2 = \left(\frac{d^2x}{ds^2}\right)^2 + \left(\frac{d^2y}{ds^2}\right)^2$$

$$\Rightarrow \qquad \frac{1}{\rho^2} = \left(\frac{d^2x}{ds^2}\right)^2 + \left(\frac{d^2y}{ds^2}\right)^2 \qquad \text{.... (5)}$$

Again,

$$\cos\Psi = \frac{dx}{ds} = \frac{dx}{d\Psi}\cdot\frac{d\Psi}{ds} = \frac{dx}{d\Psi}\cdot\frac{1}{\rho} \qquad \text{.... (6)}$$

$$\sin\Psi = \frac{dy}{ds} = \frac{dy}{d\Psi}\cdot\frac{d\Psi}{ds} = \frac{dy}{d\Psi}\cdot\frac{1}{\rho} \qquad \text{.... (7)}$$

Squaring and adding (6) and (7), we get

$$\cos^2\Psi + \sin^2\Psi = \frac{1}{\rho^2}\left\{\left(\frac{dx}{d\Psi}\right)^2 + \left(\frac{dy}{d\Psi}\right)^2\right\}$$

$$\Rightarrow \qquad 1 = \frac{1}{\rho^2}\left\{\left(\frac{dx}{d\Psi}\right)^2 + \left(\frac{dy}{d\Psi}\right)^2\right\}$$

$$\Rightarrow \qquad \rho^2 = \left(\frac{dx}{d\Psi}\right)^2 + \left(\frac{dy}{d\Psi}\right)^2 \qquad \text{.... (8)}$$

6.8. An Important Note About Parabola

The equation of the parabola is $y^2 = 4ax$. Its shape is as shown in the adjoining figure. A (0, 0) is the ***vertex*** of this parabola, X-axis is the ***axis*** of this parabola.

Y-axis is the ***tangent*** at the vertex to this parabola.

The point $S(a, 0)$ on the axis of parabola is called the ***focus*** of the parabola.

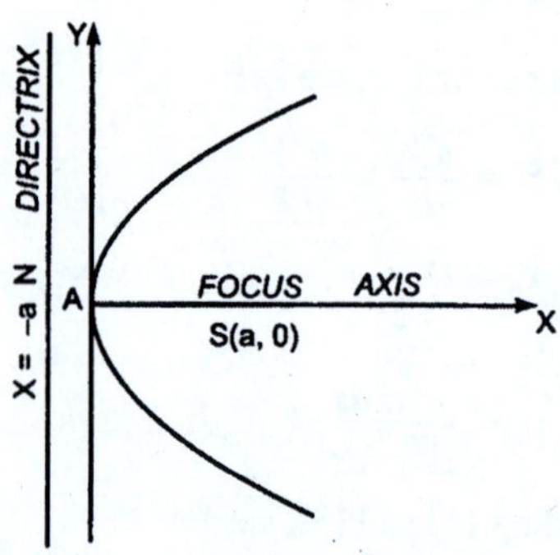

Fig. 6.5

The line $x = a$ is called the ***directrix*** of the parabola. Any chord of the parabola passing through the focus is called a ***focal chord***.

The focal chord perpendicular to the axis of the parabola is called its ***Latus Rectum***.

Length of latus rectum = $4a$, the coefficient of x.

The parametric equations for the parabola $y^2 = 4ax$ are, $x = at^2$, $y = 2at$, where, t is a parameter.

The coordinates of any point on the parabola $y^2 = 4ax$ are $(at^2, 2at)$. For brevity, the point $(at^2, 2at)$ is called point 't' on the parabola $y^2 = 4ax$.

Equation of tangent at 't' is,

$$ty = x + at^2.$$

Equation of normal at 't' is,

$$y + tx = 2at + at^3.$$

If 't_1' and 't_2' are the extremities of a focal chord of the parabola $y^2 = 4ax$, then $t_1 t_2 = -1$.

The ends of a double ordinate of the parabola $y^2 = 4ax$ are

$$(at^2, 2at) \text{ and } (at^2, -2at).$$

6.9. An Important Note About Ellipse

The equation of the ellipse is,

$$\frac{x^2}{a^2} + \frac{y^2}{b^2} = 1.$$

Its shape when $a^2 > b^2$ is as shown in the following figure —

$C(0, 0)$ is the ***centre*** of the ellipse. $AA' = 2a$ is the ***major axis*** and $BB' = 2b$ is the ***minor axis*** of the ellipse.

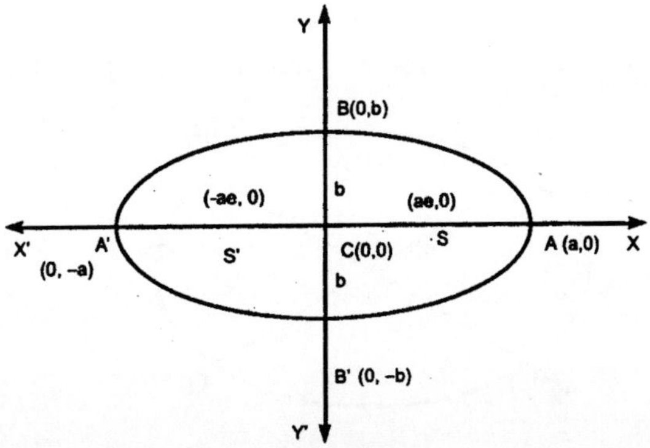

Fig. 6.6

The points S $(ae, 0)$ and S' $(-ae, 0)$ on the major axis are called the ***foci*** of the ellipse.

Eccentricity e of the ellipse is given by $b^2 = a^2(1 - e^2)$, where $0 < e < 1$. The lines $x = \pm \frac{a}{e}$ are called the ***directrices*** of the ellipse.

Any chord of the ellipse passing through either focus is called a ***focal chord***.

The focal chords perpendicular to the major axis are called ***Latera Recta.***

Length of latus rectum= $\frac{2b^2}{a}$.

The ***parametric equations*** of the ellipse $\frac{x^2}{a^2} + \frac{y^2}{b^2} = 1$ are,

$$x = a\cos\phi, \quad y = b\sin\phi;$$

which, ϕ is parameter.

The coordinates of ***any point on the ellipse*** $\frac{x^2}{a^2} + \frac{y^2}{b^2} = 1$ are $(a\cos\phi, b\sin\phi)$.

For brevity, the point $(a\cos\phi, b\sin\phi)$ is called point 'ϕ' on the ellipse. Equation of tangent at 'ϕ' is,

$$\frac{x}{a}\cos\phi + \frac{y}{b}\sin\phi = 1.$$

Equation of normal at Φ is,

$$\frac{ax}{\cos\phi} - \frac{by}{\sin\phi} = a^2 - b^2.$$

6.10. Conjugate Diameters of an Ellipse

A chord of the ellipse passing through its centre is called a *diameter* of the ellipse. Two diameters of an ellipse are said to be conjugate if each bisects

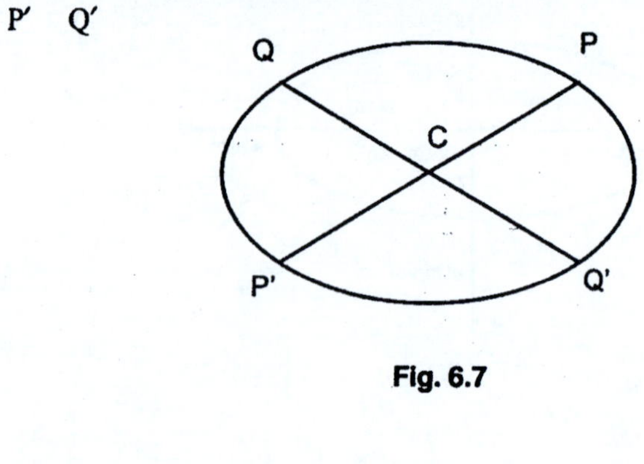

Fig. 6.7

chords parallel to the other. If, PCP' and QCQ' are two conjugate diameters of an ellipse, and P is $(a \cos \phi, b \sin \phi)$, then Q is,

$$\left[a \cos\left(\frac{\pi}{2}+\phi\right), b \sin\left(\frac{\pi}{2}+\phi\right)\right]$$

i.e., $$(-a \sin \phi, \ b \cos \phi).$$

CP and CQ are called ***conjugate semi-diameters*** of the ellipse.

ILLUSTRATIVE EXAMPLES

Example 1. *Find out the radius of curvature at the point (x, y) of the parabola $y^2 = 4ax$.*

Solution : $$y^2 = 4ax. \quad \text{.... (i)}$$

Differentiating w.r.t. x, to both sides

$$2y \frac{dy}{dx} = 4a$$

$$\Rightarrow \quad \frac{dy}{dx} = \frac{2a}{y}$$

$$= \frac{2a}{2\sqrt{ax}} \quad \text{[From (1)]}$$

$$= \frac{\sqrt{a}}{\sqrt{x}}.$$

Again, differentiating w.r.t. x,

$$\frac{d^2y}{dx^2} = \sqrt{a}\,(-\tfrac{1}{2})\, x^{-3/2} = -\frac{\sqrt{a}}{2x^{3/2}}.$$

Now, $$\rho = \frac{\left\{1+\left(\frac{dy}{dx}\right)^2\right\}^{3/2}}{d^2y/dx^2}$$

$$= \frac{\left(1+\frac{a}{x}\right)^{3/2}}{-\frac{\sqrt{a}}{2x^{3/2}}} = \frac{-2(a+x)^{3/2}}{\sqrt{a}}$$

$$= \frac{2(a+x)^{3/2}}{\sqrt{a}}. \quad \text{[Neglecting –ve sign]}$$

Example 2. *Prove that the curvature at the point (x_1, y_1) of the catenary $y = c \cosh \frac{x}{c}$ is $\frac{y^2}{c}$.*

Solution : The given equation to catenary is,

$$y = c \cosh \frac{x}{c} \qquad \text{.... (i)}$$

$$\frac{dy}{dx} = c\left(\sinh \frac{x}{c}\right)\frac{1}{c} = \sinh \frac{x}{c}$$

$$\frac{d^2y}{dx^2} = \cosh \frac{x}{c} \cdot \frac{1}{c}$$

$$\rho = \frac{\left[1+\left(\frac{dy}{dx}\right)^2\right]^{3/2}}{d^2y/dx^2}$$

$$= \frac{c\left(1+\sinh^2 \frac{x}{c}\right)^{3/2}}{\cosh \frac{x}{c}} = \frac{c \cosh^3 \frac{x}{c}}{\cosh \frac{x}{c}} = c \cosh^2 \frac{x}{c}$$

$$= c \cdot \left(\frac{y}{c}\right)^2, \qquad \text{[Using (1)]}$$

$$= \frac{y^2}{c}.$$

Example 3. *Find the radius of curvature at the point* (s, Ψ) *on the curve*

$$s = a \log (\tan \Psi + \sec \Psi) + a \tan \Psi \sec \Psi$$

Solution : $s = a \log (\tan \Psi + \sec \Psi) + a \tan \Psi \sec \Psi$

$$\therefore \quad \rho = \frac{ds}{d\Psi} = a \frac{1}{\tan \Psi + \sec \Psi}(\sec^2 \Psi + \sec \Psi \tan \Psi)$$
$$+ a(\sec^2 \Psi \sec \Psi + \tan \Psi . \sec \Psi \tan \Psi)$$

$$= \frac{a \sec \Psi (\sec \Psi + \tan \Psi)}{\tan \Psi + \sec \Psi} + a \sec \Psi (\sec^2 \Psi + \tan^2 \Psi)$$

$$= a \sec \Psi + a \sec \Psi (\sec^2 \Psi + \sec^2 \Psi - 1)$$

$$= a \sec \Psi (1 + 2 \sec^2 \Psi - 1) = 2a \sec^3 \Psi.$$

Example 4. *Prove that the radius of curvature at any point t of the cycloid* $x = a (t + \sin t)$, $y = a (1 - \cos t)$ *is given by*

$$\rho = 4a \cos \frac{t}{2}.$$

Solution :

$$x = a(t + \sin t)$$

$$\therefore \quad \frac{dx}{dt} = a(1 + \cos t)$$

$$y = a(1 - \cos t)$$

$$\therefore \quad \frac{dy}{dt} = a \sin t$$

$$\therefore \quad \frac{dy}{dx} = \frac{dy/dt}{dx/dt} = \frac{a \sin t}{a(1 + \cos t)}$$

$$= \frac{2 \sin \frac{t}{2} \cos \frac{t}{2}}{2 \cos^2 \frac{t}{2}} = \tan \frac{t}{2}$$

$$\therefore \quad \frac{d^2y}{dx^2} = \frac{d}{dx}\left(\frac{dy}{dx}\right)$$

$$= \frac{d}{dx}\left(\tan \frac{t}{2}\right)$$

$$= \frac{d}{dt}\left(\tan \frac{t}{2}\right) \frac{dt}{dx}$$

$$= \frac{\frac{d}{dt}\left(\tan \frac{t}{2}\right)}{dx/dt}$$

$$= \frac{\frac{1}{2} \sec^2 \frac{t}{2}}{a(1 + \cos t)}$$

$$= \frac{\sec^4 t/2}{4a}$$

$$\rho = \frac{(1 + y_1^2)^{3/2}}{y_2}$$

$$= \frac{\left(1 + \tan^2 \frac{t}{2}\right)^{3/2}}{\sec^4 \frac{t}{2}/4a}$$

$$= \frac{\sec^3 \frac{t}{2}}{\sec^4 \frac{t}{2}/4a} = 4a \cos \frac{t}{2}.$$

Example 5. *Prove that the radius of curvature of the point* $(a\cos^3\theta, a\sin^3\theta)$ *of the curve*

$$x^{2/3} + y^{2/3} = a^{2/3} \text{ is } 3a \sin\theta\cos\theta.$$

Solution : The parametric equation to the curve $x^{2/3} + y^{2/3} = a^{2/3}$ is

$$x = a\cos^3\theta,\ y = a\sin^3\theta$$

$$\therefore \quad \frac{dx}{d\theta} = -3a\cos^2\theta\sin\theta \text{ and } \frac{dy}{d\theta} = 3a\sin^2\theta\cos\theta$$

$$\therefore \quad \frac{dy}{dx} = \frac{3a\sin^2\theta\cos\theta}{-3a\cos^2\theta\sin\theta} = -\tan\theta$$

$$\frac{d^2y}{dx^2} = -\sec^2\theta\frac{d\theta}{dx} = -\sec^2\theta\cdot(-3a\cos^2\theta\sin\theta)$$

$$= \frac{1}{3a\cos^4\theta\sin\theta}$$

$$\rho = \frac{[1+(dy/dx)^2]^{3/2}}{d^2y/dx^2} = \frac{(1+\tan^2\theta)^{3/2}}{\dfrac{1}{(3a\cos^4\theta\sin\theta)}}$$

$$= \sec^3\theta\cdot 3a\cos^4\theta\sin\theta = 3a\cos\theta\sin\theta.$$

Example 6. *Prove that for the ellipse* $\dfrac{x^2}{a^2}+\dfrac{y^2}{b^2}=1$, $\rho = \dfrac{a^2b^2}{p^2}$, *where p is length of perpendicular from centre upon the tangent at* (x, y).

Solution : Differentiating the equation

$$\frac{x^2}{a^2} + \frac{y^2}{b^2} = 1 \qquad \text{.... (1)}$$

w.r.t. x, we get $\dfrac{2x}{a^2}+\dfrac{2y}{b^2}\dfrac{dy}{dx} = 0 \Rightarrow \dfrac{dy}{dx} = \dfrac{-b^2x}{a^2y}$

$$\frac{d^2y}{dx^2} = \frac{-b^2}{a^2}\times\frac{y - x\dfrac{dy}{dx}}{y^2}$$

$$= \frac{-b^2}{a^2y^2}\left[y - x\left(\frac{-b^2x}{a^2y}\right)\right]$$

$$= \frac{-b^2}{a^2y^2}\times\frac{a^2y^2+b^2x^2}{a^2y} = \frac{-b^2(a^2y^2+b^2x^2)}{a^4y^3}$$

$$= \frac{-b^2}{a^4 y^3} \times a^2 b^2 \left[\frac{y^2}{b^2} + \frac{x^2}{a^2}\right]$$

$$\Rightarrow \qquad \frac{d^2y}{dx^2} = \frac{-b^4}{a^2 y^3}, \qquad \text{[Using (1)]}$$

$$\therefore \qquad \rho = \frac{\left[1+\left(\frac{dy}{dx}\right)^2\right]^{3/2}}{d^2y/dx^2} = \frac{\left[1+\left(\frac{-b^2 x}{a^2 y}\right)^2\right]^{3/2}}{-b^4/a^2 y^3}.$$

Leaving negative sign

$$\rho = \frac{(a^4y^2 + b^4x^2)^{3/2}}{a^4b^4} \qquad \text{.... (2)}$$

Equation to tangent at (x, y)

$$\frac{xX}{a^2} + \frac{yY}{b^2} = 1. \qquad \text{.... (3)}$$

The length of the perpendicular from (0, 0) on (3),

$$p = \frac{1}{\sqrt{\frac{x^2}{a^4} + \frac{y^2}{b^4}}}$$

$$= \frac{a^2b^2}{\sqrt{b^4x^2 + a^4y^2}}$$

$$\frac{1}{p^3} = \frac{(b^4x^2 + a^4y^2)^{3/2}}{a^6b^6}$$

$$\Rightarrow \qquad (b^4x^2 + a^4y^2)^{3/2} = \frac{a^6b^6}{p^3}$$

Using (2), $\qquad \rho \cdot a^4b^4 = \dfrac{a^6b^6}{p^3}$

$$\therefore \qquad \rho = \frac{a^2b^2}{p^3}.$$

Example 7. *Show that the curve for which* $s = \sqrt{8ay}$ *(the cycloid) has for its intrinsic equation* $s = 4a \sin \Psi$.

Hence prove that $\qquad \rho = 4a\sqrt{1 - \dfrac{y}{2a}}.$

Solution : $\qquad s = \sqrt{8ay}$

$$\Rightarrow \qquad s^2 = 8ay. \qquad \text{.... (1)}$$

Differentiating w.r.t. s, we get

$$2s = 8a\frac{dy}{ds} = 8a\sin\Psi \qquad \left[\because \sin\Psi = \frac{dy}{ds}\right]$$

$$\Rightarrow \qquad s = 4a\sin\Psi \qquad \text{.... (2)}$$

$$\therefore \qquad \rho = \frac{ds}{d\Psi} = 4a\cos\Psi$$

$$= 4a\sqrt{1-\sin^2\Psi}$$

$$= 4a\sqrt{1-\frac{s^2}{16a^2}} \qquad \text{[From (2)]}$$

$$= 4a\sqrt{1-\frac{8ay}{16a^2}} \qquad \text{[From (1)]}$$

$$= 4a\sqrt{1-\frac{y}{2a}}.$$

Example 8. *If CP and CD be a pair of conjugate semi-diameter of an ellipse, prove that the radius of curvature at P is* $\frac{CD^3}{ab}$, *a and b being the lengths of semi-axes of the ellipse.*

Solution : Let the co-ordinates of any point P on the ellipse

$$\frac{x^2}{a^2}+\frac{y^2}{b^2} = 1 \qquad \text{.... (1)}$$

be $(a\cos\phi, b\sin\phi)$. The co-ordinates of the point Q of the conjugate diameter will be $\left\{a\cos\left(\frac{\pi}{2}+\phi\right), b\sin\left(\frac{\pi}{2}+\phi\right)\right\}$ that is

$$(-a\sin\phi, b\cos\phi).$$

The co-ordinates of the point $C = (0, 0)$.

$$CD = \sqrt{(0+a\sin\phi)^2+(0-b\cos\phi)^2}$$

$$= \sqrt{a^2\sin^2\phi+b^2\cos^2\phi}$$

$$CD^3 = (a^2\sin^2\phi+b^2\cos^2\phi)^{3/2}.$$

Now, for the point P, $x = a\cos\phi$, $y = b\sin\phi$

$$\frac{dx}{d\phi} = -a\sin\phi, \qquad \frac{dy}{d\phi} = b\cos\phi$$

$$\frac{dy}{dx} = \frac{dy/d\phi}{dx/d\phi} = \frac{b\cos\phi}{-a\sin\phi} = \frac{-b}{a}\cot\phi$$

$$\frac{d^2y}{dx^2} = \frac{b}{a}\,\text{cosec}^2\,\phi\,\frac{d\phi}{dx}$$

$$= \frac{b}{a}\,\text{cosec}^2\,\phi\left(\frac{-1}{a\sin\phi}\right) = \frac{-b}{a^2}\,\text{cosec}^3\,\phi.$$

The radius of curvature at $P\,(a\cos\phi, b\sin\phi)$ on the ellipse,

$$\rho = \frac{\left[1+\left(\frac{dy}{dx}\right)^2\right]^{3/2}}{\frac{d^2y}{dx^2}} = \frac{\left[1+\frac{b^2}{a^2}\cot^2\phi\right]^{3/2}}{-\frac{b}{a}\,\text{cosec}^3\,\phi}$$

$$= \frac{(a^2\sin^2\phi + b^2\cos^2\phi)^{3/2}}{-ab\sin^3\phi\cos^3\phi}$$

$$= \frac{(a^2\sin^2\phi + b^2\cos^2\phi)^{3/2}}{ab\sin^3\phi\cos^3\phi} \quad \text{(The numerical value)}$$

$$= \frac{CD^3}{ab}.$$

Example 9. *If ρ_1 and ρ_2 be the radii of curvature at the extremities of two conjugate diameters of an ellipse, prove that*

$$(\rho_1^{2/3} + \rho_2^{2/3})\,a^{2/3}\,b^{2/3} = a^2 + b^2.$$

Solution : Let CP and CD be the two conjugate diameters of an ellipse

$$\frac{x^2}{a^2} + \frac{y^2}{b^2} = 1.$$

Let ϕ be the eccentric angle of P, then eccentric angle D will be $\left(\frac{\pi}{2} + \phi\right)$.

∴ For the point P,

$$x = a\cos\phi, \qquad y = b\sin\phi$$

$$\frac{dx}{d\phi} = -a\sin\phi, \qquad \frac{dy}{d\Phi} = b\cos\phi$$

$$\frac{d^2x}{d\phi^2} = -a\cos\phi, \qquad \frac{d^2y}{d\phi^2} = -b\sin\phi.$$

At the end point $P\,(a\cos\phi, b\sin\phi)$ of the diameter, the radius of our vector will be

$$\rho_1 = \frac{(x'^2 + y'^2)^{3/2}}{x'y'' - x''y'}$$

$$= \frac{[a^2 \sin^2 \phi + b^2 \cos^2 \phi]^{3/2}}{-a \sin \phi (-b \sin \phi) - b \cos \phi (-a \cos \phi)}$$

$$= \frac{[a^2 \sin^2 \phi + b^2 \cos^2 \phi]^{3/2}}{ab [\sin^2 \phi + \cos^2 \phi]} = \frac{[a^2 \sin^2 \phi + b^2 \cos^2 \phi]^{3/2}}{ab}$$

$$\rho_1^{2/3} = \frac{a^2 \sin^2 \phi + b^2 \cos^2 \phi}{(ab)^{2/3}}$$

Similarly, at the end $D \left[a \cos\left(\frac{\pi}{2} + \phi\right), b \sin\left(\frac{\pi}{2} + \phi\right)\right]$ the radius of curvature,

$$\rho_2^{2/3} = \frac{a^2 \sin^2\left(\frac{\pi}{2} + \phi\right) + b^2 \cos^2\left(\frac{\pi}{2} + \phi\right)}{(ab)^{2/3}}$$

$$= \frac{a^2 \cos^2 \phi + b^2 \sin^2 \phi}{(ab)^{2/3}}.$$

Now,

$$\rho_1^{2/3} + \rho_2^{2/3} = \frac{1}{(ab)^{2/3}} [a^2 \sin^2 \phi + b^2 \cos^2 \phi + a^2 \cos^2 \phi + b^2 \sin^2 \phi]$$

$$= \frac{1}{(ab)^{2/3}} [a^2 (\sin^2 \phi + \cos^2 \phi) + b^2 (\cos^2 \phi + \sin^2 \phi)]$$

$$= \frac{a^2 + b^2}{(ab)^{2/3}}.$$

EXERCISE 6 (A)

1. Find the radius of curvature at the point (s, Ψ) on the following curves —

   (i) $s = a\Psi$ (circle)

   (ii) $s = c \tan \Psi$ (catenary)

   (iii) $s = 4a \sin \Psi$ (cycloid)

   (iv) $s = 8a \sin^2 \frac{\Psi}{6}$ (cardioid)

   (v) $s = c \log \sec \Psi$ (tractrix)

   (vi) $s = a (e^{m\Psi} - 1)$ (equiangular spiral)

   (vii) $s = c \sec^3 \Psi$

   (viii) $s = c \log \tan \left(\frac{\pi}{4} + \frac{\Psi}{2}\right)$.

2. Prove that for the curve $s = a \log \cot\left(\frac{\pi}{4} - \frac{\Psi}{2}\right) + a \sin \Psi \sec^2 \Psi$,

$$\rho = 2a \sec^3 \Psi$$

and, hence show that $\frac{d^2y}{dx^2} = \frac{1}{2a}$

and that this differential equation is satisfied by the parabola

$$x^2 = 4ay.$$

3. Prove that for the curve $\quad x = m(\sec^2 \Psi - 1),$

$$\rho = 3m \tan \Psi \sec^2 \Psi$$

and hence show that

$$3m \frac{dy}{dx} \frac{d^2y}{dx^2} = 1.$$

Also, that this differential equation is satisfied by the semi-cubical parabola $27my^2 = 8x^3$.

4. Prove that the curvature at a point of the curve $y = f(x)$ is given by $\left(\frac{d^2y}{dx^2}\right) \cdot \cos^2 \phi$, where, ϕ is the inclination of the tangent at the point to the axis of X.

5. Prove that $\frac{1}{\rho} = \frac{d}{dx}\left(\frac{dy}{ds}\right)$.

6. Prove that the radius of curvature of the curve

$$x = a \cos\left(\frac{s}{a}\right), \; y = a \sin\left(\frac{s}{a}\right) \text{ is } a.$$

7. Prove that the radius of curvature of the curve

$$y = \frac{a\left(e^{x/a} + e^{-x/a}\right)}{2} \text{ is } \frac{y^2}{a}.$$

8. Find the radius of curvature at the point (x, y) on the following curves —

(i) $xy = a^2$ (Hyperbola)

(ii) $ay^2 = x^3$

(iii) $y = a \log \sec\left(\frac{x}{a}\right)$

(iv) $x^{2/3} + y^{2/3} = a^{2/3}$ (Astroid)

(v) $a^2y = x^3 - a^3$.

9. Show that the radius of curvature of the curve $y = a \cosh(x/a)$ at any point (x, y) is equal in length to the portion of the normal intercepted between the curve and the axis of X.

10. Find the radius of curvature of the following curves at the points indicated —

(i) $\sqrt{x} + \sqrt{y} = 1$ at $\left(\frac{1}{4}, \frac{1}{4}\right)$.

(ii) $y = e^x$, at the point where it crosses the Y-axis.

(iii) $x^3 + y^3 = 3axy$ at the point $\left(\frac{3a}{2}, \frac{3a}{2}\right)$.

(iv) $\sqrt{x} + \sqrt{y} = \sqrt{a}$ at the point where the line $y = c$ cuts it.

(v) $xy = 4$ at $(2, 2)$.

(vi) $4x^2 + 9y^2 = 72$ at $(3, 2)$.

(vii) $y^2 = 16x$ at an end of latus rectum.

(viii) $a^2y = x^3 - a^3$ at the point where it crosses the X-axis.

(ix) $y = 4 \sin x - \sin 2x$ at $\left(x = \frac{\pi}{2}\right)$.

11. Find the points on the parabola $y^2 = 8x$ at which the radius of curvature is $7\frac{13}{16}$.

12. Find the radius of curvature at any point t of the following curves —

(i) $x = a \cos t,\ y = b \sin t$.

(ii) $x = a \cos^3 t, y = a \sin^3 t$.

(iii) $x = at^2,\ y = 2at$.

(iv) $x = 3a \cos t - a \cos 3t, y = 3a \sin t - a \sin 3t$.

(v) $x = a(\cos t + t \sin t), y = a(\sin t - t \cos t)$.

(vi) $x = a \sin 2t\,(1 + \cos 2t), y = a \cos 2t(1 - \cos 2t)$.

(vii) $x = c \log_e \{s + \sqrt{s + c^2}\},\ y = \sqrt{c^2 + s^2}$.

13. Show that the radius of curvature at any point (x, y) of the curve $x^{2/3} + y^{2/3} = a^{2/3}$ is three times the length of the perpendicular from the origin upon the tangent at (x, y).

14. In the ellipse $\frac{x^2}{a^2}+\frac{y^2}{b^2}=1$, show that the radius of curvature at an end of the major axis is equal to the semi-latus rectum of the ellipse.

15. Show that the radius of curvature of the lemniscate $(x^2+y^2)^2=a^3(x^2-y^2)$ at a point where the tangent is parallel to the X-axis is $\frac{\sqrt{2}a}{3}$.

ANSWERS

1. (i) a (ii) $c\sec^2\Psi$ (iii) $4a\cos\Psi$ (iv) $\frac{4}{3}a\sin\frac{\Psi}{3}$
(v) $c\tan\Psi$ (vi) $ame^{m\Psi}$ (vii) $3c\sec^2\Psi\tan\Psi$
(viii) $c\sec\Psi$.

8. (i) $\frac{(x^4+c^4)^{3/4}}{2c^2x^3}$ (ii) $(4a+9x)^{3/2}\frac{\sqrt{x}}{6a}$ (iii) $a\sec\frac{x}{a}$
(iv) $3(axy)^{1/3}$ (v) $\frac{(a^4+9x^4)^{3/2}}{6xa^4}$.

10. (i) $\frac{1}{\sqrt{2}}$ (ii) $2\sqrt{2}$ (iii) $\frac{8\sqrt{2}}{3a}$ (iv) $\frac{a}{\sqrt{2}}$
(v) $2\sqrt{2}$ (vi) $\frac{13\sqrt{13}}{12}$ (vii) $16\sqrt{2}$ (viii) $\frac{5\sqrt{10}}{3}a$
(ix) $\frac{5\sqrt{5}}{4}$.

11. $\left(\frac{9}{8}, 3\right), \left(\frac{9}{8}, -3\right)$.

12. (i) $\frac{(a^2\sin^2 t+b^2\cos^2 t)^{3/2}}{ab}$ (ii) $3a\sin t\cos t$
(iii) $2a(1+t^2)^{3/2}$ (iv) $3a\sin t$ (v) at (vi) $4a\cos 3t$
(vii) $\frac{s^2+c^2}{c}$.

6.11. Pedal Formula for Radius of Curvature

From the figure, $\Psi = \theta+\phi$.

Differentiating both sides w.r.t. s, we get

$$\frac{d\Psi}{ds}=\frac{d\theta}{ds}+\frac{d\phi}{ds}$$

$$\Rightarrow \quad \frac{1}{\rho}=\frac{1}{r}\cdot r\cdot\frac{d\theta}{ds}+\frac{d\phi}{dr}\cdot\frac{dr}{ds}$$

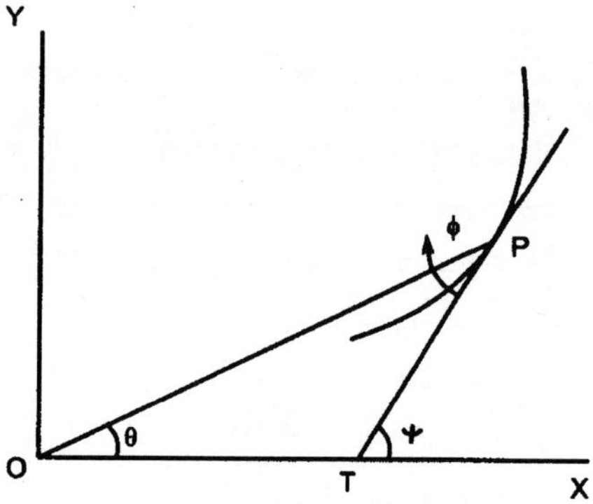

Fig. 6.8

$$\Rightarrow \qquad \frac{1}{\rho} = \frac{1}{r}\cdot \sin\phi + \frac{d\phi}{dr}\cos\phi$$

$$\Rightarrow \qquad \frac{1}{\rho} = \frac{1}{r}\left\{\sin\phi + r\cos\phi\frac{d\phi}{dr}\right\}$$

$$\Rightarrow \qquad \frac{1}{\rho} = \frac{1}{r}\cdot\frac{d}{dr}(r\sin\phi)$$

$$\Rightarrow \qquad \frac{1}{\rho} = \frac{1}{r}\cdot\frac{dp}{dr} \qquad [\because p = r\sin\phi]$$

i.e.,
$$\rho = r\frac{dr}{dp}.$$

6.12. Polar Formula for Radius of Curvature

We know that

$$\frac{1}{p^2} = \frac{1}{r^2} + \frac{1}{r^4}\left(\frac{dr}{d\theta}\right)^2. \qquad \text{.... (1)}$$

Differentiating (1) w.r.t. r, we get

$$-\frac{2}{p^3}\frac{dp}{dr} = -\frac{2}{r^3} - \frac{4}{r^5}\left(\frac{dr}{d\theta}\right)^2 + \frac{1}{r^4}\cdot 2\frac{dr}{d\theta}\cdot\frac{d^2r}{d\theta^2}\cdot\frac{d\theta}{dr}$$

$$= -\frac{2}{r^3} - \frac{4}{r^5}\left(\frac{dr}{d\theta}\right)^2 + \frac{2}{r^4}\frac{d^2r}{d\theta^2}$$

$$\Rightarrow \qquad \frac{r^5}{p^3}\frac{dp}{dr} = r^2 + 2\left(\frac{dr}{d\theta}\right)^2 - r\frac{d^2r}{d\theta^2} \qquad \text{.... (2)}$$

$$\Rightarrow \qquad \rho = r\frac{dr}{dp}$$

$$= \frac{r \cdot r^5/p^3}{r^2 + 2\left(\frac{dr}{d\theta}\right)^2 - r\frac{d^2r}{d\theta^2}} \quad \text{[From (2)]}$$

$$= \frac{r^6/p^3}{r^2 + 2\left(\frac{dr}{d\theta}\right)^2 - r\frac{d^2r}{d\theta^2}}$$

$$= r^6 \frac{\{1/p^2\}^{3/2}}{r^2 + 2\left(\frac{dr}{d\theta}\right)^2 - r\frac{d^2r}{d\theta^2}}$$

$$= r^6 \frac{\left\{\frac{1}{r^2} + \frac{1}{r^4}\left(\frac{dr}{d\theta}\right)^2\right\}^{3/2}}{r^2 + 2\left(\frac{dr}{d\theta}\right)^2 - r\frac{d^2r}{d\theta^2}} \quad \text{[From (1)]}$$

Therefore,
$$\rho = \frac{\left\{r^2 + \left(\frac{dr}{d\theta}\right)^2\right\}^{3/2}}{r^2 + 2\left(\frac{dr}{d\theta}\right)^2 - r\frac{d^2r}{d\theta^2}}$$

6.13. Tangential Polar Formula for Radius of Curvature

A relation between p and Ψ for a curve is called its tangential polar equation.

We know that

$$\frac{dr}{ds} = \cos\phi$$

and
$$\rho = \frac{ds}{d\Psi} = r\frac{dr}{dp}$$

$\therefore$
$$\frac{dp}{d\Psi} = \frac{dp}{dr} \cdot \frac{dr}{ds} \cdot \frac{ds}{d\Psi}$$

$$= \frac{dp}{dr}\cos\phi \cdot r\frac{dr}{dp}$$

$$= r\cos\phi \quad \text{.... (1)}$$

Also,
$$p = r\sin\phi \quad \text{.... (2)}$$

Squaring and adding (1) and (2), we obtain

$$p^2 + \left(\frac{dp}{d\phi}\right)^2 = r^2.$$

Differentiating w.r.t. p on both sides, we get

$$2p + 2\frac{dp}{d\Phi}\cdot\frac{d}{dp}\left(\frac{dp}{d\Psi}\right) = 2r\frac{dr}{dp}$$

$$\Rightarrow \qquad p + \frac{dp}{d\Psi}\cdot\frac{d}{d\Psi}\left(\frac{dp}{d\Psi}\right)\frac{d\Psi}{dp} = r\frac{dr}{dp}$$

$$\Rightarrow \qquad p + \frac{d^2p}{d\Psi^2} = \rho,$$

which is the required formula.

ILLUSTRATIVE EXAMPLES

Example 1. *Find the radius of curvature at the point (p, r) of cardioid*

$$r^3 = 2ap^2.$$

Solution : The equation to the cardioid

$$r^3 = 2ap^2. \qquad \text{.... (1)}$$

Differentiating (1) w.r.t. p,

$$3r^2\frac{dr}{dp} = 4ap \quad \Rightarrow r\frac{dr}{dp} = \frac{4ap}{3r}$$

$$\Rightarrow \qquad \rho = r\frac{dr}{dp} = \frac{4ap}{3r} = \frac{4a}{3r}\sqrt{\frac{r^3}{2a}} = \frac{2}{3}\sqrt{2ar}.$$

Example 2. *Find the radius of curvature at the point (p, r) on the ellipse*

$$\frac{1}{p^2} = \frac{1}{a^2} + \frac{1}{b^2} - \frac{r^2}{a^2b^2}$$

Solution : We have

$$\frac{1}{p^2} = \frac{1}{a^2} + \frac{1}{b^2} - \frac{r^2}{a^2b^2}.$$

Differentiating both sides w.r.t. r, we get

$$-\frac{2}{p^3}\frac{dp}{dr} = -\frac{2r}{a^2b^2}$$

$$\Rightarrow \qquad \frac{1}{p^3}\frac{dp}{dr} = \frac{r}{a^2b^2}$$

$$\therefore \qquad r\frac{dr}{dp} = \frac{a^2b^2}{p^3}$$

$$\therefore \qquad \rho = \frac{a^2b^2}{p^3}.$$

Example 3. *For the curve $r^2 = a^2 \cos 2\theta$, find the value of*

(i) ϕ, (ii) p, (iii) ρ.

Solution :

(i) $$r^2 = a^2 \cos 2\theta$$

$$\Rightarrow \quad 2 \log r = 2 \log a + \log \cos 2\theta$$

$$\Rightarrow \quad \frac{2}{r}\frac{dr}{d\theta} = -\frac{2 \sin 2\theta}{\cos 2\theta}$$

$$\Rightarrow \quad \frac{1}{r}\frac{dr}{d\theta} = -\tan 2\theta \Rightarrow \cot \phi = -\tan 2\theta$$

$$\Rightarrow \quad \cot \phi = \cot\left(\frac{\pi}{2} + 2\theta\right) \quad \Rightarrow \phi = \frac{\pi}{2} + 2\theta.$$

(ii) $$p = r \sin \phi \quad \Rightarrow p = r \sin\left(\frac{\pi}{2} + 2\theta\right)$$

$$\Rightarrow \quad p = r \cos 2\theta \quad \Rightarrow p = r \cdot \frac{r^2}{a^2} \quad \left[\because r^2 = a^2 \cos 2\theta\right]$$

$$\Rightarrow \quad p = \frac{r^3}{a^2}.$$

(iii) $$p = \frac{r^3}{a^2} \quad \Rightarrow \frac{dp}{dr} = \frac{3r^2}{a^2}$$

$$\therefore \quad \rho = \frac{r}{(dp/dr)} = \frac{r}{(3r^2/a^2)}$$

$$= \frac{a^2 r}{3r^2} = \frac{a^2}{3r}.$$

Example 4. *Show that the radius of curvature for the curve $r^n = a^n \cos n\theta$ is*

$$\frac{a^n}{(n+1)r^{n-1}}.$$

Solution : Taking log of both sides of

$$r^n = a^n \cos n\theta \qquad \text{.... (1)}$$

we get, $$n \log r = n \log a + \log \cos n\theta.$$

Differentiating both sides w.r.t. θ,

$$\frac{n}{r}\cdot\frac{dr}{d\theta} = 0 + \frac{1}{\cos n\theta}(-n \sin n\theta)$$

$$\Rightarrow \qquad \frac{dr}{d\theta} = -r\tan n\theta.$$

Again,
$$\frac{d^2r}{d\theta^2} = -\frac{dr}{d\theta}\tan n\theta - r\sec^2 n\theta \cdot n$$
$$= r\tan^2 n\theta - nr\sec^2 n\theta$$

$$\rho = \frac{\left[r^2 + \left(\frac{dr}{d\theta}\right)^2\right]^{3/2}}{r^2 + 2\left(\frac{dr}{d\theta}\right)^2 - r\frac{dr}{d\theta}}$$

$$= \frac{[r^2 + r^2\tan^2 n\theta]^{3/2}}{r^2 + 2r^2\tan^2 n\theta - r^2\tan^2 n\theta + nr^2\sec^2 n\theta}$$

$$= \frac{r^3\sec^3 n\theta}{r^2[(1+\tan^2 n\theta) + n\sec^2 n\theta]} = \frac{r\sec^3 n\theta}{\sec^2 n\theta + n\sec^2 n\theta}$$

$$= \frac{r\sec^3 n\theta}{(n+1)\sec^2 n\theta} = \frac{r\sec n\theta}{n+1} = \frac{r}{(n+1)\cos n\theta}$$

$$= \frac{a^n r}{(n+1)r^n} = \frac{a^n}{(n+1)r^{n-1}} \qquad \text{[Using (1)]}$$

$$\rho = \frac{a^n}{(n+1)r^{n-1}}.$$

Example 5. *Show that the pedal equation to the curve* $\frac{1}{r} = 1 + e\cos\theta$ *is*

$$\frac{1}{p^2} = \frac{1}{l^2}\left(\frac{2l}{r} - 1 + e^2\right)$$

Solution : Taking log of both sides of $\frac{l}{r} = 1 + e\cos\theta$, we get

$$\log l - \log r = \log(1 + e\cos\theta)$$

$$0 - \frac{1}{r}\frac{dr}{d\theta} = \frac{-e\sin\theta}{1+e\cos\theta}$$

$$\frac{1}{r}\frac{dr}{d\theta} = \frac{e\sin\theta}{1+e\cos\theta}$$

$$\frac{1}{p^2} = \frac{1}{r^2}\left[1 + \frac{1}{r^2}\left(\frac{dr}{d\theta}\right)^2\right]$$

$$= \frac{1}{r^2}\left[1 + \frac{e^2 \sin^2\theta}{(1 + e\cos\theta)^2}\right]$$

$$= \frac{1}{r^2}\frac{1 + 2e\cos\theta + e^2\cos^2\theta + e^2\sin^2\theta}{(1 + e\cos\theta)^2}$$

$$= \frac{1}{r^2} \times \frac{1 + 2e\cos\theta + e^2}{(1 + e\cos\theta)^2}$$

From given curve $e\cos\theta = \frac{l}{r} - 1$, we get

$$\frac{1}{p^2} = \frac{1}{r^2} \times \frac{1 + 2\left(\frac{l}{r} - 1\right) + e^2}{\left(1 + \frac{l}{r} - 1\right)^2} = \frac{1 + \frac{2l}{r} - 2 + e^2}{r^2 \times \frac{l^2}{r^2}}$$

$$= \frac{1}{l^2}\left[\frac{2l}{r} - 1 + e^2\right].$$

Example 6. *Show that the radius of curvature at any point on the cardioid $r = a(1 - \cos\theta)$ is $\frac{2}{3}\sqrt{2ar}$. Also prove that $\frac{\rho^2}{r}$ is constant.*

Solution : $r = a(1 - \cos\theta)$

$$\therefore \quad \frac{dr}{d\theta} = a\sin\theta \quad \text{and} \quad \frac{d^2r}{d\theta^2} = a\cos\theta.$$

Now,
$$\rho = \frac{\{r^2 + (dr/d\theta)^2\}^{3/2}}{r^2 + 2\left(\frac{dr}{d\theta}\right)^2 - r\left(\frac{d^2r}{d\theta}\right)^2}$$

$$= \frac{\{a^2(1 - \cos\theta)^2 + a^2\sin^2\theta\}^{3/2}}{a^2(1 - \cos\theta)^2 + 2 \cdot a^2\sin^2\theta - a(1 - \cos\theta)\,a\cos\theta}$$

$$= a \cdot \frac{(1 + \cos^2\theta - 2\cos\theta + \sin^2\theta)^{3/2}}{\{1 + \cos^2\theta - 2\cos\theta + 2\sin^2\theta - \cos\theta + \cos^2\theta\}}$$

$$= \frac{a\{2 - 2\cos\theta\}^{3/2}}{(1 + 2 - 3\cos\theta)}$$

$$= \frac{a \cdot \left(2 \cdot 2\sin^2\frac{\theta}{2}\right)^{3/2}}{3 \cdot 2\sin^2\frac{\theta}{2}} = \frac{8a \cdot \sin^3\frac{\theta}{2}}{6\sin^2\frac{\theta}{2}}$$

$$= \frac{4}{3} a \sin\frac{\theta}{2} = \frac{2a}{3}\sqrt{4\sin^2\frac{\theta}{2}}$$

$$= \frac{2a}{3}\sqrt{2\cdot(1-\cos\theta)}$$

$$= \frac{2a}{3}\sqrt{\frac{2r}{a}} = \frac{2}{3}\sqrt{2ar}.$$

Example 7. *In the equiangular spiral* $r = ae^{\theta\cot\alpha}$, *prove that* $\rho = r\operatorname{cosec}\alpha$ *and hence show that the radius of curvature subtends a right angle at the pole.*

Solution : $$r = ae^{\theta\cot\alpha} \quad ....(1)$$

Differentiating both sides w.r.t. θ, we get

$$\frac{dr}{d\theta} = ae^{\theta\cot\alpha}.\cot\alpha$$

$$\Rightarrow \quad \frac{dr}{d\theta} = r\cot\alpha$$

$$\Rightarrow \quad r\frac{d\theta}{dr} = \tan\alpha \quad ....(2)$$

$$\Rightarrow \quad \tan\phi = \tan\alpha \quad \Rightarrow \quad \phi = \alpha$$

Now, $p = r\sin\phi \quad \therefore \quad p = r\sin\alpha.$

Differentiating both sides w.r.t. r, we get

$$\frac{dp}{dr} = \sin\alpha \qquad \therefore \quad \frac{dr}{sp} = \operatorname{cosec}\alpha$$

$$\therefore \quad r\frac{dr}{dp} = r\operatorname{cosec}\alpha \qquad \therefore \quad \rho = r\operatorname{cosec}\alpha.$$

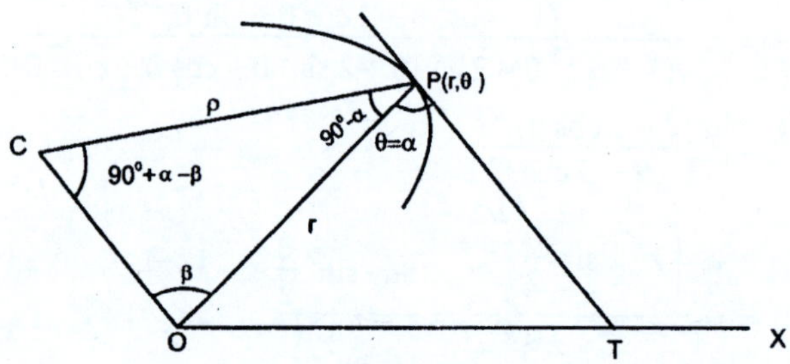

Fig. 6.9

Let $P(r, \theta)$ be any point on the curve. PT is the tangent and PC is normal to the curve at P. Let C be the centre of curvature at P. Join OC and OP, where O is the pole.

$$\therefore \quad \phi = \alpha$$

$$\therefore \quad \angle OPT = \alpha$$

$$\therefore \quad \angle OPC = 90° - \alpha \qquad [\because PC \perp PT]$$

Let the angle subtended by CP (radius of curvature) at pole, *i.e.*, $\angle POC$ be β.

Then, $$\angle OCP = 180° - [\beta + 90° - \alpha] = 90° + \alpha - \beta.$$

Now, from ΔOCP, by sine formula,

$$\frac{\rho}{\sin\beta} = \frac{r}{\sin(90° + \alpha - \beta)}$$

$$\Rightarrow \quad \frac{r \operatorname{cosec}\alpha}{\sin\beta} = \frac{r}{\cos(\alpha - \beta)}$$

$$\Rightarrow \quad \sin\alpha\sin\beta = \cos(\alpha - \beta)$$

$$\Rightarrow \quad \sin\alpha\sin\beta = \cos\alpha\cos\beta + \sin\alpha\sin\beta$$

$$\Rightarrow \quad \cos\alpha\cos\beta = 0$$

$$\Rightarrow \quad \cos\beta = 0 \qquad [\because \cos\alpha \neq 0 \text{ as } \alpha \text{ is given}]$$

$$\Rightarrow \quad \beta = \frac{\pi}{2}.$$

Hence, the radius of curvature subtends a right angle at the pole.

Example 8. *For any curve, prove that*

$$\frac{r}{\rho} = \sin\phi\left(1 + \frac{d\phi}{d\theta}\right).$$

Solution : We know that $\Psi = \theta + \phi$.

Differentiating both sides w.r.t. θ, we get

$$\frac{d\Psi}{d\theta} = 1 + \frac{d\phi}{d\theta}$$

$$\Rightarrow \quad \sin\phi\frac{d\Psi}{d\theta} = \sin\phi\left(1 + \frac{d\phi}{d\theta}\right)$$

$$\Rightarrow \quad r\frac{d\theta}{ds}\frac{d\Psi}{d\theta} = \sin\phi\left(1 + \frac{d\phi}{d\theta}\right)$$

$$\Rightarrow \quad r\frac{d\Psi}{ds} = \sin\phi\left(1 + \frac{d\phi}{d\theta}\right)$$

$$\Rightarrow \quad \frac{r}{\rho} = \sin\phi\left(1 + \frac{d\phi}{d\theta}\right).$$

EXERCISE 6 (B)

1. Find the radius of curvature at the point (p, r) on each of the following curves —

   (i) $pr = a^2$ (Hyperbola) (ii) $r^3 = a^2 p$ (Lemniscate)

   (iii) $p^3 = ar$ (Parabola) (iv) $ap = r^2$

   (v) $p^2 = \dfrac{r^4}{r^2 + a^2}$ (Archimedian spiral).

2. In the curve $a^m p = r^{m+1}$, show that radius of curvature varies inversely as the $(m-1)$th power of the radius vector.

3. Find the radius of curvature at the point (r, θ) on each of the following curves —

   (i) $r = a(1 + \cos\theta)$ (ii) $r(1 + \cos\theta) = a$

   (iii) $r^m = a^m \cos m\theta$ (iv) $r^2 \cos 2\theta = a^2$

   (v) $r^n = a^n \sin n\theta$ (vi) $\dfrac{1}{r^2} = \dfrac{\cos^2\theta}{a^2} + \dfrac{\sin^2\theta}{b^2}$.

4. Find the radius of curvature of the curve $\dfrac{2a}{r} = 1 + \cos\theta$; hence show that the square of the radius of curvature varies as the cube of the focal distance.

5. If ρ_1, ρ_2 be the radii of curvature at the extremities of any chord through the pole of $r = a(1 + \cos\theta)$, prove that

$$9(\rho_1^2 - \rho_2^2) = 16a^2.$$

6. Show that at the points in which the Archimedian spiral $r = a\theta$ intersects the reciprocal spiral $r\theta = a$, their curvature are in the ratio 3 : 1.

7. Prove that for any curve

$$\frac{d^2r}{ds^2} = \frac{\sin^2\phi}{r} - \frac{\sin\phi}{\rho}.$$

8. If the polar equation of a curve be $r = f(\theta)$ and if $u = \dfrac{1}{r}$, prove that the curvature is given by

$$\left(\frac{d^2u}{d\theta^2} + u\right)\sin^3\phi,$$

where, θ is the angle between the radius vector and the tangent at the point (r, θ).

9. Prove that $\rho = \dfrac{r\,(d\theta/ds)}{r\left(\dfrac{d\theta}{ds}\right)^2 - \dfrac{d^2r}{ds^2}}$.

10. Show that for the hypocycloid $p = a \sin b\Psi$, the radius of curvature is proportional to p.

11. Show that the radius of curvature for the curve

$$p^2 = a^2 \cos^2 \Psi + b^2 \sin^2 \Psi \text{ is } \frac{a^2b^2}{p^3}.$$

12. Find the intrinsic equation of the cardioid $r = a\,(1 + \cos \theta)$ the hence or otherwise prove that $s^2 + 9\rho^2 = 16a^2$, where ρ is the radius of curvature at any point and s is the length of the arc intercepted between the vertex and the points.

ANSWERS

1. (i) $\dfrac{r^3}{a^2}$ (ii) $\dfrac{a^2}{3r}$ (iii) $\dfrac{2r^{3/2}}{\sqrt{a}}$ (iv) $\dfrac{a}{2}$

(v) $\dfrac{(r^2+a^2)^{3/2}}{r^2+2a^2}$

3. (i) $\dfrac{2}{3}\sqrt{2ar}$ (ii) $\dfrac{\sqrt{8r^3}}{\sqrt{a}}$ (iii) $\dfrac{a^m r^{-m+1}}{m+1}$ (iv) $\dfrac{r^3}{a^2}$

(v) $\dfrac{a^n}{(n+1)\,r^{n-1}}$ (vi) $\dfrac{(b^4 \cos^2 \theta + a^4 \sin^2 \theta)^{3/2}}{u^3a^4b^4}$, where $u = \dfrac{1}{r}$

4. (i) $\sqrt{\dfrac{4r^3}{a}}$ **12.** $S = 4a \sin \dfrac{\theta}{2}$.

6.14. Radius of Curvature at the Origin

The following methods are used to find the radius of curvature at the origin.

(i) Method of substitution : We know that cartesian formula for radius of curvature is,

$$\rho = \frac{\{1 + (dy/dx)^2\}^{3/2}}{d^2y/dx^2} \qquad \text{.... (A)}$$

We find the numerical values of $\dfrac{dy}{dx}$ and $\dfrac{d^2y}{dx^2}$ at origin (0, 0) by putting $x = 0$ and $y = 0$ in their values. Then using formula (A), we find the value of radius of curvature at (0, 0).

(ii) Method of expansion : Let the equation of the curve by $y = f(x)$.
Since it passes through the origin (0, 0) $\therefore f(0) = 0$.
By Maclaurin's expansion

$$y = f(0) + xf'(0) + \frac{x^2}{2!}f''(0) + ...$$

$$\Rightarrow \qquad y = xf'(0) + \frac{x^2}{2!}f''(0) + ... \qquad .... (1)$$

$[\because f(0) = 0]$

Hence, if y can be expanded in ascending powers of x by trigonometrical or algebraical methods, then we have

$$y = px + \frac{qx^2}{2!} + ... \qquad .... (2)$$

Comparing (1) and (2), we get

$$p = f'(0) = (y_1)_0;$$

$$q = f'(0) = (y_2)_0.$$

$\therefore$ Using formula (A) above, we get

$$\rho_{\text{(at the origin)}} = \frac{(1+p^2)^{3/2}}{q}.$$

(iii) Newtonian method : If the curve passes through the origin, *i.e.*, (0, 0) and if X-axis is tangent to the curve at origin, then at $x = 0$, $y = 0$; $dy/dx = 0$.
Hence, in this case Maclaurin's expansion becomes,

$$y = 0 + 0 \cdot x + q \cdot \frac{x^2}{2!} + ...$$

Dividing by $x^2/2$ and taking the limit as $x \to 0$, we get

$$\lim_{x \to 0}\left(\frac{2y}{x^2}\right) = q;$$

[All other terms vanish in the limit since they contain x]

Now,
$$\rho_{\text{(at the origin)}} = \frac{(1+p^2)^{3/2}}{q} = \frac{1}{q}, \text{ as } p = 0$$

$$= \lim_{x \to 0}\left(\frac{x^2}{2y}\right).$$

Similarly, if the curve passes through the origin and Y-axis is the tangent at the origin, then

$$\rho_{\text{(at the origin)}} = \lim_{x \to 0}\left(\frac{y^2}{2x}\right)$$

[On interchanging x and y in the above result]

(iv) Radius of curvature at origin for polar curves : If $r=0$ when $\theta \to 0$, then initial line is a tangent at the pole and so

$$\begin{aligned}
\rho_{(\text{at the pole})} &= \lim_{x \to 0} \left(\frac{x^2}{2y}\right) \\
&= \lim_{\theta \to 0} \left(\frac{r^2 \cos^2 \theta}{2r \sin \theta}\right) = \lim_{\theta \to 0} \left(\frac{r \cos^2 \theta}{2 \sin \theta}\right) \\
&= \lim_{\theta \to 0} \left(\frac{r}{2\theta} \cdot \frac{\theta}{\sin \theta} \cdot \cos^2 \theta\right) \\
&= \lim_{\theta \to 0} \left(\frac{r}{2\theta}\right) \qquad \left[\because \lim_{\theta \to 0} \frac{\theta}{\sin \theta} = 1\right. \\
&\qquad\qquad \text{and } \lim_{\theta \to 0} \cos \theta = 1
\end{aligned}$$

If $r=0$ when $\theta = \frac{\pi}{2}$, then X-axis is the tangent to the curve at pole and so,

$$\begin{aligned}
\rho_{(\text{at the pole})} &= \lim_{x \to 0} \left(\frac{y^2}{2x}\right) \\
&= \lim_{\theta \to 0} \left(\frac{r^2 \sin^2 \theta}{2r \cos \theta}\right) = \lim_{\theta \to 0} \left(\frac{r \sin^2 \theta}{2 \cos \theta}\right) \\
&= \lim_{\theta \to 0} \left[\frac{r\theta^2}{2} \cdot \left(\frac{\sin \theta}{\theta}\right)^2 \cdot \frac{1}{\cos \theta}\right] \\
&= \lim_{\theta \to 0} \left(\frac{r\theta^2}{2}\right) \qquad \left[\because \lim_{\theta \to 0} \frac{\sin \theta}{\theta} = 1\right. \\
&\qquad\qquad \text{and } \lim_{\theta \to 0} \cos \theta = 1
\end{aligned}$$

6.15. Tangent at the Origin

The following proposition is very helpful in finding the equation(s) of tangent or tangents at the origin of the rational, integral algebraic curves.

Proposition : *"If a curve passes through the origin and its equation is given by a rational, integral algebraic function equated to zero, then the equation(s) of the tangent or tangents at the origin is (are) obtained by equating to zero the lowest degree terms in the equation of the curve."*

ILLUSTRATIVE EXAMPLES

Example 1. *Show that the radius of curvature to the curve*

$$y = 6x + 5x^2 + x^3$$

at origin is $\dfrac{37\sqrt{37}}{10}$.

Solution : $$y = 6x + 5x^2 + x^3. \quad \text{.... (1)}$$

Differentiating (1) w.r.t. x, we get

$$\frac{dy}{dx} = 6 + 10x + 3x^2 \quad \text{....(2)}$$

$$\therefore \quad \frac{dy}{dx}_{\text{(at origin)}} = 6 = (y_1)_0.$$

Differentiating (2) w.r.t. x, we get

$$\frac{d^2y}{dx^2} = 10 + 6x$$

$$\therefore \quad \frac{d^2y}{dx^2}_{\text{(at origin)}} = 10 = (y_2)_0$$

$$\therefore \quad \rho_{\text{(at origin)}} = \frac{\{1 + (y_1)_0^2\}^{3/2}}{(y_2)_0}$$

$$= \frac{\{1 + (6)^2\}^{3/2}}{10} = \frac{37\sqrt{37}}{10}.$$

Example 2. *Show that the radius of curvature of the curve*

$$y^2 = x^2\left(\frac{a+x}{a-x}\right)$$

at the origin are $\pm a\sqrt{2}$.

Solution : We have, $$y^2 = x^2\left(\frac{a+x}{a-x}\right)$$

$$\therefore \quad y = \pm x\sqrt{\frac{a+x}{a-x}}$$

$$= \pm x(a+x)^{1/2}(a-x)^{-1/2}$$

$$= \pm x\left(1 + \frac{x}{a}\right)^{1/2}\left(1 - \frac{x}{a}\right)^{-1/2}$$

$$= \pm x\left(1 + \frac{1}{2}\cdot\frac{x}{a} - \frac{1}{8}\cdot\frac{x^2}{a^2} + \ldots\right)\left(1 + \frac{1}{2}\cdot\frac{x}{a} + \frac{3}{8}\cdot\frac{x^2}{a^2} + \ldots\right)$$

$$= \pm x\left(1 + \frac{x}{a} + \ldots\right)$$

$$= \pm\left(x + \frac{x^2}{a} + \ldots\right)$$

which is of the form $$y = px + \frac{qx^2}{2} + \dots$$

Comparing, we have $$p = \pm 1 = (y_1)_0$$

and $$q = \pm\frac{2}{a} = (y_2)_0$$

$\therefore$ $$\rho_{\text{(at origin)}} = \frac{(1+p^2)^{3/2}}{q}$$

$$= \frac{(1+1)^{3/2}}{\pm 2/a} = \pm a\sqrt{2}.$$

Example 3. *Find the radius of curvature at origin for the curve*

$$5x^3 + 7y^3 + 4x^2y + xy^2 + 2x^2 + 3xy + y^2 + 4x = 0.$$

Solution : Equating to zero the terms of lowest degree we get equation to tangent at origin, that is $x = 0$ or Y-axis.

Now, dividing the equation by $2x$, we get

$$\frac{5}{2}x^2 + 6y\cdot\frac{y^2}{2x} + 2xy + \frac{y^2}{2} + x + \frac{3}{2}y + \frac{y^2}{2x} + 2 = 0.$$

If $x \to 0$, then at origin, $y \to 0$. Therefore taking limit

$$\lim_{x\to 0}\left(\frac{y^2}{2x}\right) + 2 = 0$$

$$\rho = \lim_{x\to 0}\left(\frac{y^2}{2x}\right) = 2 \quad \text{(Numerical value).}$$

Example 4. *Find the radius of curvature of the curve* $r = a\sin n\theta$ *at the pole.*

Solution : We see that when $\theta = 0$, $r = 0$ in the equation of the curve

$$r = a\sin n\theta.$$

$\therefore$ Initial line is the tangent at the pole.

Therefore $$\theta_{\text{(at pole)}} = \lim_{\theta\to 0}\left(\frac{r}{2\theta}\right) = \lim_{\theta\to 0}\left(\frac{a\sin n\theta}{2\theta}\right)$$

$$= \lim_{\theta\to 0}\left(\frac{a}{2}\frac{\sin n\theta}{n\theta} - n\right) = \frac{na}{2}.$$

$$\left[\because \text{ When } \theta \to 0, \text{ then } n\theta \to 0 \text{ and hence } \lim_{n\theta\to 0}\frac{\sin n\theta}{n\theta} = 1\right]$$

EXERCISE 6(C)

1. Find the radius of curvature at origin for the curve

$$y = x^4 - 4x^3 - 18x^2.$$

2. Find the radii of curvature at the origin for the curve

$$y^2 - 3xy + 2x^2 - x^3 + y^4 = 0.$$

3. Find the radius of curvature at the origin of the curve

$$y - x = x^2 + 2xy + y^2.$$

4. Obtain the radii of curvature for the curve $(y^2 - x^2) = x^3$ at the origin.

5. Find the radius of curvature at origin for the curve

$$x^3 + y^2 - 2x^2 + 6y = 0.$$

6. Find the radius of curvature at (r, θ) for the curve $r\theta^2 = a$.

7. Apply Newton's method to find the radius of curvature at the origin for the following curve —

(i) $x = a(\theta + \sin\theta), y = \theta(1 - \cos\theta)$, (ii) $x = 1 - t^2, y = t - t^3$.

ANSWERS

1. $-\frac{1}{36}$	2. $\frac{5\sqrt{3}}{2}, \frac{3}{2}$	3. $\frac{1}{2\sqrt{2}}$
4. $\pm\sqrt{2}\,a$	5. $\frac{3}{2}$	6. $\frac{a(\theta^2+4)^{3/2}}{\theta^3(\theta^2+2)}$
7. (i) $4a$;	(ii) $2\sqrt{2}$, for both the branches.	

6.16. Centre of Curvature

To find the coordinates of the centre of curvature for any point (x, y) of the curve $y = f(x)$.

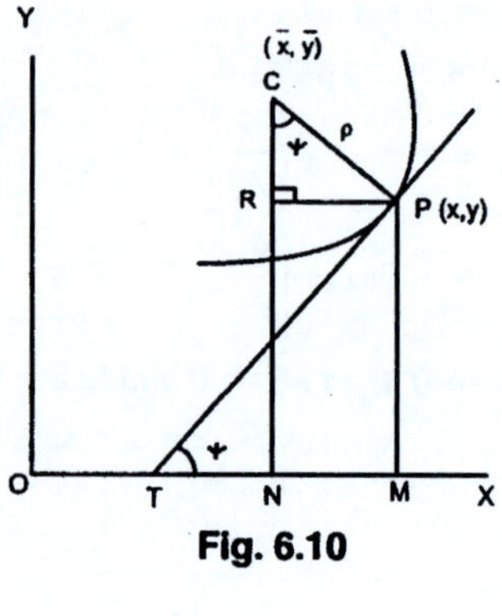

Fig. 6.10

Let $C(\bar{x}, \bar{y})$ be the centre of curvature corresponding to the point $P(x, y)$ on the curve. Then length $CP = \rho$.

Let the tangent PT at P makes an angle Ψ with the +ve direction of X-axis. Draw PM and CN perpendiculars on X-axis and $PR \perp CN$. Then,

$$\begin{aligned} \angle PCN &= 90° - \angle CPR \\ &= 90° - (90° - \angle RPT) \\ &= \angle RPT \\ &= \angle PTX = \Psi \end{aligned}$$

$$\therefore \quad \begin{aligned} \bar{x} &= ON = OM - NM \\ &= OM - RP = x - \rho \sin \Psi \\ \bar{y} &= NC = NR + RC = y + \rho \cos \Psi. \end{aligned}$$

Now, $$\tan \Psi = \frac{dy}{dx} = y_1$$

$\therefore$ $$\sin \Psi = \frac{y_1}{\sqrt{1 + y_1^2}}$$

and $$\cos \Psi = \frac{1}{\sqrt{1 + y_1^2}}.$$

Also, $$\rho = \frac{(1 + y_1^2)^{3/2}}{y_2}$$

$\therefore$ $$\bar{x} = x - \frac{(1 + y_1^2)^{3/2}}{y_2} \cdot \frac{y_1}{\sqrt{1 + y_1^2}}$$

i.e., $$\bar{x} = x - \frac{y_1 (1 + y_1^2)}{y_2}.$$

$$\bar{y} = y + \frac{(1 + y_1^2)^{3/2}}{y_2} \cdot \frac{1}{\sqrt{1 + y_1^2}}$$

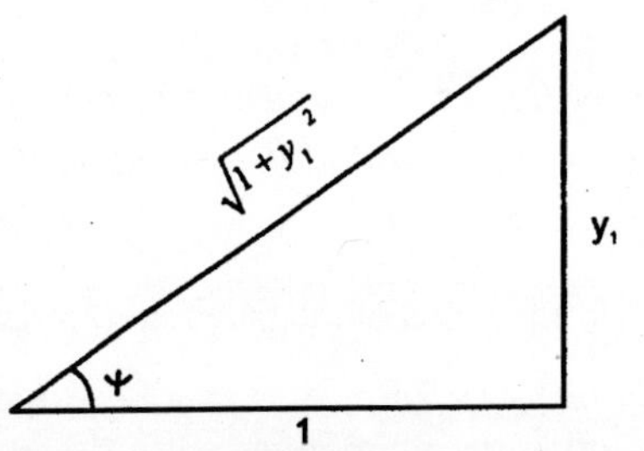

Fig. 6.11

i.e., $$\bar{y} = y + \frac{1+y_1^2}{y_2}.$$

Notes :

1. **Circle of curvature :** Its equation is given by
$$(x-\bar{x})^2 + (y-\bar{y})^2 = \rho^2.$$
2. **Evolute :** The locus of the centre of curvature is called the evolute of the curve and the curve itself is called the involute of its evolute.

6.17. To Find the Evolute of a Curve

(i) The centre of curvature $(\bar{x}, \bar{y})$ is given by
$$\bar{x} = x - \frac{y_1(1+y_1^2)}{y_2},\ \bar{y} = y + \frac{1+y_1^2}{y_2}.$$

(ii) Eliminate the parameters x and y to find a relation between $\bar{x}$ and $\bar{y}$.

(iii) Generalizing $\bar{x}$ and $\bar{y}$, we get the equation of the evolute.

ILLUSTRATIVE EXAMPLES

Example 1. *Prove that the coordinates of the centre of curvature at any point (x, y) on the curve $y = f(x)$ can be expressed in the form*
$$\left(x - \frac{dy}{d\Psi}, y + \frac{dx}{d\Psi}\right).$$

Solution: Let $(\bar{x}, \bar{y})$ be the coordinates of centre of curvature. Then
$$\bar{x} = x - \rho \sin \Psi$$
$$\Rightarrow \qquad \bar{x} = x - \frac{ds}{d\Psi} \cdot \frac{dy}{ds} = x - \frac{dy}{d\Psi}$$
and
$$\bar{y} = y + \rho \cos \Psi$$
$$= y + \frac{ds}{d\Psi} \cdot \frac{dx}{ds} = y + \frac{dx}{d\Psi}.$$

Example 2. *Find the coordinates of the centre of curvature for any point (x, y) on the parabola $y^2 = 4ax$. Also find the equation of the evolute of the parabola.*

Solution : The parametric equations of the parabola $y^2 = 4ax$ are
$$x = at^2,\ y = 2at$$
$$\frac{dx}{dt} = 2t,\ \frac{dy}{dt} = 2a$$

$$\therefore \qquad y_1 = \frac{dy}{dx} = \frac{dy/dt}{dx/dt} = \frac{2a}{2at} = \frac{1}{t}$$

$$y_2 = \frac{1}{t^2}\frac{dt}{dx} = -\frac{1/t^2}{dx/dt}$$

$$= -\frac{1/t^2}{2at} = \frac{1}{2at^3}.$$

Let $(\bar{x}, \bar{y})$ be the coordinates of the centre of curvature at 't', then,

$$\bar{x} = x - \frac{y_1(1+y_1^2)}{y_2}$$

$$= at^2 - \frac{\frac{1}{t}\left(1+\frac{1}{t^2}\right)}{-\frac{1}{2at^3}}$$

$$= at^2 + 2at^2\left(1+\frac{1}{t^2}\right)$$

$$= 3at^2 + 2a \qquad \text{.... (1)}$$

and

$$\bar{y} = y + \frac{1+y_1^2}{y_2}$$

$$= 2at + \frac{1+\frac{1}{t^3}}{-\frac{1}{2at^3}} = 2at - 2at^3\left(1+\frac{1}{t^2}\right)$$

$$= 2at - 2at^3 - 2at$$

$$= -2at^3. \qquad \text{.... (2)}$$

$\therefore$ The coordinates of the centre of curvature at any point (x, y), *i.e.*, at any point 't' are $(2a + 3at^2, -2at^3)$.

Now, evolute of a curve is the locus of its centre of curvature.

To eliminate the parameter t between (i) and (ii) :

From (1), $$t^2 = \frac{\bar{x} - 2a}{3a} \qquad \text{.... (3)}$$

From (2) $$t^3 = -\frac{\bar{y}}{2a} \qquad \text{... (4)}$$

Cubing (3), squaring (4) and equating the two values of t^3, we get

$$\left(\frac{\bar{x}-2a}{3a}\right)^3 = \left(-\frac{\bar{y}}{2a}\right)^2$$

$$\Rightarrow \qquad \frac{(\bar{x}-2a)^3}{27a^3} = \frac{\bar{y}^2}{4a^2}$$

$$\Rightarrow \qquad \frac{(\bar{x}-2a)}{27a} = \frac{\bar{y}^2}{4}$$

$$\Rightarrow \qquad 27\, a\bar{y}^2 = 4\,(\bar{x}-2a)^3.$$

$\therefore$ Locus of $(\bar{x}, \bar{y})$, *i.e.*, the evolute of the parabola is,

$$27ay^2 = 4\,(x-2a)^3.$$

Example 3. *Show that the equation of the circle of curvature at the point* $\left(\frac{a}{4}, \frac{a}{4}\right)$ *on the curve* $\sqrt{x}+\sqrt{y}=\sqrt{a}$ *is*

$$\left(x-\frac{3a}{4}\right)^2 + \left(y-\frac{3a}{4}\right)^2 = \frac{a^2}{2}.$$

Solution : We have, $\sqrt{x}+\sqrt{y} = \sqrt{a}.$ .... (1)

Differentiating w.r.t x, we get

$$\frac{1}{2}x^{-1/2} + \frac{1}{2}y^{-1/2}\frac{dy}{dx} = 0. \qquad \text{.... (2)}$$

Differentiating again w.r.t x, we get

$$-\frac{1}{4}x^{-3/2} - \frac{1}{4}y^{-3/2}\frac{dy}{dx}\cdot\frac{dy}{dx} + \frac{1}{2}y^{-1/2}\cdot\frac{d^2y}{dx^2} = 0. \qquad \text{.... (3)}$$

At the point $\left(\frac{a}{4}, \frac{a}{4}\right)$, from (2),

$$\frac{1}{2}\cdot\frac{2}{\sqrt{2}} + \frac{1}{2}\cdot\frac{2}{\sqrt{a}}\cdot y_1 = 0$$

$$\therefore \qquad y_1 = -1$$

and from (3),

$$-\frac{1}{4}\cdot\frac{4}{a}\cdot\frac{2}{\sqrt{a}} - \frac{1}{4}\cdot\frac{4}{a}\cdot\frac{2}{\sqrt{a}}\cdot(-1)^2 + \frac{1}{2}\cdot\frac{2}{\sqrt{a}}\cdot y_2 = 0$$

$$\Rightarrow \qquad -\frac{4}{a\sqrt{a}} + \frac{1}{\sqrt{a}}y_2 = 0$$

$$\Rightarrow \qquad y_2 = \frac{4}{a}.$$

$$\rho_{\text{(at the given point)}} = \frac{(1+y_1^2)^{3/2}}{y_2}$$

$$= \frac{(1+1)^{3/2}}{\frac{4}{a}} = 2\sqrt{2}\,\frac{a}{4} = \frac{a}{\sqrt{2}}.$$

Let (α, β) be the centre of curvature at $\left(\frac{a}{4}, \frac{a}{4}\right)$, then

$$\alpha = x - \frac{y_1(1+y_1^2)}{y_2}$$

$$= \frac{a}{4} - \frac{(-1)(1+1)}{4/a} = \frac{a}{4} + \frac{a}{2} = \frac{3a}{4}$$

$$\beta = y + \frac{1+y_1^2}{y_2}$$

$$= \frac{a}{4} + \frac{1+1}{4/a} = \frac{a}{4} + \frac{a}{2} = \frac{3a}{4}.$$

Equation of circle of curvature is,

$$(x-\alpha)^2 + (y-\beta)^2 = (a/\sqrt{2})^2$$

$$\Rightarrow \qquad \left(x - \frac{3a}{4}\right)^2 + \left(y - \frac{3a}{4}\right)^2 = \frac{a^2}{2}.$$

EXERCISE 6(D)

1. Find the coordinates of the centre of curvature of the curve
$$x^2 = 4ay.$$
2. Find the coordinates of the centre of curvature of the semi-cubical parabola $\qquad a^2y = x^2.$
3. Find the coordinates of the centre of curvature of the curve
$$y = c\cosh(x/c).$$
4. Find the coordinates of the centre of curvature of the curve
$$y^3 = a^2x.$$
5. Find the coordinates of the centre of curvature of the curve $xy = c^2$ and deduce the equation of its evolute.
6. Find the evolute of the curve

7. If $(\bar{x}, \bar{y})$ be the coordinates of the centre of curvature of the parabola $\sqrt{x} + \sqrt{y} = \sqrt{a}$, then prove that
$$\sqrt{x} + \sqrt{y} = 3(x+y).$$

8. Show that the evolute of the catenary $y = c \cosh (x/c)$ is
$$y = 2c \cosh \left\{\frac{4cx + y\sqrt{y^2 - 4c^2}}{4c^2}\right\}.$$

9. Find the equation of the circle of curvature at the point (0, 1) of the curve $\quad y = x^3 + 2x^2 + x + 1.$

10. Show that the circle of curvature at the origin of the parabola
$$y = mx + \frac{x^2}{a} \text{ is,}$$
$$x^2 + y^2 = a(1 + m^2)(y - mx).$$

ANSWERS

1. $\left(-\frac{x^3}{4a^2}, 2a + \frac{3x^2}{4a}\right).$

2. $\left\{\frac{x}{2}\left(1 - \frac{9x^4}{a^4}\right), \frac{5x^3}{2a^2} + \frac{a^2}{6x}\right\}.$

3. $\left\{x - y\sqrt{\frac{y^2}{c^2} - 1}, 2y\right\}.$

4. $\left(\frac{a^4 + 15y^4}{6a^2y}, \frac{a^4y - 9y^5}{2a^4}\right).$

5. $\left(\frac{3x^2 + y^2}{2x}, \frac{3y^2 + x^2}{y}\right).$

6. $(x+y)^{2/3} + (x-y)^{2/3} = a^{2/3}.$

9. $x^2 + y^2 + x - 3y - 2 = 0.$

6.18. Chord of Curvature

Let PQ be any chord through P making an angle α with CP, the radius of curvature at P. Then from right angle triangle PQD, chord of curvature $PQ = PD \cos \alpha = 2\rho \cos \alpha$.

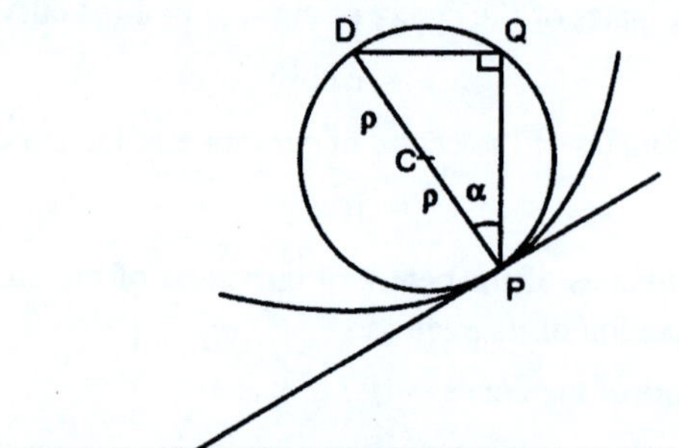

Fig. 6.12

6.19. Chord of Curvature Parallel to the Axes

Let the tangent at $P(x, y)$ makes an angle Ψ with the X-axis. Then chord of curvature PA parallel to X-axis makes angle $90° - \Psi$ with CP, the radius of curvature at P and the chord of curvature PB, parallel to Y-axis, make angle Ψ with CP.

Now, the length of the chord of curvature PA, parallel to the X-axis

$$= 2\rho \cos(90° - \Psi) = 2\rho \sin \Psi$$

and, length of the chord of curvature PB, parallel to the Y-axis

$$= 2\rho \cos \Psi$$

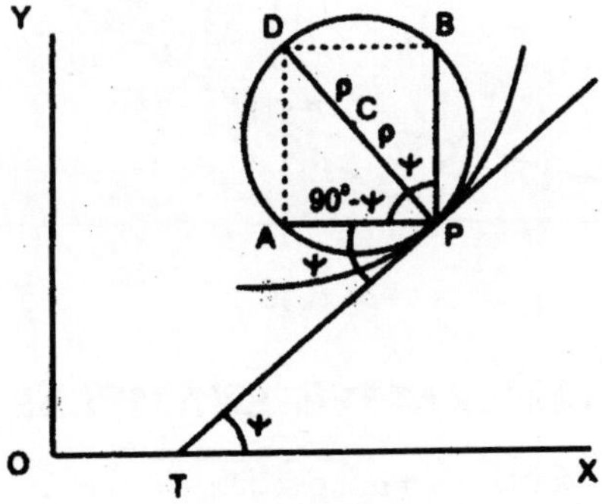

Fig. 6.13

6.20. Chord of Curvature through Pole (Origin)

Let the tangent at P makes an angle ϕ with the radius vector OP. PQ is the chord of curvature through pole O. It makes an angle $90° - \phi$ with CP, the radius of curvature at P.

$\therefore$ Length of the chord of curvature PQ, through pole $= 2\rho \sin \phi$.

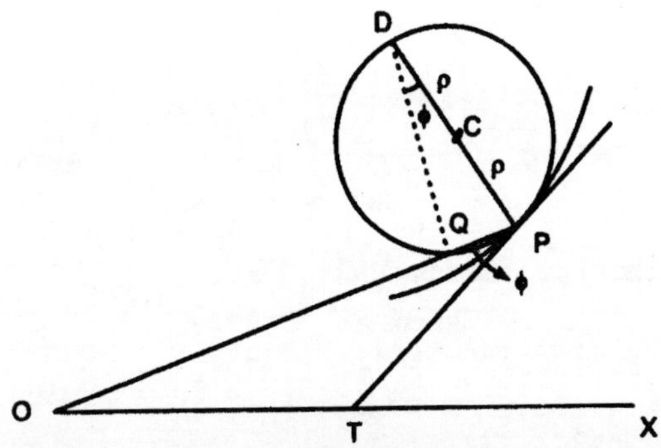

Fig. 6.14

6.21. Chord of Curvature Perpendicular to the Radius Vector

The chord of curvature perpendicular to the radius vector is PR. It makes an angle ϕ with CP, the radius of curvature at P.

$\therefore$ The length of $PR = 2\rho \cos \phi$.

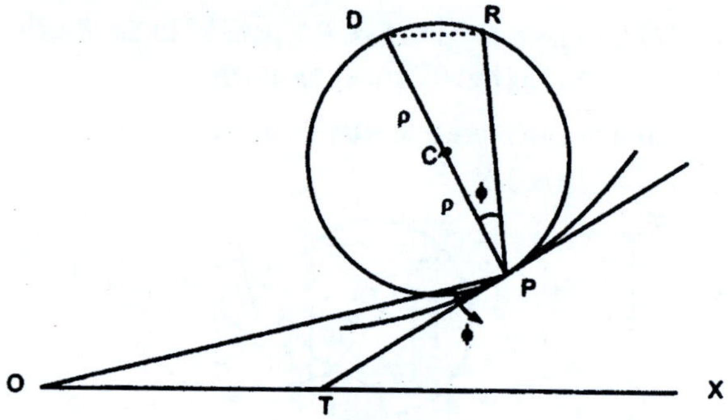

Fig. 6.15

ILLUSTRATIVE EXAMPLES

Example 1. *For the curve* $y = a \log \sec (x/a)$ *the length of chord of curvature parallel to Y-axis is constant.*

Solution : Equation to the curve

$$y = a \log \sec (x/a)$$

$$\frac{dy}{dx} = \frac{a}{\sec (x/a)} \cdot \sec \frac{x}{a} \tan \frac{x}{a} \times \frac{1}{a} = \tan \frac{x}{a}$$

$$\frac{d^2y}{dx^2} = \sec^2 \frac{x}{a} \cdot \frac{1}{a} = \frac{1}{a} \sec^2 \frac{x}{a}$$

$$\therefore \qquad \rho = \frac{\left(1 + \tan^2 \frac{x}{a}\right)^{3/2}}{\frac{1}{a} \sec^2 \frac{x}{a}} = \frac{a \sec^3 \frac{x}{a}}{\sec^2 \frac{x}{a}} = \sec \frac{x}{a}.$$

The length of chord curvature parallel to Y-axis

$$= 2\rho \cos x$$

$$= 2a \sec \frac{x}{a} \cdot \cos \frac{x}{a} = 2a = \text{constant.}$$

Example 2. *If c_x and c_y be the chord of curvature parallel to the axis of X and Y respectively at any point of the curve $y = ae^{x/a}$, prove that*

$$\frac{1}{c_x^2} + \frac{1}{c_y^2} = \frac{1}{2ac_x}.$$

Solution : $y = ae^{x/a}$

$\therefore \quad y_1 = ae^{x/a} \cdot \frac{1}{a} = e^{x/a} = \frac{y}{a}$

$\therefore \quad y_2 = \frac{y_1}{a} = \frac{y}{a^2}$

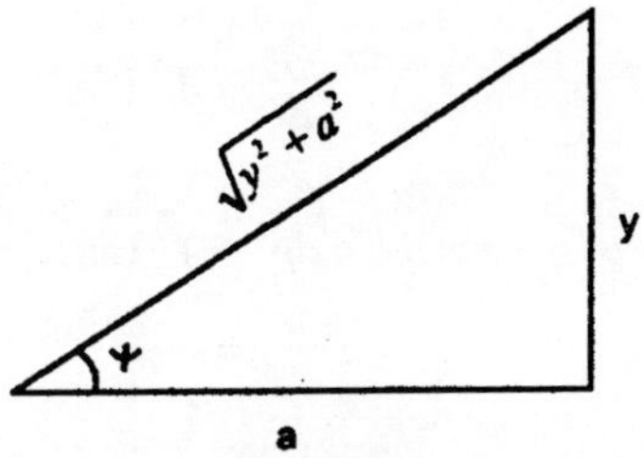

Fig. 6.16

$$\rho = \frac{(1+y_1)^{3/2}}{y_2}$$

$$= \frac{\left(1+\frac{y^2}{a^2}\right)^{3/2}}{y/a^2}$$

$$= \frac{(a^2+y^2)^{3/2}}{ay}$$

$$\tan\Psi = y_1 = \frac{y}{a}$$

$\therefore \quad \sin\Psi = \frac{y}{\sqrt{y^2+a^2}}$

and $\quad \cos\Psi = \frac{a}{\sqrt{y^2+a^2}}$

$$c_x = 2\rho \sin \Psi$$

$$= 2\frac{(a^2+y^2)^{3/2}}{ay} \cdot \frac{y}{\sqrt{y^2+a^2}} = \frac{2 \cdot (a^2+y^2)}{a}.$$

$$c_y = 2\rho \cos \Psi$$

$$= 2 \cdot \frac{(a^2+y^2)^{3/2}}{ay} \cdot \frac{a}{\sqrt{y^2+a^2}} = \cdot \frac{2(a^2+y^2)}{a}.$$

$\therefore \quad \frac{1}{c_x^2} + \frac{1}{c_y^2} = \frac{a^2}{4(a^2+y^2)^2} + \frac{y^2}{4(a^2+y^2)^2}$

$$= \frac{a^2+y^2}{4(a^2+y^2)^2} = \frac{1}{4(a^2+y^2)}$$

$$= \frac{1}{2a.c_x}. \qquad \left[\because c_x = \frac{2(a^2+y^2)}{a}\right]$$

Example 3. *Show that the chord of curvature through the pole for curve*

$$p = f(r) \text{ is } \frac{2f(r)}{f'(r)}.$$

Solution : $\therefore \quad p = f(r)$

$$\therefore \quad \frac{dp}{dr} = f'(r)$$

$$\therefore \quad \frac{dr}{dp} = \frac{1}{f'(r)}$$

$$\therefore \quad r\frac{dr}{dp} = \frac{r}{f'(r)}$$

$$\therefore \quad \rho = \frac{r}{f'(r)}$$

$$\because \quad p = r\sin\phi \qquad \therefore \ \phi = \frac{p}{r}.$$

Now, chord of curvature through the pole = $2\rho \sin\phi$

$$= 2\frac{r}{f'(r)}\frac{p}{r} = 2\frac{r}{f'(r)}\frac{f(r)}{r} = 2\frac{f(r)}{f'(r)}.$$

Example 4. *Find the length of chord of curvature passing through pole for the curve* $r^n = a^n \cos n\theta$.

Solution : The equation to the curve is

$$r^n = a^n \cos n\theta \qquad \text{.... (1)}$$

Taking log of both sides,

$$n\log r = n\log a + \log\cos n\theta.$$

Differentiating it w.r.t. θ,

$$\frac{n}{r}\frac{dr}{d\theta} = 0 + \frac{-\sin n\theta}{\cos n\theta}\cdot n$$

$$\Rightarrow \quad \frac{1}{r}\frac{dr}{d\theta} = -\tan n\theta$$

$$\Rightarrow \quad r\frac{d\theta}{dr} = -\cot n\theta$$

$$\Rightarrow \quad \tan\phi = \tan\left(\frac{\pi}{2} + n\theta\right)$$

$$\phi = \frac{\pi}{2} + n\theta$$

$$\sin \phi = \sin\left(\frac{\pi}{2} + n\theta\right) = \cos n\theta$$

$$\sin \phi = \frac{r^n}{a^n}. \qquad \text{[Using (1)]}$$

The pedal equation to (1) is $p = r \sin \theta$

$$p = r \cdot \frac{r^n}{a^n} = \frac{r^{n+1}}{a^n}$$

$$a^n p = r^{n+1}.$$

Differentiating w.r.t. r,

$$a^n \frac{dp}{dr} = (n+1)\, r^n$$

$$\Rightarrow \qquad \rho = r\frac{dr}{dp} = r\frac{a^n}{(n+1)\, r^n} = \frac{a^n}{(n+1)\, r^{n-1}}.$$

The length of chord of curvature passing through origin

$$= 2\rho \sin \theta = 2 \cdot \frac{a^n}{(n+1)\, r^{n-1}} \cos n\theta$$

$$= \frac{2a^n}{(n+1)\, r^{n-1}} \times \frac{r^n}{a^n} \qquad \text{[Using (i)]}$$

$$= \frac{2r}{n+1}.$$

EXERCISE 6(E)

1. Show that the length of chord of curvature parallel to Y-axis at the origin of the parabola $y = mx + \frac{x^2}{a}$ is $(1 + m^2)\, a$.

2. If c_x, c_y be the chords of curvature parallel to the axes of X and Y respectively at any point of the curve $y = c \cosh (x/c)$, prove that
$$4c^2 (c_x^2 + c_y^2) = c_y^4.$$

3. Show that the chord of curvature parallel to the axis of Y for the curve $y = c \cosh (x/c)$ is double the ordinate whereas the chord parallel to X-axis is of the length $c \sinh (2x/c)$.

4. Find the chord of curvature through the pole of the cardioid
$$r = a(1 - \cos \theta).$$

5. Find the chord of curvature through the pole of the curve $r^2 \cos 2\theta = a^2$.

6. Find the chord of curvature through the pole of the curve $r^2 = a^2 \cos 2\theta$.

7. Find the chord of curvature through the pole of the curve $r^n = a^n \sin n\theta$.

8. Show that the chord of curvature through the pole of the equiangular spiral $r = ae^{\theta \cot \alpha}$ is $2r$.

9. Show that in a parabola the chord of curvature through the focus of the parabola is four times the focal distance of the point, and the chord of curvature parallel to the axis has the same length.

ANSWERS

4. $\frac{4}{3}r$ 5. $2r$ 6. $\frac{2r}{3}$ 7. $\frac{2r}{n+1}$

OBJECTIVE EXERCISE

Write the correct answers.

1. The cartesian formula for curvature is —

(a) $\frac{(1+y_1^2)^{3/2}}{y_2}$ (b) $\frac{(1+y_1^2)^{3/2}}{y_1}$

(c) $\frac{y_1}{(1+y_1^2)^{3/2}}$ (d) $\frac{y_2}{(1+y_1^2)^{3/2}}$

2. The radius of curvature at the point (2, 2) for the curve $xy = 4$ is —

(a) $\sqrt{2}$ (b) 2

(c) $1/\sqrt{2}$ (d) $2/\sqrt{2}$

3. The centre of curvature at (1, 2) for the curve $y^2 = 4x$ is —

(a) (2, 5) (b) (5, −2)

(c) (2, −5) (d) (−5, −2).

4. If $s = c \sec \Psi$ then ρ is —

(a) $c \sec \Psi$ (b) $c \sec^2 \Psi \tan \Psi$

(c) $c \sec \Psi \tan \Psi$ (d) None.

5. Curvature of circle of radius r is —

(a) 0 (b) r

(c) $1/r$ (d) $2r$.

ANSWERS

1. (d) 2. (d) 3. (b) 4. (c) 5. (c)

7

Envelopes and Evolutes

7.1. Family of Curves

Consider the equation $x^2 + y^2 = k^2$. It is the equation of a circle whose centre is (0, 0) and radius is k. Put $k = 1$, we get, $x^2 + y^2 = 1^2$. Put $k = 2$, we get $x^2 + y^2 = 2^2$. Put $k = 3$, we get $x^2 + y^2 = 3^2, \ldots\ldots$, and so on. Thus we see that for different values of k, we get different circles. Also we see that the value of k is constant for one member but varies from member to member. Such a quantity is known as *Parameter*. Thus *a quantity which is constant for the same curve but different for different curves is called the Parameter*. The equation $x^2 + y^2 = k^2$ which involves a parameter k is called a family of circles. Symbolically a family of curves is written as $f(x, y, \alpha) = 0$, where α is a parameter.

7.2. Envelope

Definition: Envelope is always related to a family of curves and in itself is a curve which touches each member of the family of curves and at each point is touches by some member of the family.

From coordinate geometry, we know that the equation of a tangent line to the circle $x^2 + y^2 = a^2$ in terms of m is $y = mx + a\sqrt{1 + m^2}$ where m is a parameter and a in a constant whatsoever be the value of m, this line always touches the circle $x^2 + y^2 = a^2$. Moreover, the equation of the tangent at any point of the circle $x^2 + y^2 = a^2$ can be put in the form $y = mx + a\sqrt{1 + m^2}$ for a corresponding particular value of m. Thus we conclude that the envelope of the family of the straight lines $y = mx + a\sqrt{1 + m^2}$ is the circle $x^2 + y^2 = a^2$.

7.3. Envelope: Another definition

The envelope of a family of curves is the locus of the limiting position of the point of intersection of any two consecutive members of the family, when one of them tends to conicide with the other which is kept fixed.

Let $f(x, y, C) \leq 0$ be a family of curves, C being the parameter. Let $f(x, y, \alpha) = 0 \ldots (1)$ and

$f(x, y, \alpha + \delta\alpha) = 0 \ldots (2)$ be two consecutive members of the family corresponding to the values α & $\alpha + \delta\alpha$ of the parameter C.

Then the envelope is the locus of the point of intersection of (1) and (2) when $\delta\alpha \to 0$.

$\therefore$ The point of intersection of (1) and (2) satisfies both (1) and (2), therefore it will satisfy the equation

$$f(x, y, \alpha + \delta\alpha) - f(x, y, \alpha) = 0$$

i.e.,
$$\frac{f(x, y, \alpha + \delta\alpha) - f(x, y, \alpha)}{\delta\alpha} = 0$$

Proceeding to the limit as $\delta\alpha \to 0$,

we get
$$\frac{\partial f}{\partial \alpha} = 0 \qquad \ldots (3)$$

Thus the coordinate of the points on the envelope satisfy the equations (1) and (3).

Let the elimination of α between (1) and (3) leads to an equation

$$\phi(x, y) = 0$$

This is then the required envelop.

Remember: The equation of the envelope of the family of curves $f(x, y, \alpha) = 0$ where α is the parameter, is obtained by eliminating α in between the equations

$$f(x, y, \alpha) = 0$$

and
$$\frac{\partial}{\partial \alpha} f(x, y, \alpha) = 0$$

Similarly, the equation of the envelope of the family of curves $f(r, \theta, \alpha) = 0$ where α is the parameter is obtained by eliminating α in between the equations

$$f(r, \theta, \alpha) = 0$$

and
$$\frac{\partial}{\partial \alpha} f(r, \theta, \alpha) = 0$$

7.4. Envelope in Case $f(x, y, \alpha) = 0$ is a quadratic in α

Let the given family of curves be a quadratic in α of the form

$A\alpha^2 + B\alpha + C = 0 \ldots\ldots (1)$ where *A, B, C* are the functions of *x* and *y*. Differentiating (1) partially w.r.t. α, we get

$$2A\alpha + B = 0$$

$$\Rightarrow \alpha = -\frac{B}{2A} \qquad (2)$$

Eliminating α between (1) and (2),

we get,

$$A\left(-\frac{B}{2A}\right)^2 + B\left(-\frac{B}{2A}\right) + C = 0$$

$$\Rightarrow \frac{B^2}{4A} - \frac{B^2}{2A} + C = 0$$

$$\Rightarrow -\frac{B^2}{4A} + C = 0$$

$$\Rightarrow \boxed{B^2 = 4AC}$$

which is the equation of the envelope.

Remember : If the equation of a family of curves be a quadratic in parameter α, then the envelope of the family is

Discriminant = 0

i.e., $B^2 = 4AC$

7.5. Envelop when $f(x, y, \alpha) = 0$ is of the form $A \cos\alpha + B \sin\alpha = C$

Equation of the family of curves is

$$A \cos \alpha + B \sin \alpha = C \qquad \ldots.(1)$$

where α is the parameter and A, B, C are functions of x and y.

Differentiating (1) partially w.r.t. α, we get

$$-A \sin \alpha + B \cos \alpha = 0 \qquad \ldots.(2)$$

To eliminate α between (1) and (2), square and add, we get

$$\boxed{A^2 + B^2 = C^2}$$

which gives the equation of the envelope.

7.6. The Envelope Touches each Member of the Family

Let $\phi(x, y, \alpha) = 0$ .... (1) be the given family of curves, where α is the parameter.

The envelope of (1) in the result of elimination between (1) and $\frac{\partial}{\partial\alpha}\phi(x, y, \alpha) = 0$.... (2)

Solving (1) and (2) for x and y in terms of α, we get

$$x = f(\alpha)$$
$$y = \psi(\alpha) \qquad \ldots.(3)$$

Equations (3) represent the parametric equations of the envelope.

Clearly the values of x and y given by equations (3) satisfy equation (1) identically i.e., for all values of α.

Now considering x, y as the functions of α, differentiating (1) w.r.t. α, we get

$$\frac{\partial\phi}{\partial\alpha}\frac{dx}{d\alpha} + \frac{\partial\phi}{\partial y}\frac{dy}{d\alpha} + \frac{\partial\phi}{\partial\alpha}\frac{d\alpha}{d\alpha} = 0$$

$$\Rightarrow \quad \frac{\partial\phi}{\partial x}\frac{dx}{d\alpha} + \frac{\partial\phi}{\delta y}\frac{dy}{d\alpha} + \frac{\partial\phi}{\partial\alpha} = 0$$

$$\Rightarrow \quad \frac{\partial\phi}{\partial x}f'(x) + \frac{\partial\phi}{\partial y}\psi'(\alpha) = 0 \qquad \text{From (2) and (3)}$$

$$\Rightarrow \quad \frac{-\partial\phi/\partial x}{\partial\phi/\partial y} = \frac{\psi'(\alpha)}{f'(\alpha)} \qquad \ldots.(4)$$

The gradient of the tangent line drawn at any point (x, y) of any member $\varphi(x, y, \alpha) = 0$ [where α is a constant, as α is constant for an individual member] of the family is $\frac{-\partial\phi/\partial x}{\partial\phi/\partial y}$ which is equal to $\frac{\psi'(\alpha)}{f'(\alpha)}$ from (4).

Also the gradient of the tangent line drawn to the envelope at the same point is $\frac{dy}{dx}$ i.e. $\frac{dy/d\alpha}{dx/d\alpha}$ i.e. $\frac{\psi'(\alpha)}{f'(\alpha)}$ from (3).

Since the two gradients are the same, hence the envelope and the curve have a common tangent at their common point *i.e.,* they touch each other.

ILLUSTRATIVE EXAMPLES

Example 1. *Find the envelope of the family of curves*

$y = mx + am^p$, *where m is the parameter.*

Solution :

The equation of the family is

$$y = mx + am^p \qquad \ldots.(1)$$

Differentiating (1) partially w.r.t. m, we get

$$0 = x + apm^{p-1}$$
$$\Rightarrow 0 = mx + apm^p$$

$$\Rightarrow am^p = -\frac{mx}{p} \quad \ldots (2)$$

Substituting the value of am^p in (1), we get,

$$y = mx - \frac{mx}{p}$$

$$\Rightarrow py = mx\,(p-1) \quad \ldots (3)$$

Raising both sides to power p,

$$p^p y^p = m^p x^p\,(p-1)^p$$

$$\Rightarrow p^p y^p = \frac{-mx}{ap} \,.\, x^p \,.\, (p-1)^p \quad \text{From (2)}$$

$$= -\frac{py}{p-1} \,.\, \frac{1}{ap} \,.\, x^p \,.\, (p-1)^p \quad \text{From (3)}$$

$$\Rightarrow a\, p^p\, y^{p-1} = -x^p\,(p-1)^{p-1}$$

$$\Rightarrow a\, p^p\, y^{p-1} + x^p\,(p-1)^{p-1} = 0$$

which is the required envelope.

Example 2. *Find the envelope of the family of curves*

$x \cos m\alpha + y \sin m\alpha = a\,(\cos n\alpha)^{m/n}$, *where α is the parameter.*

Solution : The given family of curves is

$$x \cos m\alpha + y \sin m\alpha = a\,(\cos n\alpha)^{m/n} \quad \ldots (1)$$

Differentiating w.r.t. α, we get

$$-mx \sin m\alpha + my \cos m\alpha = a \,.\, \frac{m}{n}\,(\cos n\alpha)^{\frac{m}{n}-1} \,.\, (-\sin n\alpha) \,.\, n$$

$$\Rightarrow x \sin m\alpha - y \cos m\alpha = a \sin n\alpha\,(\cos n\alpha)^{\frac{m-n}{n}} \quad \ldots (2)$$

Squaring and adding (1) and (2), we get

$$x^2 + y^2 = a^2 \cos^{\frac{2m}{n}} n\alpha + a^2 \sin^2 n\alpha \cos^{\frac{2(m-n)}{n}} n\alpha$$

$$= a^2 \cos^{\frac{2m}{n}} n\alpha \left[1 + \frac{\sin^2 n\alpha}{\cos^2 n\alpha}\right]$$

$$= a^2 \cos^{2\left(\frac{m-n}{n}\right)} n\alpha \quad \ldots (3)$$

Multiplying (1) by $\sin n\alpha$, we get

$$x \cos m\alpha \sin n\alpha + y \sin m\alpha \sin n\alpha = a \sin n\alpha\,(\cos n\alpha)^{m/n} \quad \ldots (4)$$

Multiplying (2) by $\cos n\alpha$, we get

$$x \sin m\alpha \cos n\alpha - y \cos n\alpha \cdot \cos m\alpha =$$

$$a \sin n\alpha \,.\, \cos n\alpha\, (\cos n\alpha)^{\frac{m-n}{n}}$$

$$= a \sin n\alpha\, (\cos n\alpha)^{m/n} \qquad \text{.... (5)}$$

Substracting (5) from (4), we get,

$$x \sin (n-m)\,\alpha + y \cos (n-m)\, \alpha = 0$$

$$\Rightarrow \quad -x \sin (m-n)\, \alpha + y \cos (m-n)\, \alpha = 0$$

$$\Rightarrow \quad x \sin (m-n)\, \alpha = y \cos (m-n)\, \alpha$$

$$\Rightarrow \tan (m-n)\, \alpha = \frac{y}{x}$$

put $x = r \cos \theta$

& $y = r \sin \theta$

$$\therefore \quad \tan (m-n)\, \alpha = \tan \theta$$

$$\Rightarrow \quad (m-n)\, \alpha = \theta$$

$$\Rightarrow \quad \alpha = \frac{\phi}{m-n}$$

Putting in (3), we get

$$x^2 + y^2 = a^2 \left[\cos\left(\frac{n\phi}{m-n}\right)\right]^{\frac{2(m-n)}{n}}$$

$$\Rightarrow \quad r^2 = a^2 \left[\cos\left(\frac{n\phi}{m-n}\right)\right]^{\frac{2(m-n)}{n}}$$

$$\Rightarrow \quad r = a \left[\cos\left(\frac{n\phi}{m-n}\right)\right]^{\frac{(m-n)}{n}}$$

$$\Rightarrow \quad r^{\frac{n}{m-n}} = a^{\frac{n}{m-n}} \cos\left(\frac{n\phi}{m-n}\right)$$

which gives the required envelope where (r, α) are the polar coordinates of (x, y).

Example 3. *Find the envelope of the family of lines* $y = mx + \dfrac{a}{m}$ *where m is a parameter.*

Solution : The equation of the family of lines is

$$y = mx + \frac{a}{m}$$

$$\Rightarrow xm^2 - ym + a = 0$$

This is a quadratic in parameter m. Therefore equating to zero the discriminant of the quadratic, we have the equation of the envelope as

$$B^2 - 4AC = 0$$

$$\Rightarrow (-y)^2 - 4\ (x)\ (a) = 0$$

$$\Rightarrow y^2 = 4\,ax$$

Example 4. *Find the envelope of the circles* $x^2 + y^2 - 2ax \cos \alpha - 2ay \sin \alpha = c^2$ *where* α *is the parameter. Interpret the result geometrically.*

Solution : The given family is

$$x^2 + y^2 - 2ax \cos \alpha - 2ay \sin \alpha = c^2 \qquad \text{.... (1)}$$

Differentiating (1) partially w.r.t. α, we have,

$$2\,ax \sin \alpha - 2ay \cos \alpha = 0 \qquad \text{.... (2)}$$

(1) can be written as

$$2ax \cos \alpha + 2ay \sin \alpha = x^2 + y^2 - c^2 \qquad \text{... (3)}$$

To eliminate α, squaring and adding (2) and (3), we get,

$$4\,a^2\,x^2 + 4\,a^2\,y^2 = (x^2 + y^2 - c^2)^2$$

$$\Rightarrow\ 4\,a^2\,(x^2 + y^2) = (x^2 + y^2 - c^2)^2 \qquad \text{.... (4)}$$

which is the required envelope.

Geometrical Interpretation

Equation (4) can be written as

$$(x^2 + y^2)^2 - 2\ (2a^2 + c^2)\ (x^2 + y^2) + c^4 = 0$$

[quadratic in $x^2 + y^2$]

$$\therefore x^2 + y^2 = \frac{2\,(2a^2 + c^2) \pm \sqrt{4\ (2a^2 + c^2)^2 - 4\,c^4}}{2}$$

$$= 2a^2 + c^2 \pm \sqrt{(2a^2 + c^2)^2 - c^4}$$

$$= 2a^2 + c^2 \pm \sqrt{4a^4 + 4a^2\,c^2}$$

$$= 2a^2 + c^2 \pm 2a\,\sqrt{a^2 + c^2}$$

$$= a^2 + c^2 \pm 2a\,\sqrt{a^2 + c^2} + a^2$$

$$= (\sqrt{a^2 + c^2} \pm a)^2$$

This shows that the required envelope consists of two circles with centre as origin and radii as $\sqrt{a^2 + c^2} \pm a$.

Example 5. *Obtain the envelope of the family of curves*

$\dfrac{x^2}{\alpha^2} + \dfrac{y^2}{k^2 - \alpha^2} = 1$ *where α is the parameter and interpret the result.*

Solution :

$$\frac{x^2}{\alpha^2} + \frac{y^2}{k^2 - \alpha^2} = 1$$

$$\Rightarrow \quad (k^2 - \alpha^2)\, x^2 + \alpha^2 y^2 = \alpha^2\, (k^2 - \alpha^2)$$

$$\Rightarrow \quad \alpha^4 - (x^2 - y^2 + k^2)\, \alpha^2 + k^2 x^2 = 0$$

which is quadratic in α^2. Hence equating to zero the discriminant of this equation, the required envelope is,

$$(x^2 - y^2 + k^2)^2 - 4k^2 x^2 = 0$$

$$\Rightarrow (x^2 - y^2 + k^2 + 2kx)\, (x^2 - y^2 + k^2 - 2kx) = 0$$

$$\Rightarrow [(x + k)^2 - y^2]\, [(x - k)^2 - y^2] = 0$$

$$\Rightarrow \quad (x + y + k)\, (x - y + k)\, (x + y - k)\, (x - y - k) = 0$$

which represents the four lines

$x + y + k = 0;$

$x + y - k = 0;$

$x - y + k = 0;$

$x - y - k = 0$

which form a square.

Hence the required envelope of the given family of curves is a square.

Example 6. *Find the envelope of the family of curves*

$\dfrac{x^3}{a^3 \cos\theta} + \dfrac{y^3}{b^3 \sin\theta} = 1,$ *θ being the parameter.*

Solution : Equation of the family of curves is

$$\frac{x^3}{a^3} \sec\theta + \frac{y^3}{b^3} \operatorname{cosec}\theta = 1 \qquad \text{.... (1)}$$

Differentiating partially w.r.t. θ, we have,

$$\frac{x^3}{a^3} \sec\theta \tan\theta \frac{-y^3}{b^3} \operatorname{cosec}\theta \cot\theta = 0$$

$$\Rightarrow \frac{x^3 \sin\theta}{a^3 \cos^2\theta} = \frac{y^3 \cos\theta}{b^3 \sin^2\theta}$$

$$\Rightarrow \tan^3\theta = \frac{a^3 y^3}{b^3 x^3}$$

$$\Rightarrow \tan\theta = \frac{ay}{bx} = \frac{y/b}{x/a}$$

whence $\sec\theta = \dfrac{\sqrt{\dfrac{x^2}{a^2}+\dfrac{y^2}{b^2}}}{\dfrac{x}{a}}$

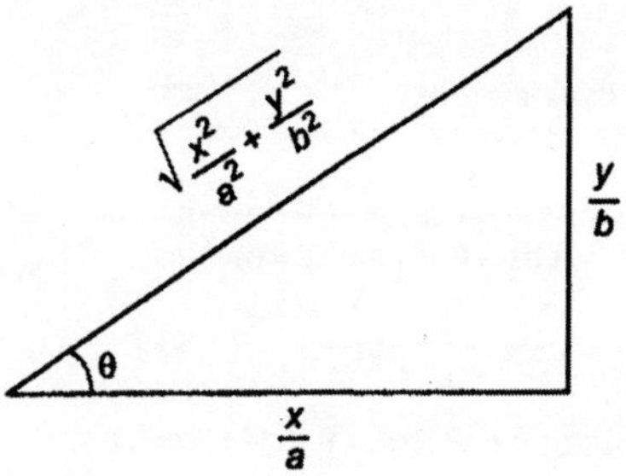

Fig 7.1

whence $\operatorname{cosec}\theta = \dfrac{\sqrt{\dfrac{x^2}{a^2}+\dfrac{y^2}{b^2}}}{\dfrac{y}{b}}$

Substituting in (1), we have

$$\frac{x^3}{a^3} = \frac{\sqrt{\dfrac{x^2}{a^2}+\dfrac{y^2}{b^2}}}{\dfrac{x}{a}} + \frac{y^3}{b^3}\;\frac{\sqrt{\dfrac{x^2}{a^2}+\dfrac{y^2}{b^2}}}{\dfrac{y}{b}} = 1$$

$$\Rightarrow \sqrt{\frac{x^2}{a^2}+\frac{y^2}{b^2}}\left(\frac{x^2}{a^2}+\frac{y^2}{b^2}\right) = 1$$

$$\Rightarrow \left(\frac{x^2}{a^2}+\frac{y^2}{b^2}\right)^{3/2} = 1$$

$$\Rightarrow \frac{x^2}{a^2}+\frac{y^2}{b^2} = 1$$

which is the required envelop.

Example 7. *Find the envelope of the straight lines*

$$y\sin t + x\cos t = a + a\cos t\log\tan\frac{t}{2}$$

Solution : The equation of the straight line is

$$y\sin t + x\cos t = a + a\cos t\log\tan\frac{t}{2}$$

$$\Rightarrow x + y\tan t = a\sec t + a\log\tan\frac{t}{2} \quad \text{....(1)}$$

Dividing by $\cos t$

Differentiating (1) partially w.r.t.t, we have

$$y\sec^2 t = a\sec t\tan t + a\,\frac{\sec^2\dfrac{t}{2}}{\tan\dfrac{t}{2}}\;\frac{1}{2}$$

$$= \frac{a \sin t}{\cos^2 t} + \frac{a}{\sin t}$$

$$= \frac{a}{\sin t \cos^2 t}$$

$$\Rightarrow \quad \frac{y}{\cos^2 t} = \frac{a}{\sin t \cos^2 t}$$

$$\Rightarrow \quad y = \frac{a}{\sin t} \qquad \text{....(2)}$$

Putting the value of y in (1), we have

$$x + a \sec t = a \sec t + a \log \tan \frac{t}{2}$$

$$\Rightarrow \quad x = a \log \tan \frac{t}{2}$$

$$\Rightarrow \quad \tan \frac{t}{2} = e^{x/a} \text{ and } \cot \frac{t}{2} = e^{-x/a}$$

$$\therefore \quad e^{x/a} + e^{x/a} = \tan \frac{t}{2} + \cot \frac{t}{2}$$

$$\Rightarrow \quad \frac{e^{x/a} + e^{-x/a}}{2} = \frac{1}{\sin t} = \frac{y}{a} \qquad \text{[From (2)]}$$

$$\Rightarrow \quad \cosh\left(\frac{x}{a}\right) = \frac{x}{a}$$

$$\Rightarrow \quad y = a \cosh\left(\frac{x}{a}\right)$$

Which is the required envelope.

Example 8. *From any point on the ellipse* $\frac{x^2}{a^2} + \frac{y^2}{b^2} = 1$, *perpendiculars are drawn to the axes. Show that the line joining the feet of the perpendiculars always touches the curve* $\left(\frac{x}{a}\right)^{2/3} + \left(\frac{y}{b}\right)^{2/3} = 1$.

Solution :

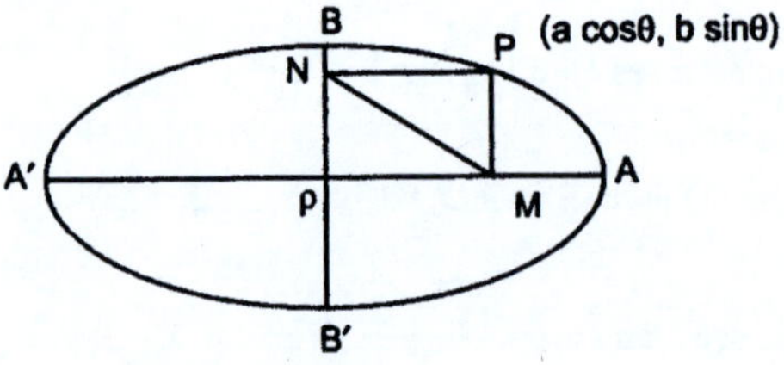

Fig 7.2

Let P $(a \cos \theta, b \sin \theta)$ by any point on the ellipse $\frac{x^2}{a^2} + \frac{y^2}{b^2} = 1$

Let PM and PN be the perpendiculars from P on the axis of x and y respectively.

Then the coordinates of M are $(a \cos \theta, 0)$ and those of N are $(0, b \sin \theta)$.

$\therefore$ Equation of line MN is

$$y - 0 = \frac{b \sin \theta - 0}{0 - a \cos \theta}(x - a \cos \theta)$$

$$\Rightarrow \frac{y}{b \sin \theta} = -\frac{x}{a \cos \theta} + 1$$

$$\Rightarrow \frac{x}{a \cos \theta} + \frac{y}{b \sin \theta} = 1 \qquad \text{.... (1)}$$

Differentiating (1) partially w.r.t. θ, we have,

$$\frac{x}{a \cos^2 \theta} \sin \theta - \frac{y}{b \sin^2 \theta} \cos \theta = 0$$

$$\Rightarrow \frac{x}{a \cos^3 \theta} = \frac{y}{b \sin^3 \theta}$$

$$\Rightarrow \frac{\cos^3 \theta}{\frac{x}{a}} = \frac{\sin^3 \theta}{\frac{y}{b}}$$

$$\Rightarrow \frac{\cos \theta}{\left(\frac{x}{a}\right)^{\frac{1}{3}}} = \frac{\sin \theta}{\left(\frac{y}{b}\right)^{\frac{1}{3}}} = \frac{\sqrt{\cos^2 \theta + \sin^2 \theta}}{\sqrt{\left(\frac{x}{a}\right)^{2/3} + \left(\frac{y}{b}\right)^{2/3}}} = \frac{1}{\sqrt{\left(\frac{x}{a}\right)^{2/3} + \left(\frac{y}{b}\right)^{2/3}}}$$

so that

$$\cos \theta = \frac{\left(\frac{x}{a}\right)^{\frac{1}{3}}}{\sqrt{\left(\frac{x}{a}\right)^{2/3} + \left(\frac{y}{b}\right)^{2/3}}}$$

$$\text{and } \sin \theta = \frac{\left(\frac{y}{b}\right)^{\frac{1}{3}}}{\sqrt{\left(\frac{x}{a}\right)^{2/3} + \left(\frac{y}{b}\right)^{2/3}}}$$

Putting in (1), we have

$$\frac{x}{a} \cdot \frac{\sqrt{\left(\frac{x}{a}\right)^{2/3} + \left(\frac{y}{b}\right)^{2/3}}}{\left(\frac{x}{a}\right)^{\frac{1}{3}}} + \frac{y}{b} \cdot \frac{\sqrt{\left(\frac{x}{a}\right)^{2/3} + \left(\frac{y}{b}\right)^{2/3}}}{\left(\frac{y}{b}\right)^{\frac{1}{3}}} = 1$$

$$\Rightarrow \sqrt{\left(\frac{x}{a}\right)^{2/3} + \left(\frac{y}{b}\right)^{2/3}} \left[\left(\frac{x}{a}\right)^{2/3} + \left(\frac{y}{b}\right)^{2/3}\right] = 1$$

$$\Rightarrow \left[\left(\frac{x}{a}\right)^{2/3} + \left(\frac{y}{b}\right)^{2/3}\right]^{3/2} = 1$$

$$\Rightarrow \left(\frac{x}{a}\right)^{2/3} + \left(\frac{y}{b}\right)^{2/3} = 1$$

which is the required envelope.

Example 9. *Show that the envelope of the family of curves*

$A\lambda^3 + 3B\lambda^2 + 3C\lambda + D = 0$ where λ is the parameter and A, B, C, D are functions of x and y, is $(BC - AD)^2 = 4\,(BD - C^2)\,(AC - B^2)$.

Solution : The equation of the family of curves is

$$A\lambda^3 + 3B\lambda^2 + 3C\lambda + D = 0 \qquad \text{.... (1)}$$

Differentiating partially w.r.t. λ, we have

$$3A\lambda^2 + 6B\lambda + 3C = 0$$

$$\Rightarrow A\lambda^2 + 2B\lambda + C = 0 \qquad \text{... (2)}$$

The required envelope is obtained by eliminating λ between (1) and (2).

Multiplying (2) by λ and subtracting from (1), we have

$$B\lambda^2 + 2C\lambda + D = 0 \qquad \text{.... (3)}$$

From (2) and (3)

$$\frac{\lambda^2}{2BD - 2C^2} = \frac{\lambda}{BC - AD} = \frac{1}{2AC - 2B^2}$$

whence $\lambda^2 = \dfrac{BD - C^2}{AC - B^2}$

and $\lambda = \dfrac{BC - AD}{2\,(AC - B^2)}$

Now, $\lambda^2 = (\lambda)^2$

$$\Rightarrow \quad \frac{BD - C^2}{AC - B^2} = \frac{(BC - AD)^2}{4\,(AC - B^2)^2}$$

$$\Rightarrow (BC - AD)^2 = 4\,(BD - C^2)\,(AC - B^2)$$

which is the required envelope.

Example 10. *Show that the radius of curvature of the envelope of the line* $x \cos \alpha + y \sin \alpha = f(\alpha)$ *is* $f(\alpha) + f''(\alpha)$.

Solution : The equation of the line is

$$x \cos \alpha + y \sin \alpha = f(\alpha) \qquad \text{.... (1)}$$

Differentiating (1) partially w.r.t. parameter α, we get,

$$-x \sin \alpha + y \cos \alpha = f'(\alpha) \qquad \text{.... (2)}$$

Since the envelope is obtained by eliminating α from (1) and (2), so on solving (1) and (2), we get

$$x = f(\alpha) - \cos \alpha - f'(\alpha) \sin \alpha \qquad \text{.... (3)}$$

$$y = f(\alpha) \sin \alpha + f'(\alpha) \cos \alpha \qquad \text{.... (4)}$$

Equations (3) and (4) taken together represent the envelope of lines (1) in the parametric form. We are required to find out the radius of curvature of the curve given by (3) and (4).

From (3),

$$\frac{dx}{d\alpha} = f'(\alpha) \cos \alpha - f(\alpha) \sin \alpha - f'(\alpha) \cos \alpha - f''(\alpha) \sin \alpha$$

$$= -\sin \alpha\,[f(\alpha) + f''(\alpha)] \qquad \text{...... (5)}$$

From (4)

$$\frac{dy}{d\alpha} = f(\alpha) \cos \alpha + f'(\alpha) \sin \alpha + f''(\alpha) \cos \alpha - f'(\alpha) \sin \alpha$$

$$= \cos \alpha\,[f(\alpha) + f''(\alpha)] \qquad \text{..... (6)}$$

$$\therefore \quad \frac{dy}{dx} = \frac{dy/d\alpha}{dx/d\alpha} = -\cot \alpha$$

$$\frac{d^2y}{dx^2} = \frac{d}{dx}\left(\frac{dy}{dx}\right)$$

$$= \frac{d}{dx}(-\cot \alpha)$$

$$= \frac{d}{dx}(-\cot \alpha) \cdot \frac{d\alpha}{dx}$$

$$= \frac{\operatorname{cosec}^2 \alpha}{-\sin \alpha\ [f(\alpha) + f''(\alpha)]}$$

$$= -\frac{\operatorname{cosec}^3 \alpha}{f(\alpha) + f''(\alpha)}$$

$\therefore$ ρ = radius of curvature

$$= \frac{\left[1 + \left(\frac{dy}{dx}\right)^2\right]^{3/2}}{\frac{d^2y}{dx^2}}$$

$$= \frac{[1 + \cot^2 \alpha]^{3/2}}{\frac{-\operatorname{cosec}^3 \alpha}{f(\alpha) + f''(\alpha)}}$$

$$= -[f(\alpha) + f''(\alpha)]$$

$$= f(\alpha) + f''(\alpha) \quad \text{numerically.}$$

EXERCISE 7 (A)

1. Find the envelopes of the following:

(a) $x \cos \alpha + y \sin \alpha = p$, where α is the parameter.

(b) $x \operatorname{cosec} \theta - y \cot \theta = c$, where θ is the parameter.

(c) $x^2 \sin \phi + y^2 \cos \phi = a^2$ where ϕ is the parameter.

(d) $ax \sec t - by \operatorname{cosec} t = a^2 - b^2$ where t is the parameter.

(e) $x \cos^n \theta + y \sin^n \theta = b$, where θ is the parameter.

(f) $x \sin \theta - y \cos \theta = a\theta$, where θ is the parameter.

2. Find the envelopes of the following:

(a) $y = mx + \sqrt{a^2 m^2 + b^2}$

(b) $y = mx - a\sqrt{1 + m^2}$

(c) $y^2 = 2a(x - a)$

(d) $(x - \alpha)^2 + y^2 = 4\alpha$

(e) $(x - \alpha)^2 + (y - \alpha)^2 = 2\alpha$

(f) $cy^2 = x(x + c)^2$, where c is the parameter.

3. Find the envelopes of the following family of curves:

(a) $tx^3 + t^2 y = a$, t being the parameter.

(b) $y = t^2 (x - t)$, t being the parameter.

4. Find the envelope of the straight lines
$x \cos \alpha + y \sin \alpha = \alpha \sin \alpha \cos \alpha$
where the angle α is the parameter. Give the geometrical interpretation.

5. Find the envelope of the ellipses
$x = a \sin (\theta - \alpha)$
$y = b \cos \theta$
where α is the parameter.

6. Find the envelope of the family of curves
$\frac{a^2}{x} \cos \theta - \frac{b^2}{y} \sin \theta = \frac{c^2}{a}$, for different values of θ.

7. Find the envelope of
$x \cos^3 \theta + y \sin^3 \theta = c$ for different values of θ.

8. Find the envelopes of the family of trajectories
$y = x \tan \alpha - \frac{1}{2} \frac{g x^2}{x^2 \cos^2 \alpha}$, where α is the parameter.

9. Find the envelope of the family of curves (lines)
$x \cos \theta - y \sin \theta = a (1 + \cos \theta)$, where θ is the parameter.

10. Find the envelope of the family of lines
$ax \cos \theta + by \cot \theta = a^2 + b^2$, where θ is the parameter.

11. Show that the envelope of the straight lines $y \cos \theta - x \sin \theta = a - a \sin \theta \log_e \tan \left(\frac{\pi}{4} + \frac{\theta}{2}\right)$, where θ is the parameter, is the catenary $y = a \cos h \left(\frac{x}{a}\right)$.

12. Find the envelope of the circles, whose centres lies on the parabola $y^2 = 4ax$ and which pass through its vertex.

13. Show that the envelope of the family of the circles whose diameters are the double ordinates of the parabola $y^2 = 4ax$ is the parabola $y^2 = 4a (x + a)$.

14. Find the envelope of the circles described on the radii vectors of the parabola $y^2 = 4ax$ as diameters.

15. Find the envelope of the circles which pass through the centre of the ellipse $\frac{x^2}{a^2} + \frac{y^2}{b^2} = 1$ and have their centres upon its circumference.

16. Find the envelope of the circles whose diameters are the double ordinates of the ellipse $\frac{x^2}{a^2} + \frac{y^2}{b^2} = 1$.

17. Find the envelope of the circles drawn upon the central radii vectors of the ellipse $\frac{x^2}{a^2} + \frac{y^2}{b^2} = 1$ as diameter.

18. Show that the envelope of the circles described on the central radii of the rectangular hyperbola $x^2 - y^2 = a^2$ is the lemniscate $r^2 = a^2 \cos 2\theta$.

19. Show that the line joining the extremities of a pair of semi-conjugate diameters of the ellipse $\frac{x^2}{a^2} + \frac{y^2}{b^2} = 1$, always touches the similar ellipse $\frac{x^2}{a^2} + \frac{y^2}{b^2} = \frac{1}{2}$.

20. Show that the envelope of the polars of the point on the ellipse $\frac{x^2}{h^2} + \frac{y^2}{k^2} = 1$ with respect to the ellipse $\frac{x^2}{a^2} + \frac{y^2}{b^2} = 1$ is $\frac{h^2 x^2}{a^4} + \frac{k^2 y^2}{b^4} = 1$.

21. From any point on the ellipse $\frac{x^2}{l^2} + \frac{y^2}{m^2} = 1$, tangents are drawn to the ellipse $\frac{x^2}{a^2} + \frac{y^2}{b^2} = 1$. Show that the envelope of the chords of contact is $\frac{l^2 x^2}{a^4} + \frac{m^2 y^2}{b^4} = 1$.

22. Show that the envelope of the straight line of a given length l, which slides with its extremities on two fixed straight lines at right angles, is $x^{2/3} + y^{2/3} = l^{2/3}$.

23. A straight line of given length slides with its extremities on two fixed straight lines at right angles. Find the envelope of the circle drawn on the sliding line as diameter.

24. From any point on the curve $y = ax^3$, perpendiculars are drawn to the axes, and the feet of these perpendiculars are joined. Show that the straight line thus obtained always touches the curve $27y + 4ax^3 = 0$.

25. A series of circles have their centres on a given straight line, and their radii are proportional to the distance of their corresponding centres from a given point in that line. Find the envelope.

ANSWERS

1. (a) $x^2 + y^2 = p^2$

(b) $x^2 - y^2 = c^2$

(c) $x^4 + y^4 = a^4$

(d) $(ax)^{2/3} + (by)^{2/3} = (a^2 - b^2)^{2/3}$

(e) $x^{\frac{2}{2-n}} + y^{\frac{2}{2-n}} = a^{\frac{2}{2-n}}$

(f) $x = a \cos \theta + a\theta \sin \theta$
$y = a \sin \theta - a\theta \cos \theta$

2. (a) $\dfrac{x^2}{a^2} + \dfrac{y^2}{b^2} = 1$

(b) $x^2 + y^2 = a^2$

(c) $y = \pm x\sqrt{2}$

(d) $y^2 - 4x - 4 = 0$

(e) $(x + y + 1)^2 = 2(x^2 + y^2)$

(f) $y^2 = 4x^2$

3. (a) $x^6 + 4ay = 0$

(b) $4x^3 = 27y$

4. $x^{2/3} + y^{2/3} = l^{2/3}$ (Astroid)

5. $x = \pm a$

6. $\dfrac{a^4}{x^2} + \dfrac{b^4}{y^2} = c^2$

7. $\dfrac{1}{x^2} + \dfrac{1}{y^2} = \dfrac{1}{c^2}$

8. $y + \frac{1}{2}\frac{gx^2}{u^2} = \frac{u^2}{2g}$

9. $x^2 + y^2 = 2\,ax$

10. $(ax)^{2/3} - (by)^{2/3} = (a^2 + b^2)^{2/3}$

12. $x^3 + y^2\,(x + 2a) = 0$

14. $ay^2 + x\,(x^2 + y^2) = 0$

15. $4\,(a^2x^2 + b^2y^2) = (x^2 + y^2)^2$

16. $b^2x^2 + (a^2 + b^2)\,y^2 = b^2\,(a^2 + b^2)$

17. $a^2x^2 + b^2y^2 = (x^2 + y^2)^2$

23. $x^2 + y^2 = l^2$

25. $k^2\,x^2 = (1 - k^2)\,y^2$ where k is the constant of proportionality.

7.7. Two Parameters Connected by a Relation

Let the equation of the family of curves be

$$f\,(x,\,y,\,a,\,b) = 0 \qquad \text{...(1)}$$

where the two parameters a and b are connected by the relation

$$\phi\,(a,\,b) = 0 \qquad \text{...(2)}$$

Now in order to find out the envelope of (1) under (2), we eliminate either a or b in between (1) and (2). Then the given equation (1) of the family of curves reduces to the equation of a family involving either of the two parameters a and b, whose envelope can be determined by the usual method.

Example 1. *Find the envelope of a system of concentric and coaxial ellipses of constant area.*

Solution : The given family is

$$\frac{x^2}{a^2} + \frac{y^2}{b^2} = 1 \qquad \text{...(1)}$$

where,

$$\text{area} = \text{constant}$$

$$\Rightarrow \quad \pi ab = \text{constant}$$

$$\Rightarrow ab = \text{constant} = \lambda \text{ (say)} \qquad \text{...(2)}$$

From (2),

$$b = \frac{\lambda}{a}$$

put the value of b in (1),

$$\frac{x^2}{a^2} + \frac{y^2}{\frac{\lambda^2}{a^2}} = 1$$

$$\Rightarrow \frac{x^2}{a^2} + \frac{a^2 y^2}{\lambda^2} = 1$$

$$\Rightarrow \lambda^2 x^2 + a^4 y^2 = a^2 \lambda^2$$

$$\Rightarrow a^4 y^2 - a^2 \lambda^2 + \lambda^2 x^2 = 0$$

This is a quadratic in a^2. Hence the envelope is given by

$$B^2 = 4AC$$

$$\Rightarrow (-\lambda^2)^2 = 4y^2 \lambda^2 x^2$$

$$\Rightarrow \lambda^2 = 4x^2 y^2$$

$$\Rightarrow 2xy = \pm \lambda$$

7.8. The above process of finding out the envelopes becomes very combersome in some cases. In such case, we proceed to follow the method as explained in the following examples.

Example 2. *Find the envelope of the curves* $\frac{x^m}{a^m} + \frac{y^m}{b^m} = 1$, *where the parameters a and b are connected by the relation* $a^p + b^p = c^p$.

Solution : Let us suppose that a and b both are the functions of some third arbitrary parameters t. Then, differentiating

$$\frac{x^m}{a^m} + \frac{y^m}{b^m} = 1 \text{ w.r.t. } t, \text{ we get,} \quad \text{.... (1)}$$

$$(-m) \frac{x^m}{a^{m+1}} \frac{da}{dt} + (-m) \frac{y^m}{b^{m+1}} \frac{db}{dt} = 0$$

$$\Rightarrow \frac{x^m}{a^{m+1}} \frac{da}{dt} + \frac{y^m}{b^{m+1}} \frac{db}{dt} = 0$$

$$\Rightarrow \frac{da/dt}{db/dt} = -\frac{y^m}{b^{m+1}} \cdot \frac{a^{m+1}}{x^m} \quad \text{... (2)}$$

Differentiating $a^p + b^p = c^p$ w.r.t. t, we get ... (3)

$$p\, a^{p-1} \frac{da}{dt} + p\, b^{p-1} \frac{db}{dt} = 0$$

$$\Rightarrow a^{p-1} \frac{da}{dt} + b^{p-1} \frac{db}{dt} = 0$$

$$\Rightarrow \frac{da/dt}{db/dt} = -\frac{b^{p-1}}{a^{p-1}} \qquad \text{... (4)}$$

Equating the values of $\frac{da/dt}{db/dt}$ from equations (3) and (4), we get,

$$\frac{-y^m\, a^{m+1}}{b^{m+1}\, x^m} = -\frac{b^{p-1}}{a^{p-1}}$$

$$\Rightarrow y^m\, a^{m+p} = x^m\, b^{m+p}$$

$$\Rightarrow \frac{x^m}{a^{m+p}} = \frac{y^m}{b^{m+p}}$$

$$\Rightarrow \frac{\frac{x^m}{a^m}}{a^p} = \frac{\frac{y^m}{b^m}}{b^p} = \frac{\frac{x^m}{a^m} + \frac{y^m}{b^m}}{a^p + b^p}$$

$$\Rightarrow \frac{x^m}{a^{m+p}} = \frac{y^m}{b^{m+p}} = \frac{1}{c^p} \qquad \text{[From (1) and (3)]}$$

$$\Rightarrow a^{m+p} = c^p\, x^m$$

$$\& \quad b^{m+p} = c^p\, y^m$$

$$\Rightarrow a = c^{\frac{p}{m+p}}\, x^{\frac{m}{m+p}}$$

$$\& \quad b = c^{\frac{p}{m+p}}\, y^{\frac{m}{m+p}}$$

Putting these values of a and b in (3), we get

$$c^{\frac{p^2}{m+p}} \cdot x^{\frac{mp}{m+p}} + c^{\frac{p^2}{m+p}}\, y^{\frac{mp}{m+p}} = c^p$$

$$\Rightarrow x^{\frac{mp}{m+p}} + y^{\frac{mp}{m+p}} = c^{p - \frac{p^2}{m+p}} = c^{\frac{mp}{m+p}}$$

which is the required envelope.

Example 3. *Find the envelope of the family of curves* $\sqrt{\frac{x}{a}} + \sqrt{\frac{y}{b}} = 1$ *when* $ab = c^2$.

Solution : Let a and b be the functions of some third arbitrary parameter t.

Then, differentiating $\sqrt{\frac{x}{a}} + \sqrt{\frac{y}{b}} = 1$ w.r.t. t, we get, ... (1)

$$-\frac{1}{2}\frac{\sqrt{x}}{a^{3/2}}\frac{da}{dt}-\frac{1}{2}\frac{\sqrt{y}}{b^{3/2}}\frac{db}{dt}=0$$

$$\Rightarrow \quad \frac{\sqrt{x}}{a^{3/2}}\frac{da}{dt}=-\frac{\sqrt{y}}{b^{3/2}}\frac{db}{dt}$$

$$\Rightarrow \quad \frac{da/dt}{db|dt}=-\frac{\sqrt{y}}{b^{3/2}}\cdot\frac{a^{3/2}}{\sqrt{x}} \qquad \text{... (2)}$$

Differentiating $ab = c^2$ w.r.t. t, we get, ... (3)

$$a\frac{db}{dt}+\frac{da}{dt}b=0$$

$$\Rightarrow \quad \frac{\frac{da}{dt}}{\frac{db}{dt}}=-\frac{a}{b} \qquad \text{... (4)}$$

Equating the two values of $\dfrac{da/dt}{db/dt}$ from (3) and (4), we get,

$$\frac{\sqrt{y}\,a^{3/2}}{b^{3/2}\sqrt{x}}=\frac{a}{b}$$

$$\Rightarrow \quad \sqrt{y}\,a^{\frac{1}{2}}=\sqrt{x}\,b^{\frac{1}{2}}$$

$$\Rightarrow \quad \frac{\sqrt{\frac{x}{a}}}{1}=\frac{\sqrt{\frac{y}{b}}}{1}=\frac{\sqrt{\frac{x}{a}}+\sqrt{\frac{y}{b}}}{1+1}=\frac{1}{2} \qquad \text{[From (1)]}$$

$$\Rightarrow \quad \sqrt{\frac{x}{a}}=\frac{1}{2}$$

$$\& \quad \sqrt{\frac{y}{b}}=\frac{1}{2}$$

$$\Rightarrow \quad \sqrt{a}=2\sqrt{x}$$

$$\Rightarrow \quad a=4x$$

$$\& \quad \sqrt{b}=2\sqrt{y}$$

$$\Rightarrow \quad b=4y$$

Putting these values of a and b in (3), we get,

$$4x\,.\,4y=c^2$$

$$\Rightarrow 16xy=c^2$$

which is the required envelope.

EXERCISE 7 (B)

1. Find the envelope of the straight lines $\frac{x}{a} + \frac{y}{b} = 1$ when the parameters a and b are connected by the relation

   (a) $a + b = c$

   (b) $a^2 + b^2 = c^2$ i.e., length intercepted between the axes is c.

   (c) $ab = c^2$

   (d) $a^3 + b^3 = c^3$

   (e) $a^n + b^n = c^n$

   (f) $a^m\ b^m = c^{m+n}$

   where c is a constant.

2. Prove that the envelope of the ellipses having the axes of coordinates as the principal axes and the sum of their semi-axes constant and equal to c, is the astroid $x^{2/3} + y^{2/3} = c^{2/3}$.

3. Find the envelope of the ellipse $\frac{x^2}{a^2} + \frac{y^2}{b^2} = 1$ when

   (a) $a^2 + b^2 = c^2$

   (b) $a^n + b^n = c^n$

   (c) $a^m\ b^n = c^{m+n}$ where a and b are the parameters and c is any constant.

4. Find the envelope of the family of parabola $\sqrt{\frac{x}{a}} + \sqrt{\frac{y}{b}} = 1$ when

   (a) $a + b = c$

   (b) $a^n + b^n = c^n$ where c is any constant.

ANSWERS

1. (a) $\sqrt{x} + \sqrt{y} = \sqrt{c}$

   (b) $x^{2/3} + y^{2/3} = a^{2/3}$

   (c) $4xy = c^2$

   (d) $x^{3/4} + y^{3/4} = c^{3/4}$

   (e) $x^{\frac{n}{n+1}} + y^{\frac{n}{n+1}} = c^{\frac{n}{n+1}}$

   (f) $(m + n)^{m+n} \cdot x^m\ y^n = m^m\ n^n\ c^{m+n}$

3. (a) $x \pm y = \pm c$

(b) $x^{\frac{2n}{n+2}} + y^{\frac{2n}{n+2}} = c^{\frac{2n}{n+2}}$

(c) $\dfrac{(m+n)^{m+n}}{m^m n^n} x^{2m} y^{2n} = c^{2m+2n}$

4. (a) $x^{\frac{1}{3}} + y^{\frac{1}{3}} = c^{\frac{1}{3}}$

(b) $x^{\frac{n}{2n+1}} + y^{\frac{n}{2n+1}} = c^{\frac{n}{2n+1}}$

7.9. Polar Coordinates

ILLUSTRATIVE EXAMPLES

Example 1. *Find the envelope of the straight lines at right angles to the radii vectors of the curve $r = ae^{\theta \cot \alpha}$ through their extremities.*

Solution

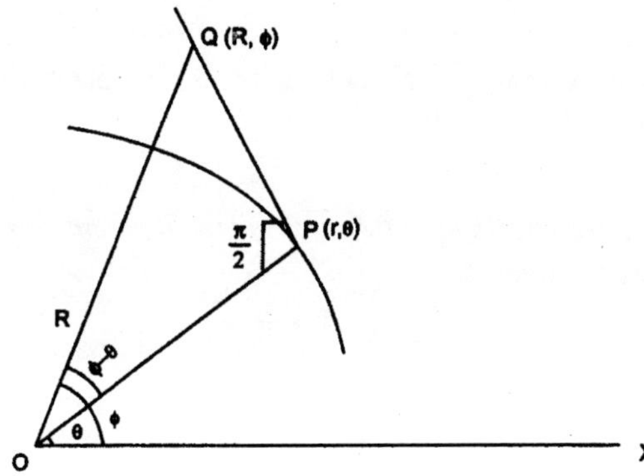

Fig. 7.3

Let $P\ (r,\ \theta)$ be any point on the curve $r = a\,e^{\theta \cot \alpha}$. Let $(R,\ \theta)$ be any point on the line drawn at right angles to the radius vector OP through the extremity P.

From right angle triangle OPQ,

$$R \cos (\phi - \theta) = r$$

$$\Rightarrow \quad R \cos (\phi - \theta) = a\, e^{\theta \cot \alpha} \qquad \ldots (1)$$

which is the equation of the line PQ. Taking log, we get

$$\log R + \log \cos (\phi - \theta) = \log a + \theta \cot \alpha$$

Differentiating partially w.r.t. the parameter θ, we get

$$\tan(\phi - \theta) = \cot\alpha = \tan\left(\frac{\pi}{2} - \alpha\right)$$

$$\Rightarrow \quad \phi - \theta = \frac{\pi}{2} - \alpha$$

$$\Rightarrow \theta = \phi + \alpha - \frac{\pi}{2}$$

Substituting this value of θ in (1), we get,

$$R \cos\left\{\theta - \theta - \alpha + \frac{\pi}{2}\right\} = a\, e^{(\theta + \alpha - \frac{\pi}{2})\cot\alpha}$$

$$\Rightarrow \quad R \sin\alpha = a\, e^{\left(\alpha - \frac{\pi}{2}\right)\cot\alpha} \cdot e^{\theta\cot\alpha}$$

$$\Rightarrow R = \frac{a\, e^{\left(\alpha - \frac{\pi}{2}\right)\cot\alpha}}{\sin\alpha}\, e^{\theta\cot\alpha}$$

$$\Rightarrow \quad R = A\, e^{\varphi\cot\alpha} \text{ where}$$

$$A = \frac{a\, e^{\left(\alpha - \frac{\pi}{2}\right)\cot\alpha}}{\sin\alpha}$$

It we take the current coordinates as (r, θ) in place of (R, θ), then required equation of the envelope is

$$r = A\, e^{\theta\cot\alpha}$$

Example 2. *Find the envelope of the circles described on the radii vectors of the following curve as diameters*

$$r = a(1 + \cos\theta)$$

Solution :

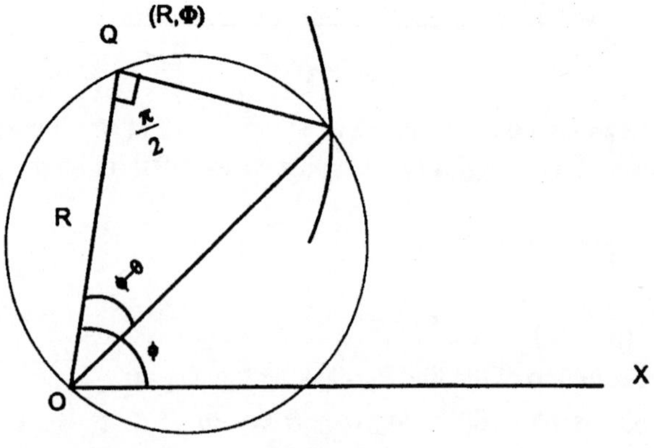

Fig. 7.4

Let $P(r, \theta)$ be any point on the cardioid $r = a(1 + \cos\theta)$. Let $Q(R, \phi)$ be any point or the circle described on the radius vector OP of the cardioid as diameter. Then, from right angled triangle OPQ,

$$R = r\cos(\phi - \theta)$$

$$\Rightarrow R = a(1 + \cos\theta)\cos(\phi - \theta) \qquad \text{...(1)}$$

This is the equation of the given circle where θ is the parameter.

Taking log, we get,

$$\log R = \log a + \log(1 + \cos\theta) + \log\cos(\phi - \theta)$$

Differentiating partially w.r.t. θ, the parameter, we get,

$$0 = 0 - \frac{\sin\theta}{1 + \cos\theta} + \frac{\sin(\phi - \theta)}{\cos(\phi - \theta)}$$

$$\Rightarrow \frac{\sin\theta}{1 + \cos\theta} = \frac{\sin(\phi - \theta)}{\cos(\phi - \theta)}$$

$$\Rightarrow \frac{2\sin\frac{\theta}{2}\cos\frac{\theta}{2}}{2\cos^2\frac{\theta}{2}} = \tan(\phi - \theta)$$

$$\Rightarrow \tan\frac{\theta}{2} = \tan(\phi - \theta)$$

$$\Rightarrow \frac{\theta}{2} = \phi - \theta$$

$$\Rightarrow \theta = \frac{3\theta}{2}$$

$$\Rightarrow \theta = \frac{2}{3}\phi$$

Substituting this value of θ in (1), we get,

$$R = a\left(1 + \cos\frac{2}{3}\phi\right)\cos\left(\phi - \frac{2}{3}\phi\right)$$

$$\Rightarrow R = a \,.\, 2\cos^3\frac{\phi}{3} \,.\, \cos\frac{\phi}{3}$$

$$\Rightarrow R = 2a\cos^3\frac{\phi}{3}$$

If we take the current coordinates as (r, θ) in place of (R, ϕ), then the required equation of the envelope is

$$r = 2a\cos^3\frac{\theta}{3}.$$

EXERCISE 7 (C)

1. Find the envelopes of the straigth lines, drawn through the extremities of, and at right angles to the radii vectors of the following curves:

   (a) $r = a\,(1 + \cos\theta)$

   (b) $r^n = a^n \cos n\theta$

   (c) $r = a + b\cos\theta$

2. Find the envelopes of the circles described on the radii vectors of the following curves as diameters:

   (a) $r = 2a\cos\theta$

   (b) $r^2 = a^2 \cos 2\theta$

   (c) $r^n = a^n \cos n\theta$

   (d) $r \cos^n\left(\dfrac{\theta}{n}\right) = a$

   (e) $\dfrac{l}{r} = 1 + e\cos\theta$

ANSWERS

1. (a) $r = 2a\cos\theta$

   (b) $r^{\frac{n}{1-n}} = a^{\frac{n}{1-n}} \cos\left\{\dfrac{n}{1-n}\,\theta\right\}$

   (c) $r^2 - 2br\cos\theta + (b^2 - a^2) = 0$

2. (a) $r = a\,(1 + \cos\theta)$

   (b) $r^{2/3} = a^{2/3} \cos\dfrac{2\theta}{3}$

   (c) $r^{\frac{n}{n+1}} = a^{\frac{n}{n+1}} \cos\left(\dfrac{n\theta}{n+1}\right)$

   (d) $r \cos^{n-1}\left(\dfrac{\theta}{n-1}\right) = a$

   (e) $r^2(e^2 - 1) - 2\,l\,e\,r\cos\theta + l^2 = 0$

7.10. Evolute

Evolute of a curve is the envelope of its normals.

We know that the locus of the centre of curvature is called the evolute of the curve and the curve in itself is called the involute of its evolute. But the centre of curvature is the limiting position of the point of inter-section of two consecutive normals, hence evolute may be defined as the envelope of the normals of that curve.

ILLUSTRATIVE EXAMPLES

Example 1. *Find the evolute of the parabola* $y^2 = 4ax$ or *Prove that normals of the curve* $y^2 = 4ax$ *touch the curve* $27ay^2 = 4(x - 2a)^3$.

Solution: Equation of any normal to the parabola $y^2 = 4ax$ is $y = mx - 2am - am^3$ where m is the parameter. ... (1)

Differentiating partially (1) w.r.t. m, we get,

$$0 = x - 2a - 3am^2$$

$$\Rightarrow m = \left(\frac{x - 2a}{3a}\right)^{1/2}$$

Substituting this value of m in (1), we get,

$$y = \left(\frac{x - 2a}{3a}\right)^{\frac{1}{2}}\left[x - 2a - a \cdot \frac{x - 2a}{3a}\right]$$

$$= \left(\frac{x - 2a}{3a}\right)^{\frac{1}{2}}(x - 2a) \cdot \frac{2}{3}$$

$$= \frac{2(x - 2a)^{3/2}}{3\sqrt{3a}}$$

Squaring, we get,

$$27ay^2 = 4(x - 2a)^3$$

which is the required evolute.

Example 2. *Find the evolute of the ellipse*

$$\frac{x^2}{a^2} + \frac{y^2}{b^2} = 1.$$

Solution : Equation of any normal to the ellipse is

$ax \sec\theta - by \operatorname{cosec}\theta = a^2 - b^2$ ... (1) where θ is the parameter.

Differentiating (1) partially w.r.t. θ, we get,

$$ax \sec\theta \tan\theta + by \operatorname{cosec}\theta \cot\theta = 0$$

$$\Rightarrow ax\frac{\tan\theta}{\cos\theta} + \frac{by}{\sin\theta\tan\theta} = 0$$

$$\Rightarrow ax\tan^3\theta + by = 0$$

$$\Rightarrow \tan^3 \theta = -\frac{by}{ax}$$

$$\Rightarrow \tan \theta = \frac{-(by)^{1/3}}{(ax)^{1/3}}$$

whence $\sin \theta = \dfrac{-(by)^{1/3}}{\sqrt{(ax)^{2/3} + (by)^{2/3}}}$

and $\cos \theta = \dfrac{(ax)^{1/3}}{\sqrt{(ax)^{2/3} + (by)^{2/3}}}$

$\therefore$ From (1)

$$ax \left\{\frac{\sqrt{(ax)^{2/3} + (by)^{2/3}}}{(ax)^{1/3}}\right\} - by \left\{\frac{\sqrt{(ax)^{2/3} + (by)^{2/3}}}{-(by)^{1/3}}\right\} = a^2 - b^2$$

$$\Rightarrow \sqrt{(ax)^{2/3} + (by)^{2/3}} \left\{(ax)^{2/3} + (by)^{2/3}\right\} = a^2 - b^2$$

$$\Rightarrow \left\{(ax)^{2/3} + (by)^{2/3}\right\}^{3/2} = a^2 - b^2$$

$$\Rightarrow (ax)^{2/3} + (by)^{2/3} = (a^2 - b^2)^{2/3}$$

which is the required evolute.

EXERCISE 7 (D)

1. Prove that the ellipses
   $b^2 x^2 + a^2 y^2 = a^2 b^2$ and $a^2 x^2 \sec^4 \alpha + b^2 y^2 \operatorname{cosec}^4 \alpha = (a^2 - b^2)^2$
   are so related that the evolute of the first is the envelope of the second for all values of α.
2. Prove that the evolute of the hyperbola
   $2xy = a^2$ is
   $(x + y)^{2/3} - (x - y)^{2/3} = 2a^{2/3}$
3. Prove that the evolute of the hyperbola $\dfrac{x^2}{a^2} - \dfrac{y^2}{b^2} = 1$ is
   $(ax)^{2/3} - (by)^{2/3} = (a^2 + b^2)^{2/3}$.
4. Prove that the evolute of astroid $x^{2/3} + y^{2/3} = a^{2/3}$,
   $\{$or $x = a \cos^3 t,\ y = a \sin^3 t\}$ is
   $(x + y)^{2/3} + (x - y)^{1/2} = 2a^{2/3}$.

8 Partial Differentiation

8.1. Partial and Total Increment of a Function

Let $z = f(x, y)$ be a function of two independent variables. Then change in one variable does not affect the other. So keeping y constant, an increment Δx in x will produce corresponding increment in z. This increment in z is called partial increment with respect to x and it is denoted by $\Delta_x z$. So

$$\Delta_x z = f(x + \Delta x, y) - f(x, y).$$

Similarly, keeping x constant, increment Δy in y will produce increment in z. This increment in z is called partial increment w.r.t. y and it is denoted by $\Delta_y z$. Hence

$$\Delta_y z = f(x, y + \Delta y) - f(x, y).$$

If increment Δx in x and Δy in y will produce an increment Δz in z, then Δz is called total increment in z.

$$\therefore \qquad \Delta z = f(x + \Delta x, y + \Delta y) - f(x, y).$$

8.2. Partial Derivaties

The ordinary derivative of a function of two or more variables with respect to one of the independent variables, keeping all other independent variables constant is called the partial derivative of the function w.r.t. the variable.

Let $z = f(x, y)$ be a function of two independent variables x and y. Then the ordinary derivative of $f(x, y)$ with respect to x, keeping y constant is called partial derivative of the function $f(x, y)$ with respect to the x.

Generally, it is denoted by $\frac{\partial f}{\partial x}$ or f_x or $f_x(x, y)$.

$$\therefore \qquad \frac{\partial f}{\partial x} = \lim_{\delta_x \to 0} \frac{f(x + \delta x, y) - f(x, y)}{\delta x}$$

provided this limit exists and is unique.

Similarly, the ordinary derivative of $f(x, y)$ w.r.t. y, keeping x constant is called partial derivative of the function $f(x, y)$ w.r.t. y. It is denoted by $\frac{\partial f}{\partial x}$ or f_y or $f_y(x, y)$.

$$\therefore \qquad \frac{\partial f}{\partial y} = \lim_{\delta_y \to 0} \frac{f(x, y+\delta y) - f(x, y)}{\delta y}$$

provided this limit exists and is unique.

The partial derivatives at a particular point (a, b) are

$$f_x(a, b) = \left(\frac{\partial f}{\partial x}\right)_{(a, b)} = \lim_{h \to 0} \frac{f(a+h, b) - f(a, b)}{h}$$

and

$$f_y(a, b) = \left(\frac{\partial f}{\partial y}\right)_{(a, b)} = \lim_{k \to 0} \frac{f(a, b+k) - f(a, b)}{k}$$

8.3. Partial Derivaties of Function of Three Variables

Let $u = f(x, y, z)$ be a function of three independent variables x, y, z and let $f(x, y, z)$ be defined in some neighbourhood of (x, y, z). Then partial derivative of $f(x, y, z)$ with respect to x, y, z are defined as follows :

(i) If $\dfrac{\partial f}{\partial x} = f_x = \lim_{\delta_x \to 0} \dfrac{f(x+\delta x, y, z) - f(x, y, z)}{\delta x}$

exists, then f_x is said to be partial derivative of $f(x, y, z)$, w.r.t. 'x'.

(ii) If $\dfrac{\partial f}{\partial y} = f_y = \lim_{\delta_y \to 0} \dfrac{f(x, y+\delta y, z) - f(x, y, z)}{\delta y}$

exists, then f_y is said to be partial derivative of $f(x, y, z)$ with respect to y.

(iii) If $\dfrac{\partial f}{\partial z} = f_z = \lim_{\delta_z \to 0} \dfrac{f(x, y, z+\delta z) - f(x, y, z)}{\delta z}$

exists, then f_z is said to be partial derivative of $f(x, y, z)$ w.r.t. z.

ILLUSTRATIVE EXAMPLES

Example 1. *If $f(x, y) = x + y^2$ find $f_x(x, y)$ and $f_y(x, y)$ from definition.*

Solution : We know,

$$f_x(x, y) = \frac{\partial f}{\partial x} = \lim_{\delta_x \to 0} \frac{f(x+\delta x, y) - f(x, y)}{\delta x}$$

$$= \lim_{\delta_x \to 0} \frac{\{(x+\delta x)^3 + y^2\} - (x^3 + y^2)}{\delta x}$$

$$= \lim_{\delta_x \to 0} \frac{\{x^3 + 3x^2\delta x + 3x(\delta x)^2 + (\delta x)^3 + y^2 - x^3 - y^2\}}{\delta x}$$

$$= \lim_{\delta_x \to 0} \frac{3x^2\,\delta x + 3x\,(\delta x)^2 + (\delta x)^3}{\delta x}$$

$$= \lim_{\delta_x \to 0} \{3x^2 + 3x\,\delta x + (\delta x)^2\}$$

$$= 3x^2.$$

$$f_y(x, y) = \frac{\partial f}{\partial y} = \lim_{\delta_y \to 0} \frac{f(x, y + \delta y) - f(x, y)}{\delta y}$$

$$= \lim_{\delta_y \to 0} \frac{\{x^3 + (y + \delta y)^2\} - (x^3 + y^2)}{\delta y}$$

$$= \lim_{\delta_y \to 0} \frac{x^3 + y^2 + 2y\,\delta y + (\delta y)^2 - x^3 - y^2}{\delta y}$$

$$= \lim_{\delta_y \to 0} \frac{2y\,\delta y + (\delta y)^2}{\delta y}$$

$$= \lim_{\delta_y \to 0} \{2y + \delta y\}$$

$$= 2y.$$

Example 2. *If* $f(x, y) = \dfrac{x^2 - y^2}{x^2 + y^2 + 1}$ *find by definition* $f_x(0, 0)$ *and* $f_y(0, 0)$.

Solution :

$$f_x(0, 0) = \frac{\partial f}{\partial x} = \lim_{\delta_x \to 0} \frac{f(0 + \delta x, 0) - f(0, 0)}{\delta x}$$

$$= \lim_{\delta_x \to 0} \left\{ \frac{\dfrac{(\delta x)^2 - 0}{(\delta x)^2 + 0 + 1} - \dfrac{0}{1}}{\delta x} \right\} = \lim_{\delta_x \to 0} \frac{\dfrac{(\delta x)^2}{(\delta x)^2 + 1}}{\delta x}$$

$$= \lim_{\delta_x \to 0} \frac{(\delta x)^2\,\delta x}{(\delta x)^2 + 1}$$

$$= \frac{0}{1} = 0.$$

Similarly,

$$f_y(0, 0) = \frac{\partial f}{\partial y} = \lim_{\delta_y \to 0} \frac{f(0, 0 + \partial y) - f(0, 0)}{\delta y}$$

$$= \lim_{\delta_y \to 0} \frac{\dfrac{0 - (0 + \delta y)^2}{0 + (0 + \delta y)^2 + 1} - \dfrac{0}{1}}{\delta y}$$

$$= \lim_{\delta_y \to 0} \frac{-\delta y}{(\delta y)^2 + 1}$$

$$= \frac{0}{1} = 0.$$

Hence, $f_x(0, 0) = 0$

and $f_y(0, 0) = 0$

Example 3. *If $u = f(y/x)$, then prove that*

$$x\frac{\partial u}{\partial x} + y\frac{\partial u}{\partial y} = 0.$$

Solution : $u = f\left(\frac{y}{x}\right)$

$\therefore$ $\frac{\partial u}{\partial x} = f'\left(\frac{y}{x}\right)\left(-\frac{y}{x^2}\right)$

$\therefore$ $x\frac{\partial u}{\partial x} = -\frac{y}{x}f'\left(\frac{y}{x}\right)$ .... (i)

Again, $\frac{\partial u}{\partial y} = f'\left(\frac{y}{x}\right)\frac{1}{x}$

$\therefore$ $y\frac{\partial u}{\partial y} = \frac{y}{x}f'\left(\frac{y}{x}\right)$ .... (ii)

Adding (i) and (ii), we get

$$x\frac{\partial u}{\partial x} + y\frac{\partial u}{\partial y} = 0$$

Example 4. *If* $f(x, y) = \begin{cases} 1 & \text{if } (x, y) \neq (0, 0) \\ 0 & \text{if } (x, y) = (0, 0) \end{cases}$ *show that $f_x(0, 0)$ and $f_y(0, 0)$ does not exist.*

Solution : $f_x(0, 0) = \frac{\partial f(0, 0)}{\partial x}$

$$= \lim_{\delta_x \to 0}\frac{f(0 + \delta x, 0) - f(0, 0)}{\delta x}$$

$$= \lim_{\delta_x \to 0}\frac{1 - 0}{\delta x} = \lim_{\delta_x \to 0}\frac{1}{\delta x}$$

The limit is not finite. Hence, $f_x(0, 0)$ does not exist.

Similarly, $f_y(0, 0) = \frac{\partial}{\partial y}f(0, 0)$

$$= \lim_{\delta_y \to 0}\frac{f(0, 0 + \delta y) - f(0, 0)}{\delta y}$$

$$= \lim_{\delta_y \to 0} \frac{1-0}{\delta y} = \lim_{\delta_y \to 0} \frac{1}{\delta y}$$

This limit is also not finite. Hence, $f_y(0, 0)$ also does not exist.

Example 5. *If* $f(x, y) = \begin{cases} \dfrac{xy}{x^2+y^2}, & (x, y) \neq (0, 0) \\ 0 & (x, y) = (0, 0) \end{cases}$

Show that both partial derivatives $f_x(0, 0)$ *and* $f_y(0, 0)$ *exist but the function is not continuous at origin.*

Solution : Using the definition of partial derivatives

$$f_x(0, 0) = \lim_{h \to 0} \frac{f(0+h, 0) - f(0, 0)}{h} = \lim_{h \to 0} \frac{0}{h} = 0$$

and $$f_y(0, 0) = \lim_{k \to 0} \frac{f(0, 0+k) - f(0, 0)}{k} = \lim_{k \to 0} \frac{0}{k} = 0$$

Thus, partial derivatives exist at origin.

Now, to test the continuity at origin, if we let the path of approach along the line $y = mx$, we get

$$\lim_{(x, y) \to (0, 0)} f(x, y) = \lim_{x \to 0} \frac{x \cdot mx}{x^2 + m^2x^2} = \frac{m}{1+m^2}.$$

So that the limit depends on the value of m. Therefore, limit is not unique and is different for the different paths. It follows that limit does not exist. Hence, function $f(x, y)$ is not continuous at origin.

Example 6. *If* $f(x, y) = \begin{cases} \dfrac{x^3+y^3}{x-y}, & x \neq y \\ 0 & x = y \end{cases},$

then show that the function is discontinuous at the origin but possesses partial derivatives f_x *and* f_y *at every point including the origin.*

Solution : Let $(a, b) \in \mathbf{R}^2$ be an arbitrary point. Then by definition of partial derivatives

$$f_x(a, b) = \lim_{h \to 0} \frac{f(a+h, b) - f(a, b)}{h}$$

$$= \lim_{h \to 0} \left[\frac{\dfrac{(a+h)^3 + b^3}{a+h-b} - \dfrac{(a^3+b^3)}{a-b}}{h} \right]$$

$$= \lim_{h \to 0} \left[\frac{(a-b)(a^3+b^3)+(a-b)(h^3+3ah^2+3a^2h) - (a-b)(a^3+b^3) - h(a^3+b^3)}{h(a+h-b)(a-b)} \right]$$

$$= \lim_{h \to 0} \frac{(a-b)(h^2+3ah+3a^2)-(a^3+b^3)}{(a+h-b)(a-b)}$$

$$= \frac{(a-b)\,3a^2-(a^3+b^3)}{(a-b)^2} = \frac{2a^3-3a^2b-b^3}{(a-b)^2}$$

Also, $$f_y(a, b) = \lim_{k \to 0} \frac{f(a, b+k)-f(a, b)}{k}$$

$$= \lim_{k \to 0} \left[\frac{\dfrac{a^3+(b+k)^3}{a-(b+k)} - \dfrac{a^3+b^3}{a-b}}{k} \right]$$

$$= \lim_{k \to 0} \left[\frac{(a-b)(k^3+3bk^2+3b^2k)+k(a^3+b^3)}{k(a-b-k)(a-b)} \right]$$

$$= \frac{(a-b)\,3b^2+(a^3+b^3)}{(a-b)^2} = \frac{a^3+3ab^2-2b^3}{(a-b)^2}$$

Again, $$f_x(0, 0) = \lim_{h \to 0} \frac{f(0+h, 0)-f(0, 0)}{h}$$

$$= \lim_{h \to 0} \frac{h^3}{h^2} = 0$$

$$f_y(0, 0) = \lim_{k \to 0} \frac{f(0, 0+k)-f(0, 0)}{k}$$

$$= \lim_{k \to 0} \left(\frac{-k^3}{k^2} \right) = 0.$$

Hence, partial derivatives f_x and f_y exist at every point of $\mathbf{R}^2$ including origin.

To test the continuity of the function at origin, see the previous chapter.

EXERCISE 8 (A)

1. $f(x, y) = xy^2 + x^2y$, find f_x and f_y by definition.
2. $f(x, y) = \sqrt{x^4+y^2+1}$, find the value of $f_x(1, 2)$ and $f_y(1, 2)$.
3. $$f(x, y) = \begin{cases} \dfrac{x^2-xy}{x+y}, & \text{if } (x, y) \neq (0, 0) \\ 0 & \text{if } (x, y) = (0, 0) \end{cases}$$

   find the value of $f_x(0, 0)$ and $f_y(0, 0)$.

4. $$f(x, y) = \begin{cases} \tan \dfrac{xy}{x^2 + y^2}, & \text{if } (x, y) \neq (0, 0) \\ 0 & \text{if } (x, y) = (0, 0) \end{cases}$$

find the value of $f_x(0, 0)$ and $f_y(0, 0)$.

5. If $f(x, y) = x^3 y + e^{xy^2}$, then find f_x and f_y by the definition of partial derivatives.

6. If $f(x, y) = xy \left(\dfrac{x^2 - y^2}{x^2 + y^2}\right)$, when $(x, y) \neq (0, 0)$ and $f(0, 0) = 0$, show that

$$f_x(x, 0) = 0 = f_y(0, y)$$
$$f_x(0, y) = -y, \; f_y(x, 0) = x.$$

7. If $$f(x, y) = \begin{cases} xy \tan\left(\dfrac{y}{x}\right), & (x, y) \neq (0, 0) \\ 0, & (x, y) = (0, 0). \end{cases}$$

Show that $$xf_x + yf_y = 2f.$$

8. Calculate $f_x, f_y, f_x(0, 0), f_y(0, 0)$ for the following functions :

(i) $$f(x, y) = \begin{cases} \dfrac{x^3 - y^3}{x^2 + y^2}, & x \neq 0, \; y \neq 0 \\ 0 & x = 0 = y \end{cases}$$

(ii) $$f(x, y) = \begin{cases} \dfrac{xy}{\sqrt{x^2 + y^2}}, & \text{if } x^2 + y^2 \neq 0 \\ 0 & \text{if } x = y = 0 \end{cases}$$

9. Let $$f(x, y) = \begin{cases} \sin\left(\dfrac{xy}{x^2 + y^2}\right), & \text{if } (x, y) \neq (0, 0) \\ 0 & \text{if } (x, y) = (0, 0) \end{cases}$$

Evaluate $f_x(0, 0)$ and $f_y(0, 0)$.

Show that $f(x, y)$ is not continuous at the origin.

10. If $f(x, y) = \sqrt{|xy|}$, find $f_x(0, 0)$ and $f_y(0, 0)$.

ANSWERS

1. $y^2 + 2xy, \; 2xy + x^2$
2. $\sqrt{2}/3, \; 8\sqrt{2}/3$
3. $1, 0$
4. $0, 0$
5. $f_x = 3x^2 y + y^2 e^{xy^2}, f_y = x^3 + 2xy\, e^{xy^2}$
9. $0, 0$
10. $0, 0.$

8.4. Higher Order Partial Derivatives

If a function $f(x, y)$ has partial derivatives of the first order at each point (x, y) of a given region, then f_x, f_y are themselves functions of x and y. These derivatives can be further differentiated partially w.r.t. x, y and called second order partial derivatives.

Thus, partial derivatives of $\frac{\partial f}{\partial x}$ w.r.t. x and y are $\frac{\partial}{\partial x}\left(\frac{\partial f}{\partial x}\right)$ and $\frac{\partial}{\partial y}\left(\frac{\partial f}{\partial x}\right)$ respectively.

Also,
$$\frac{\partial}{\partial x}\left(\frac{\partial f}{\partial x}\right) = \frac{\partial^2 f}{\partial x^2} = f_{xx}$$

and
$$\frac{\partial}{\partial y}\left(\frac{\partial f}{\partial x}\right) = \frac{\partial^2 f}{\partial y\, \partial x} = f_{yx}$$

Similarly, partial derivatives of $\frac{\partial f}{\partial y}$ w.r.t. x and y are

$$\frac{\partial}{\partial x}\left(\frac{\partial f}{\partial y}\right) = \frac{\partial^2 f}{\partial x\, \partial y} = f_{xy}$$

and
$$\frac{\partial}{\partial y}\left(\frac{\partial f}{\partial y}\right) = \frac{\partial^2 f}{\partial y^2} = f_{yy} \text{ respectively.}$$

8.5. Partial Derivatives of Second Order

Partial Derivatives of second order are illustrated as follows :

At a point (a, b)

(i) $$f_{xx}(a, b) = \left[\frac{\partial^2 f}{\partial x^2}\right]_{(a,\, b)} = \lim_{h \to 0} \frac{f_x(a+h, b) - f_x(a, b)}{h}$$

(ii) $$f_{yy}(a, b) = \left[\frac{\partial^2 f}{\partial y^2}\right]_{(a,\, b)} = \lim_{h \to 0} \frac{f(a+2h, b) - 2f(a+h, b) + f(a, b)}{(h)^2}$$

$$= \lim_{k \to 0} \frac{f_y(a, b+k) - f_y(a, b)}{k}$$

$$= \lim_{k \to 0} \frac{f(a, b+2k) - 2f(a, b+k) + f(a, b)}{(k)^2}$$

(iii) $$f_{xy}(a, b) = \left(\frac{\partial^2 f}{\partial x\, \partial y}\right)_{(a,\, b)}$$

$$= \lim_{h \to 0} \frac{f_y(a+h, b) - f_y(a, b)}{h}$$

$$= \lim_{h\to 0}\frac{1}{h}\left[\lim_{k\to 0}\frac{f(a+h,b+k)-f(a+h,b)}{k}\right.$$

$$\left. - \lim_{k\to 0}\frac{f(a,b+k)-f(a,b)}{k}\right]$$

$$= \lim_{h\to 0}\lim_{k\to 0}\frac{f(a+h,b+k)-f(a+h,b)-f(a,b+k)+f(a,b)}{hk}$$

(iv) $f_{yx}(a,b) = \left(\dfrac{\partial^2 f}{\partial y\,\partial x}\right)_{(a,b)}$

$$= \lim_{k\to 0}\frac{f_x(a,b+k)-f_x(a,b)}{k}$$

$$= \lim_{k\to 0}\frac{1}{k}\left[\lim_{h\to 0}\frac{f(a+h,b+k)-f(a,b+k)}{h}\right.$$

$$\left. - \lim_{h\to 0}\frac{f(a+h,b)-f(a,b)}{h}\right]$$

$$= \lim_{k\to 0}\lim_{h\to 0}\frac{f(a+h,b+k)-f(a+h,b)-f(a,b+k)+f(a,b)}{hk}$$

Thus, from (iii) and (iv) we see that $f_{xy}(a,b)$ and $f_{yx}(a,b)$ are the repeated limits of the same expression taken in different orders. Hence, $f_{xy}(a,b)$ and $f_{yx}(a,b)$ may or may not be equal.

In a similar manner higher order partial derivatives are defined. For example

$\dfrac{\partial^3 f}{\partial x\,\partial x\,\partial y} = f_{xxy}$ and so on.

Schwarz's Theorem 1. ***If*** (a, b) ***be a point of the domain*** $D \subset R^2$ ***of a real valued function*** $f(x, y)$ ***such that***

(i) f_x ***exists in a certain neighbourhood of*** (a, b)

(ii) f_{xy} ***is continuous at*** (a, b),

then $f_{yx}(a,b)$ ***exists and is equal to*** $f_{xy}(a,b)$, ***i.e.,*** $f_{yx}(a,b) = f_{xy}(a,b)$.

Proof : Under the given conditions f_x, f_y and f_{xy} exist in a certain neighbourhood of (a, b). Let $(a+h, b+k)$ be any point of this neighbourhood. Consider

$$\phi(h,k) = f(a+h,b+k)-f(a+h,b)-f(a,b+k)+f(a,b)$$

and $\quad g(y) = f(a+h,y)-f(a,y)$

so that $\phi(h,k) = g(a+h)-g(a)$ ...(1)

Since f_y exists in a neighbourhood of (a, b), the function $g(y)$ is derivable in open interval $]b, b+k[$ and therefore, by Lagrange's mean value theorem, we get from (1)

$$\begin{aligned}\phi(h, k) &= k\,g'(b+\theta k), \quad 0<\theta<1\\ &= k\,[f_y(a+h, b+\theta k) - f_y(a, b+\theta k)] \quad \text{... (2)}\end{aligned}$$

Again, since f_{xy} exists in a neighbourhood of (a, b), the function f_y is differentiable with respect to x in $]\,a, a+h[$ and therefore, by Lagrange's mean value theorem, we get from (2)

$$\phi(h, k) = h\,k\,f_{xy}(a+\theta' h, b+\theta k) \;\ldots\; 0 < \theta' < 1$$

or
$$\begin{aligned}f(a+h, b+k) - f(a+h, b) - f(a, b+k) + f(a, b)\\ = hk\,f_{xy}(a+\theta' h, b+\theta k), \qquad 0<\theta<1 \text{ and } 0<\theta'<1\end{aligned}$$

$$\Rightarrow \frac{1}{k}\left\{\frac{f(a+h, b+k) - f(a, b+k)}{h} - \frac{f(a+h, b) - f(a, b)}{h}\right\}$$
$$= f_{xy}(a+\theta' h, b+\theta k).$$

Since f_{xy} exists in a neighbourhood of (a, b), this gives when $h \to 0$

$$\frac{f_x(a, b+k) - f_x(a, b)}{k} = \lim_{h\to 0} f_{xy}(a+\theta' h, b+\theta k)$$

Again, since f_{xy} is continuous at (a, b) so that taking limit when $k \to 0$, we get

$$\lim_{k\to 0}\frac{f_x(a, b+k) - f_x(a, b)}{k} = \lim_{k\to 0}\lim_{h\to 0} f_{xy}(a+\theta' h, b+\theta k)$$

or
$$f_{yx}(a, b) = f_{xy}(a, b).$$

Note : This theorem lays down sufficient conditions for the equality of f_{xy} and f_{yx} at a point (a, b).

8.6. Corollary

If f_{xy} and f_{yx} are both continuous at (a, b), then

$$f_{xy}(a, b) = f_{yx}(a, b).$$

ILLUSTRATIVE EXAMPLES

Example 1. *If* $f(x, y) = x^2 + xy - y^2$ *find* $f_x, f_y, f_{xx}, f_{xy}, f_{yy}$ and $f_{yx}(1, 2)$ *by definition and show that* $f_{xy} = f_{yx}$.

Solution :

$$f_x(1, 2) = \frac{\partial}{\partial x} f(1, 2) = \lim_{\delta x\to 0}\frac{f(1+\delta x, 2) - f(1, 2)}{\delta x}$$

$$= \lim_{\delta x\to 0}\frac{\{(1+\delta)\,x^2 + 2\,(1+\partial x) - 4\} - (-1)}{\partial x}$$

[as $f(1, 2) = -1$]

$$= \lim_{\delta x \to 0} \frac{1 + (\delta x)^2 + 2\delta x + 2 + 2\delta x - 4 + 1}{\delta x}$$

$$= \lim_{\delta x \to 0} \{4 + \delta x\} = 4.$$

$$f_y(1, 2) = \frac{\partial}{\partial y} f(1, 2) = \lim_{\delta y \to 0} \frac{f(1, 2 + \delta y) - f(1, 2)}{\delta y}$$

$$= \lim_{\delta y \to 0} \frac{1 + (2 + \delta y) - (2 + \delta)^2 - (-1)}{\delta y}$$

$$= \lim_{\delta y \to 0} \frac{1 + 2 + \delta y - \{4 + (\delta y)^2 + 4\delta y\} + 1}{\delta y}$$

$$= \lim_{\delta y \to 0} \frac{-(\delta y)^2 - 3\delta y}{\delta y}$$

$$= \lim_{\delta y \to 0} \{-\partial y - 3\} = -3.$$

$$f_{xx}(1, 2) = \frac{\partial^2}{\partial x^2} f(1, 2) = \lim_{\delta x \to 0} \frac{f(1 + 2\delta x, 2) - 2f(1 + \delta x, 2) + f(1, 2)}{(\partial x)^2}$$

$$= \lim_{\delta x \to 0} \frac{\{(1 + 2\delta x)^2 + 2(1 + 2\delta x) - 4\} - 2\{(1 + \delta x)^2 + 2(1 + \delta x) - 4 + (-1)}{(\delta x)^2}$$

$$= \lim_{\delta x \to 0} \frac{1 + 4(\delta x)^2 + 4\delta x + 2 + 4\delta x - 4 - 2\{1 + (\delta x)^2 + 2\delta x + 2 + 2\delta x - 4\} - 1}{(\delta x)^2}$$

$$= \lim_{\delta x \to 0} \frac{2(\delta x)^2}{(\delta x)^2}$$

$$= \lim_{\delta x \to 0} 2 = 2.$$

$$f_{yy}(1, 2) = \frac{\partial^2}{\partial y^2} f(1, 2)$$

$$= \lim_{\delta y \to 0} \frac{f(1, 2 + 2\delta y) - 2f(1, 2 + \delta y) + f(1, 2)}{(\delta y)^2}$$

$$= \lim_{\delta y \to 0} \frac{\{1 + (2 + 2\delta y) - (2 + 2\delta y)^2\} - 2\{1 + (2 + \delta y) - (2 + \delta y)^2\} + (-1)}{(\delta y)^2}$$

$$= \lim_{\delta y \to 0} \frac{1 + 2 + 2\delta y - \{4 + 4(\delta y)^2 + 8\delta y\} - 6 - 2\delta y + 2\{4 + (\delta y)^2 + 4\delta y\} - 1}{(\delta y)^2}$$

$$= \lim_{\delta y \to 0} \frac{-2(\delta y)^2}{(\delta y)^2}$$

$$= \lim_{\delta y \to 0} -2 = -2.$$

$$f_{xy}(1, 2) = \frac{\partial^2}{\partial x\, \partial y} f(1, 2)$$

$$= \lim_{\delta y \to 0} \lim_{\delta x \to 0} \frac{f(1 + \delta x, 2 + \delta y) - f(1 + \delta x, 2) - f(1, 2 + \delta y) + f(1, 2)}{\delta y\, \delta x}$$

$$= \lim_{\delta y \to 0} \lim_{\delta x \to 0} \frac{[\{(1 + \delta x)^2 + (1 + \delta x)(2 + \delta y) - (2 + \delta y)^2\} - \{(1 + \delta x)^2 + 2(1 + \delta x) - 4\} - \{1 + (2 + \delta y) - (2 + \delta y)^2 - 1\}]}{\delta y\, \delta x}$$

$$= \lim_{\delta y \to 0} \lim_{\delta x \to 0} \frac{[\{1 + (\delta x)^2 + 2\delta x + 2 + 2\delta x + \delta y + \delta x\, \delta y - 4 + (\delta y)^2 + 4\delta y\} - \{1 + (\delta x)^2 + 2\delta x + 2 + 2\delta x - 4\} - \{1 + 2 + \delta y - 4 - (\delta y)^2 - 4\delta y\} - 1]}{\delta y\, \delta x}$$

$$= \lim_{\delta y \to 0} \lim_{\delta x \to 0} \frac{4\delta y\, \delta x}{\delta y\, \delta x}$$

$$= \lim_{\delta y \to 0} \lim_{\delta x \to 0} 4 = 4.$$

$$f_{yx}(1, 2) = \frac{\partial^2}{\partial y\, \partial x} f(1, 2)$$

$$= \lim_{\delta x \to 0} \lim_{\delta y \to 0} \frac{f(1 + \delta x, 2 + \delta y) - f(1 + \delta x, 2) - f(1, 2 + \delta y) + f(1, 2)}{\delta x\, \delta y}$$

$$= \lim_{\delta x \to 0} \lim_{\delta y \to 0} \frac{4\delta y\, \delta x}{\delta x\, \delta y} \qquad \text{[Solving as above]}$$

$$= \lim_{\delta x \to 0} \lim_{\delta y \to 0} 4 = 4.$$

Hence, it is clear that

$$\frac{\partial^2 f}{\partial x\, \partial y} = \frac{\partial^2 f}{\partial y\, \partial x}$$

Example 2. *Given that*

$$f(x, y) = \begin{cases} \dfrac{xy(x^2 - y^2)}{x^2 + y^2} & \text{if } (x, y) \neq (0, 0) \\ = 0, & \textit{otherwise} \end{cases}$$

Evaluate

(i) $f_x(0, 0)$, $f_y(0, 0)$, $f_{xx}(0, 0)$, $f_{yy}(0, 0)$ and $f_{xy}(0, 0)$

(ii) Show that $f_{xy}(0, 0) \neq f_{yx}(0, 0)$.

Solution :

(i)

$$f_x(0,0) = \frac{\partial f(0,0)}{\partial x}$$

$$= \lim_{\delta x \to 0} \frac{(0+\delta x, 0) - f(0,0)}{\delta x}$$

$$= \lim_{\delta x \to 0} \frac{0-0}{\delta x} = 0.$$

$$f_y(0,0) = \frac{\partial f(0,0)}{\partial y}$$

$$= \lim_{\delta y \to 0} \frac{(0+\partial y, 0) - f(0,0)}{\delta y}$$

$$= \lim_{\delta y \to 0} \frac{0-0}{\delta y} = 0.$$

$$f_{xx}(0,0) = \frac{\partial^2 f(0,0)}{\partial x^2}$$

$$= \lim_{\delta x \to 0} \frac{f(0+2\delta x, 0) - 2f(0+\delta x, 0) + f(0,0)}{(\delta x)^2}$$

$$= \lim_{\delta x \to 0} \frac{f(2\delta x, 0) - 2f(\delta x, 0) + f(0,0)}{(\delta x)^2}$$

$$= \lim_{\delta x \to 0} \frac{0-0+0}{(\delta x)^2} = 0.$$

$$f_{yy}(0,0) = \frac{\partial^2 f(0,0)}{\partial y^2}$$

$$= \lim_{\delta y \to 0} \frac{f(0, 0+2\delta y) - 2f(0, 0+\delta y) + f(0,0)}{(\delta y)^2}$$

$$= \lim_{\delta y \to 0} \frac{f(0, 2\delta y) - 2f(0, \delta y) + f(0,0)}{(\delta y)^2}$$

$$= \lim_{\delta y \to 0} \frac{0-0+0}{(\delta y)^2} = 0.$$

$$f_{xy}(0,0) = \frac{\partial^2 f(0,0)}{\partial x\, \partial y}$$

$$[f(0+\delta x, 0+\delta y) - f(0+\delta x, 0)$$

$$= \lim_{\delta x \to 0} \lim_{\delta y \to 0} \frac{-f(0, 0 + \delta y) + f(0, 0)]}{\delta x\, \delta y}$$

$$= \lim_{\delta x \to 0} \lim_{\delta y \to 0} \frac{\delta x\, \delta y \{ (\delta x)^2 - (\delta y)^2 \}}{\{ (\delta x)^2 + (\delta y)^2 \}\, \delta x\, \delta y}$$

$$= \lim_{\delta x \to 0} \lim_{\delta y \to 0} \frac{\{ (\delta x)^2 - (\delta y)^2 \}}{(\delta x)^2 + (\delta y)^2}$$

$$= \lim_{\delta x \to 0} \frac{(\delta x)^2}{(\delta x)^2} = \lim_{\delta x \to 0} 1 = 1. \qquad \text{... (A)}$$

$$f_{yx}(0, 0) = \frac{\partial^2 f(0, 0)}{\partial y\, \partial x}$$

$$= \lim_{\delta y \to 0} \lim_{\delta x \to 0} \frac{f(0 + \delta x, \delta y) - f(\delta x, 0) - f(0, \delta y) + f(0, 0)}{\delta y\, \delta x}$$

$$= \lim_{\delta y \to 0} \lim_{\delta x \to 0} \frac{\delta x\, \delta y\ \{ (\delta x)^2 - (\delta y)^2 \}}{\delta y\, \delta x\, \{ (\delta x)^2 + (\delta y)^2 \}}$$

$$= \lim_{\delta y \to 0} \lim_{\delta x \to 0} \frac{(\delta x)^2 - (\delta y)^2}{(\delta x)^2 + (\delta y)^2}$$

$$= \lim_{\delta y \to 0} \frac{-(\delta y)^2}{(\delta y)^2} = \lim_{\delta y \to 0} -1 = -1. \qquad \text{... (B)}$$

(ii) Hence, from relation (A) and (B) $f_{xy}(0, 0) \neq f_{yx}(0, 0)$.

Example 3. *If* $u = x^y$ *show that*

$$\frac{\partial^2 u}{\partial y\, \partial x} = \frac{\partial^2 u}{\partial x\, \partial y}.$$

Solution :

$$u = x^y$$

$$\therefore \quad \frac{\partial u}{\partial y} = x^y \log x$$

$$\therefore \quad \frac{\partial}{\partial x}\left(\frac{\partial u}{\partial y}\right) = \frac{\partial^2 u}{\partial x\, \partial y}$$

$$= y\, x^{y-1} \log x + x^y \cdot \frac{1}{x}$$

$$= y\, x^{y-1} \log x + x^{y-1} \qquad \text{... (i)}$$

Again,

$$\frac{\partial u}{\partial x} = y\, x^{y-1}$$

$$\therefore \quad \frac{\partial^2 u}{\partial y\, \partial x} = \frac{\partial}{\partial y}\left(\frac{\partial u}{\partial x}\right)$$

$$= \frac{\partial}{\partial y}(y\,x^{y-1})$$

$$= x^{y-1} + y\,x^{y-1}\log x \qquad \text{... (ii)}$$

From (i) and (ii), we see

$$\frac{\partial^2 u}{\partial x\,\partial y} = \frac{\partial^2 u}{\partial y\,\partial x}.$$

Example 4. *If* $V = (x^2 + y^2 + z^2)^{-1/2}$, *prove that*

(1) $x\dfrac{\partial V}{\partial x} + y\dfrac{\partial V}{\partial y} + z\dfrac{\partial V}{\partial z} = -V$

(2) $\dfrac{\partial^2 V}{\partial x^2} + \dfrac{\partial^2 V}{\partial y^2} + \dfrac{\partial^2 V}{\partial z^2} = 0.$

Solution : $V = (x^2 + y^2 + z^2)^{-1/2}$ ... (i)

$$\therefore \quad \frac{\partial V}{\partial x} = -\frac{1}{2}(x^2 + y^2 + z^2)^{-3/2}.2x$$

$$= -x\,(x^2 + y^2 + z^2)^{-3/2} \qquad \text{... (ii)}$$

$$\therefore \quad \frac{\partial^2 V}{\partial x^2} = [(x^2 + y^2 + z^2)^{-3/2} + x.\left(-\frac{3}{2}\right)(x^2 + y^2 + z^2)^{-5/2}.2x]$$

$$= -[(x^2 + y^2 + z^2)^{-3/2} - 3x^2(x^2 + y^2 + z^2)^{-5/2}]$$

$$= -(x^2 + y^2 + z^2)^{-5/2}[1 - 3x^2(x^2 + y^2 + z^2)^{-1}]$$

$$= -(x^2 + y^2 + z^2)^{-5/2}\left[1 - \frac{3x^2}{x^2 y^2 + z^2}\right]$$

$$= -(x^2 + y^2 + z^2)^{-5/2}[y^2 + z^2 - 2x^2]$$

$$= \frac{2x^2 - y^2 - z^2}{(x^2 + y^2 + z^2)^{5/2}} \qquad \text{... (iii)}$$

Similarly,

$$\frac{\partial V}{\partial y} = -y\,(x^2 + y^2 + z^2)^{-3/2} \qquad \text{... (iv)}$$

$$\frac{\partial^2 V}{\partial y^2} = \frac{2y^2 - z^2 - x^2}{(x^2 + y^2 + z^2)^{3/2}} \qquad \text{... (v)}$$

$$\frac{\partial V}{\partial z} = -z\,(x^2 + y^2 + z^2)^{-3/2} \qquad \text{... (vi)}$$

$$\frac{\partial^2 V}{\partial z^2} = \frac{2z^2 - x^2 - y^2}{(x^2 + y^2 + z^2)^{5/2}} \qquad \text{... (vii)}$$

Hence, (1) $x\dfrac{\partial V}{\partial x}+y\dfrac{\partial V}{\partial y}+z\dfrac{\partial V}{\partial z} = \dfrac{-x^2-y^2-z^2}{(x^2+y^2+z^2)^{3/2}} = \dfrac{-(x^2+y^2+z^2)}{(x^2+y^2+z^2)^{3/2}}$

$$= -(x^2+y^2+z^2)^{-1/2} = -V.$$

(2) From relations (iii), (v) and (vii)

$$\frac{\partial^2 V}{\partial x^2}+\frac{\partial^2 V}{\partial y^2}+\frac{\partial^2 V}{\partial z^2} = \frac{2x^2-y^2-z^2+2y^2-z^2-x^2+2z^2-x^2-y^2}{(x^2+y^2+z^2)^{5/2}}$$

$$= 0.$$

Example 5. *If* $z = x^2\tan^{-1}\left(\dfrac{y}{x}\right) - y^2\tan^{-1}\left(\dfrac{x}{y}\right)$, *then prove that*

$$\frac{\partial^2 z}{\partial y\,\partial x} = \frac{x^2-y^2}{x^2+y^2}.$$

Solution : $z = x^2\tan^{-1}\left(\dfrac{y}{x}\right) - y^2\tan^{-1}\left(\dfrac{x}{y}\right)$

$$\therefore \quad \frac{\partial z}{\partial x} = 2x\tan^{-1}\frac{y}{x} + x^2\frac{1}{1+\dfrac{y^2}{x^2}}\left(-\frac{y}{x^2}\right) - y^2\frac{1}{1+\dfrac{x^2}{y^2}}\cdot\frac{1}{y}$$

$$= 2x\tan^{-1}\frac{y}{x} - \frac{x^2 y}{x^2+y^2} - \frac{y^3}{x^2+y^2}$$

$$\therefore \quad \frac{\partial^2 z}{\partial y\,\partial x} = \frac{\partial}{\partial y}\left(\frac{\partial z}{\partial x}\right)$$

$$= 2x\cdot\frac{1}{1+\dfrac{y^2}{x^2}}\cdot\frac{1}{x} - x^2\left[\frac{(x^2+y^2).1-y.2y}{(x^2+y^2)^2}\right]$$

$$- \frac{(x^2+y^2).3y^2-y^3.2y}{(x^2+y^2)^2}$$

$$= \frac{2x^2}{x^2+y^2} - x^2\left\{\frac{x^2-y^2}{(x^2+y^2)^2}\right\} - \frac{y^2.[3x^2+3y^2-2y^2]}{(x^2+y^2)^2}$$

$$= \frac{2x^2}{x^2+y^2} - x^2\left\{\frac{x^2-y^2}{(x^2+y^2)^2}\right\} - y^2\left\{\frac{3x^2+y^2}{(x^2+y^2)^2}\right\}$$

$$= \frac{2x^2(x^2+y^2) - x^2(x^2-y^2) - y^2(3x^2+y^2)}{(x^2+y^2)^2}$$

$$= \frac{2x^4(x^2+y^2) - x^4 + x^2y^2 - 3x^2y^2 - y^4}{(x^2+y^2)^2}$$

$$= \frac{x^4 - y^4}{(x^2 + y^2)^2} = \frac{(x^2 - y^2)(x^2 + y^2)}{(x^2 + y^2)^2}$$

$$= \frac{x^2 - y^2}{x^2 + y^2}.$$

Example 6. *If* $x^x y^y z^z = c$, *then show that*

$$\frac{\partial^2 z}{\partial x\, \partial y} = -(x \log ex)^{-1}, \text{ when } x = y = z.$$

Solution : $x^x y^y z^z = c$

Taking logarithm, we get

$$x \log x + \log y + z \log z = \log c.$$

Differentiating partially w.r.t., x [Noting that z is a function of x and y here], we get

$$\left(x.\frac{1}{x} + 1.\log x\right) + \left(z.\frac{1}{z} + 1.\log z\right)\frac{\partial z}{\partial x} = 0$$

$$\Rightarrow \quad (1 + \log x) + (1 + \log z)\frac{\partial z}{\partial x} = 0$$

$$\Rightarrow \quad \frac{\partial z}{\partial x} = -\frac{1 + \log x}{1 + \log z}.$$

Similarly, $$\frac{\partial z}{\partial y} = -\frac{1 + \log y}{1 + \log z}$$

Now, $$\frac{\partial^2 z}{\partial x\, \partial y} = \frac{\partial}{\partial x}\left(\frac{\partial x}{\partial y}\right)$$

$$= \frac{\partial}{\partial x}\left(-\frac{1 + \log y}{1 + \log z}\right)$$

$$= -(1 + \log y)\frac{\partial}{\partial x}\left(\frac{1}{1 + \log z}\right)$$

$$= -(1 + \log y)\left(-\frac{1}{(1 + \log z)^2} \cdot \frac{1}{z}\frac{\partial z}{\partial x}\right)$$

$$= \frac{1 + \log y}{(1 + \log z)^2} . \frac{1}{z}\frac{\partial z}{\partial x}$$

$$= \frac{1 + \log y}{(1 + \log z)^2}\left\{-\frac{1}{z}\left(\frac{1 + \log x}{1 + \log z}\right)\right\}$$

$$= -\frac{(1 + \log y)(1 + \log x)}{(z + 1 + \log z)^3}$$

$$= -\frac{(1+\log x)(1+\log x)}{x(1+\log x)^3} \qquad [\text{at } x = y = z]$$

$$= -\frac{1}{x(1+\log x)}$$

$$= -\frac{1}{x(\log e + \log x)}$$

$$= -\frac{1}{x \log (ex)}$$

$$= -[x \log (ex)]^{-1}.$$

Example 7. *If* $u = e^{xyz}$, *show that*

$$\frac{\partial^3 u}{\partial x\, \partial y\, \partial z} = (1 + 3xyz + x^2y^2z^2)\, e^{xyz}.$$

Solution :

$$u = e^{xyz}$$

$$\therefore \quad \frac{\partial u}{\partial z} = e^{xyz} xy$$

$$\therefore \quad \frac{\partial^2 u}{\partial y\, \partial z} = \frac{\partial}{\partial y}\left(\frac{\partial u}{\partial z}\right) = \frac{\partial}{\partial y}[e^{xyz} xy]$$

$$= x \frac{\partial}{\partial y}[ye^{xyz}]$$

$$= x[e^{xyz} + ye^{xyz} xz]$$

$$= x(1 + xyz)\, e^{xyz}$$

$$= (x + x^2yz)\, e^{xyz}$$

$$\therefore \quad \frac{\partial}{\partial x}\frac{\partial}{\partial y}\left(\frac{\partial u}{\partial z}\right) = \frac{\partial^3 u}{\partial x\, \partial y\, \partial z}$$

$$= (1 + 2xyz)\, e^{xyz} + e^{xyz}.yz\,(x + x^2\, yz)$$

$$= e^{xyz}(1 + 2xyz + xyz + x^2y^2z^2)$$

$$= e^{xyz}(1 + 3xyz + x^2y^2z^2).$$

Example 8. *If* $u = \log (x^3 + y^3 + z^3 - 3xyz)$, *show that,*

(1) $\dfrac{\partial u}{\partial x} + \dfrac{\partial u}{\partial y} + \dfrac{\partial u}{\partial z} = \dfrac{3}{x+y+z}.$

(2) $\left(\dfrac{\partial}{\partial x} + \dfrac{\partial}{\partial y} + \dfrac{\partial}{\partial z}\right)^2 u = -\dfrac{9}{(x+y+z)^2}.$

(3) $$\frac{\partial^2 u}{\partial x^2} + \frac{\partial^2 u}{\partial y^2} + \frac{\partial^2 u}{\partial z^2} = -\frac{3}{(x+y+z)^2}.$$

Solution :

(1) Given

$$u = \log(x^3 + y^3 + z^3 - 3xyz) \qquad \text{... (i)}$$

$$\frac{\partial u}{\partial x} = \frac{3x^2 - 3yz}{x^3 + y^3 + z^3 - 3xyz} \qquad \text{... (ii)}$$

$$\frac{\partial u}{\partial y} = \frac{3y^2 - 3xz}{x^3 + y^3 + z^3 - 3xyz} \qquad \text{... (iii)}$$

$$\frac{\partial u}{\partial z} = \frac{3z^2 - 3xy}{x^3 + y^3 + z^3 - 3xyz} \qquad \text{... (iv)}$$

Adding (ii), (iii) and (iv)

$$\frac{\partial u}{\partial x} + \frac{\partial u}{\partial y} + \frac{\partial u}{\partial z} = \frac{3(x^3 + y^3 + z^3 - xy - yz - xz)}{x^3 + y^3 + z^3 - 3xyz}$$

$$= \frac{3(x^2 + y^2 + z^2 - xy - yz - xz)}{(x+y+z)(x^2 + y^2 + z^2 - xy - yz - xz)}$$

$$= \frac{3}{x+y+z}.$$

(2) $$\left(\frac{\partial}{\partial x} + \frac{\partial}{\partial y} + \frac{\partial}{\partial z}\right)^2 u$$

$$= \left(\frac{\partial}{\partial x} + \frac{\partial}{\partial y} + \frac{\partial}{\partial z}\right)\left(\frac{\partial}{\partial x} + \frac{\partial}{\partial y} + \frac{\partial}{\partial z}\right) u$$

$$= \left(\frac{\partial}{\partial x} + \frac{\partial}{\partial y} + \frac{\partial}{\partial z}\right)\left(\frac{\partial u}{\partial x} + \frac{\partial u}{\partial y} + \frac{\partial u}{\partial z}\right)$$

$$= \left(\frac{\partial}{\partial x} + \frac{\partial}{\partial y} + \frac{\partial}{\partial z}\right)\left(\frac{3}{x+y+z}\right)$$

$$= \frac{\partial}{\partial x}\left(\frac{3}{x+y+z}\right) + \frac{\partial}{\partial y}\left(\frac{3}{x+y+z}\right) + \frac{\partial}{\partial z}\left(\frac{3}{x+y+z}\right)$$

$$= -\frac{3}{(x+y+z)^2} - \frac{3}{(x+y+z)^2} - \frac{3}{(x+y+z)^2}$$

$$= \frac{9}{(x+y+z)^2}.$$

(3) We know from algebra

$$x^3 + y^3 + z^3 - 3xyz = (x+y+z)(x + y\omega + z\omega^2)(x + y\omega^2 + z\omega)$$

where ω is imaginary cube root of unity where $1+\omega+\omega^2=0$ and $\omega^3=1$. Now,

$$u = \log(x^3+y^3+z^3-3xyz)$$

$$= \log(x+y+z)+\log(x+y\omega+z\omega^2) + \log(x+y\omega^2+z\omega) \quad \text{... (v)}$$

$$\frac{\partial u}{\partial x} = \frac{1}{x+y+z}+\frac{1}{x+y\omega+z\omega^2}+\frac{1}{x+y\omega^2+z\omega}$$

$$\frac{\partial^2 u}{\partial x^2} = -\frac{1}{(x+y+z)^2}-\frac{1}{(x+y\omega+z\omega)^2}-\frac{1}{(x+y\omega^2+z\omega)^2} \quad \text{... (vi)}$$

$$\frac{\partial u}{\partial y} = \frac{1}{x+y+z}+\frac{\omega}{x+y\omega+z\omega^2}+\frac{\omega^2}{x+y\omega^2+z\omega}$$

$$\frac{\partial^2 u}{\partial y^2} = -\frac{1}{(x+y+z)^2}-\frac{\omega^2}{(x+y\omega+z\omega^2)^2}-\frac{\omega^4}{(x+y\omega^2+z\omega)^2} \quad \text{... (vii)}$$

$$\frac{\partial u}{\partial z} = \frac{1}{x+y+z}+\frac{\omega^2}{x+y\omega+z\omega^2}+\frac{\omega}{x+y\omega^2+z\omega}$$

$$\frac{\partial^2 u}{\partial z^2} = -\frac{1}{(x+y+z)^2}-\frac{\omega^4}{(x+y\omega+z\omega^2)^2}-\frac{\omega^2}{(x+y\omega^2+z\omega)^2} \quad \text{... (viii)}$$

Adding (vi), (vii) and (viii)

$$\frac{\partial^2 u}{\partial x^2}+\frac{\partial^2 u}{\partial y^2}+\frac{\partial^2 u}{\partial z^2}$$

$$= -\frac{3}{(x+y+z)^2}-\frac{1+\omega^2+\omega^4}{(x+y\omega+z\omega^2)^2}-\frac{1+\omega^4+\omega^2}{(x+y\omega^2+z\omega)^2}$$

$$= -\frac{3}{(x+y+z)^3}-\frac{1+\omega^2+\omega}{(x+y\omega+z\omega^2)^2}-\frac{1+\omega+\omega^2}{(x+y\omega^2+z\omega)^2}$$

$$= -\frac{3}{(x+y+z)^2}. \qquad [\because 1+\omega^2+\omega=0]$$

Example 9. *If* $u = (1-2xy+y^2)^{-1/2}$ *prove that*

$$\frac{\partial}{\partial x}\left\{(1-x^2)\frac{\partial u}{\partial x}\right\}+\frac{\partial}{\partial y}\left\{u^2\frac{\partial u}{\partial y}\right\} = 0.$$

Solution : $u = (1-2xy+y^2)^{-1/2}$

$$\frac{\partial u}{\partial x} = -\frac{1}{2}(1-2xy+y^2)^{-3/2}(-2y)$$

$$= y(1-2xy+y^2)^{-3/2} \quad \text{...(i)}$$

$$\frac{\partial^2 u}{\partial x^2} = -\frac{3}{2} y (1 - 2xy + y^2)^{-5/2} (-2y)$$

$$= 3y^2 (1 - 2xy + y^2)^{-5/2} \quad \text{... (ii)}$$

$$\frac{\partial u}{\partial y} = -\frac{1}{2} (1 - 2xy + y^2)^{-3/2} (-2x + 2y)$$

$$= (x - y)(1 - 2xy + y^2)^{-3/2} \quad \text{... (iii)}$$

$$\frac{\partial^2 u}{\partial y^2} = -\frac{3}{2} (1 - 2xy + y^2)^{-5/2} (x - y)(-2x + 2y)$$

$$= (1 - 2xy + y^2)^{-3/2} (-1)$$

$$= \frac{3(x-y)^2}{(1 - 2xy + y^2)^{5/2}} - \frac{1}{(1 - 2xy + y^2)^{3/2}} \quad \text{... (iv)}$$

Now, L.H.S. $= \dfrac{\partial}{\partial x}\left\{(1 - x^2)\dfrac{\partial u}{\partial x}\right\} + \dfrac{\partial}{\partial y}\left(y^2 \dfrac{\partial u}{\partial y}\right)$

$$= (1 - x^2)\frac{\partial^2 u}{\partial x^2} - 2x\frac{\partial u}{\partial x} + y^2 \frac{\partial^2 u}{\partial y^2} + 2y \frac{\partial u}{\partial y}$$

Putting the values of $\dfrac{\partial u}{\partial x}, \dfrac{\partial u}{\partial y}, \dfrac{\partial^2 u}{\partial x^2}$ and $\dfrac{\partial^2 u}{\partial y^2}$

From (i), (ii), (iii) and (iv)

$$\text{L.H.S.} = \frac{(1 - x^2)\, 3y^2}{(1 - 2xy + y^2)^{5/2}} + \frac{3y^2 (x - y)^2}{(1 - 2xy + y^2)^{5/2}}$$

$$- \frac{y^2}{(1 - 2xy + y^2)^{3/2}} - \frac{2xy}{(1 - 2xy + y^2)^{3/2}} + \frac{2y(x - y)}{(1 - 2xy + y^2)^{3/2}}$$

$$= \frac{3y^2 (1 - x^2 + x^2 + y^2 - 2xy)}{(1 - 2xy + y^2)^{5/2}} + \frac{-y^2 - 2xy + 2xy - 2y^2}{(1 - 2xy + y^2)^{3/2}}$$

$$= \frac{3y^2 (1 - 2xy + y^2)}{(1 - 2xy + y^2)^{3/2}} - \frac{3y^2}{(1 - 2xy + y^2)^{3/2}}$$

$$= \frac{3y^2}{(1 - 2xy + y^2)^{3/2}} - \frac{3y^2}{(1 - 2xy + y^2)^{3/2}}$$

$= 0 =$ R.H.S.

Example 10. *If $u = f(r)$ where $r^2 = x^2 + y^2$, show that*

$$\frac{\partial^2 u}{\partial x^2} + \frac{\partial^2 u}{\partial y^2} = f''(r) + \frac{f'(r)}{r}$$

Solution : Given

$$r^2 = x^2 + y^2 \qquad \text{... (i)}$$

$$2r\frac{\partial r}{\partial x} = 2x \Rightarrow \frac{\partial r}{\partial x} = \frac{x}{r}$$

$$2r\frac{\partial r}{\partial y} = 2y \Rightarrow \frac{\partial r}{\partial x} = \frac{y}{r}$$

$$u = f(r) \qquad \text{... (ii)}$$

$$\frac{\partial u}{\partial x} = f'(r)\frac{\partial r}{\partial x} = f'(r)\frac{x}{r} \qquad \text{... (iii)}$$

Differentiating partially with respect to x,

$$\frac{\partial^2 u}{\partial x^2} = \frac{\partial}{\partial x}\left\{f'(r)\frac{x}{r}\right\}$$

$$= \frac{\left\{f''\frac{\partial r}{\partial x}x + f'(r)\right\}r - \frac{\partial r}{\partial x}xf'(r)}{r^2}$$

$$= \frac{1}{r^2}\left\{(f''(r)x^2 + f'(r))r - \frac{x^2}{r}f'(r)\right\}$$

$$= \frac{1}{r^3}\{f''(r)x^2r + f'(r)r^2 - x^2f'(r)\} \qquad \text{... (iv)}$$

Differentiating (iii) partially with respect to y

$$\frac{\partial u}{\partial y} = f'(r)\frac{\partial r}{\partial y}$$

$$= f'(r)\frac{y}{r} \qquad \text{... (v)}$$

Differentiating (v) partially with respect to y

$$\frac{\partial^2 u}{\partial y^2} = \frac{\partial}{\partial y}\left\{f'(r)\frac{y}{r}\right\}$$

$$= \frac{\left\{f''(r)\frac{\partial r}{\partial y}y + f'(r)\right\} - r\frac{\partial r}{\partial y}yf'(r)}{r^2}$$

$$= \frac{\left\{f''(r)\frac{y^2}{r} + f'(r)\right\}r - \frac{y^2}{r}f'(r)}{r^2}$$

$$= \frac{1}{r^3}\{f''(r)y^2r + f'(r)r^2 - y^2f'(r)\} \qquad \text{... (vi)}$$

Adding (iv) and (vi)

$$\frac{\partial^2 u}{\partial x^2} + \frac{\partial^2 u}{\partial y^2} = \frac{1}{r^3}\{f''(r)\, r(x^2+y^2) + 2f'(r)\, r^2 - f'(r)\,(x^2+y^2)\}$$

$$= \frac{1}{r^3}\{f''(r)\, r^3 + 3f'(r)\, r^2 - f'(r)\, r^2\}$$

$$= f''(r) + \frac{f'(r)}{r}.$$

Example 11. *If* $\dfrac{x^2}{a^3+u} + \dfrac{y^2}{b^2+u} + \dfrac{z^2}{c^2+u} = 1$, *prove that*

$$u_x^2 + u_y^2 + u_z^2 = 2\,\{xu_x + yu_y + zu_z\}.$$

Solution : Given that

$$\frac{x^2}{a^3+u} + \frac{y^2}{b^2+u} + \frac{z^2}{c^2+u} = 1 \qquad \text{... (i)}$$

Differentiating both sides partially with respect to x

$$\frac{2x}{a^2+u} - \left\{\frac{x^2}{(a^2+u)^2} + \frac{y^2}{(b^2+u)^2} + \frac{z^2}{(c^2+u)^2}\right\}\frac{\partial u}{\partial x} = 0$$

$$\therefore \qquad \frac{\partial u}{\partial x} = \frac{\dfrac{2x}{a^2+u}}{\dfrac{x^2}{(a^2+u)^2} + \dfrac{y^2}{(b^2+u)^2} + \dfrac{z^2}{(c^2+u)^2}} \qquad \text{... (ii)}$$

$$u_x^2 = \left(\frac{\partial u}{\partial x}\right)^2 = \frac{\dfrac{4x^2}{(a^2+u)^2}}{\left\{\dfrac{x^2}{(a^2+u)^2} + \dfrac{y^2}{(b^2+u)^2} + \dfrac{z^2}{(c^2+u)^2}\right\}}.$$

Hence,

$$u_x^2 + u_y^2 + u_z^2 = \frac{4}{\left\{\dfrac{x^2}{(a^2+u)^2} + \dfrac{y^2}{(b^2+u)^2} + \dfrac{z^2}{(c^2+u)^2}\right\}} \qquad \text{... (iii)}$$

From (ii), we get

$$2x\,u_x = \frac{\dfrac{4x^2}{a^2+u}}{\dfrac{x^2}{(a^2+u)^2} + \dfrac{y^2}{(b^2+u)^2} + \dfrac{z^2}{(c^2+u)^2}} \qquad \text{... (iv)}$$

$$\text{Similarly, } 2y\,u_y = \frac{\dfrac{4y^2}{b^2+u}}{\dfrac{x^2}{(a^2+u)^2} + \dfrac{y^2}{(b^2+u)^2} + \dfrac{z^2}{(c^2+u)^2}}$$

$$2z\,u_z = \frac{\dfrac{4z^2}{c^4+u}}{\dfrac{x^2}{(a^2+u)^2} + \dfrac{y^2}{(b^2+u)^2} + \dfrac{z^2}{(c^2+u)^2}} \qquad \text{...(v)}$$

Hence,

$2\{x\,u_x + y\,u_y + z\,u_z\}$

$$= \frac{4\left\{\dfrac{x^2}{a^2+u} + \dfrac{y^2}{b^2+u} + \dfrac{z^2}{c^2+u}\right\}}{\left\{\dfrac{x^2}{(a^2+u)^2} + \dfrac{y^2}{(b^2+u)^2} + \dfrac{z^2}{(c^2+u)^2}\right\}} \qquad \text{... (vi)}$$

$$= \frac{4}{\left\{\dfrac{x^2}{(a^2+u)^2} + \dfrac{y^2}{(b^2+u)^2} + \dfrac{z^2}{(c^2+u)^2}\right\}} \qquad \text{... (vii)}$$

putting the values from (i), (iii) and (vii) it is obvious

$$u_x^2 + u_y^2 + u_z^2 = 2\{x\,u_x + y\,u_y + z\,u_z\}.$$

8.7. Differentiability

Let $f(x, y)$ be a function of two independent variables x and y, defined in some neighbourhood of a point (a, b). Then the function $f(x, y)$ is said to be differentiable at (a, b), if there exists two constants α and β depending on f and (a, b) such that

$$\lim_{(h,k)\to(0,0)} \frac{f(a+h, b+k) - f(a, b) - \alpha h - \beta k}{\sqrt{h^2+k^2}} = 0 \qquad \text{... (1)}$$

From (1) when $k = 0$, we get

$$\lim_{h\to 0} \frac{f(a+h, b) - f(a, b) - \alpha h}{h} = 0$$

or $$\lim_{h\to 0} \frac{f(a+h, b) - f(a, b)}{h} = \alpha$$

or $$f_x(a, b) = \alpha$$

Again, from (1), when $h = 0$, we get

$$\lim_{k\to 0} \frac{f(a, b+k) - f(a, b)}{k} - \beta = 0$$

$$\Rightarrow \qquad \lim_{k \to 0} \frac{f(a, b+k) - f(a, b)}{k} = \beta$$

or
$$f_y(\alpha, \beta) = \beta.$$

Thus, the constants α and β are respectively the partial derivatives of f with respect to x and y. Hence, a function which is differentiable at a point possesses the first order partial derivatives threat.

Theorem 2. ***If $f(x, y)$ is differentiable at a point (a, b), then it is continuous at (a, b) but converse need not be true.***

Proof : Since $f(x, y)$ is differentiable at (a, b), then we have

$$\lim_{(h, k) \to (0, 0)} \frac{f(a+h, b+k) - f(a, b) - h f_x(a, b) - k f_y(a, b)}{\sqrt{h^2 + k^2}} = 0$$

$$\Rightarrow \quad \lim_{(h, k) \to (0, 0)} \{f(a+h, b+k) - f(a, b) - h f_x(a, b) - k f_y(a, b)\} = 0$$

$$\Rightarrow \quad \lim_{(h, k) \to (a, b)} f(a+h, b+k) = f(a, b).$$

Hence, $f(x, y)$ is continuous at (a, b).

For the converse part see the example (1).

ILLUSTRATIVE EXAMPLE

Example 1. *Let* $f(x, y) = \begin{cases} \dfrac{xy}{\sqrt{x^2 + y^2}}, & (x, y) \neq (0, 0) \\ 0, & (x, y) = (0, 0) \end{cases}$

Show that function $f(x, y)$ is continuous and partial derivatives $f_x(0, 0)$, $f_y(0, 0)$ exist but not differentiable at $(0, 0)$.

Solution : To observe continuity of the function, see the previous chapter.

Now, by the definition of partial derivatives

$$f_x(0, 0) = \lim_{h \to 0} \frac{f(0+h, 0) - f(0, 0)}{h} = \lim_{h \to 0} \left(\frac{0}{h}\right) = 0$$

$$\text{and } f_y(0, 0) = \lim_{k \to 0} \frac{f(0, 0+k) - f(0, 0)}{k} = \lim_{k \to 0} \left(\frac{0}{k}\right) = 0$$

Therefore,

$$\lim_{(h, k) \to (0, 0)} \frac{f(0+h, 0+k) - f(0, 0) - h f_x(0, 0) - k f_y(0, 0)}{\sqrt{h^2 + k^2}}$$

$$= \lim_{(h, k) \to (0, 0)} \frac{f(h, k)}{\sqrt{h^2 + k^2}}$$

$$= \lim_{(h, k) \to (0, 0)} \frac{hk}{\sqrt{h^2 + k^2}\,\sqrt{(h^2 + k^2)}} = \lim_{(h, k) \to (0, 0)} \frac{hk}{\sqrt{(h^2 + k^2)}}$$

[let $(h, k) \to (0, 0)$ along the straight line $k = mh$]

$$\lim_{h \to 0} \frac{m}{1+m^2} = \frac{m}{1+m^2}.$$

Since this value depends on m and when $m \neq 0$, $\frac{m}{1+m} \neq 0$. Hence, $f(x, y)$ is not differentiable at (0, 0). Thus, $f(x, y)$ is continuous and $f_x(0, 0)$, $f_y(0, 0)$ exist at origin but $f(x, y)$ is not differentiable at (0, 0).

Example 2. *Show that the function $f(x, y) = \sin x + \cos y$ is differentiable every where in $\mathbf{R}^2$.*

Solution : Let (a, b) be an arbitrary point R^2. Then

$$f_x(a, b) = \lim_{h \to 0} \frac{f(a+h, b) - f(a, b)}{h}$$

$$= \lim_{h \to 0} \frac{\sin(a+h) + \cos b - \sin a - \cos b}{h}$$

$$= \lim_{h \to 0} \frac{\sin(a+h) - \sin a}{h}$$

$$= \lim_{h \to 0} 2\cos\left(a + \frac{h}{2}\right) \sin\frac{h}{2}$$

$$= \lim_{h \to 0} \cos\left(a + \frac{h}{2}\right) \cdot \left(\frac{\sin h/2}{h/2}\right)$$

$$= \cos a.$$

Similarly, $f_y(a, b) = -\sin b$

Now

$$\lim_{(h, k) \to (0, 0)} \left[\frac{f(a+h, b+k) - f(a, b) - h f_x(a, b) - k f_y(a, b)}{\sqrt{h^2 + k^2}}\right]$$

$$= \lim_{(h, k) \to (0, 0)} \frac{[\sin(a+h) + \cos(b+k) - \sin a - \cos b - h\cos a + k\sin b]}{\sqrt{h^2 + k^2}}$$

$$= \lim_{(h, k) \to (0, 0)} \frac{h}{\sqrt{h^2 + k^2}} \left[\frac{\sin(a+h) - \sin a}{h} - \cos a\right]$$

$$+ \lim_{(h, k) \to (0, 0)} \frac{k}{\sqrt{h^2 + k^2}} \left[\frac{\cos(b+k) - \cos b}{k} + \sin b\right]$$

$= 0 + 0 = 0.$

Hence, the given function is differentiable at every point $(a, b) \in R^2$.

Example 3. *Show that the function* $f(x, y)$ *defined by*

$$f(x, y) = \begin{cases} \dfrac{x^3 - y^3}{x^2 + y^2}, & \text{if } (x, y) \neq (0, 0) \\ 0 & \text{if } (x, y) = (0, 0) \end{cases}$$

is continuous but not differentiable at the origin.

Solution : To observe continuity, we shall show that for given $\varepsilon > 0$, there exists δ (depending on ε) > 0 s.t.

$$|f(x, y) - f(0, 0)| < \varepsilon \text{ whenever } |x - 0| < \delta \text{ and } |y - 0| < \delta.$$

Now taking $x = r\cos\theta$ and $y = r\sin\theta$, we have

$$\begin{aligned} |f(r\cos\theta, r\sin\theta) - f(0, 0)| &= \left|\frac{r^3(\cos^3\theta - \sin^3\theta)}{r^2(\cos^2\theta + \sin^2\theta)}\right| \\ &= r|\cos^3\theta - \sin^3\theta| \\ &\leq r(|\cos^3\theta| + |\sin^3\theta|) \\ &\leq 2r, \text{ for all values of } \theta. \end{aligned}$$

Choosing $r < \delta = \varepsilon/2$, we have

$|f(r\cos\theta, r\sin\theta) - f(0, 0)| < \varepsilon$ when $\sqrt{(x-0)^2 + (y-0)^2} < \delta$

or $|f(r\cos 0, r\sin\theta) - f(0, 0)| < \varepsilon$ when $|x - 0| < \delta$ and $|y - 0| < \delta$

Hence, $f(x, y)$ is continuous at $(0, 0)$.

Now, by definition of partial derivatives,

$$f_x(0, 0) = \lim_{h\to 0} \frac{f(0+h, 0) - f(0, 0)}{h} = \lim_{h\to 0}\left(\frac{h}{h}\right) = 1$$

and $$f_y(0, 0) = \lim_{k\to 0} \frac{f(0, 0+k) - f(0, 0)}{k} = \lim_{k\to 0}\left(\frac{-k}{k}\right) = -1$$

$$\therefore \lim_{(h,k)\to(0,0)} \frac{f(0+h, 0+k) - f(0, 0) - hf_x(0, 0) - kf_y(0, 0)}{\sqrt{h^2 + k^2}}$$

$$= \lim_{(h,k)\to(0,0)} \left[\frac{h^3 - k^3}{h^2 + k^2} - h + k\right] \cdot \frac{1}{\sqrt{h^2 + k^2}}$$

$$= \lim_{(h,k)\to(0,0)} \left[\frac{(h-k)hk}{(h^2 + k^2)^{3/2}}\right]$$

$$= \lim_{h\to 0} \frac{h^3 m(1-m)}{h^3(1+m^2)^{3/2}}$$

[taking $(h, k) \to (0, 0)$ along the line $k = mh$]

$$= \frac{m(1-m)}{(1+m^2)^{3/2}} \neq 0.$$

Since the value of the limit depends upon m, so this limit does not exist.

Hence, the given function is not differentiable at origin.

Example 4. *Prove that* $f(x, y) = \sqrt{|xy|}$ *is not differentiable at* (0, 0) *although partial derivatives* f_x *and* f_y *both exist at* (0, 0).

Solution : By the definition of partial derivatives,

$$\left(\frac{\partial f}{\partial x}\right)_{(0,0)} = \lim_{h \to 0} \frac{f(0+h, 0) - f(0, 0)}{h}$$

$$= \lim_{h \to 0} \frac{0}{h} = 0$$

and $$\left(\frac{\partial f}{\partial y}\right)_{(0,0)} = \lim_{k \to 0} \frac{f(0, 0+k) - f(0, 0)}{k} = \lim_{k \to 0} \left(\frac{0}{k}\right) = 0.$$

Now, $$\lim_{(h,k) \to (0,0)} \frac{f(0+h, 0+k) - f(0, 0) - h f_x(0, 0) - k f_y(0, 0)}{\sqrt{h^2 + k^2}}$$

$$= \lim_{(h,k) \to (0,0)} \frac{\sqrt{|hk|}}{\sqrt{h^2 + k^2}}$$

putting $h = r\cos\theta, k = r\sin\theta$

$$= \lim_{r \to 0} \frac{\sqrt{|r\cos\theta \cdot r\sin\theta|}}{\sqrt{r^2\cos^2\theta + r^2\sin^2\theta}} = \lim_{r \to 0} \sqrt{|\cos\theta \sin\theta|}$$

$$= |\cos\theta \sin\theta|^{1/2}$$

$$\neq 0 \text{ for arbitrary } \theta.$$

Hence, function is not differentiable at origin.

Example 5. *Show that the function* $f(x, y)$ *defined by*

$$f(x, y) = \begin{cases} x^2 \sin\dfrac{1}{x} + y^2 \sin\dfrac{1}{y}, & \text{when } x \neq 0, y \neq 0 \\ 0, & \text{when } x = 0, y = 0 \end{cases}$$

is differentiable at (0, 0).

Solution : By definition of partial derivatives,

$$f_x(0, 0) = \lim_{h \to 0} \frac{f(h, 0) - f(0, 0)}{h}$$

$$= \lim_{h \to 0} \frac{h^2 \sin\frac{1}{h}}{h} = \lim_{h \to 0} h \sin\frac{1}{h} = 0$$

and $$f_y(0, 0) = \lim_{k \to 0} \frac{k^2 \sin 1/k}{k} = \lim_{k \to 0} k \sin\frac{1}{k} = 0$$

Now, $\lim_{(h, k) \to (0, 0)} \dfrac{f(0+h, 0+k) - h f_x(0, 0)\, k f_y(0, 0)}{\sqrt{h^2+k^2}}$

$$= \lim_{(h, k) \to (0, 0)} \frac{f(0+h, 0+k)}{\sqrt{h^2+k^2}}$$

[taking $h = r\cos\theta$, $k = r\sin\theta$]

$$= \lim_{r \to 0} \frac{\left[r^2 \cos^2\theta \sin\left(\dfrac{1}{r\sin\theta}\right) + r^2 \sin^2\theta \sin\left(\dfrac{1}{r}\sin\theta\right)\right]}{r}$$

$$= \lim_{r \to 0} r\left[\cos^2\theta \sin\left(\frac{1}{r\cos\theta}\right) + \sin^2\theta \sin\left(\frac{1}{r\sin\theta}\right)\right]$$

$$= 0.$$

Hence, $f(x, y)$ is differentiable at (0, 0).

EXERCISE 8 (B)

1. If $z = \dfrac{x^2+y^2}{x+y}$, show that

$$\left(\frac{\partial z}{\partial x} - \frac{\partial z}{\partial y}\right)^2 = 4\left(1 - \frac{\partial z}{\partial x} - \frac{\partial z}{\partial y}\right).$$

2. If $u = \sin^{-1}\dfrac{y}{x} + \tan^{-1}\dfrac{y}{x}$, show that

$$x\frac{\partial u}{\partial x} + y\frac{\partial u}{\partial y} = 0.$$

3. If $u = \sin^{-1}\dfrac{\sqrt{x} - \sqrt{y}}{\sqrt{x} + \sqrt{y}}$, show that

$$x\frac{\partial u}{\partial x} + y\frac{\partial u}{\partial y} = 0.$$

4. If $u = \tan^{-1}\dfrac{xy}{\sqrt{1+x^2+y^2}}$, show that

$$\frac{\partial^2 u}{\partial x\, \partial y} = \frac{1}{(x^2+y^2+z^2)^{3/2}}.$$

5. If $u = x^2+y^2+z^2$, show that

$$x\frac{\partial u}{\partial x} + y\frac{\partial u}{\partial y} + z\frac{\partial u}{\partial z} = 2u.$$

6. If $u = x^2(y-z)+y^2(z-x)+z^2(x-y)$, show that

$$\frac{\partial u}{\partial x}+\frac{\partial u}{\partial y}+\frac{\partial u}{\partial z} = 0.$$

7. If $u = (x^2+y^2+z^2)^{1/2}$, show that

$$\frac{\partial^2 u}{\partial x^2}+\frac{\partial^2 u}{\partial y^2}+\frac{\partial^2 u}{\partial z^2} = \frac{2}{u}.$$

8. If $u = e^x\{x\cos y - y\sin y\}$, show that

$$\frac{\partial^2 u}{\partial x^2}+\frac{\partial^2 u}{\partial y^2} = 0.$$

9. If $u = 3(lx+my+nz)^2 - (x^2+y^2+z^2)$, show that

$$\frac{\partial^2 u}{\partial x^2}+\frac{\partial^2 u}{\partial y^2}+\frac{\partial^2 u}{\partial z^2} = 0, \text{ if } l^2+m^2+n^2 = 1.$$

10. If $u - \sin(\sqrt{x}+\sqrt{y})$, show that

$$x\frac{\partial u}{\partial x}+y\frac{\partial u}{\partial y} = \frac{1}{2}(\sqrt{x}+\sqrt{y})\cos(\sqrt{x}+\sqrt{y}).$$

11. If $u = \log_e\sqrt{x^2+y^2+z^2}$, show that

$$(x^2+y^2+z^2)\left(\frac{\partial^2 u}{\partial x^2}+\frac{\partial^2 u}{\partial y^2}+\frac{\partial^2 u}{\partial z^2}\right) = 1.$$

12. If $u = \log_e(x^2+y^2+z^2)$, show that

$$x\frac{\partial^2 u}{\partial y\,\partial z} = y\frac{\partial^2 u}{\partial z\,\partial x} = z\frac{\partial^2 u}{\partial x\,\partial y}$$

13. If $u = xf(x+y) + y\psi(x+y)$, show that

$$\frac{\partial^2 u}{\partial x^2}-2\frac{\partial^2 u}{\partial x\,\partial y}+\frac{\partial^2 u}{\partial y^2} = 0.$$

14. If $u = f(x^2+y^2+z^2)$, show that

$$\frac{\partial^2 u}{\partial x^2}+\frac{\partial^2 u}{\partial y^2}+\frac{\partial^2 u}{\partial z^2} = 4(x^2+y^2+z^2)\,f''(x^2+y^2+z^2)+6f'(x^2+y^2+z^2).$$

15. If $a^2x^2+b^2y^2-c^2z^2 = 0$, show that

$$\frac{\partial^2 z}{\partial x^2}+\frac{\partial^2 z}{\partial y^2}+\frac{a^2b^2}{c^4z^3}(x^2+y^2).$$

16. If $u = \begin{vmatrix} x^2 & y^2 & z^2 \\ x & y & z \\ 1 & 1 & 1 \end{vmatrix}$, show that

$$u_x + u_y + u_z = 0.$$

$$\left[\textbf{Hint} : u_x = \frac{\partial u}{\partial x},\ u_y = \frac{\partial u}{\partial y},\ u_z = \frac{\partial u}{\partial z} \right]$$

17. If $u = At^{-1/2} e^{-x^2/4a^2t}$, prove that

$$\frac{\partial u}{\partial t} = a^2 \frac{\partial^2 u}{\partial x^2}.$$

18. Verify that the following functions are differentiable at the points indicated against each :

(a) $f(x, y) = xy$ at $(3, 2)$

(b) $f(x, y) = e^{x+y}$ at $(1, 3)$

(c) $f(x, y) = \cos(xy)$ at $(0, 0)$.

19. Examine the differentiability of the following functions at the origin :

(i) $f(x, y) = |x^2 - y^2|$

(ii) $f(x, y) = y^2 \sin x/y$, where $f(0, 0) = 0$.

20. Show that the function $f(x, y)$ defined by

$$f(x, y) = \begin{cases} \dfrac{x^3 - y^3}{x^2 + y^2}, & (x, y) \neq (0, 0) \\ 0, & \text{otherwise} \end{cases}$$

is continuous but not differentiable at (0, 0).

21. Show that the function $f(x, y)$ defined by

$$f(x, y) = \begin{cases} \dfrac{xy^2}{x^2 + y^2}, & (x, y) \neq (0, 0) \\ 0, & \text{otherwise} \end{cases}$$

is continuous but not differentiable.

22. Examine the differentiability of the following function at the origin :

$$f(x, y) = \begin{cases} xy \dfrac{x^2 - y^2}{\sqrt{x^2 + y^2}}, & x \neq 0,\ y \neq 0 \\ 0, & x = 0 = y \end{cases}$$

9

Euler's Theorem on Homogeneous Functions and Total Derivative

9.1. Homogeneous Functions

A function in which every term is of the same degree is known as a homogeneous function of that degree.

For instance, let

$$f(x, y) = a_0 x^n + a_1 x^{n-1} y + a_2 x^{n-2} y^2 + ... + a_{n-1} xy^{n-1} + a_n y^n \quad ...(1)$$

be a function of x and y. We see that every term of this function has the same degree; *i.e.*, n. Hence, it is a homogeneous function of x and y of degree n.

Now, (1) can be rewritten as

$$\begin{aligned} f(x, y) &= x^n \left[a_0 + a_1 \left(\frac{y}{x}\right) + a_2 \left(\frac{y}{x}\right)^2 + ... + a_{n-1} \left(\frac{y}{x}\right)^{n-1} + a_n \left(\frac{y}{x}\right)^n \right] \\ &= x^n F\left(\frac{y}{x}\right). \end{aligned}$$

Thus, every homogeneous function of x and y of degree n can be expressed in the form

$$x^n F\left(\frac{y}{x}\right).$$

whether F may have any value, for example, $x^5 \sin \frac{y}{x}$; $x^4 \tan^{-1} \frac{y}{x}$; $x^3 \log \frac{x}{y}$; are homogeneous functions of degree 5, 4, 3 respectively.

9.2. A Test for Homogeneity of a Function of x and y

Let $f(x, y)$ be any function of x and y. Then $f(x, y)$ is said to be a homogeneous function of x and y of degree n, if

$$f(tx, ty) = t^n f(x, y).$$

9.3. Euler's Theorem on Homogeneous Functions

Statement : *If u is a homogeneous function of* x and y *of degree n, then*

$$x\frac{\partial u}{\partial x} + y\frac{\partial u}{\partial y} = nu.$$

Proof : Since u is a homogeneous function of x and y of degree n, hence we can write

$$u = x^n F\left(\frac{y}{x}\right) \quad \text{... (1)}$$

$\therefore$
$$\frac{\partial u}{\partial x} = nx^{n-1} F\left(\frac{y}{x}\right) + x^n F'\left(\frac{y}{x}\right)\left(-\frac{y}{x^2}\right)$$

$\Rightarrow$
$$x\frac{\partial u}{\partial x} = nx^n F\left(\frac{y}{x}\right) - x^{n-1} y F'\left(\frac{y}{x}\right)$$

$\Rightarrow$
$$x\frac{\partial u}{\partial x} = nu - x^{n-1} y F'\left(\frac{y}{x}\right) \quad \text{... (2)}$$

and
$$\frac{\partial u}{\partial y} = x^n F'\left(\frac{y}{x}\right)\cdot\frac{1}{x}$$

$\Rightarrow$
$$y\frac{\partial u}{\partial y} = x^{n-1} y F'\left(\frac{y}{x}\right). \quad \text{... (3)}$$

Adding (2) and (3), we obtain

$$x\frac{\partial u}{\partial x} + y\frac{\partial u}{\partial y} = nu.$$

Note : Euler's theorem can be extended to a homogeneous function of any number of variables. Thus, if $f(x_1, x_2, ..., x_n)$ be a homogeneous function of n variables, say, $x_1, x_2, ..., x_n$, then

$$x_1\frac{\partial f}{\partial x_1} + x_2\frac{\partial f}{\partial x_2} + ... + x_n\frac{\partial f}{\partial x_n} = nf.$$

9.4. Corollary

Statement : *If u is a homogeneous function of degree n, then*

(i) $x\dfrac{\partial^2 u}{\partial x^2} + y\dfrac{\partial^2 u}{\partial x\,\partial y} = (n-1)\dfrac{\partial u}{\partial x}$

(ii) $x\dfrac{\partial^2 u}{\partial x\,\partial y} + y\dfrac{\partial^2 u}{\partial y^2} = (n-1)\dfrac{\partial u}{\partial y}$

(iii) $x^2 \dfrac{\partial^2 u}{\partial x^2} + 2xy \dfrac{\partial^2 u}{\partial x\, \partial y} + y^2 \dfrac{\partial^2 u}{\partial y^2} = n\ (n-1)\ u.$

Proof : Since u is a homogeneous function of x and y of degree n, therefore, by Euler's theorem, we have

$$x \frac{\partial u}{\partial x} + y \frac{\partial u}{\partial y} = nu. \qquad \text{...(1)}$$

Differentiating both sides of (1) partially w.r.t. x, we get

$$x \frac{\partial^2 u}{\partial x^2} + \frac{\partial u}{\partial x} + y \frac{\partial^2 u}{\partial x\, \partial y} = n \frac{\partial u}{\partial x}$$

$$\Rightarrow \qquad x \frac{\partial^2 u}{\partial x^2} + y \frac{\partial^2 u}{\partial x\, \partial y} = (n-1) \frac{\partial u}{\partial x} \qquad \text{...(2)}$$

Again differentiating both sides of (1) partially with respect to y, we get

$$x \frac{\partial^2 u}{\partial y\, \partial x} + \frac{\partial u}{\partial y} + y \frac{\partial^2 u}{\partial y^2} = n \frac{\partial u}{\partial y}$$

$$\Rightarrow \qquad x \frac{\partial^2 u}{\partial y\, \partial x} + y \frac{\partial^2 u}{\partial y^2} = (n-1) \frac{\partial u}{\partial y}$$

$$\Rightarrow \qquad x \frac{\partial^2 u}{\partial x\, \partial y} + y \frac{\partial^2 u}{\partial y^2} = (n-1) \frac{\partial u}{\partial y} \qquad \text{...(3)}$$

$$\left[\because \frac{\partial^2 u}{\partial y\, \partial x} = \frac{\partial^2 u}{\partial x\, \partial y}\right]$$

Now, multiplying (2) by x and (3) by y and adding, we get

$$x^2 \frac{\partial^2 u}{\partial x^2} + 2xy \frac{\partial^2 u}{\partial x\, \partial y} + y^2 \frac{\partial^2 u}{\partial y^2} = (n-1) \left[x \frac{\partial u}{\partial x} + y \frac{\partial u}{\partial y}\right]$$

$$= n\,(n-1)\,u \qquad \text{[using (1)]}$$

9.5. An Important Deduction from Euler's Theorem

Let $$u = \phi\,(T_n) \qquad \text{...(1)}$$

where, T_n is a homogeneous function of degree n.

Again, suppose that (1) implies

$$T_n = f(u).$$

Then using Euler's theorem on homogeneous functions, we get

$$x \frac{\partial}{\partial x} f(u) + y \frac{\partial}{\partial y} f(u) = nf(u)$$

$$\Rightarrow \qquad xf'(u)\frac{\partial u}{\partial x} + yf'(u)\frac{\partial u}{\partial y} = nf(u)$$

$$\Rightarrow \qquad \boxed{x\frac{\partial u}{\partial x} + y\frac{\partial u}{\partial y} = n\frac{f(u)}{f'(u)}.}$$

Illustrative Examples

Example 1. *If* $u = \dfrac{x^{1/4}+y^{1/4}}{x^{1/5}+y^{1/5}}$, *then verify Euler's theorem for u.*

Solution :
$$u = \frac{x^{1/4}+y^{1/4}}{x^{1/5}+y^{1/5}}$$

$$= \frac{x^{1/4}\,[1+(y/x)^{1/4}]}{x^{1/5}\,[1+(y/x)^{1/5}]} = x^{1/20}\,\frac{1+(y/x)^{1/4}}{1+(y/x)^{1/5}}$$

$$= x^{1/20}\,F\left(\frac{y}{x}\right)$$

$\Rightarrow$ u is a homogeneous function of x and y of degree $\dfrac{1}{20}$.

Hence, to verify Euler's theorem for u, we have to prove that

$$x\frac{\partial u}{\partial x} + y\frac{\partial u}{\partial y} = \frac{1}{20}u.$$

Now,
$$u = \frac{x^{1/4}+y^{1/4}}{x^{1/5}+y^{1/5}}$$

$$\Rightarrow \qquad \log u = \log(x^{1/4}+y^{1/4}) - \log(x^{1/5}+y^{1/5}) \qquad \text{...(i)}$$

Differentiating (i) partially w.r.t. x, we get

$$\frac{1}{u}\frac{\partial u}{\partial x} = \frac{\frac{1}{4}x^{-3/4}}{x^{1/4}+y^{1/4}} - \frac{\frac{1}{2}x^{-4/5}}{x^{1/5}+y^{1/5}}$$

$$\Rightarrow \qquad \frac{1}{u}x\frac{\partial u}{\partial y} = \frac{\frac{1}{4}x^{1/4}}{x^{1/4}+y^{1/4}} - \frac{\frac{1}{5}x^{1/5}}{x^{1/5}+y^{1/5}}. \qquad \text{... (ii)}$$

Differentiating (i) partially w.r.t. y, we get

$$\frac{1}{u}\frac{\partial u}{\partial y} = \frac{\frac{1}{4}y^{-3/4}}{x^{1/4}+y^{1/4}} - \frac{\frac{1}{5}y^{-4/5}}{x^{1/5}+y^{1/5}}$$

$$\Rightarrow \qquad \frac{1}{u}\, y\, \frac{\partial u}{\partial y} = \frac{\frac{1}{4} y^{1/4}}{x^{1/4}+y^{1/4}} - \frac{\frac{1}{5} y^{1/5}}{x^{1/5}+y^{1/5}}. \qquad \text{... (iii)}$$

Adding (ii) and (iii), we get

$$\frac{1}{u}\left[x\,\frac{\partial u}{\partial x} + y\,\frac{\partial u}{\partial y}\right] = \frac{1}{4}\left[\frac{x^{1/4}+y^{1/4}}{x^{1/4}+y^{1/4}}\right] - \frac{1}{5}\left[\frac{x^{1/5}+y^{1/5}}{x^{1/5}+y^{1/5}}\right]$$

$$= \frac{1}{4} - \frac{1}{5} = \frac{1}{20}$$

$$\Rightarrow \qquad x\,\frac{\partial u}{\partial y} + u\,\frac{\partial u}{\partial y} = \frac{u}{20}.$$

Hence, we verify Euler's theorem of the given function.

Example 2. *If* $u = \log\left(\frac{x^4+y^4}{x+y}\right)$, *then show that*

$$x\,\frac{\partial u}{\partial x} + y\,\frac{\partial u}{\partial y} = 3.$$

Solution :
$$u = \log\left(\frac{x^4+y^4}{x+y}\right)$$

$$\Rightarrow \qquad e^u = \frac{x^4+y^4}{x+y}$$

$$\Rightarrow \qquad z = \frac{x^4+y^4}{x+y} \qquad [\text{where } z = e^u]$$

Clearly, z is a homogeneous function of x and y of degree 3. Hence, by Euler's theorem on homogeneous functions, we have

$$x\,\frac{\partial z}{\partial x} + y\,\frac{\partial z}{\partial y} = 3z.$$

$$\Rightarrow \qquad x\,\frac{\partial}{\partial x}(e^u) + y\,\frac{\partial}{\partial y}(e^u) = 3e^u$$

$$\Rightarrow \qquad xe^u\,\frac{\partial u}{\partial x} + ye^u\,\frac{\partial u}{\partial y} = 3e^u$$

$$\Rightarrow \qquad x\,\frac{\partial u}{\partial x} + y\,\frac{\partial u}{\partial y} = 3.$$

Example 3. *If* $u = \sin^{-1}\frac{x+y}{\sqrt{x}+\sqrt{y}}$, *then show that*

$$x\,\frac{\partial u}{\partial x} + y\,\frac{\partial u}{\partial y} = \frac{1}{2}\tan u.$$

Solution : $$u = \sin^{-1}\left(\frac{x+y}{\sqrt{x}+\sqrt{y}}\right)$$

$$\Rightarrow \qquad \sin u = \frac{x+y}{\sqrt{x}+\sqrt{y}}$$

$$\Rightarrow \qquad z = \frac{x+y}{\sqrt{x}+\sqrt{y}} \qquad \text{[where } z = \sin u\text{]}$$

Now z is a homogeneous function of x and y of degree $\frac{1}{2}$. Hence, by Euler's theorem on homogeneous functions, we get

$$x\frac{\partial z}{\partial x} + y\frac{\partial z}{\partial y} = \frac{1}{2}z.$$

$$\Rightarrow \qquad x\frac{\partial}{\partial x}(\sin u) + y\frac{\partial}{\partial y}(\sin u) = \frac{1}{2}\sin u$$

$$\Rightarrow \qquad x\cos u\frac{\partial u}{\partial x} + y\cos u\frac{\partial u}{\partial y} = \frac{1}{2}\sin u$$

$$\Rightarrow \qquad x\frac{\partial u}{\partial x} + y\frac{\partial u}{\partial y} = \frac{1}{2}\tan u.$$

Example 4. *If* $u = x\,\phi\left(\frac{y}{x}\right) + \psi\left(\frac{y}{x}\right)$, *prove that*

$$x^2\frac{\partial^2 u}{\partial x^2} + 2xy\frac{\partial^2 u}{\partial x\,\partial y} + y^2\frac{\partial^2 u}{\partial y^2} = 0.$$

Solution : $$u = x\,\phi\left(\frac{y}{x}\right) + \psi\left(\frac{y}{x}\right)$$

$$\Rightarrow \qquad u = v + w, \text{ where } v = x\,\phi\left(\frac{y}{x}\right),\ w = \psi\left(\frac{y}{x}\right)$$

Now, v is a homogeneous function of x and y of degree 1. Hence, by Euler's theorem on homogeneous functions, we get

$$x^2\frac{\partial^2 v}{\partial x^2} + 2xy\frac{\partial^2 v}{\partial x\,\partial y} + y^2\frac{\partial^2 v}{\partial y^2} = 0. \qquad \text{... (i)}$$

[using 9.4 (iii)]

Again, w is a homogeneous function of x and y of degree 0, hence, by Euler's theorem, we get

$$x^2\frac{\partial^2 w}{\partial x^2} + 2xy\frac{\partial^2 w}{\partial x\,\partial y} + y^2\frac{\partial^2 w}{\partial y^2} = 0. \qquad \text{...(ii)}$$

[using 9.4 (iii)]

Adding (i) and (ii), we get

$$x^2 \frac{\partial^2}{\partial x^2}(v+w) + 2xy \frac{\partial^2}{\partial x\,\partial y}(v+w) + y^2 \frac{\partial^2}{\partial y^2}(v+w) = 0$$

$$\Rightarrow \qquad x^2 \frac{\partial^2 u}{\partial x^2} + 2xy \frac{\partial^2}{\partial x\,\partial y} + y^2 \frac{\partial^2 u}{\partial y^2} = 0$$

[using $u = v + w$]

Second Method : *Question can easily be evaluated without applying Euler's theorem.*

$$u = x\,\phi\left(\frac{y}{x}\right) + \psi\left(\frac{y}{x}\right) \qquad \text{... (i)}$$

$$\frac{\partial u}{\partial x} = \phi\left(\frac{y}{x}\right) + x\,\phi'\left(\frac{y}{x}\right)\left(-\frac{y}{x^2}\right) + \psi'\left(\frac{y}{x}\right)\left(-\frac{y}{x^2}\right)$$

$$\Rightarrow \qquad x\frac{\partial u}{\partial x} = x\,\phi\left(\frac{y}{x}\right) - y\,\phi'\left(\frac{y}{x}\right) - \frac{y}{x}\psi'\left(\frac{y}{x}\right) \qquad \text{... (ii)}$$

$$\frac{\partial u}{\partial y} = x\,\phi'\left(\frac{y}{x}\right)\frac{1}{x} + \psi'\left(\frac{y}{x}\right)\frac{1}{x}$$

$$\Rightarrow \qquad y\frac{\partial u}{\partial y} = y\,\phi'\left(\frac{y}{x}\right) + \frac{y}{x}\psi'\left(\frac{y}{x}\right) \qquad \text{... (iii)}$$

Adding (ii) and (iii),

$$x\frac{\partial u}{\partial x} + y\frac{\partial u}{\partial y} = x\,\phi\left(\frac{y}{x}\right) - y\,\phi'\left(\frac{y}{x}\right) - \frac{y}{x}\psi'\left(\frac{y}{x}\right) + y\,\phi'\left(\frac{y}{x}\right) + \frac{y}{x}\psi'\left(\frac{y}{x}\right)$$

$$= x\,\phi\left(\frac{y}{x}\right) \qquad \text{... (iv)}$$

Differentiating (iv) partially with respect to x,

$$x\frac{\partial^2 u}{\partial x^2} + \frac{\partial u}{\partial x} + y\frac{\partial^2 u}{\partial x\,\partial y} = \phi\left(\frac{y}{x}\right) + x\phi'\left(\frac{y}{x}\right)\left(-\frac{y}{x^2}\right) \qquad \text{... (v)}$$

Differentiating (iv) partially with respect to y,

$$x\frac{\partial^2 y}{\partial x\,\partial y} + \frac{\partial^2 u}{\partial y^2} + \frac{\partial u}{\partial y} = x\,\phi'\left(\frac{y}{x}\right)\frac{1}{x} \qquad \text{... (vi)}$$

Multiplying (iv) by x and (v) by y and adding,

$$x^2\frac{\partial^2 u}{\partial x^2} + 2xy\frac{\partial^2 u}{\partial x\,\partial y} + y^2\frac{\partial^2 u}{\partial y^2} + x\frac{\partial u}{\partial x} + y\frac{\partial u}{\partial y}$$

$$= x\,\phi\left(\frac{y}{x}\right) - y\,\phi'\left(\frac{y}{x}\right) + y\,\phi'\left(\frac{y}{x}\right) \qquad \text{... (vii)}$$

Putting the value from (iv) in (vii),

$$x^2 \frac{\partial^2 u}{\partial x^2} + 2xy \frac{\partial^2 u}{\partial x \partial y} + y^2 \frac{\partial^2 u}{\partial y^2} + x\,\phi\left(\frac{y}{x}\right) = x\phi\left(\frac{y}{x}\right)$$

$$\Rightarrow \qquad x^2 \frac{\partial^2 u}{\partial x^2} + 2xy \frac{\partial^2 u}{\partial x\, \partial y} + y^2 \frac{\partial^2 u}{\partial y^2} = 0.$$

Example 5. *If* $V = \log_e \sin \left\{\frac{\pi (2x^2 + y^2 + xz)^{1/2}}{2 (x^2 + xy + 2yz + z^2)^{1/3}}\right\}$, *find the value of*

$$x\frac{\partial V}{\partial x} + y\frac{\partial V}{\partial y} + z\frac{\partial V}{\partial z} \quad at\ x = 0,\ y = 1,\ z = 2.$$

Solution :

$$V = \log_e \sin\left[\frac{\pi (2x^2 + y^2 + xz)^{1/2}}{2 (x^2 + xy + 2yz + z^2)^{1/3}}\right] \qquad \text{... (i)}$$

$$\Rightarrow \qquad V = \log_e \sin u$$

where

$$u = \frac{\pi (2x^2 + y^2 + xz)^{1/2}}{2 (x^2 + xy + 2yz + z^2)^{1/3}}$$

From (ii),

$$\frac{\partial V}{\partial x} = \cot u \frac{\partial u}{\partial x}$$

$$\frac{\partial V}{\partial y} = \cot u \frac{\partial u}{\partial y}$$

$$\frac{\partial V}{\partial z} = \cot u \frac{\partial u}{\partial z}$$

$$\therefore\ x\frac{\partial V}{\partial x} + y\frac{\partial V}{\partial y} + z\frac{\partial V}{\partial z} = \cot u\left[x\frac{\partial u}{\partial x} + y\frac{\partial u}{\partial y} + z\frac{\partial u}{\partial z}\right] \qquad \text{... (iii)}$$

But u is homogeneous function of x, y and z of degree $\frac{1}{3}$. Hence,

$$x\frac{\partial u}{\partial x} + y\frac{\partial u}{\partial y} + z\frac{\partial u}{\partial z} = \frac{1}{3} u \qquad \text{[using Euler's theorem]}$$

$\therefore$ From (iii),

$$x\frac{\partial V}{\partial x} + y\frac{\partial V}{\partial y} + z\frac{\partial V}{\partial z} = \frac{1}{3} u \cot u$$

When $x = 0$, $y = 1$ and $z = 2$, we have $u = \pi/4$

$$\therefore \qquad x\frac{\partial V}{\partial x} + y\frac{\partial V}{\partial y} + z\frac{\partial V}{\partial z} = \frac{1}{3} \cdot \frac{\pi}{4} \cdot \cot\frac{\pi}{4} = \frac{\pi}{12}$$

Example 6. *If* $u = \sin^{-1}\left(\dfrac{x^{1/3}+y^{1/3}}{x^{1/2}+y^{1/2}}\right)$, *prove that*

$$x^2\frac{\partial^2 u}{\partial x^2} + 2xy\frac{\partial^2 u}{\partial x\,\partial y} + y^2\frac{\partial^2 u}{\partial y^2} = \frac{\tan u}{144}(13+\tan^2 u).$$

Solution :

$$u = \sin^{-1}\left(\frac{x^{1/3}+y^{1/3}}{x^{1/2}+y^{1/2}}\right)^{1/2}$$

$$\Rightarrow \quad \sin u = \left(\frac{x^{1/3}+y^{1/3}}{x^{1/2}+y^{1/2}}\right)^{1/2}$$

$$\Rightarrow \quad z = \left(\frac{x^{1/3}+y^{1/3}}{x^{1/2}+y^{1/2}}\right)^{1/2}, \quad \text{where } z = \sin u$$

Now z is a homogeneous function of x and y of degree $\frac{1}{12}$. Hence, by Euler's theorem on homogeneous functions, we get

$$x\frac{\partial z}{\partial x} + y\frac{\partial z}{\partial y} = \frac{1}{12}z$$

$$\Rightarrow \quad x\frac{\partial}{\partial x}(\sin u) + y\frac{\partial}{\partial y}(\sin u) = -\frac{1}{12}\sin u$$

$$\Rightarrow \quad x\cos u\frac{\partial u}{\partial x} + y\cos u\frac{\partial u}{\partial y} = -\frac{1}{12}\sin u$$

$$\Rightarrow \quad x\frac{\partial u}{\partial x} + y\frac{\partial u}{\partial y} = -\frac{1}{12}\tan u \quad \text{...(i)}$$

Differentiation (i) partially with respect to x, we get

$$x\frac{\partial^2 u}{\partial x^2} + \frac{\partial u}{\partial x} + y\frac{\partial^2 u}{\partial x\,\partial y} = -\frac{1}{12}\sec^2 u\frac{\partial u}{\partial x}$$

$$\Rightarrow \quad x\frac{\partial^2 u}{\partial x^2} + y\frac{\partial^2 u}{\partial x\,\partial y} = -\left(\frac{1}{12}\sec^2 u + 1\right)\frac{\partial u}{\partial x} \quad \text{... (ii)}$$

Differentiating (ii) partially with respect to y, we get

$$x\frac{\partial^2 u}{\partial u\,\partial y} + y\frac{\partial^2 u}{\partial y^2} + \frac{\partial u}{\partial y} = -\frac{1}{12}\sec^2 u\frac{\partial u}{\partial y}$$

$$\Rightarrow \quad y\frac{\partial^2 u}{\partial y^2} + x\frac{\partial^2 u}{\partial x\,\partial y} = -\left(\frac{1}{12}\sec^2 u + 1\right)\frac{\partial u}{\partial y}. \quad \text{...(iii)}$$

Multiplying (ii) and (i) by x and y respectively and adding, we get

$$x^2\frac{\partial^2 u}{\partial x^2} + 2xy\frac{\partial^2 u}{\partial x\,\partial y} + y^2\frac{\partial^2 u}{\partial y^2} = -\left(\frac{1}{12}\sec^2 u + 1\right)\left(x\frac{\partial u}{\partial x} + y\frac{\partial u}{\partial y}\right)$$

$$= -\left(\frac{1}{12}\sec^2 u + 1\right)\left(-\frac{1}{12}\tan u\right) \quad \text{[from (i)]}$$

$$= \frac{1}{144}(\sec^2 u + 12)\tan u$$

$$= \frac{1}{144}(1 + \tan^2 u + 12)\tan u$$

$$= \frac{1}{144}\tan u\,(13 + \tan^2 u).$$

EXERCISE 9 (A)

1. Verify Euler's theorem for the following functions :

(a) $u = ax^2 + 2hxy + by^2$ (b) $u = \dfrac{x^2 y^2}{x^2 + y^2}$

(c) $u = \dfrac{x(x^3 - y^3)}{x^3 + y^3}$ (d) $u = axy + byz + czx.$

2. Verify Euler's theorem for the following functions :

(a) $u = x^4 \log (y/x)$ (b) $u = x^n \sin (y/x)$

(c) $u = (x^{1/2} + y^{1/2})(x^n + y^n).$

3. If $u = \sin^{-1}(x/y) + \tan^{-1}(y/x)$, prove that

$$x\frac{\partial u}{\partial x} + y\frac{\partial u}{\partial y} = 0.$$

4. If $z = xy\,f(y/x)$, show that

$$x\frac{\partial z}{\partial x} + y\frac{\partial z}{\partial y} = 2z.$$

5. If $u = x\sin^{-1}(y/x)$, prove that

$$x^2\frac{\partial^2 u}{\partial x^2} + 2xy\frac{\partial^2 u}{\partial x\,\partial y} + y^2\frac{\partial^2 u}{\partial y^2} = 0.$$

6. If $u = \dfrac{x^2 y^2}{x^2 + y^2}$, show that

(a) $x\dfrac{\partial^2 u}{\partial x^2} + y\dfrac{\partial^2 u}{\partial y\,\partial x} = \dfrac{\partial u}{\partial x}$

(b) $x \dfrac{\partial^2 u}{\partial x \, \partial y} + y \dfrac{\partial^2 u}{\partial y^2} = \dfrac{\partial u}{\partial y}$

(c) $x^2 \dfrac{\partial^2 u}{\partial x^2} + 2xy \dfrac{\partial^2 u}{\partial x \, \partial y} + y^2 \dfrac{\partial^2 u}{\partial y^2} = 2u.$

7. (a) If $u = \tan^{-1}\left(\dfrac{x^3 + y^3}{x - y}\right)$, prove that

$$x \frac{\partial u}{\partial x} + y \frac{\partial u}{\partial y} = \sin 2u.$$

(b) If $u = \sec^{-1}\left(\dfrac{x^3 + y^3}{x + y}\right)$, prove that

$$x \frac{\partial u}{\partial x} + y \frac{\partial u}{\partial y} = 2 \cot u.$$

(c) If $u = \cos^{-1}\left(\dfrac{x + y}{\sqrt{x} + \sqrt{y}}\right)$, prove that

$$x \frac{\partial u}{\partial x} + y \frac{\partial u}{\partial y} + \frac{1}{2} \cot u = 0.$$

(d) If $u = \sin^{-1}\left(\dfrac{x^2 + y^2}{x + y}\right)$, prove that

$$x \frac{\partial u}{\partial x} + y \frac{\partial u}{\partial y} = \tan u.$$

8. If $u = (\sqrt{x} + \sqrt{y})^5$, prove that

$$x^2 \frac{\partial^2 u}{\partial x^2} + 2xy \frac{\partial^2 u}{\partial x \, \partial y} + y^2 \frac{\partial^2 u}{\partial y^2} = \frac{15}{4} u.$$

9. If $u = \tan^{-1}\left(\dfrac{x^2 + y^2}{x - y}\right)$, prove that

$$x^2 \frac{\partial^2 u}{\partial x^2} + 2xy \frac{\partial^2 u}{\partial x \, \partial y} + y^2 \frac{\partial^2 u}{\partial y^2} = (1 - 4 \sin^2 u) \sin 2u.$$

10. If $u = \sin^{-1}\left(\dfrac{x + y}{\sqrt{x} + \sqrt{y}}\right)$, show that

$$x^2 \frac{\partial^2 u}{\partial x^2} + 2xy \frac{\partial^2 u}{\partial x \, \partial y} + y^2 \frac{\partial^2 u}{\partial y^2} = -\frac{\sin u \cos 2u}{4 \cos^2 u}.$$

11. If $u = f(x, y, z)$, be a homogeneous function of degree n in x, y, z; prove that

$$x^2 \frac{\partial^2 u}{\partial x^2} + y^2 \frac{\partial^2 u}{\partial y^2} + z^2 \frac{\partial^2 u}{\partial z^2} + 2yz \frac{\partial^2 u}{\partial y\, \partial z} + 2zx \frac{\partial^2 u}{\partial z\, \partial x} + 2xy \frac{\partial^2 u}{\partial x\, \partial y}$$

$$= n(n-1)\, u.$$

9.6. An Important Convention

Let (x, y) be the cartesian coordinates of any point P referred to OX and OY as coordinate axes. Let (r, θ) be the polar coordinates of P referred to O as pole and OX as initial line. If $\angle POX = \theta$, then we know that

$$x = r\cos\theta \qquad \text{...(1)}$$

$$y = r\sin\theta \qquad \text{...(2)}$$

$$r^2 = x^2 + y^2 \qquad \text{...(3)}$$

$$\tan\theta = \frac{y}{x}$$

$$\Rightarrow \qquad \theta = \tan^{-1}\frac{y}{x}. \qquad \text{...(4)}$$

It is noteworthy here, that

the operator $\frac{\partial}{\partial x}$ always means $\left(\frac{\partial}{\partial x}\right)_{y = \text{const.}}$

the operator $\frac{\partial}{\partial y}$ always means $\left(\frac{\partial}{\partial y}\right)_{x = \text{const.}}$

the operator $\frac{\partial}{\partial r}$ always means $\left(\frac{\partial}{\partial r}\right)_{\theta = \text{const.}}$

the operator $\frac{\partial}{\partial \theta}$ always means $\left(\frac{\partial}{\partial \theta}\right)_{r = \text{const.}}$

Thus, from (1),

$$\frac{\partial x}{\partial r} = \cos\theta$$

$$\frac{\partial x}{\partial \theta} = -\, r\sin\theta.$$

From (2),

$$\frac{\partial y}{\partial r} = \sin\theta$$

$$\frac{\partial y}{\partial \theta} = r\cos\theta.$$

Now from (1), $\qquad r = x\sec\theta$

If we write
$$\frac{\partial r}{\partial x} = \sec\theta,$$
then it is not correct because we have determined $\left(\frac{\partial r}{\partial x}\right)_{\theta\,=\,\text{const.}}$ here, whereas we want $\left(\frac{\partial r}{\partial x}\right)_{y\,=\,\text{const.}}$. To solve the problem, we write
$$r^2 = x^2 + y^2$$
and then
$$2r\frac{\partial r}{\partial x} = 2x$$
$$\Rightarrow \qquad \frac{\partial r}{\partial x} = \frac{x}{r} = \cos\theta.$$
Also,
$$2r\frac{\partial r}{\partial y} = 2y$$
$$\Rightarrow \qquad \frac{\partial r}{\partial y} = \frac{y}{r} = \sin\theta.$$
Again from (4),
$$\frac{\partial\theta}{\partial x} = \frac{1}{1+\frac{y^2}{x^2}}\left(-\frac{y}{x^2}\right) = -\frac{y}{x^2+y^2}$$
$$= -\frac{r\sin\theta}{r^2} = -\frac{\sin\theta}{r}$$
and
$$\frac{\partial\theta}{\partial y} = \frac{1}{1+\frac{y^2}{x^2}}\frac{1}{x} = \frac{x}{x^2+y^2}$$
$$= \frac{r\cos\theta}{r^2} = \frac{\cos\theta}{r}.$$

9.7. Composite Function and Total Differential Coefficient

If u is given to be a function of the variables x and y and these variables themselves are given to be the functions of the variable t, then u is said to be a composite function of the variable t.

Thus, the relations
$$u = f(x, y);$$
$$x = \phi(t);$$
$$y = \psi(t)$$
define u as a composite function of t.

Similarly, the relations $z = f(x, y); x = \phi(t, r)\, y = \psi(t, r)$ define z as a composite function of t and r.

$\frac{du}{dt}$ is called the total differential coefficient of u with respect to t.

By substituting the values of x and y in terms of t, we can express u as a function of t and then in the usual way, we can find $\frac{du}{dt}$.

But sometimes, the elimination of x and y becomes very tedious and therefore, we require a new advice to find $\frac{du}{dt}$.

9.8. Theorem

Statement : *If u is a composite function of t, defined by the relations*

$$u = f(x, y); \; x = \phi(t); \; y = \psi(t),$$

where, u possesses continuous first order partial derivatives w.r.t x and y, and x and y have continuous derivatives w.r.t. 't', then

$$\frac{du}{dt} = \frac{\partial u}{\partial x}\frac{dx}{dt} + \frac{\partial u}{\partial y}\frac{dy}{dt}$$

Proof :

$$u = f(x, y). \qquad \text{... (1)}$$

Let t receive an increment δt and let the corresponding increments in x, y, u be $\delta x, \delta y, \delta u$ respectively. Then

$$u + \delta u = f(x + \delta x, y + \delta y). \qquad \text{... (2)}$$

Subtracting (1) from (2), we get

$$\delta u = f(x + \delta x, y + \delta y) - f(x, y).$$

Subtracting and adding $f(x, y + \delta y)$ on R.H.S., we get

$$\delta u = [f(x + \delta x, y + \delta y) - f(x, y + \delta y)] + [f(x, y + \delta y) - f(x, y)]$$

$$\Rightarrow \quad \frac{\delta u}{\delta t} = \frac{f(x+\delta x, y+\delta y) - f(x, y+\delta y)}{\delta y} \cdot \frac{\delta y}{\delta t} + \frac{f(x, y+\delta y) - f(x, y)}{\delta y} \cdot \frac{\delta y}{\delta t} \qquad \text{... (3)}$$

Now, proceeding to limits as $\delta t \to 0$ and consequently δx and δy also tend to zero, we have

$$\lim_{\delta t \to 0} \frac{\delta u}{\delta t} = \frac{du}{dt}$$

$$\lim_{\delta t \to 0} \frac{\delta x}{\delta t} = \frac{dx}{dt}$$

$$\lim_{\delta t \to 0} \frac{\delta y}{\delta t} = \frac{dy}{dt}$$

and $$\lim_{\delta x \to 0} \frac{f(x + \delta x, y + \delta y) - f(x, y + \delta y)}{\delta x} = \frac{\partial u}{\partial x}$$

[because while x changes to $x + \delta x$, $y + \delta y$ remains unchanged]

Similarly, $$\lim_{\delta x \to 0} \frac{f(x, y + \delta y) - f(x, y)}{\delta y} = \frac{\partial u}{\partial y}$$

Hence, (3) gives

$$\frac{du}{dt} = \frac{\partial u}{\partial x}\frac{dx}{dt} - \frac{\partial u}{\partial y}\frac{dy}{dt} \quad \text{... (4)}$$

Generalization : In general, if $u = f(x_1, x_2, x_3, ..., x_n)$ where $x_1, x_2, x_3, ..., x_n$ all are functions of t, we have

$$\frac{du}{dt} = \frac{\partial u}{\partial x_1}\frac{dx_1}{dt} + \frac{\partial u}{\partial x_2} \cdot \frac{dx_2}{dt} + \frac{\partial u}{\partial x_3}\frac{dx_3}{dt} + ... + \frac{\partial u}{\partial x_n}\frac{dx_n}{dt}$$

9.9. Deduction

Again, if $u = f(x, y)$, where, $x = \phi(t, r)$; $y = \psi(t, r)$,

then from (4), we have by writing in place of ordinary differential coefficients, the corresponding partial differential coefficients,

$$\frac{\partial u}{\partial t} = \frac{\partial u}{\partial x} \cdot \frac{\partial x}{\partial t} + \frac{\partial u}{\partial y} \cdot \frac{\partial y}{\partial t}$$

and $$\frac{\partial u}{\partial r} = \frac{\partial u}{\partial x} \cdot \frac{\partial x}{\partial r} + \frac{\partial u}{\partial y} \cdot \frac{\partial y}{\partial r}$$

Generalization : If $u = f(x, y, z, ...)$; $x = \phi(t, r)$; $y = \psi(t, r)$ ... then

$$\frac{\partial u}{\partial t} = \frac{\partial u}{\partial x}\frac{\partial x}{\partial t} + \frac{\partial u}{\partial y}\frac{\partial y}{\partial t} + \frac{\partial u}{\partial z}\frac{\partial z}{\partial t} + ...$$

and $$\frac{\partial u}{\partial r} = \frac{\partial u}{\partial x}\frac{\partial x}{\partial r} + \frac{\partial u}{\partial y}\frac{\partial y}{\partial r} + \frac{\partial u}{\partial z}\frac{\partial z}{\partial r} + ...$$

Again, from $x = \phi(t, r)$; $y = \psi(t, r)$

we can find the values of t and r in terms of x and y, *i.e.*,

$$t = f_1(x, y); \; r = f_2(x, y)$$

and clearly $u = f(t, r)$, therefore, we have

$$\frac{\partial u}{\partial x} = \frac{\partial u}{\partial t}\frac{\partial t}{\partial x} + \frac{\partial u}{\partial r}\frac{\partial r}{\partial x}$$

and $$\frac{\partial u}{\partial y} = \frac{\partial u}{\partial t}\frac{\partial t}{\partial y} + \frac{\partial u}{\partial r}\frac{\partial r}{\partial y}.$$

A special case : Let $u = f(x, y)$; $x = x$; $y = \phi(x)$

i.e., u is a function of x and y, y is a function of x, x is of course, a function of x, then u is a composite function of x.

$$\therefore \qquad \frac{du}{dx} = \frac{\partial u}{\partial x}\frac{dx}{dx} + \frac{\partial u}{\partial y}\frac{dy}{dx}$$

$$\Rightarrow \qquad \frac{du}{dx} = \frac{\partial u}{\partial x} + \frac{\partial u}{\partial y}\frac{dy}{dx}.$$

Moreover : If here $u = f(x, y) =$ constant, then we get

$$0 = \frac{\partial u}{\partial x} + \frac{\partial u}{\partial y}\frac{dy}{dx}$$

$$\Rightarrow \qquad \frac{dy}{dx} = -\frac{\frac{\partial u}{\partial x}}{\frac{\partial u}{\partial y}} = -\frac{p}{q}$$

where $p = \dfrac{\partial u}{\partial x}$ and $q = \dfrac{\partial u}{\partial y}$.

Thus, we can easily determine *the first differential coefficient of an implicit function.*

9.10. Second Differential Coefficient of an Implicit Function

Let the implicit function be $f(x, y) = c$, then we have

$$\frac{dy}{dx} = -\frac{p}{q}. \qquad \text{...(1)}$$

[see Art. 8.7]

We want to find out $\dfrac{d^2y}{dx^2}$.

Let us denote $\dfrac{\partial f}{\partial x}, \dfrac{\partial f}{\partial y}, \dfrac{\partial^2 f}{\partial x^2}, \dfrac{\partial^2 f}{\partial x\,\partial y}$ and $\dfrac{\partial^2 f}{\partial y^2}$ by p, q, r, s and t respectively.

From (1),

$$\frac{d^2y}{dx^2} = \frac{d}{dx}\left(-\frac{p}{q}\right) = -\frac{q\dfrac{dp}{dx} - p\dfrac{dq}{dx}}{q^2} \qquad \text{...(2)}$$

But
$$\frac{dp}{dx} = \frac{\partial p}{\partial x} + \frac{\partial p}{\partial y}\frac{dy}{dx} = \frac{\partial^2 f}{\partial x^2} + \frac{\partial^2 f}{\partial y\,\partial x}\frac{dy}{dx}$$

$$= r + s\left(-\frac{p}{q}\right)$$

$$= \frac{rq - sp}{q}$$

and
$$\frac{dp}{dx} = \frac{\partial q}{\partial x} + \frac{\partial q}{\partial y}\frac{dy}{dx} = \frac{\partial^2 f}{\partial x\, \partial y} + \frac{\partial^2 f}{\partial y^2}\frac{dy}{dx}$$

$$= s + t\left(-\frac{p}{q}\right)$$

$$= \frac{sq - pt}{q}.$$

Putting for $\frac{dp}{dx}$ and $\frac{dq}{dx}$ in (1) and simplifying, we get

$$\frac{d^2y}{dx^2} = -\frac{q^2 r - 2pqs + p^2 t}{q^3}$$

Illustrative Examples

Example 1. *If* $x^y + y^x = c$, *find* $\frac{dy}{dx}$.

Solution : Let $f(x, y) = x^y + y^x - c = 0$

$$\therefore \quad \frac{\partial f}{\partial x} = yx^{y-1} + y^x \log y$$

$$\frac{\partial f}{\partial y} = x^y \log x + xy^{x-1}$$

$$\therefore \quad \frac{dy}{dx} = -\frac{\partial f/\partial x}{\partial f/\partial y}$$

$$= -\frac{yx^{y-1} + y^x \log y}{x^y \log x + xy^{x-1}}.$$

Example 2. If $f(x, y) = 0$, $\phi(y, z) = 0$, *show that*

$$\frac{\partial f}{\partial y} \cdot \frac{\partial \phi}{\partial z} \cdot \frac{dz}{dx} = \frac{\partial f}{\partial x} \cdot \frac{\partial \phi}{\partial y}.$$

Solution : $\because$
$$f(x, y) = 0$$

$$\therefore \quad \frac{dy}{dx} = -\frac{\partial f/\partial x}{\partial f/\partial y} \qquad \text{... (i)}$$

$$\because \quad \phi(y, z) = 0$$

$$\therefore \qquad \frac{dz}{dy} = -\frac{\partial\phi/\partial y}{\partial\phi/\partial z}. \qquad \text{... (ii)}$$

Multiplying (i) and (ii), we get

$$\frac{dy}{dx}\frac{dz}{dy} = \left\{-\left(\frac{\partial f/\partial x}{\partial f/\partial y}\right)\right\} - \left(-\left\{\frac{\partial\phi/\partial y}{\partial\phi/\partial z}\right\}\right)$$

$$\Rightarrow \qquad \frac{dz}{dx} = \frac{\partial f/\partial x}{\partial f/\partial y}\frac{\partial\phi/\partial y}{\partial\phi/\partial z}$$

$$\Rightarrow \qquad \frac{\partial f}{\partial y}\frac{\partial f}{\partial z}\frac{dz}{dx} = \frac{\partial f}{\partial x}\frac{\partial\phi}{\partial y}.$$

Example 3. *If* $z = \sqrt{x^2+y^2}$ and $x^3+y^3+3axy = 5a^2$, *find the value of* $\frac{dz}{dx}$ *when* $x = a, y = a$.

Solution : We have

$$x^3 + y^3 + 3axy = 5a^2$$

$$\Rightarrow \qquad f \equiv x^3 + y^3 + 3axy - 5a^2 = 0$$

$$\therefore \qquad \frac{dy}{dx} = -\frac{\partial f/\partial x}{\partial f/\partial y} = -\frac{3x^2+3ay}{3y^2+3ax}$$

$$\Rightarrow \qquad \frac{dy}{dx} = -\frac{x^2+ay}{y^2+ax}$$

Now,
$$z = \sqrt{x^2+y^2}$$

$$\therefore \qquad \frac{\partial z}{\partial y} = \frac{1}{2}(x^2+y^2)^{-1/2}\cdot 2x = \frac{x}{\sqrt{x^2+y^2}},$$

and
$$\frac{\partial z}{\partial y} = \frac{1}{2}(x^2+y^2)^{-1/2}\cdot 2y = \frac{y}{\sqrt{x^2+y^2}}$$

$$\therefore \qquad \frac{dz}{dx} = \frac{\partial z}{\partial x} + \frac{\partial z}{\partial y}\frac{dy}{dx}$$

$$= \frac{x}{\sqrt{x^2+y^2}} + \frac{y}{\sqrt{x^2+y^2}}\left(-\frac{x^2+ay}{y^2+ax}\right)$$

$$= \frac{a}{\sqrt{a^2+a^2}} + \frac{a}{\sqrt{a^2+a^2}}\left(-\frac{a^2+a^2}{a^2+a^2}\right) \qquad \text{at } x = a, y = a$$

$$= \frac{1}{\sqrt{2}} - \frac{1}{\sqrt{2}}$$

$$= 0.$$

Example 4. *Find* $\frac{d^2y}{dx^2}$ if $ax^2 + 2hxy + by^2 = 1$.

Solution : Let $$f \equiv ax^2 + 2hxy + by^2 - 1 = 0$$

$$\therefore \quad p = \frac{\partial f}{\partial x} = 2(ax + hy)$$

$$q = \frac{\partial f}{\partial y} = 2(hx + by)$$

$$r = \frac{\partial^2 f}{\partial x^2} = 2a$$

$$s = \frac{\partial^2 f}{\partial x\, \partial y} = 2h$$

$$t = \frac{\partial^2 f}{\partial y^2} = 2b$$

$$\therefore \quad \frac{d^2y}{dx^2} = -\frac{q^2r - 2pqs + p^2t}{q^3}$$

$$= \left[\frac{(hx + by)^2 a - 2(ax + hy)(hx + by)h + (ax + hy)^2 b}{(hx + by)^2}\right]$$

$$= \frac{h^2 - ab}{(hx + by)^2} \qquad \text{[after simplification]}$$

Example 5. *If* $u = f(x - y, y - z, z - x)$, *prove that*

$$\frac{\partial u}{\partial x} + \frac{\partial u}{\partial y} + \frac{\partial u}{\partial z} = 0.$$

Solution : Let $X = x - y,\ Y = y - z,\ Z = z - x$, then

$$\frac{\partial X}{\partial x} = 1, \qquad \frac{\partial X}{\partial y} = -1, \qquad \frac{\partial X}{\partial z} = 0$$

$$\frac{\partial Y}{\partial x} = 0, \qquad \frac{\partial Y}{\partial y} = 1, \qquad \frac{\partial Y}{\partial z} = -1$$

$$\frac{\partial Z}{\partial x} = -1, \qquad \frac{\partial Z}{\partial y} = 0, \qquad \frac{\partial Z}{\partial z} = 1$$

and $$u = f(X, Y, Z)$$

$$\therefore \quad \frac{\partial u}{\partial x} = \frac{\partial u}{\partial X}\frac{\partial X}{\partial x} + \frac{\partial u}{\partial Y}\frac{\partial Y}{\partial x} + \frac{\partial u}{\partial Z}\frac{\partial Z}{\partial x}$$

$$= \frac{\partial u}{\partial X}(1) + \frac{\partial u}{\partial Y}(0) + \frac{\partial u}{\partial Z}(-1)$$

$$= \frac{\partial u}{\partial X} - \frac{\partial u}{\partial Z} \qquad \text{... (i)}$$

$$\frac{\partial u}{\partial y} = \frac{\partial u}{\partial X}\frac{\partial X}{\partial y} + \frac{\partial u}{\partial Y}\frac{\partial Y}{\partial y} + \frac{\partial u}{\partial Z}\frac{\partial Z}{\partial y}$$

$$= \frac{\partial u}{\partial X}(-1) + \frac{\partial u}{\partial Y}(1) + \frac{\partial u}{\partial Z}(0)$$

$$= -\frac{\partial u}{\partial X} + \frac{\partial u}{\partial Y} \qquad \text{... (ii)}$$

and
$$\frac{\partial u}{\partial z} = \frac{\partial u}{\partial X}\frac{\partial X}{\partial z} + \frac{\partial u}{\partial Y}\frac{\partial Y}{\partial z} + \frac{\partial u}{\partial Z}\frac{\partial Z}{\partial z}$$

$$= \frac{\partial u}{\partial X}(0) + \frac{\partial u}{\partial Y}(-1) + \frac{\partial u}{\partial Z}(1)$$

$$= -\frac{\partial u}{\partial Y} + \frac{\partial u}{\partial Z} \qquad \text{... (iii)}$$

Adding (i), (ii) and (iii), we get

$$\frac{\partial u}{\partial x} + \frac{\partial u}{\partial y} + \frac{\partial u}{\partial z} = 0.$$

Example 6. *If* $u = f(x, y)$, *where,* $x = r\cos\theta$ and $y = r\sin\theta$, *prove that the equation* $\frac{\partial^2 u}{\partial x^2} + \frac{\partial^2 u}{\partial y^2} = 0$ *transforms into*

$$\frac{\partial^2 u}{\partial r^2} + \frac{1}{r}\frac{\partial u}{\partial r} + \frac{1}{r^2}\frac{\partial^2 u}{\partial \theta^2} = 0.$$

Solution : We have

$$x = r\cos\theta, \qquad y = r\sin\theta$$

$$\therefore \quad r^2 = x^2 + y^2, \qquad \theta = \tan^{-1}\left(\frac{y}{x}\right)$$

$$\therefore \quad \frac{\partial r}{\partial x} = \frac{x}{r} = \cos\theta$$

$$\frac{\partial r}{\partial y} = \frac{y}{r} = \sin\theta$$

$$\frac{\partial \theta}{\partial x} = -\frac{y}{x^2+y^2} - \frac{r\sin\theta}{r^2} = \frac{\sin\theta}{r}$$

$$\frac{\partial \theta}{\partial y} = \frac{x}{x^2+y^2} = \frac{r\cos\theta}{r^2} = \frac{\cos\theta}{r}.$$

Now, u may be considered as a function of r, θ where r, θ are functions of x and y.

$$\therefore \quad \frac{\partial u}{\partial x} = \frac{\partial u}{\partial r}\frac{\partial r}{\partial x} + \frac{\partial u}{\partial \theta}\frac{\partial \theta}{\partial x}$$

$$= \frac{\partial u}{\partial r}\cdot\cos\theta + \frac{\partial u}{\partial \theta}\cdot\left(-\frac{\sin\theta}{r}\right)$$

$$\therefore \quad \frac{\partial^2 u}{\partial x^2} = \frac{\partial}{\partial x}\left(\frac{\partial u}{\partial x}\right)$$

$$= \frac{\partial}{\partial x}\left\{\frac{\partial u}{\partial r}\cos\theta - \frac{\sin\theta}{r}\frac{\partial u}{\partial \theta}\right\}$$

$$\Rightarrow \quad \frac{\partial^2 u}{\partial x^2} = \frac{\partial}{\partial r}\left\{\frac{\partial u}{\partial r}\cos\theta - \frac{\sin\theta}{r}\frac{\partial u}{\partial \theta}\right\}\cos\theta$$

$$- \frac{\sin\theta}{r}\frac{\partial}{\partial \theta}\left\{\frac{\partial u}{\partial r}\cos\theta - \frac{\sin\theta}{r}\frac{\partial u}{\partial \theta}\right\}$$

$$= \frac{\partial^2 u}{\partial r^2}\cos^2\theta - \sin\theta\cos\theta\left\{\frac{1}{r}\frac{\partial^2 u}{\partial r\,\partial \theta} - \frac{1}{r^2}\frac{\partial u}{\partial \theta}\right\}$$

$$- \frac{\sin\theta}{r}\left\{-\frac{\partial u}{\partial r}\sin\theta + \frac{\partial^2 u}{\partial \theta\,\partial r}\cos\theta - \frac{1}{r}\left(\cos\theta\frac{\partial u}{\partial \theta} + \sin\theta\frac{\partial^2 u}{\partial \theta^2}\right)\right\}$$

$$= \frac{\partial^2 u}{\partial r^2}\cos^2\theta - \frac{\sin\theta\cos\theta}{r}\frac{\partial^2 u}{\partial r\,\partial \theta} + \frac{\sin\theta\cos\theta}{r^2}\frac{\partial u}{\partial \theta}$$

$$+ \frac{\sin^2\theta}{r}\frac{\partial u}{\partial r} - \frac{\sin\theta\cos\theta}{r}\frac{\partial^2 u}{\partial \theta\,\partial r}$$

$$+ \frac{\sin\theta\cos\theta}{r^2}\frac{\partial u}{\partial \theta} + \frac{\sin^2\theta}{r^2}\frac{\partial^2 u}{\partial \theta^2}$$

$$= \frac{\partial^2 u}{\partial r^2}\cos^2\theta - \frac{2\sin\theta\cos\theta}{r}\frac{\partial^2 u}{\partial r\,\partial \theta} + \frac{2\sin\theta\cos\theta}{r^2}\frac{\partial u}{\partial \theta}$$

$$+ \frac{\sin^2\theta}{r}\frac{\partial u}{\partial r} + \frac{\sin^2\theta}{r^2}\frac{\partial^2 u}{\partial \theta^2}. \quad \text{... (i)}$$

Similarly, we have

$$\frac{\partial^2 u}{\partial y^2} = \frac{\partial^2 u}{\partial r^2}\sin^2\theta + \frac{2\sin\theta\cos\theta}{r}\frac{\partial^2 u}{\partial r\,\partial\theta} - \frac{2\sin\theta\cos\theta}{r^2}\frac{\partial u}{\partial\theta} + \frac{\cos^2\theta}{r}\frac{\partial u}{\partial r} + \frac{\cos^2\theta}{r^2}\frac{\partial^2 u}{\partial\theta^2} \quad \text{... (ii)}$$

Adding (i) and (ii), we get

$$\frac{\partial^2 u}{\partial x^2} + \frac{\partial^2 u}{\partial y^2} = \frac{\partial^2 u}{\partial r^2} + \frac{1}{r}\frac{\partial u}{\partial r} + \frac{1}{r^2}\frac{\partial^2 u}{\partial\theta^2}.$$

EXERCISE 9 (B)

1. Find the total differential coefficient of x^2y with respect to x, when x and y are connected by the relation

$$x^2 + xy + y^2 = 1$$

2. Find $\frac{dy}{dx}$ in the following cases :

(a) $x^y = y^x$, (b) $x^y + y^x = (x+y)^{x+y}$, (c) $(\cos x)^y = (\sin y)^x$.

3. Find $\frac{du}{dx}$, if $u = \sin(x^2 + y^2)$, where $a^2x^2 + b^2y^2 = c^2$.

4. If $u = e^{mx}(y - z)$, $y = m \sin x$ and $z = \cos x$, find $\frac{du}{dx}$.

5. If $ax^2 + by^2 + cz^2 = 1$ and $lx + my + nz = 0$, find the values of $\frac{dy}{dx}$ and $\frac{dz}{dx}$.

6. If the curves $f(x, y) = 0$ and $\phi(x, y) = 0$ touch, show that at the point of contact,

$$\frac{\partial f}{\partial x}\cdot\frac{\partial\phi}{\partial y} - \frac{\partial f}{\partial y}\frac{\partial\phi}{\partial x} = 0.$$

7. If x, y, z are connected by the equations $\phi(x, y, z) = 0$ and $\psi(x, y, z) = 0$, find $\frac{dy}{dx}$.

8. If $xyf\left(\frac{y}{x}\right) = c$, show that

$$\frac{f'\left(\frac{y}{x}\right)}{f\left(\frac{y}{x}\right)} = \frac{x\left(y + x\frac{dy}{dx}\right)}{y\left(y - x\frac{dy}{dx}\right)}.$$

9. if $ax^2 + 2hxy + by^2 + 2gx + 2fy + c = 0$, prove that

$$\frac{d^2y}{dx^2} = \frac{abc + 2fgh - af^2 - bg^2 - ch^2}{(hx + by + f)^3}$$

10. If z is a function of x and y, prove that, if $x = e^u + e^{-v}$, $y = e^{-u} - e^v$,

$$\frac{\partial z}{\partial u} - \frac{\partial z}{\partial v} = x\frac{\partial z}{\partial x} - y\frac{\partial z}{\partial y}.$$

11. If $z = f(u, v)$, $u = x^2 - 2xy - y^2$, $v = y$, show that

$$(x + y)\frac{\partial z}{\partial x} + (x - y)\frac{\partial z}{\partial y} = 0$$

is equivalent to $\dfrac{\partial z}{\partial v} = 0.$

12. $x = \xi \cos\alpha - \eta \sin\alpha$, $y = \xi \sin\alpha + \eta\cos\alpha$, show that

$$\frac{\partial^2 u}{\partial x^2} + \frac{\partial^2 u}{\partial y^2} = \frac{\partial^2 u}{\partial \xi^2} + \frac{\partial^2 u}{\partial \eta^2}.$$

13. If $x = r\cos\theta$, $y = r\sin\theta$, prove that

(i) $\dfrac{\partial r}{\partial x} = \dfrac{\partial x}{\partial r}.$

(ii) $\dfrac{1}{r}\dfrac{\partial x}{\partial \theta} = r\dfrac{\partial \theta}{\partial x}.$

(iii) $\dfrac{\partial^2 \theta}{\partial x^2} + \dfrac{\partial^2 \theta}{\partial y^2} = 0.$

14. If $x = r\cos\theta$, $y = r\sin\theta$, where r and θ are functions of t, prove that

(a) $\dfrac{dx}{dt} = \cos\theta\dfrac{dr}{dt} - r\sin\theta\dfrac{d\theta}{dt}.$

(b) $\dfrac{dy}{dt} = \sin\theta\dfrac{dr}{dt} + r\cos\theta\dfrac{d\theta}{dt}.$

(c) $\dfrac{d^2x}{dt^2} = \cos\theta\dfrac{d^2r}{dt^2} - 2\sin\theta\dfrac{dr}{dt}\dfrac{d\theta}{dt} - r\cos\theta\left(\dfrac{d\theta}{dt}\right)^2 - r\sin\theta\dfrac{d^2\theta}{dt^2}.$

(d) $\dfrac{d^2y}{dt^2} = \sin\theta\dfrac{d^2r}{dt^2} + 2\cos\theta\dfrac{dr}{dt}\dfrac{d\theta}{dt} - r\cos\theta\left(\dfrac{d\theta}{dt}\right)^2 + r\cos\theta\dfrac{d^2\theta}{dt^2}.$

(e) $x\dfrac{dy}{dt} - y\dfrac{dx}{dt} = r^2\dfrac{d\theta}{dt}.$

15. If $x = r\cos\theta,\ y = r\sin\theta$, prove that

(a) $\left(\dfrac{\partial r}{\partial x}\right)^2 + \left(\dfrac{\partial r}{\partial y}\right)^2 = 1.$

(b) $\dfrac{\partial^2 r}{\partial x^2}\cdot\dfrac{\partial^2 r}{\partial y^2} = \left(\dfrac{\partial^2 r}{\partial y\,\partial x}\right)^2.$

(c) $\dfrac{\partial^2 r}{\partial x^2} + \dfrac{\partial^2 r}{\partial y^2} = \dfrac{1}{r}\left\{\left(\dfrac{\partial r}{\partial x}\right)^2 + \left(\dfrac{\partial r}{\partial y}\right)^2\right\}.$

16. If V is a function of r alone and $r^2 = x^2 + y^2 + z^2$, prove that

$$\frac{\partial^2 V}{\partial x^2} + \frac{\partial^2 V}{\partial y^2} + \frac{\partial^2 V}{\partial z^2} = \frac{\partial^2 V}{\partial r^2} + \frac{2}{r}\frac{dV}{dr}.$$

17. If $u = f(y - x, z - y)$ prove that

$$\frac{\partial u}{\partial x} + \frac{\partial u}{\partial y} + \frac{\partial u}{\partial z} = 0.$$

ANSWERS

1. $2xy - \dfrac{x^2(2x+y)}{x+2y}.$

2. (a) $-\dfrac{yx^{y-1} - y^x\log y}{x^y\log x - xy^{x-1}}.$

(b) $-\dfrac{y^x\log y + yx^{y-1} - (x+y)^{x+y}\ \{1 + \log(x+y)\}}{x^y\log x + xy^{x-1} - (x+y)^{x+y}\ \{1 + \log(x+y)\}}$

(c) $\dfrac{\log\sin y + y\tan x}{\log\cos x - x\cot y}.$

3. $2x\ [\cos(x^2+y^2)]\left(1 - \dfrac{a^2}{b^2}\right).$

4. $e^{mx}(m^2 + 1)\sin x.$

5. $\dfrac{dy}{dx} = \dfrac{clz - anx}{bny - cmz},\ \dfrac{dz}{dx} = \dfrac{amx - bly}{bny - cmz}.$

7. $\dfrac{\dfrac{\partial\phi}{\partial z}\dfrac{\partial\psi}{\partial x} - \dfrac{\partial\psi}{\partial z}\dfrac{\partial\phi}{\partial x}}{\dfrac{\partial\psi}{\partial z}\dfrac{\partial\phi}{\partial y} - \dfrac{\partial\phi}{\partial z}\dfrac{\partial\psi}{\partial y}}.$

10
Change of Variables

10.1. Introduction

The difference between the dependent and independent variables is well known to us. Sometimes, it is desirable, particularly dealing with the solutions of the differential equations, to change the independent variable into the dependent variable or into another variable with which it is connected by a relation. A similar transformation may be affected in case of several independent variables which are connected to a set of other variables by means of given relations.

10.2. Change of Independent Variable into Dependent Variable

Let $\qquad y = f(x)$, then

$$\frac{dy}{dx} = \frac{1}{\frac{dx}{dy}} = \left(\frac{dx}{dy}\right)^{-1} \qquad \text{... (1)}$$

$$\frac{d^2y}{dx^2} = \frac{d}{dx}\left(\frac{dy}{dx}\right) = \frac{d}{dx}\left(\frac{dx}{dy}\right)^{-1}$$

$$= \frac{d}{dy}\left(\frac{dx}{dy}\right)^{-1} \frac{dy}{dx} = -\left(\frac{dx}{dy}\right)^{-2} \frac{d^2x}{dy^2}\frac{dy}{dx}$$

$$= -\left(\frac{dx}{dy}\right)^{-2} \frac{d^2x}{dy^2}\left(\frac{dx}{dy}\right)^{-1}$$

$$= -\left(\frac{dx}{dy}\right)^{-3} \frac{d^2x}{dy^2}. \qquad \text{... (2)}$$

$$\frac{d^3y}{dx^3} = \frac{d}{dx}\left(\frac{d^2y}{dx^2}\right) = \frac{d}{dx}\left[-\left(\frac{dx}{dy}\right)^{-3} \frac{d^2x}{dy^2}\right]$$

$$= -\frac{d}{dy}\left[\left(\frac{dx}{dy}\right)^{-3} \frac{d^2x}{dy^2}\right]\frac{dy}{dx}$$

$$= -\left[\left(\frac{dx}{dy}\right)^{-3}\frac{d^3x}{dy^3} - 3\left(\frac{dx}{dy}\right)^{-4}\left(\frac{d^2x}{dy^2}\right)^2\right]\left(\frac{dx}{dy}\right)^{-1}$$

$$= -\left(\frac{dx}{dy}\right)^{-4}\frac{d^3x}{dy^3} + 3\left(\frac{dx}{dy}\right)^{-5}\left(\frac{d^2x}{dy^2}\right)^2 \qquad \text{... (3)}$$

and so on.

10.3. To Change the Independent Variable x into Another Variable t where $x = f(t)$

$$\frac{dy}{dx} = \frac{dy}{dt}\frac{dt}{dx} = \frac{dy}{dt}\left(\frac{dx}{dt}\right)^{-1} = \left(\frac{dx}{dt}\right)^{-1}\frac{dy}{dt}. \qquad \text{...(1)}$$

$$\Rightarrow \qquad \frac{d}{dx} \equiv \left(\frac{dx}{dt}\right)^{-1}\frac{d}{dt}$$

$$\frac{d^2y}{dx^2} = \frac{d}{dx}\left(\frac{dy}{dx}\right) = \left(\frac{dx}{dt}\right)^{-1}\frac{d}{dt}\left[\left(\frac{dx}{dt}\right)^{-1}\frac{dy}{dt}\right]$$

$$= \left[\frac{dx}{dt}\right]^{-1}\left[\left(\frac{dx}{dt}\right)^{-1}\frac{d^2y}{dt^2} + \left(\frac{dy}{dt}\right)(-1)\left(\frac{dx}{dt}\right)^{-2}\frac{d^2x}{dt^2}\right]$$

$$= \left(\frac{dx}{dt}\right)^{-3}\left[\frac{dx}{dt}\frac{d^2y}{dt^2} - \frac{d^2x}{dt^2}\frac{dy}{dt}\right]. \qquad \text{... (2)}$$

$$\frac{d^3y}{dx^3} = \frac{d}{dx}\left(\frac{d^2y}{dx^2}\right) = \left(\frac{dx}{dt}\right)^{-1}\frac{d}{dt}\left[\frac{\frac{dx}{dt}\frac{d^2y}{dt^2} - \frac{d^2x}{dt^2}\frac{dy}{dt}}{\left(\frac{dx}{dt}\right)^3}\right]$$

$$= \frac{\left(\frac{dx}{dt}\right)^3\left[\frac{dx}{dt}\frac{d^3y}{dt^3} - \frac{d^3x}{dt^3} - \frac{dy}{dt}\right] - \left[\frac{dx}{dt}\frac{d^2y}{dt^2} - \frac{d^2x}{dt^2} - \frac{dy}{dt}\right]3\left(\frac{dx}{dt}\right)^2\frac{d^2x}{dt^2}}{\left(\frac{dx}{dt}\right)\left(\frac{dx}{dt}\right)^6}$$

$$= \left(\frac{dx}{dt}\right)^{-5}\left[\left(\frac{dx}{dt}\frac{d^3y}{dt^3} - \frac{d^3x}{dt^3}\frac{dy}{dt}\right)\frac{dx}{dt} - 3\left(\frac{dx}{dt}\frac{d^2y}{dt^2} - \frac{d^2x}{dt^2}\frac{dy}{dt}\right)\frac{d^2x}{dt^2}\right] \qquad \text{... (3)}$$

and so on.

ILLUSTRATIVE EXAMPLES

Example 1. *Show that the equation $d^2y/dx^2 = a$, may be written in the form*

$$\frac{d^2x}{dy^2} + a\left(\frac{dx}{dy}\right)^3 = 0.$$

Solution :

$$\frac{d^2y}{dx^2} = a$$

$$\Rightarrow \quad -\left(\frac{dx}{dy}\right)^{-3}\frac{d^2x}{dy^2} = a$$

$$\Rightarrow \quad -\frac{d^2x/dy^2}{(dx/dy)^3} = a$$

$$\Rightarrow \quad \frac{d^2x}{dy^2} + a\left(\frac{dx}{dy}\right)^3 = 0.$$

Example 2. *Change the independent variable from x to y in*

$$\rho = \frac{\{1 + (dy/dx)^2\}^{3/2}}{d^2y/dx^2}$$

Solution :

$$\rho = \frac{\left\{1 + \left(\frac{dy}{dx}\right)^2\right\}^{3/2}}{\frac{d^2y}{dx^2}}$$

$$\Rightarrow \quad \rho = \frac{\{1 + [(dx/dy)^{-1}]^2\}^{3/2}}{-\left(\frac{dx}{dy}\right)^{-3}\frac{d^2x}{dy^2}} = \frac{\{1 + (dx/dy)^{-2}\}^{3/2}}{-\left(\frac{dx}{dy}\right)^{-3}\frac{d^2x}{dy^2}}$$

$$= \frac{\{1 + (dx/dy)^2\}^{3/2}}{-\left(\frac{dx}{dy}\right)^3\left(\frac{dx}{dy}\right)^{-3}\frac{d^2x}{dy^2}}$$

$$= -\frac{\{1 + (dx/dy)^2\}^{3/2}}{d^2x/dy^2}.$$

Example 3. *Transform the equation*

$$\sin^2 2z \frac{d^2y}{dz^2} + \sin 4z \frac{dy}{dz} + 4y = 0$$

by putting $\tan z = e^x$.

Solution : $\tan z = e^x$

$$\Rightarrow \quad sec^2 z \frac{dz}{dx} = e^x = \tan z$$

$$\Rightarrow \quad \frac{dz}{dx} = \frac{\tan z}{sec^2 z} = \sin z \cos z$$

$$\Rightarrow \quad \frac{dz}{dx} = \frac{1}{2} \sin 2z$$

$$\therefore \quad \frac{dy}{dz} = \frac{dy}{dx}\frac{dx}{dz}$$

$$= \frac{dy}{dx}\frac{2}{\sin 2z}$$

$$= \frac{2}{\sin 2z}\frac{dy}{dx}$$

$$\Rightarrow \quad \frac{d}{dz} = \frac{2}{\sin 2z}\frac{d}{dx}$$

$$\therefore \quad \frac{d^2y}{dz^2} = \frac{d}{dz}\left(\frac{dy}{dz}\right) = \frac{2}{\sin 2z}\frac{d}{dx}\left(\frac{2}{\sin 2z}\frac{dy}{dx}\right)$$

$$= \frac{2}{\sin 2z}\left[\frac{2}{\sin 2z}\frac{d^2y}{dx^2} - \frac{4\cos 2z}{\sin^2 2z}\frac{dz}{dx}\frac{dy}{dx}\right]$$

$$= \frac{2}{\sin 2z}\left[\frac{2}{\sin 2z}\frac{d^2y}{dx^2} - \frac{4\cos 2z}{\sin^2 2z}\frac{\sin 2z}{2}\frac{dy}{dx}\right]$$

$$= \frac{4}{\sin^2 2z}\frac{d^2y}{dx^2} - \frac{4\cos 2z}{\sin^2 2z}\frac{dy}{dx}.$$

Substituting these values, the given equation becomes

$$\sin^2 2z\left[\frac{4}{\sin^2 2z}\frac{d^2y}{dx^2}-\frac{4\cos 2z}{\sin^2 2z}\frac{dy}{dx}\right]+\sin 4z\frac{2}{\sin 2z}\frac{dy}{dx}+4y=0$$

$$\Rightarrow \qquad 4\frac{d^2y}{dx^2}-4\cos 2z\frac{dy}{dx}+\frac{4\sin 2z\cos 2z}{\sin 2z}\frac{dy}{dx}+4y=0$$

$[\because \sin 4z = 2\sin 2z\cos 2z]$

$$\Rightarrow \qquad \frac{d^2y}{dx^2}+y=0.$$

Example 4. *Transform the equation*

$$\frac{d^2y}{dx^2}+\cot x\frac{dy}{dx}+4y\operatorname{cosec}^2 x=0$$

by the substitution $z = 2\log\tan x/2$.

Solution : $z = 2\log\tan x/2$

$$\therefore \qquad \frac{dz}{dx}=\frac{2}{\tan x/2}\sec^2\frac{x}{2}\cdot\frac{1}{2}=\frac{2}{\sin x}$$

$$\therefore \qquad \frac{dy}{dx}=\frac{dy}{dz}\frac{dz}{dx}$$

$$=\frac{dy}{dz}\frac{2}{\sin x}=\frac{2}{\sin x}\frac{dy}{dz}$$

$$\Rightarrow \qquad \frac{d}{dx}\equiv\frac{2}{\sin x}\frac{d}{dz}$$

$$\frac{d^2y}{dx^2}=\frac{d}{dx}\left(\frac{dy}{dx}\right)=\frac{2}{\sin x}\frac{d}{dz}\left[\frac{2}{\sin x}\frac{dy}{dz}\right]$$

$$=\frac{2}{\sin x}\left[\frac{2}{\sin x}\frac{d^2y}{dz^2}-\frac{2\cos x}{\sin^2 x}\frac{dx}{dz}\frac{dy}{dz}\right]$$

$$=\frac{2}{\sin x}\left[\frac{2}{\sin x}\frac{d^2y}{dz^2}-\frac{2\cos x}{\sin^2 x}\frac{\sin x}{2}\cdot\frac{dy}{dz}\right]$$

$$=\frac{4}{\sin^2 x}\frac{d^2y}{dz^2}-\frac{2\cos x}{\sin^2 x}\frac{dy}{dz}.$$

Substituting these values in the given equation, we get

$$\frac{4}{\sin^2 x}\frac{d^2y}{dz^2} - \frac{2\cos x}{\sin^2 x}\frac{dy}{dz} + \cot x \cdot \frac{2}{\sin x}\frac{dy}{dz} + 4y\ \text{cosec}^2 x = 0$$

$$\Rightarrow \quad 4\ \text{cosec}^2 x \frac{d^2y}{dz^2} + 4y\ \text{cosec}^2 x = 0$$

$$\Rightarrow \quad \frac{d^2y}{dz^2} + y = 0$$

which is the required transformed equation.

Example 5. *If* $x^2 + z^2 = 1$, *show that the equation*

$$\frac{d}{dx}\left\{(1-x^2)\frac{dy}{dx}\right\} + n(n+1)\, y = 0$$

becomes $z(z^2-1)\dfrac{d^2y}{dz^2} + (2z^2-1)\dfrac{dy}{dz} - n(n+1)\, zy = 0.$

Solution :

$$x^2 + z^2 = 1$$

$$\therefore \quad 2x + 2z\frac{dz}{dx} = 0$$

$$\therefore \quad \frac{dz}{dx} = -\frac{x}{z}$$

$$\therefore \quad \frac{dy}{dx} = \frac{dy}{dz}\frac{dz}{dx} = -\frac{x}{z}\frac{dy}{dz}$$

$$\Rightarrow \quad \frac{d}{dx} \equiv -\frac{x}{z}\frac{d}{dx}.$$

Hence, the given equation becomes,

$$-\frac{x}{z}\cdot\frac{d}{dz}\left[(1-x^2)\left(-\frac{x}{z}\frac{dy}{dz}\right)\right] + n(n+1)y = 0$$

$$\Rightarrow \quad -\frac{x}{z}\frac{d}{dz}\left[z^2\left(-\frac{x}{z}\frac{dy}{dz}\right)\right] + n(n+1)y = 0$$

$$\Rightarrow \quad -\frac{x}{z}\frac{d}{dz}\left(-zx\frac{dy}{dz}\right) + n(n+1)y = 0$$

$$\Rightarrow \quad \frac{x}{z}\left[zx\frac{d^2y}{dz^2} + \left(z\frac{dx}{dz} + x\cdot 1\right)\frac{dy}{dz}\right] + n(n+1)y = 0$$

$$\Rightarrow \quad x^2\frac{d^2y}{dz^2} + \frac{x}{z}\left[z\frac{dx}{dz} + x\right]\frac{dy}{dz} + n(n+1)y = 0$$

$$\Rightarrow \qquad x^2\frac{d^2y}{dz^2} + \frac{x}{z}\left[z\left(-\frac{z}{x}\right)+x\right]\frac{dy}{dz} + n(n+1)y = 0$$

$$\Rightarrow \qquad x^2\frac{d^2y}{dz^2} + \frac{1}{z}(x^2-z^2)\frac{dy}{dz} + n(n+1)y = 0$$

$$\Rightarrow \qquad (1-z^2)\frac{d^2y}{dz^2} + \frac{1-2z^2}{z}\frac{dy}{dz} + n(n+1)y = 0$$

$[\because x^2+z^2 = 1]$

$$\Rightarrow \qquad z(z^2-1)\frac{d^2y}{dz^2} + (2z^2-1)\frac{dy}{dz} - n(n+1)zy = 0.$$

10.4. Change of Both the Dependent and Independent Variables

The relations between cartesian coordinates (x, y) and polar coordinates (r, θ) of any point are,

$$x = r\cos\theta$$

$$y = r\sin\theta.$$

$\therefore$ We have

$$\frac{dy}{dx} = \frac{dy/d\theta}{dx/d\theta}.$$

$$= \frac{(dr/d\theta)\sin\theta + r\cos\theta}{(dr/d\theta)\cos\theta - r\sin\theta}$$

$$\frac{d^2y}{dx^2} = \frac{d}{dx}\left(\frac{dy}{dx}\right) = \frac{d}{d\theta}\left(\frac{dy}{dx}\right)\frac{d\theta}{dx}$$

$$= \frac{\frac{d}{d\theta}\left(\frac{dy}{dx}\right)}{\frac{dx}{d\theta}}$$

$$= \frac{\frac{d}{d\theta}\left[\frac{(dr/d\theta)\sin\theta + r\cos\theta}{(dr/d\theta)\cos\theta - r\sin\theta}\right]}{\frac{dx}{d\theta}}$$

$$= \frac{r^2 + 2\left(\frac{dr}{d\theta}\right)^2 - r\frac{d^2r}{d\theta^2}}{\left(\cos\theta\frac{dr}{d\theta} - r\sin\theta\right)^3}. \qquad \text{[on simplification]}$$

10.5. Theorem

Show that,
$$x\frac{dx}{dt} + y\frac{dy}{dt} = r\frac{dr}{dt} \quad \text{... (1)}$$

$$x\frac{dy}{dt} - y\frac{dx}{dt} = r^2\frac{d\theta}{dt} \quad \text{... (2)}$$

$$\left(\frac{dx}{dt}\right)^2 + \left(\frac{dy}{dt}\right)^2 = \left(\frac{dr}{dt}\right)^2 + r^2\left(\frac{d\theta}{dt}\right)^2 \quad \text{... (3)}$$

Proof : $\because$
$$x^2 + y^2 = r^2$$

$\therefore$
$$2x\frac{dx}{dt} + 2y\frac{dy}{dt} = 2r\frac{dr}{dt}$$

$\Rightarrow$
$$x\frac{dx}{dt} + y\frac{dy}{dt} = r\frac{dr}{dt}. \quad \text{... (1)}$$

$$\frac{y}{x} = \tan\theta$$

$\Rightarrow$
$$\frac{x\dfrac{dy}{dt} - y\dfrac{dx}{dt}}{x^2} = \sec^2\theta\,\frac{d\theta}{dt}$$

$\Rightarrow$
$$x\frac{dy}{dt} - y\frac{dx}{dt} = x^2\sec^2\theta\,\frac{d\theta}{dt} = r^2\cos^2\theta\sec^2\theta\,\frac{d\theta}{dt}$$

$$= r^2\frac{d\theta}{dt}. \quad \text{... (2)}$$

Squaring and adding (1) and (2), we get
$$\left(\frac{dx}{dt}\right)^2 + \left(\frac{dy}{dt}\right)^2 = \left(\frac{dr}{dt}\right)^2 + r^2\left(\frac{d\theta}{dt}\right)^2. \quad \text{... (3)}$$

Example 6. *Transform to polar form the formula*
$$p = \frac{x\dfrac{dy}{dx} - y}{\sqrt{1 + (dy/dx)^2}}.$$

Solution : $\because$
$$p = \frac{x\dfrac{dy}{dx} - y}{\sqrt{1 + (dy/dx)^2}}$$

$\therefore$
$$p = \frac{x\dfrac{dy}{dt} - y\dfrac{dx}{dt}}{\sqrt{(dx/dt)^2 + (dy/dt)^2}}$$

[multiplying the numerator and denominator by dx/dt]

$$= \frac{r^2 \dfrac{d\theta}{dt}}{\sqrt{(dr/dt)^2 + r^2 (d\theta/dt)^2}}$$

$\therefore$
$$\frac{1}{p^2} = \frac{(dr/dt)^2 + r^2 (d\theta/dt)^2}{\left(r^2 \dfrac{d\theta}{dt}\right)^2} = \frac{1}{r^2} + \frac{1}{r^4}\left(\frac{dr}{dt}\right)^2 \times \left(\frac{dt}{d\theta}\right)^2.$$

$\Rightarrow$
$$\frac{1}{p^2} = \frac{1}{r^2} + \frac{1}{r^4}\left(\frac{dr}{d\theta}\right)^2.$$

EXERCISE 10 (A)

1. Change the equation $\dfrac{d^2y}{dx^2} - \left(\dfrac{dy}{dx}\right)^2 - y\left(\dfrac{dy}{dx}\right)^3 = 0$ so that the variable y may be considered the independent variable instead of x.

2. Change the independent variable from x to y in the equation
$$3\left(\frac{d^2y}{dx^2}\right)^2 - \frac{dy}{dx}\frac{d^3y}{dx^3} - \frac{d^2y}{dx^2}\left(\frac{dy}{dx}\right)^2 = 0.$$

3. Change the equation $\dfrac{dy}{dx}\dfrac{d^3y}{dx^3} - 3\left(\dfrac{d^2y}{dx^2}\right)^2 = 0$, so that y may be considered the independent variable instead of x. Hence show that the solution of the equation may be given by
$$y = a \pm \sqrt{bx + c}.$$

4. Change the independent variable from x to z in the equation
$$x^2\frac{d^2y}{dx^2} + x\frac{dy}{dx} + y = 0, \text{ where } x = 1/z.$$

5. Change the independent variable from x to z in the equation
$$x^4\frac{d^2y}{dx^2} + a^2y = 0, \text{ where } x = 1/z.$$

6. Transform the equation $x^4\dfrac{d^2y}{dx^2} + 2x^3\dfrac{dy}{dx} + n^2y = 0$ by the substitution $x = 1/z$.

7. Transform the equation $x^2\dfrac{d^2y}{dx^2} + x\dfrac{dy}{dx} + y = 0$ by the substitution $x = e^z$.

8. Transform the equation $(1 - x^2)\dfrac{d^2y}{dx^2} - x\dfrac{dy}{dx} = 0$, where $x = \cos\theta$.

9. Transform the equation $(1 - x^2)\dfrac{d^2y}{dx^2} - x\dfrac{dy}{dx} + a^2y = 0$, where $x = \sin\theta$.

10. Change the independent variable from x to θ in the equation

$$(1 - x^2)\frac{d^2y}{dx^2} - x\frac{dy}{dx} + a^2y = 0, \text{ where } x = \cos\theta.$$

11. Transform the equation $(1 + x^2)^2\dfrac{d^2y}{dx^2} + 2x(1 + x^2)\dfrac{dy}{dx} + y = 0$ into one in which z is the independent variable, given $x = \tan z$.

12. Transform the equation

$$\frac{d^2y}{dx^2} + \left(1 - \frac{1}{x}\right)\frac{dy}{dx} + 4x^2ye^{-2x} = 4(x^2 + x^3)e^{-2x}$$

by the substitution $z = (1 + x)e^{-x}$.

13. Transform the equation $\dfrac{d^2y}{dx^2} + xy\dfrac{dy}{dx} + \sec^2 x = 0$, where $x = \tan^{1} z$.

14. Transform to Cartesian formula $p = r\sin\phi$, where $\tan\phi = r\dfrac{d\theta}{dr}$.

15. Transform the formula for radius of curvature

$$\rho = \frac{[1 + (dy/dx)^2]^{3/2}}{d^2y/dx^2}$$

into polar coordinates.

16. Transform $\tan\phi = r\dfrac{d\theta}{dr}$ into Cartesian form.

17. Transform the equation

$$\left(\frac{d^2y}{dx^2}\right)\cos x + \frac{dy}{dx}\sin x - 2y\cos^3 x = 2\cos^5 x$$

into one in which z is independent variable given by $z = \sin x$.

ANSWERS

1. $\dfrac{d^2y}{dx^2} - \dfrac{dx}{dy} + y = 0.$

2. $\dfrac{d^3x}{dy^3} + \dfrac{d^2x}{dy^2} = 0.$

3. $\dfrac{d^3x}{dy^3} = 0.$

4. $z^2 \dfrac{d^2y}{dz^2} + z\dfrac{dy}{dz} + y = 0.$

5. $\dfrac{d^2y}{dz^2} + \dfrac{2}{z}\dfrac{dy}{dz} + a^2y = 0.$

6. $\dfrac{d^2y}{dz^2} + n^2y = 0.$

7. $\dfrac{d^2y}{dx^2} + y = 0.$

8. $\dfrac{d^2y}{d\theta^2} = 0.$

9. $\dfrac{d^2y}{d\theta^2} + a^2y = 0.$

10. $\dfrac{d^2y}{d\theta^2} + a^2y = 0.$

11. $\dfrac{d^2y}{dz^2} + y = 0.$

12. $\dfrac{d^2y}{dz^2} + 4y = 4z.$

13. $(1 + z^2)\dfrac{d^2y}{dz^2} + (2z + y \tan^{-1} z)\dfrac{dy}{dz} + 1 = 0.$

14. $p = \dfrac{-y + x\, dy/dx}{\sqrt{1 + (dy/dx)^2}}$

15. $\rho = \dfrac{[r^2 + (dr/d\theta)^2]^{3/2}}{r + 2\left(\dfrac{dr}{d\theta}\right)^2 - r\dfrac{d^2r}{d\theta^2}}.$

16. $\tan\phi = \dfrac{x\,(dy/dx) - y}{x + y\,(dy/dx)}.$

17. $\dfrac{d^2y}{dz^2} - 2y = 2(1 - z^2).$

10.6. Change of Two Independent Variables

Let
$$u = f(x, y); \quad \text{... (1)}$$
where
$$x = F_1(t_1, t_2); \quad \text{... (2)}$$
$$y = F_2(t_1, t_2). \quad \text{... (3)}$$

Then we know that

$$\frac{\partial u}{\partial t_1} = \frac{\partial u}{\partial x}\frac{\partial x}{\partial t_1} + \frac{\partial u}{\partial y}\frac{\partial y}{\partial t_1} \quad \text{... (4)}$$

and

$$\frac{\partial u}{\partial t_2} = \frac{\partial u}{\partial x}\frac{\partial x}{\partial t_2} + \frac{\partial u}{\partial y}\frac{\partial y}{\partial t_2} \quad \text{... (5)}$$

Now, if we substitute the values of x and y from (2) and (3) respectively in (1), then

$$u = \psi(t_1, t_2) \quad \text{... (6)}$$

Also, from (2) and (3),

$$t_1 = \phi_1(x, y) \quad \text{... (7)}$$

and $$t_2 = \phi_2(x, y). \qquad \text{... (8)}$$

So we have

$$\frac{\partial u}{\partial x} = \frac{\partial u}{\partial t_1}\frac{\partial t_1}{\partial x} + \frac{\partial u}{\partial t_2}\frac{\partial t_2}{\partial x} \qquad \text{... (9)}$$

and $$\frac{\partial u}{\partial y} = \frac{\partial u}{\partial t_1}\frac{\partial t_1}{\partial y} + \frac{\partial u}{\partial t_2}\frac{\partial t_2}{\partial y} \qquad \text{... (10)}$$

ILLUSTRATIVE EXAMPLES

Example 1. If $x = e^\theta$, $y = e^\phi$, *transform the equation*

$$e^{2\theta}\frac{\partial^2 V}{\partial x^2} + e^{2\phi}\frac{\partial^2 V}{\partial y^2} + e^\theta\frac{\partial V}{\partial x} + e^\phi\frac{\partial V}{\partial y} = 0$$

so that θ *and* ϕ *are the independent variables.*

Solution : We have $x = e^\theta \Rightarrow \theta = \log x$

and $$y = e^\phi \Rightarrow \phi = \log y$$

$$\therefore \quad \frac{\partial V}{\partial x} = \frac{\partial V}{\partial \theta}\frac{\partial \theta}{\partial x} + \frac{\partial V}{\partial \phi}\frac{\partial \phi}{\partial x}$$

$$= \frac{\partial V}{\partial \theta}\frac{1}{x} + \frac{\partial V}{\partial \phi}\cdot 0 = \frac{1}{x}\frac{\partial V}{\partial \theta}$$

$$\Rightarrow \quad x\frac{\partial V}{\partial x} = \frac{\partial V}{\partial \theta}$$

$$\Rightarrow \quad x\frac{\partial}{\partial x} \equiv \frac{\partial}{\partial \theta}.$$

Now, $$x\frac{\partial}{\partial x}\left(x\frac{\partial V}{\partial x}\right) = \frac{\partial}{\partial \theta}\left(\frac{\partial V}{\partial \theta}\right)$$

$$\Rightarrow \quad x\left[x\frac{\partial^2 V}{\partial x^2} + \frac{\partial V}{\partial x}\right] = \frac{\partial^2 V}{\partial \theta^2}$$

$$\Rightarrow \quad x^2\frac{\partial^2 V}{\partial x^2} + x\frac{\partial V}{\partial x} = \frac{\partial^2 V}{\partial \theta^2}$$

$$\Rightarrow \quad e^{2\theta}\frac{\partial^2 V}{\partial x^2} + e^\theta\frac{\partial V}{\partial x} = \frac{\partial^2 V}{\partial \theta^2} \qquad [\because x = e^\theta]$$

Similarly,

$$e^{2\phi}\frac{\partial^2 V}{\partial y^2}+e^{\phi}\frac{\partial V}{\partial y}=\frac{\partial^2 V}{\partial \phi^2}.$$

Adding the two, we get

$$e^{2\theta}\frac{\partial^2 V}{\partial x^2}+e^{2\phi}\frac{\partial^2 V}{\partial y^2}+e^{\theta}\frac{\partial V}{\partial x}+e^{\phi}\frac{\partial V}{\partial y}=\frac{\partial^2 V}{\partial \theta^2}+\frac{\partial^2 V}{\partial \phi^2}.$$

Example 2. *If* $x = e^v \sec u$, $y = e^v \tan u$ *and* ϕ *is a function of* x *and* y, *show that*

$$\cos u\left(\frac{\partial^2\phi}{\partial u\,\partial v}-\frac{\partial\phi}{\partial u}\right)=xy\left(\frac{\partial^2\phi}{\partial x^2}+\frac{\partial^2\phi}{\partial y^2}\right)+(x^2+y^2)\frac{\partial^2\phi}{\partial x\,\partial y}.$$

Solution :

$$\frac{\partial\phi}{\partial u}=\frac{\partial\phi}{\partial x}\frac{\partial x}{\partial u}+\frac{\partial\phi}{\partial y}\frac{\partial y}{\partial u}=\frac{\partial\phi}{\partial x}e^v\sec u\tan u+\frac{\partial\phi}{\partial y}e^v\sec^2 u$$

$$\Rightarrow\quad \cos u\frac{\partial\phi}{\partial u}=e^v\tan u\frac{\partial\phi}{\partial x}+e^v\sec u\frac{\partial\phi}{\partial y}$$

$$=y\frac{\partial\phi}{\partial x}+x\frac{\partial\phi}{\partial y}\qquad\text{... (i)}$$

$$\frac{\partial\phi}{\partial v}=\frac{\partial\phi}{\partial x}\frac{\partial x}{\partial v}+\frac{\partial\phi}{\partial y}\frac{\partial y}{\partial v}=\frac{\partial\phi}{\partial x}e^v\sec u+\frac{\partial\phi}{\partial y}e^v\tan u$$

$$=x\frac{\partial\phi}{\partial x}+y\frac{\partial\phi}{\partial y}\qquad\text{... (ii)}$$

From (i),

$$\cos u\frac{\partial}{\partial u}\equiv y\frac{\partial}{\partial x}+x\frac{\partial}{\partial y}$$

$$\therefore\quad \cos u\frac{\partial}{\partial u}\left(\frac{\partial\phi}{\partial v}\right)=\left(y\frac{\partial}{\partial x}+x\frac{\partial}{\partial y}\right)\frac{\partial\phi}{\partial v}$$

$$\Rightarrow\quad \cos u\frac{\partial^2\phi}{\partial u\,\partial v}=\left(y\frac{\partial}{\partial x}+x\frac{\partial}{\partial y}\right)\left(x\frac{\partial\phi}{\partial x}+y\frac{\partial\phi}{\partial y}\right)$$

$$=y\left(x\frac{\partial^2\phi}{\partial x^2}+\frac{\partial\phi}{\partial x}+y\frac{\partial^2\phi}{\partial x\,\partial y}\right)+x\left(x\frac{\partial^2\phi}{\partial y\,\partial x}+y\frac{\partial^2\phi}{\partial y^2}+\frac{\partial\phi}{\partial y}\right)$$

$$=xy\left(\frac{\partial^2\phi}{\partial x^2}+\frac{\partial^2\phi}{\partial y^2}\right)+(x^2+y^2)\frac{\partial^2\phi}{\partial x\,\partial y}+y\frac{\partial\phi}{\partial x}+x\frac{\partial\phi}{\partial y}$$

$$=xy\left(\frac{\partial^2\phi}{\partial x^2}+\frac{\partial^2\phi}{\partial y^2}\right)+(x^2+y^2)\frac{\partial^2\phi}{\partial x\,\partial y}+\cos u\frac{\partial\phi}{\partial u}$$

$$\Rightarrow \qquad \cos u \left\{\frac{\partial^2 \phi}{\partial u\, \partial v} - \frac{\partial \phi}{\partial u}\right\} = xy \left\{\frac{\partial^2 \phi}{\partial x^2} + \frac{\partial^2 \phi}{\partial y^2}\right\} + (x^2 + y^2) \frac{\partial^2 \phi}{\partial x\, \partial y}$$

Example 3. *If* $u = f(x, y), x^2 = \xi\eta$ and $y^2 = \xi/\eta$, *change the independent variables to* ξ, η *in the equation*

$$x^2 \frac{\partial^2 u}{\partial x^2} - 2xy \frac{\partial^2 u}{\partial x\, \partial y} + y^2 \frac{\partial^2 u}{\partial y^2} + 2y \frac{\partial u}{\partial y} = 0.$$

Solution : We have, $\quad x^2 = \xi\eta, y^2 = \dfrac{\xi}{\eta}$

$$\Rightarrow \qquad \xi = xy \text{ and } \eta = \frac{y}{x} \quad \text{[on solving]}$$

$$\frac{\partial u}{\partial x} = \frac{\partial u}{\partial \xi}\frac{\partial \xi}{\partial x} + \frac{\partial u}{\partial \eta}\frac{\partial \eta}{\partial x}$$

$$= \frac{\partial u}{\partial \xi} \cdot y + \frac{\partial u}{\partial \eta}\left\{\frac{-y}{x^2}\right\}$$

$$= y\frac{\partial u}{\partial \xi} - \frac{y}{x^2}\frac{\partial u}{\partial \eta}$$

$$\Rightarrow \qquad x\frac{\partial u}{\partial x} = xy\frac{\partial u}{\partial \xi} - \frac{y}{x}\frac{\partial u}{\partial \eta}$$

$$\Rightarrow \qquad x\frac{\partial u}{\partial x} = \xi\frac{\partial u}{\partial \xi} - \eta\frac{\partial u}{\partial \eta}. \qquad \text{... (i)}$$

$$\frac{\partial u}{\partial y} = \frac{\partial u}{\partial \xi}\frac{\partial \xi}{\partial y} + \frac{\partial u}{\partial \eta}\frac{\partial \eta}{\partial y}$$

$$= \frac{\partial u}{\partial \xi} \cdot x + \frac{\partial u}{\partial \eta} \cdot \frac{1}{x}$$

$$\Rightarrow \qquad y\frac{\partial u}{\partial y} = xy\frac{\partial u}{\partial \xi} + \frac{y}{x}\frac{\partial u}{\partial \eta}$$

$$\Rightarrow \qquad y\frac{\partial u}{\partial y} = \xi\frac{\partial u}{\partial \xi} + \eta\frac{\partial u}{\partial \eta}. \qquad \text{... (ii)}$$

From (i) and (ii),

$$x\frac{\partial}{\partial x} - y\frac{\partial}{\partial y} = -2\eta\frac{\partial}{\partial \eta}.$$

Now, $\left(x\dfrac{\partial}{\partial x} - y\dfrac{\partial}{\partial y}\right)\left(x\dfrac{\partial u}{\partial x} - y\dfrac{\partial u}{\partial y}\right)$

$$= x^2\frac{\partial^2 u}{\partial x^2} + x\frac{\partial u}{\partial x} - xy\frac{\partial^2 u}{\partial x\,\partial y} - xy\frac{\partial^2 u}{\partial x\,\partial y} + y\frac{\partial u}{\partial y} + y^2\frac{\partial^2 u}{\partial^2 y}$$

$$= x^2\frac{\partial^2 u}{\partial x^2} - 2xy\frac{\partial^2 u}{\partial x\,\partial y} + y^2\frac{\partial^2 u}{\partial y^2} + x\frac{\partial u}{\partial x} + y\frac{\partial u}{\partial y}$$

$$= x^2\frac{\partial^2 u}{\partial x^2} - 2xy\frac{\partial^2 u}{\partial x\,\partial y} + y^2\frac{\partial^2 u}{\partial y^2} + 2y\frac{\partial u}{\partial y} + \left(x\frac{\partial u}{\partial x} - y\frac{\partial u}{\partial y}\right)$$

$$\Rightarrow\quad x^2\frac{\partial^2 u}{\partial x^2} - xy\frac{\partial^2 u}{\partial x\,\partial y} + y^2\frac{\partial^2 u}{\partial y^2} + 2y\frac{\partial u}{\partial y}$$

$$= \left(x\frac{\partial}{\partial x} - y\frac{\partial}{\partial y}\right)\left(x\frac{\partial u}{\partial x} - y\frac{\partial u}{\partial y}\right) - \left(x\frac{\partial u}{\partial x} - y\frac{\partial u}{\partial y}\right)$$

$$= \left(-2\eta\frac{\partial}{\partial \eta}\right)\left(-2\eta\frac{\partial u}{\partial \eta}\right) - \left(-2\eta\frac{\partial u}{\partial \eta}\right)$$

$$= 4\eta^2\frac{\partial^2 u}{\partial \eta^2} + 2\eta\frac{\partial u}{\partial \eta}.$$

Hence, the given equation becomes,

$$4\eta^2\frac{\partial^2 u}{\partial \eta^2} + 2\eta\frac{\partial u}{\partial \eta} = 0$$

$$\Rightarrow\quad 2\eta\frac{\partial^2 u}{\partial \eta^2} + \frac{\partial u}{\partial \eta} = 0.$$

Example 4. *If* $x+y = 2e^{\theta}\cos\phi$ and $x-y = 2ie^{\theta}\sin\phi$, *prove that*

$$\frac{\partial^2 v}{\partial \theta^2} + \frac{\partial^2 v}{\partial \phi^2} = 4xy\frac{\partial^2 v}{\partial x\,\partial y}$$

Solution : We have, $x + y = 2e^{\theta}\cos\phi$

and $x - y = 2ie^{\theta}\sin\phi.$

Adding and subtracting, we get

$$x = e^{\theta}(\cos\phi + i\sin\phi) = e^{\theta + i\phi}$$

and

$$y = e^{\theta}(\cos\phi + i\sin\phi) = e^{\theta - i\phi}$$

$$\frac{\partial v}{\partial \theta} = \frac{\partial v}{\partial x}\frac{\partial x}{\partial \theta} + \frac{\partial v}{\partial y}\frac{\partial y}{\partial \theta}$$

$$= \frac{\partial v}{\partial x}e^{\theta + i\phi} + \frac{\partial v}{\partial y}e^{\theta - i\phi}$$

$$= \frac{\partial v}{\partial x}x + \frac{\partial v}{\partial y}y$$

$$= x\frac{\partial v}{\partial x} + y\frac{\partial v}{\partial y} \qquad \text{... (i)}$$

$$\frac{\partial v}{\partial \phi} = \frac{\partial v}{\partial x}\frac{\partial x}{\partial \phi} + \frac{\partial v}{\partial y}\frac{\partial y}{\partial \phi}$$

$$= \frac{\partial v}{\partial x} e^{\theta + i\phi}(i) + \frac{\partial v}{\partial y} e^{\theta - i\phi}(-i)$$

$$= i\left[x\frac{\partial v}{\partial x} - y\frac{\partial v}{\partial y}\right]. \qquad \text{... (ii)}$$

But
$$\frac{\partial^2 v}{\partial \theta^2} + \frac{\partial^2 v}{\partial \phi^2} = \left[\frac{\partial}{\partial \theta} + i\frac{\partial}{\partial \phi}\right]\left[\frac{\partial}{\partial \theta} - i\frac{\partial}{\partial \phi}\right] v$$

$$= \left(2y\frac{\partial}{\partial y}\right)\left(2x\frac{\partial v}{\partial x}\right) = 4xy\frac{\partial^2 v}{\partial x\,\partial y}.$$

Example 5. *If $x = r\cos\theta$, $y = r\sin\theta$ and $r = e^z$, prove that*

$$x^2\frac{\partial^2 u}{\partial x^2} + 2xy\frac{\partial^2 u}{\partial x\,\partial y} + y^2\frac{\partial^2 u}{\partial y^2} = r\frac{\partial}{\partial r}\left(r\frac{\partial}{\partial r} - 1\right)u = \frac{\partial}{\partial z}\left(\frac{\partial}{\partial z} - 1\right)u.$$

Solution : We have, $x = r\cos\theta$, $y = r\sin\theta$

$$\therefore \quad \frac{\partial u}{\partial r} = \frac{\partial u}{\partial x}\frac{\partial x}{\partial r} + \frac{\partial u}{\partial y}\frac{\partial y}{\partial r}$$

$$= \frac{\partial u}{\partial x}\cos\theta + \frac{\partial u}{\partial y}\sin\theta$$

$$= \cos\theta\frac{\partial u}{\partial x} + \sin\theta\frac{\partial u}{\partial y}$$

$$\Rightarrow \quad r\frac{\partial u}{\partial r} = r\cos\theta\frac{\partial u}{\partial x} + r\sin\theta\frac{\partial u}{\partial y}$$

$$\Rightarrow \quad r\frac{\partial u}{\partial r} = x\frac{\partial u}{\partial x} + y\frac{\partial u}{\partial y}. \qquad \text{... (i)}$$

Thus, $r\dfrac{\partial}{\partial r}\left(r\dfrac{\partial}{\partial r} - 1\right)u$

$$= \left(x\frac{\partial}{\partial x} + y\frac{\partial}{\partial y}\right)\left(x\frac{\partial}{\partial x} + y\frac{\partial}{\partial y} - 1\right)u$$

$$= x^2\frac{\partial^2 u}{\partial x^2} + y^2\frac{\partial^2 u}{\partial y^2} + 2xy\frac{\partial^2 u}{\partial x\,\partial y} + x\frac{\partial u}{\partial x} + y\frac{\partial u}{\partial y} - x\frac{\partial u}{\partial x} - y\frac{\partial u}{\partial y}$$

$$= x^2\frac{\partial^2 u}{\partial x^2} + y^2\frac{\partial^2 u}{\partial y^2} + 2xy\frac{\partial^2 u}{\partial x\,\partial y}.$$

Again, $r = e^z$

$$\Rightarrow \qquad \frac{\partial r}{\partial z} = e^z = r.$$

$$\therefore \qquad \frac{\partial u}{\partial r} = \frac{\partial u}{\partial z}\frac{\partial z}{\partial r} = \frac{\partial u}{\partial z}\frac{1}{r}$$

$$\Rightarrow \qquad \frac{\partial u}{\partial r} = \frac{1}{r}\frac{\partial u}{\partial z}$$

$$\Rightarrow \qquad r\frac{\partial u}{\partial r} = \frac{\partial u}{\partial z}$$

$$\Rightarrow \qquad r\frac{\partial}{\partial r} \equiv \frac{\partial}{\partial z}.$$

Hence,
$$r\frac{\partial}{\partial r}\left(r\frac{\partial}{\partial r} - 1\right)u = \frac{\partial}{\partial z}\left(\frac{\partial}{\partial z} - 1\right)u.$$

Example 6. *Transform the equation*

$$\frac{\partial^2 z}{\partial x^2} + 2xy^2\frac{\partial z}{\partial x} + 2(y - y^3)\frac{\partial z}{\partial y} + x^2y^2z = 0$$

by the substitution $x = uv,\ y = 1/v$ *and hence, show that* z *is the same function of* u *and* v *as of* x *and* y.

Solution : We have, $x = uv,\ y = 1/v$

$$\Rightarrow \qquad u = xy,\ v = 1/y$$

$$\therefore \qquad \frac{\partial z}{\partial x} = \frac{\partial z}{\partial u}\frac{\partial u}{\partial x} + \frac{\partial z}{\partial v}\frac{\partial v}{\partial x}$$

$$= \frac{\partial z}{\partial u}y + \frac{\partial z}{\partial v}\cdot 0 = \frac{\partial z}{\partial u}\cdot\frac{1}{v}$$

$$= \frac{1}{v}\cdot\frac{\partial z}{\partial u}. \qquad \text{... (i)}$$

$$\therefore \qquad \frac{\partial^2 z}{\partial x^2} = \frac{\partial}{\partial x}\left(\frac{\partial z}{\partial x}\right) = \frac{1}{v}\frac{\partial}{\partial u}\left(\frac{1}{v}\frac{\partial z}{\partial u}\right)$$

$$= \frac{1}{v^2}\frac{\partial^2 z}{\partial u^2}. \qquad \text{... (ii)}$$

$$\frac{\partial z}{\partial y} = \frac{\partial z}{\partial u}\frac{\partial u}{\partial y} + \frac{\partial z}{\partial v}\frac{\partial v}{\partial y}$$

$$= \frac{\partial z}{\partial u} \cdot x + \frac{\partial z}{\partial v}\left(-\frac{1}{y^2}\right)$$

$$\Rightarrow \qquad y\frac{\partial z}{\partial y} = xy\frac{\partial z}{\partial u} - \frac{1}{y}\frac{\partial z}{\partial v}$$

$$= u\frac{\partial z}{\partial u} - v\frac{\partial z}{\partial v}. \qquad \text{...(iii)}$$

Putting these values in the given equation, we get,

$$\frac{1}{v^2}\frac{\partial^2 z}{\partial u^2} + 2u\frac{1}{v}\left(\frac{1}{v}\frac{\partial z}{\partial u}\right) + 2\left(1 - \frac{1}{v^2}\right)\left(u\frac{\partial z}{\partial u} - v\frac{\partial z}{\partial v}\right) + u^2 z = 0$$

$$\Rightarrow \qquad \frac{\partial^2 z}{\partial u^2} + \frac{\partial z}{\partial u}[2u + 2u(v^2 - 1)] - 2v(v^2 - 1)\frac{\partial z}{\partial v} + u^2v^2z = 0$$

$$\Rightarrow \qquad \frac{\partial^2 z}{\partial u^2} + 2uv^2\frac{\partial z}{\partial u} + 2(v - v^2)\frac{\partial z}{\partial v} + u^2v^2z = 0.$$

This equation is exactly similar to the given equation which shows that z is the same function of u and v as of x and y.

Example 7. *If $u + iv = f(x + iy)$ where x and y are independent and u, v, x, y are all real, prove that*

$$\frac{\partial^2 V}{\partial x^2} + \frac{\partial^2 V}{\partial y^2} = \left(\frac{\partial^2 V}{\partial u^2} + \frac{\partial^2 V}{\partial v^2}\right)f'(x + iy) \cdot f'(x - iy).$$

Solution : We have, $u + iv = f(x + iy)$. ... (i)

$$u - iv = f(x - iy). \qquad \text{... (ii)}$$

Differentiating (i) partially with respect to x and y.

$$\frac{\partial u}{\partial x} + i\frac{\partial v}{\partial x} = f'(x + iy)$$

$$\frac{\partial u}{\partial y} + i\frac{\partial v}{\partial y} = if'(x + iy)$$

$$\therefore \qquad \frac{\partial u}{\partial y} + i\frac{\partial v}{\partial y} = i\left[\frac{\partial u}{\partial x} + i\frac{\partial v}{\partial u}\right]$$

$$= -\frac{\partial v}{\partial x} + i\frac{\partial u}{\partial x}.$$

Equating real and imaginary parts, we have

$$\frac{\partial u}{\partial y} = -\frac{\partial v}{\partial x}$$

and $$\frac{\partial v}{\partial y} = \frac{\partial u}{\partial x}. \qquad \text{... (iii)}$$

Again, differentiating (i) partially with respect to u and v, we get

$$1 = f'(x+iy)\left[\frac{\partial x}{\partial u} + i\frac{\partial y}{\partial u}\right]$$

and $$i = f'(x+iy)\left[\frac{\partial x}{\partial v} + i\frac{\partial y}{\partial v}\right]$$

$\Rightarrow$ $$i\left[\frac{\partial v}{\partial u} + i\frac{\partial y}{\partial u}\right] = \frac{\partial x}{\partial v} + i\frac{\partial y}{\partial v}.$$

Again, equating real and imaginary parts, we get

$$\left.\begin{aligned} \frac{\partial x}{\partial u} &= \frac{\partial y}{\partial v} \\ \frac{\partial y}{\partial u} &= -\frac{\partial x}{\partial v} \end{aligned}\right\} \qquad \text{... (iv)}$$

Now, $$\frac{\partial V}{\partial x} = \frac{\partial V}{\partial u}\frac{\partial u}{\partial x} + \frac{\partial V}{\partial v}\frac{\partial v}{\partial x}$$

$\Rightarrow$ $$\frac{\partial}{\partial x} \equiv \frac{\partial u}{\partial x}\frac{\partial}{\partial u} + \frac{\partial v}{\partial x}\frac{\partial}{\partial v}$$

$$\frac{\partial V}{\partial y} = \frac{\partial V}{\partial u}\frac{\partial u}{\partial y} + \frac{\partial V}{\partial v}\frac{\partial v}{\partial y}$$

$$= -\frac{\partial V}{\partial u}\frac{\partial v}{\partial x} + \frac{\partial V}{\partial v}\frac{\partial u}{\partial x} \qquad \text{[from (iii)]}$$

$\Rightarrow$ $$\frac{\partial}{\partial y} \equiv -\frac{\partial v}{\partial x}\frac{\partial}{\partial u} + \frac{\partial u}{\partial x}\frac{\partial}{\partial v}.$$

Hence, $$\frac{\partial}{\partial x} + i\frac{\partial}{\partial y} = \left(\frac{\partial u}{\partial x}\frac{\partial}{\partial u} + \frac{\partial v}{\partial x}\frac{\partial}{\partial v}\right) - i\frac{\partial v}{\partial x}\frac{\partial}{\partial u} + i\frac{\partial u}{\partial x}\frac{\partial}{\partial v}$$

$$= \left(\frac{\partial u}{\partial x} - i\frac{\partial v}{\partial x}\right)\left(\frac{\partial}{\partial u} + i\frac{\partial}{\partial v}\right).$$

Putting $-i$ for i, we get

$$\frac{\partial}{\partial x} - i\frac{\partial}{\partial y} = \left(\frac{\partial u}{\partial x} + i\frac{\partial v}{\partial x}\right)\left(\frac{\partial}{\partial u} - i\frac{\partial}{\partial v}\right).$$

Thus,

$$\frac{\partial^2 V}{\partial x^2} + \frac{\partial^2 V}{\partial y^2} = \left(\frac{\partial}{\partial x} + i\,\partial y\right)\left(\frac{\partial}{\partial x} - i\frac{\partial}{\partial y}\right)V$$

$$= \left(\frac{\partial u}{\partial x} - i\frac{\partial v}{\partial x}\right)\left(\frac{\partial}{\partial u} + i\frac{\partial}{\partial v}\right)\left(\frac{\partial u}{\partial x} + i\frac{\partial v}{\partial x}\right)\left(\frac{\partial u}{\partial u} - i\frac{\partial V}{\partial v}\right)$$

$$= f'(x-iy)\left(\frac{\partial}{\partial u}+i\frac{\partial}{\partial v}\right)\left[f'(x+iy)\left(\frac{\partial V}{\partial u}-i\frac{\partial V}{\partial v}\right)\right]$$

$$= f'(x-iy)f'(x+iy)\left(\frac{\partial}{\partial u}+i\frac{\partial}{\partial v}\right)\left(\frac{\partial V}{\partial u}-i\frac{\partial V}{\partial v}\right)$$

$$+ f'(x-iy)\left(\frac{\partial V}{\partial u}-i\frac{\partial V}{\partial v}\right)\left(\frac{\partial}{\partial u}+i\frac{\partial}{\partial v}\right)f'(x+iy)$$

$$= f'(x-iy)f'(x+iy)\left(\frac{\partial^2 V}{\partial u^2}+\frac{\partial^2 V}{\partial v^2}\right)$$

$$+ f'(x-iy)f'(x+iy)\left(\frac{\partial^2 V}{\partial u^2}+\frac{\partial^2 V}{\partial v^2}\right)$$

[using (iv)]

$$= f'(x-iy)f'(x+iy)\left[\frac{\partial^2 V}{\partial u^2}+\frac{\partial^2 V}{\partial v^2}\right].$$

EXERCISE 10 (B)

1. If $x = u + v$ and $v = uv$, then prove that,

$$\frac{\partial^2 V}{\partial u^2} - 2\frac{\partial^2 V}{\partial u\,\partial v} + \frac{\partial^2 V}{\partial v^2} = (x^2-4y)\frac{\partial^2 V}{\partial y^2} - 2\frac{\partial V}{\partial y}.$$

2. If $x = e^u + e^{-v}$ and $y = e^{-u} - e^v$ and $z = f(x, y)$, then prove that,

$$\frac{\partial z}{\partial u} - \frac{\partial z}{\partial v} = x\frac{\partial z}{\partial x} - y\frac{\partial z}{\partial y}.$$

3. If $\xi = y^2 + z^{-2}$, $\eta = z^2 + x^{-2}$, $\zeta = x^2 + y^{-2}$, and $u = f(x, y, z)$, then prove that,

$$x\frac{\partial u}{\partial x} + y\frac{\partial u}{\partial y} + z\frac{\partial u}{\partial z} + 2\left(\xi\frac{\partial u}{\partial \xi} + \eta\frac{\partial u}{\partial \eta} + \zeta\frac{\partial u}{\partial \zeta}\right) = 4\left(y^2\frac{\partial u}{\partial \xi} + z^2\frac{\partial u}{\partial \eta} + x^2\frac{\partial u}{\partial \zeta}\right).$$

4. If $x = e^u \sec v$ and $y = e^u \tan v$, show that,

(i) $$\left(\frac{\partial f}{\partial x}\right)^2 - \left(\frac{\partial f}{\partial y}\right)^2 = e^{-2u}\left[\left(\frac{\partial f}{\partial u}\right)^2 - \cos^2 v\left(\frac{\partial f}{\partial v}\right)^2\right].$$

(ii) $$\frac{\partial^2 f}{\partial x^2} - \frac{\partial^2 f}{\partial y^2} = e^{-2u}\left[\frac{\partial^2 f}{\partial x^2} - \cos^2 v\frac{\partial^2 f}{\partial v^2} + \sin u\cos u\frac{\partial f}{\partial v}\right].$$

5. If the equation $\frac{\partial^2 \phi}{\partial x^2} + \frac{\partial^2 \phi}{\partial y^2} = 0$, is satisfied when ϕ is a function of x and y, show that it will also be satisfied when ϕ is the same function of u and v where

$$u = \frac{1}{2} \log (x^2 + y^2) \text{ and } v = \tan^{-1} (y/x).$$

6. If $x = ae^{\theta} \cos \phi$ and $y = ae^{\theta} \sin \phi$, prove that,

$$y^2 \frac{\partial^2 u}{\partial x^2} - 2xy \frac{\partial^2 u}{\partial x \, \partial y} + x^2 \frac{\partial^2 u}{\partial y^2} = \frac{\partial^2 u}{\partial \phi^2} + \frac{\partial u}{\partial \theta}.$$

7. If $x = r \cos \theta$, $y = r \sin \theta$ and $r = e^z$, then prove that,

$$x^2 \frac{\partial^2 u}{\partial y^2} - 2xy \frac{\partial^2 u}{\partial x \, \partial y} + y^2 \frac{\partial^2 u}{\partial x^2} = \frac{\partial^2 u}{\partial \theta^2} + r \frac{\partial u}{\partial r} = \frac{\partial^2 u}{\partial \theta^2} + \frac{\partial u}{\partial z}.$$

8. If $u = c \cosh x \cos y$, $v = c \sinh x \sin y$, prove that,

$$\frac{\partial^2 V}{\partial x^2} + \frac{\partial^2 V}{\partial y^2} = \frac{c^2}{2} (\cosh 2x - \cos 2v) \left(\frac{\partial^2 V}{\partial u^2} + \frac{\partial^2 V}{\partial v^2} \right).$$

9. If $x = \cosh u \cos v$, $y = \sinh u \sin v$, prove that,

(i) $\coth u \frac{\partial \phi}{\partial u} + \cot V \frac{\partial \phi}{\partial v} - (\cosh^2 u - \cos^2 v) \frac{1}{y} \frac{\partial \phi}{\partial y}.$

(ii) $\frac{\partial^2 \phi}{\partial u^2} + \frac{\partial^2 \phi}{\partial v^2} = (\cosh^2 u - \cos^2 v) \left(\frac{\partial^2 \phi}{\partial x^2} + \frac{\partial^2 \phi}{\partial y^2} \right).$

Deduce the result that $\phi = \log (\tanh u/2)$, is a solution of the equation

$$\frac{\partial^2 \phi}{\partial x^2} + \frac{\partial^2 \phi}{\partial y^2} + \frac{1}{y} \frac{\partial \phi}{\partial y} = 0.$$

10. If $x + iy = f(\xi + i\eta)$, prove that the equation

$$\frac{\partial^2 V}{\partial x^2} + \frac{\partial^2 V}{\partial y^2} = 0$$

becomes
$$\frac{\partial^2}{\partial \xi^2} + \frac{\partial^2 V}{\partial \eta^2} = 0.$$

11. If $f(x, y)$ is transformed into $g(u, v)$ by transformations $x = u^2 - v^2$, $y = 2uv$ then prove that,

$$\frac{\partial^2 f}{\partial x^2} + \frac{\partial^2 f}{\partial y^2} = \frac{1}{4(u^2 + v^2)} \left(\frac{\partial^2 g}{\partial u^2} + \frac{\partial^2 g}{\partial v^2} \right).$$

12. If (x, y) be the rectangular and (r, θ) be the polar coordinates of a point, then prove that,

$$\frac{\partial^2 V}{\partial x^2}\frac{\partial^2 V}{\partial y^2} - \left(\frac{\partial^2 V}{\partial x\,\partial y}\right)^2 = \frac{1}{r^2}\frac{\partial^2 V}{\partial r^2}\cdot\frac{\partial^2 V}{\partial \theta^2} + \frac{1}{r}\frac{\partial V}{\partial r}\cdot\frac{\partial^2 V}{\partial r^2} - \frac{1}{r^2}\left[\frac{\partial^2 V}{\partial r \partial \theta} - \frac{1}{r}\frac{\partial V}{\partial \theta}\right].$$

13. If $x = r\cos\theta,\ y = r\sin\theta$, prove that,

$$r\frac{\partial^2 u}{\partial r\,\partial \theta} - \frac{\partial u}{\partial \theta} = xy\left(\frac{\partial^2 u}{\partial y^2} - \frac{\partial^2 u}{\partial x^2}\right) + (x^2 - y^2)\frac{\partial u^2}{\partial x\,\partial y}.$$

14. Change the variables x, y, z in the equation

$$x\frac{\partial u}{\partial x} + y\frac{\partial u}{\partial y} + z\frac{\partial u}{\partial z} = nu$$

to ξ, η, ζ where $\xi = x/z,\ \eta = y/z,\ \zeta = z$.

Deduce that u is of the form $\xi^n F(\xi, \eta)$ *i.e.*, $z^n F(x/z, y/z)$, and is therefore homogeneous.

15. Given that $f(x, y)$ has continuous and partial derivatives of the first two orders and $x + y = (u + v)^3,\ x = y = (u - v)^3$, show that,

$$9(x^2 - y^2)\left(\frac{\partial^2 f}{\partial x^2} - \frac{\partial^2 f}{\partial y^2}\right) = (u^2 - v^2)\left(\frac{\partial^2 f}{\partial u^2} - \frac{\partial^2 f}{\partial v^2}\right).$$

11
Functions of Two Variables : Taylor's Theorem

11.1. Taylor's Theorem

If a function $f(x, y)$ of two independent variables x and y, possesses continuous partial derivatives of order n in any domain D of a point (a, b) and let $(a+h, b+k)$ be any point of D, then there exists a real number θ such that

$$f(a+h, b+k)$$

$$= f(a, b) + \left(h\frac{\partial}{\partial x} + k\frac{\partial}{\partial y}\right) f(a, b) + \frac{1}{2!}\left(h\frac{\partial}{\partial x} + k\frac{\partial}{\partial y}\right)^2 f(a, b)$$

$$+ \ldots + \frac{1}{(n-1)!}\left(h\frac{\partial}{\partial x} + k\frac{\partial}{\partial y}\right)^{n-1} f(a, b) + R$$

where, $R_n = \dfrac{1}{n!}\left(h\dfrac{\partial}{\partial x} + k\dfrac{\partial}{\partial y}\right)^n f(a+\theta h, b+\theta k),\ 0 < \theta < 1$

is called remainder after n-terms.

Proof : Let (x, y) be a point in the neighbourhood of (a, b) such that

$x = a + th, y = b + tk$, where, $0 \le t \le 1$ is a parameter.

Let we define a composite function.

$$F(t) = f(x, y) = f(a+th, b+tk)$$

Since the partial derivatives of $f(x, y)$ of order n are continuous in the domain D. Therefore, the nth derivative $F^n(t)$ of composite function $F(t)$ is continuous in [0, 1].

Hence, by Maclaurin's theorem for the function $F(t)$, on [0, 1], we get

$$F(t) = F(0) + tF'(0) + \frac{t^2}{2!}F''(0) + \ldots + \frac{t^{n-1}}{(n-1)!}F^{n-1}(0) + \frac{t^n}{n!}F^n(\theta t) \quad \ldots (1)$$

where, $0 < \theta < 1$.

Now putting $t = 1$, we obtain

$$F(1) = F(0) + F'(0) + \frac{1}{2!}F''(0) + \ldots + \frac{F^{n-1}(0)}{(n-1)!} + \frac{F^n(0)}{n!} \quad \ldots (2)$$

But $F'(t) = \dfrac{dF}{dt} = \dfrac{d}{dt}(f) = \dfrac{\partial f}{\partial x}\dfrac{dx}{dt} + \dfrac{\partial f}{\partial y}\dfrac{dy}{dt}$

$$= h\frac{\partial f}{\partial x} + k\frac{\partial f}{\partial y}$$

$$= \left(h\frac{\partial}{\partial x} + k\frac{\partial}{\partial y}\right) f$$

Again, $$F'' = \frac{d}{dt}F'(t) = \frac{d}{dt}\left(h\frac{\partial f}{\partial x} + k\frac{\partial f}{\partial y}\right)$$

$$= \frac{\partial}{\partial x}\left(h\frac{\partial f}{\partial x} + k\frac{\partial f}{\partial y}\right)\frac{dx}{dt} + \frac{\partial}{\partial y}\left(h\frac{\partial f}{\partial x} + k\frac{\partial f}{\partial y}\right)\frac{dy}{dt}$$

$$= \left(h\frac{\partial^2 f}{\partial x^2} + k\frac{\partial f}{\partial x\,\partial y}\right)\frac{dx}{dt} + \left(h\frac{\partial f}{\partial x\,\partial y} + k\frac{\partial^2 f}{\partial y^2}\right)\frac{dy}{dt}$$

$$= \left(h\frac{\partial^2 f}{\partial x^2} + k\frac{\partial^2 f}{\partial x\,\partial y}\right)h + \left(h\frac{\partial^2 f}{\partial x\,\partial y} + k\frac{\partial^2 f}{\partial y^2}\right)k$$

$$= h^2\frac{\partial^2 f}{\partial x^2} + 2hk\frac{\partial^2 f}{\partial x\,\partial y} + k^2\frac{\partial^2 f}{\partial y^2}$$

$$= \left(h\frac{\partial}{\partial x} + k\frac{\partial}{\partial y}\right)^2 f.$$

Similarly, $$F'''(t) = \left(h\frac{\partial}{\partial x} + k\frac{\partial}{\partial y}\right)^3 f.$$

In general, $$F^n(t) = \left(h\frac{\partial}{\partial x} + k\frac{\partial}{\partial y}\right)^n f.$$

Now, for $t = 0$,

$$F(0) = f(a, b)$$

$$F'(0) = \left(h\frac{\partial}{\partial x} + k\frac{\partial}{\partial y}\right) f(a, b)$$

$$F''(0) = \left(h\frac{\partial}{\partial x} + k\frac{\partial}{\partial y}\right)^2 f(a, b)$$

... ...

... ...

$$F^{n-1}(0) = \left(h\frac{\partial}{\partial x} + k\frac{\partial}{\partial y}\right)^{n-1} f(a, b)$$

and $$F^n(\theta) = \left(h\frac{\partial}{\partial x} + k\frac{\partial}{\partial y}\right)^n f(a + \theta h, b + \theta k).$$

Substituting these values in (2), we get

$$f(a+h, b+k) = f(a, b) + \left(h\frac{\partial}{\partial x} + k\frac{\partial}{\partial y}\right) f(a, b) + \frac{1}{2!}\left(h\frac{\partial}{\partial x} + k\frac{\partial}{\partial y}\right)^2 f(a, b) + \ldots + \frac{1}{(n-1)!}\left(h\frac{\partial}{\partial x} + k\frac{\partial}{\partial y}\right)^{n-1} f(a, b) + R_n$$

where, $R_n = \dfrac{1}{n!}\left(h\dfrac{\partial}{\partial x} + k\dfrac{\partial}{\partial y}\right)^n f(a+\theta h, b+\theta k),\ 0<\theta<1.$

This is Taylor's theorem with remainder for the function of two variables.

11.2. Maclaurin's Theorem

If we put $a = b = 0$, $h = x$, $k = y$ in the Taylor's theorem, we get

$$f(x, y) = f(0, 0) + \left(x\frac{\partial}{\partial x} + y\frac{\partial}{\partial y}\right) f(0, 0) + \frac{1}{2!}\left(x\frac{\partial}{\partial x} + y\frac{\partial}{\partial y}\right)^2 f(0, 0) + \ldots + \frac{1}{(n-1)!}\left(x\frac{\partial}{\partial x} + y\frac{\partial}{\partial y}\right)^{n-1} f(0, 0) + R_n$$

where, $R_n = \dfrac{1}{n!}\left(x\dfrac{\partial}{\partial x} + y\dfrac{\partial}{\partial y}\right)^n f(\theta x, \theta y),\ 0<\theta<1.$

This form of expansion is called Maclaurin's theorem or Maclaurin's expansion in finite form.

11.3. Another Form of Taylor's Theorem

Putting $\quad a + h = x \Rightarrow h = x - a$

and $\quad b + k = y \Rightarrow k = y - b$

in Taylor's theorem, we obtain

$$f(x, y) = f(a, b) + \left[(x-a)\frac{\partial}{\partial x} + (y-b)\frac{\partial}{\partial y}\right] f(a, b) + \frac{1}{2!}\left[(x-a)\frac{\partial}{\partial x} + (y-b)\frac{\partial}{\partial y}\right]^2 f(a, b) + \ldots + \frac{1}{(n-1)!}\left[(x-a)\frac{\partial}{\partial x} + (y-b)\frac{\partial}{\partial y}\right]^{n-1} f(a, b) + R_n$$

$$\text{where, } R_n = \frac{1}{n\,!}\left[(x-a)\frac{\partial}{\partial x} + (y-b)\frac{\partial}{\partial y}\right]^n f(a+(x-a)\,\theta, b+(y-b)\,\theta),$$

$$0 < \theta < 1$$

This is also called the Taylor's expansion in finite form of the function $f(x, y)$ about the point (a, b) in the powers of $(x-a)$ and $(y-b)$.

11.4. Taylor's Series

If partial derivatives of all orders of $f(x, y)$ exist and continuous in domain D and $R_n \to 0$ as $n \to \infty$, where

$$R_n = \frac{1}{n\,!}\left((x-a)\frac{\partial}{\partial x} + (y-b)\frac{\partial}{\partial y}\right)^n f(a+\theta\,(x-a), b+\theta\,(y-b)),$$

$$0 < \theta < 1$$

is Taylor's remainder after n terms, then we obtain Taylor's series,

$$f(x, y) = f(a, b) + \left\{(n-a)\frac{\partial}{\partial x} + (y-b)\frac{\partial}{\partial y}\right\} f(a, b)$$

$$+ \frac{1}{2\,!}\left\{(x-a)\frac{\partial}{\partial x} + (y-b)\frac{\partial}{\partial y}\right\}^2 f(a, b) + \ldots$$

11.5. Application of Taylor's Theorem : Approximate Calculations

Let $u = f(x, y)$ be a function of x and y. Let δu be an increment in u corresponding to increments δx and δy in x and y respectively.

$$\therefore \quad u + \delta u = f(x + \delta y,\ y + \delta y) - f(x, y)$$

$$= \left[f(x, y) + \delta x \frac{\partial u}{\partial x} + \delta y \frac{\partial u}{\partial y} + \ldots\right] - f(x, y)$$

$$\text{or} \quad \delta u = \frac{\partial u}{\partial x}\,\delta x + \frac{\partial u}{\partial y}\,\delta y \text{ (approximately)}$$

[neglecting second and higher order derivatives]

This formula gives the small error in u due to small errors δx and δy in x and y respectively.

Note :

$$\left(h\frac{\partial}{\partial x} + k\frac{\partial}{\partial y}\right) f = h^n \frac{\partial^n f}{\partial x^n} + nh^{n-1} k \frac{\partial^n f}{\partial x^{n-1}\,\partial y}$$

$$+ \frac{n\,(n-1)}{2\,!} h^{n-2} k^2 \frac{\partial^n f}{\partial x^{n-2}\,\partial y^2}$$

$$+ \ldots + nhk^{n-1} \frac{\partial^n f}{\partial x\,\partial y^{n-1}} + k^n \frac{\partial^n f}{\partial y^n}.$$

ILLUSTRATIVE EXAMPLES

Example 1. *If* $f(x, y) = e^{x+y}$, *then find Taylor's expansion at the point* $(0, 0)$ *for* $n = 3$.

Solution : For $n = 3$ at the point $(0, 0)$ Taylor's expansion is given as

$$f(x, y) = f(0, 0) + \left(x\frac{\partial}{\partial x} + y\frac{\partial}{\partial y}\right) f(0, 0) + \frac{1}{2\,!}\left(x\frac{\partial}{\partial x} + y\frac{\partial}{\partial y}\right)^2 f(0, 0) + R_3$$

$$\text{where, } R_3 = \frac{1}{3\,!}\left(x^3\frac{\partial^3}{\partial x^3} + 3x^2y\frac{\partial^3}{\partial x^2\,\partial y} + 3xy^2\frac{\partial^3}{\partial x\,\partial y^2} + y^3\frac{\partial^3}{\partial y^3}\right) \times f(\theta x, \theta y),$$

$$0 < \theta < 1$$

We have given,

$$f(x, y) = e^{x+y}, \ \therefore\ f(0, 0) = 1.$$

$$\text{Now, } \frac{\partial f}{\partial x} = e^{x+y}, \frac{\partial f}{\partial y} = e^{x+y}, \frac{\partial^2 f}{\partial x^2} = e^{x+y}, \frac{\partial^2 f}{\partial x\,\partial y} = e^{x+y}, \frac{\partial^2 f}{\partial y^2} = e^{x+y}$$

$$\text{Similarly, } \frac{\partial^2 f}{\partial x^3} = \frac{\partial^3 f}{\partial x^2\,\partial y} = \frac{\partial^3 f}{\partial x\,\partial y^2} = \frac{\partial^2 f}{\partial y^3} = e^{x+y}$$

Therefore, at the point $(0, 0)$

$$f_x = f_y = 1$$

$$f_{xx} = f_{xy} = f_{yy} = 1$$

and at the point $(\theta x, \theta y)$

$$\frac{\partial^2 f}{\partial x^3} = \frac{\partial^3 f}{\partial x^2\,\partial y} = \frac{\partial^3 f}{\partial x\,\partial y^2} = \frac{\partial^2 f}{\partial y^3} = e^{\theta(x+y)}$$

Thus,

$$e^{x+y} = 1 + (x+y) + \frac{1}{2\,!}(x^2 + 2xy + y^2)$$

$$+ \frac{1}{3\,!}(x^3 + 3x^2y + 3xy^2 + y^3)\,e^{\theta(x+y)}.$$

Example 2. *Obtain Taylor's expansion of* $f(x, y) = \cos(x+y)$ *at the point* $(0, 0)$ *for* $n = 3$.

Solution : For $n = 3$, Taylor's expansion is

$$f(x, y) = f(0, 0) + \left(x\frac{\partial}{\partial x} + y\frac{\partial}{\partial y}\right) f(0, 0)$$

$$+ \frac{1}{2\,!}\left(x^2\frac{\partial^2}{\partial x^2} + 2xy\frac{\partial^2}{\partial x\,\partial y} + y^2\frac{\partial}{\partial y^2}\right) f(0, 0) + R_3$$

where,

$$R_3 = \frac{1}{3\,!}\left(x^3 \frac{\partial^3}{\partial x^3} + 3x^2 y \frac{\partial^3}{\partial x^2\,\partial y} + 3xy^2 \frac{\partial^3}{\partial x\,\partial y^2} + y^3 \frac{\partial^3}{\partial y^3}\right) \times f(\theta x, \theta y),$$

$$0 < \theta < 1$$

Since, it is given

$$f(x, y) = \cos(x + y) \quad \therefore f(0, 0) = 1$$

Now, $\dfrac{\partial f}{\partial x} = \dfrac{\partial f}{\partial y} = -\sin(x + y),\ \dfrac{\partial^2 f}{\partial x^2} = \dfrac{\partial^2 f}{\partial x\,\partial y} = \dfrac{\partial^2 f}{\partial y^2} = -\cos(x + y)$

$$\frac{\partial^3 f}{\partial x^3} = \frac{\partial^3 f}{\partial x^2\,\partial y} = \frac{\partial^3 f}{\partial x\,\partial y^2} = \frac{\partial^3 f}{\partial y^3} = \sin(x + y).$$

Therefore, at (0, 0)

$$\frac{\partial f}{\partial x} = \frac{\partial f}{\partial y} = 0,\ \frac{\partial^2 f}{\partial x^2} = \frac{\partial^2 f}{\partial x\,\partial y} = \frac{\partial^2 f}{\partial y^2} = -1.$$

Also, at the point $(\theta x, \theta y)$

$$\frac{\partial^3 f}{\partial x^3} = \frac{\partial^3 f}{\partial x^2\,\partial y} = \frac{\partial^2 f}{\partial x\,\partial y^2} = \frac{\partial^3 f}{\partial y^3} = \sin\theta\,(x + y).$$

Hence, $\cos(x + y) = 1 + 0 + \dfrac{1}{2\,!}(x^2 + 2xy + y^2)(-1)$

$$+ \frac{1}{3\,!}(x^3 + 3x^2 y + 3xy^2 + y^3)\sin\theta\,(x + y)$$

$$= 1 - \frac{(x + y)^2}{2\,!} + \frac{(x + y)^3}{3\,!}\sin\theta\,(x + y).$$

Example 3. *If* $f(x, y) = e^{xy}$, *then expand* $f(x, y)$ *by Taylor's series about the point* (1, 1).

Solution : We have given

$$f(x, y) = e^{xy}$$

$$\therefore \quad f(1, 1) = e$$

Now, $\dfrac{\partial f}{\partial x} = ye^{xy},\ \dfrac{\partial f}{\partial y} = xe^{xy}$

$$\frac{\partial^2 f}{\partial x^2} = y^2 e^{xy},\ \frac{\partial^2 f}{\partial x\,\partial y} = e^{xy} + e^{xy}\cdot xy = e^{xy}(1 + xy)$$

$$\frac{\partial^2 f}{\partial y^2} = x^2 e^{xy}$$

$$\frac{\partial^3 f}{\partial x^3} = y^3 e^{xy}, \frac{\partial^3 f}{\partial y^3} = x^3 e^{xy}, \frac{\partial^3 f}{\partial x^2 \partial y} = \frac{\partial}{\partial x}\left(\frac{\partial^2 f}{\partial x \partial y}\right) = (2y + xy^2) e^{xy}$$

$$\frac{\partial^3 f}{\partial x \partial y^2} = (2x + yx^2) e^{xy}$$

At the point (1, 1)

$$f_x(1, 1) = f_y(1, 1) = e$$

$$f_{xx}(1, 1) = f_{yy}(1, 1) = e$$

$$f_{xy}(1, 1) = (1 + 1) e = 2e$$

$$f_{xxx}(1, 1) = f_{yyy}(1, 1) = e$$

$$f_{xxy} = f_{xyy}(1, 1) = (2 + 1) e = 3e.$$

Therefore, by Taylor's series expansion about (1, 1),

$$f(x, y) = f(1, 1) + \left\{(x-1)\frac{\partial}{\partial x} + (y-1)\frac{\partial}{\partial y}\right\} f(1, 1)$$

$$+ \frac{1}{2!}\left\{(x-1)\frac{\partial}{\partial x} + (y-1)\frac{\partial}{\partial y}\right\}^2 f(1, 1) + \ldots$$

$$\therefore \quad e^{xy} = e + \{(x-1) e + (y-1) e\}$$

$$+ \frac{1}{2!}\{(x-1)^2 e + 2(x-1)(y-1) e + (y-1)^2 e\} + \ldots$$

$$= e\,[1 + (x-1) + (y-1) + \frac{1}{2}(x-1)^2 + (x-1)(y-1)$$

$$+ \frac{1}{2}(y-1)^2 + \ldots].$$

Example 4. *Expand function* $f(x, y) = x^2 + xy + y^2$ *in powers of* $(x-2)$ *and* $(y-3)$.

Solution : Here $\quad f(x, y) = x^2 + xy + y^2$

$\therefore \quad f(2, 3) = 19$

Now, $f_x(x, y) = 2x + y \quad \therefore \quad f_x(2, 3) = 2 \times 2 + 3 = 4 + 3 = 7$

$f_y(x, y) = x + 2y, \quad \therefore \quad f_y(2, 3) = 8$

$f_{xx}(x, y) = 2 \quad \therefore \quad f_{xx}(2, 3) = 2$

$f_{xy}(x, y) = 2 \quad \therefore \quad f_{xy}(2, 3) = 2$

$f_{yy}(x, y) = 2 \quad \therefore \quad f_{yy}(2, 3) = 2$

$$f_{xxy}(x, y) = 0 \qquad \therefore \quad f_{xxy}(2, 3) = 0$$

$$f_{xyy}(x, y) = 0 \qquad \therefore \quad f_{xyy}(2, 3) = 0$$

$$f_{xxx}(x, y) = 0 \qquad \therefore \quad f_{xxx}(2, 3) = 0$$

$$f_{yyy}(x, y) = 0 \qquad \therefore \quad f_{yyy}(2, 3) = 0.$$

Thus, we see that third and higher order partial derivatives are zero.

Hence, from Taylor's theorem,

$$f(x, y) = f(2, 3) + \left\{(x-2)\frac{\partial}{\partial x} + (y-3)\frac{\partial}{\partial y}\right\} f(2, 3)$$
$$+ \frac{1}{2!}\left\{(x-2)^2 \frac{\partial^2}{\partial x^2} + 2(x-2)(y-3)\frac{\partial^2}{\partial x \partial y} + (y-3)\frac{\partial^2}{\partial y^2}\right\}$$
$$f(2, 3) + \ldots$$

or $f(x, y) = f(2, 3) + (x-2) f_x(2, 3) + (y-3) f_y(2, 3)$

$$+ \frac{1}{2!}\{(x-2)^2 f_{xx}(2, 3) + 2(x-2)(y-3) f_{xy}(2, 3)$$
$$+ (y-3)^2 f_{yy}(2, 3)\} + 0$$

$$= 19 + 7(x-2) + 8(y-3) + \frac{1}{2}\{(x-2)^2 \cdot 2 + 2(x-1)(y-3)$$
$$+ 2(y-3)^2\}$$

$$= 19 + 7(x-2) + 8(y-3) + (x-2)^2 + (x-2)(y-3)$$
$$+ (y-3)^2.$$

Example 5. *Expand $x^2 y + 3y - 2$ in powers of $(x-1)$ and $(y+2)$.*

Solution : Let $f(x, y) = x^2 y + 3y - 2$

Then

$$f_x = 2xy$$

$$f_y = x^2 + 3$$

$$f_{xx} = 2y$$

$$f_{xy} = 2x$$

$$f_{yy} = 0$$

$$f_{xxx} = 0$$

$$f_{xxy} = 2$$

$$f_{xyy} = 0$$
$$f_{yyy} = 0$$

and other higher order partial derivatives are zero.

At the point (1, 2)

$$f_x(1, 2) = 4$$
$$f_y(1, 2) = 4$$
$$f_{xx}(1, 2) = 4$$
$$f_{xy}(1, 2) = 2$$

using Taylor's theorem with $h = x - 1$ and $k = y + 2$.

Also, $a = 1$ and $b = -2$, we have

$$f(x, y) = f(1, -2) + \left\{(x-1)\frac{\partial}{\partial x} + (y+2)\frac{\partial}{\partial y}\right\} f(1, -2)$$
$$+ \frac{1}{2\,!}\left\{(x-1)^2\frac{\partial^2}{\partial x^2} + 2(x-1)(y+2)\frac{\partial^2}{\partial x\,\partial y} + (y+2)^2\frac{\partial^2}{\partial y^2}\right\}$$
$$f(1, -2) + \frac{1}{3\,!}\left\{(x-1)^3\frac{\partial^3}{\partial x^3} + 3(x-1)^2(y+2)\frac{\partial^3}{\partial x^2\,\partial y}\right.$$
$$\left. + 3(x-1)(y+2)^2\frac{\partial^3}{\partial x\,\partial y^2} + (y-2)^3\frac{\partial^3}{\partial y^3}\right\} \times f(1, -2)$$

$$\Rightarrow x^2y + 3y - 2 = -10 + \{-4(x-1) + 4(y+2)\}$$
$$+ \frac{1}{2\,!}\{-4(x-1)^2 + 4(x-1)(y+2)\}$$
$$+ \frac{1}{3\,!}\{3(x-1)^2(y+2)(2)\} + 0$$
$$= -10 - 4(x-1) + 4(y+2) - 2(x-1)^2$$
$$+ 2(x-1)(y+2) + (x-1)^2(y+2).$$

Example 6. *Prove that*

$$\sin x \sin y = xy - \frac{1}{6}\{(x^3 + 3xy^2)\cos\theta x \sin\theta y + (y^3 + 3yx^2)\sin\theta x \cos\theta y\}$$

where, $0 < \theta < 1$.

Solution : From the Taylor's theorem,

$$f(x, y) = f(0, 0) + \left(x\frac{\partial}{\partial x} + y\frac{\partial}{\partial y}\right) f(0, 0)$$

$$+\frac{1}{2!}\left(x^2\frac{\partial^2}{\partial x^2}+2xy\frac{\partial^2}{\partial x\,\partial y}+y^2\frac{\partial^2}{\partial y^2}\right)f(0,0)$$

$$+\frac{1}{3!}\left(x^3\frac{\partial^3}{\partial x^3}+3x^2y\frac{\partial^3}{\partial x^2\,\partial y}+3xy^2\frac{\partial^2}{\partial x\,\partial y^2}+y^3\frac{\partial^3}{\partial y^3}\right)f(\theta x,\theta y)$$

where, $0<\theta<1$.

Here $f(x,y)=\sin x\sin y$

$\therefore \quad f(0,0)=0$

$$\therefore \quad \frac{\partial f}{\partial x}=\cos x\sin y,\ \frac{\partial f}{\partial y}=\sin x\cos y,\ \frac{\partial^2 f}{\partial x^2}=-\sin x\sin y$$

$$\frac{\partial^2 f}{\partial y^2}=-\sin x\sin y,\ \frac{\partial^2 f}{\partial x\,\partial y}=\cos x\cos y,\ \frac{\partial^2 f}{\partial x^2}=-\cos x\sin y$$

$$\frac{\partial^3 f}{\partial x^2\,\partial y}=-\sin x\cos y,\ \frac{\partial^3 f}{\partial x\,\partial y^2}=-\cos x\sin y,\ \frac{\partial^3 f}{\partial y^3}=-\sin x\cos y$$

$\therefore$ At (0, 0)

$$\frac{\partial f}{\partial x}=\frac{\partial f}{\partial y}=0,\ \frac{\partial^2 f}{\partial x^2}=\frac{\partial^2 f}{\partial y^2}=0,\ \frac{\partial^2 f}{\partial x\,\partial y}=1.$$

Also at the point $(\theta x,\theta y)$,

$$\frac{\partial^3 f}{\partial x^3}=\frac{\partial^3 f}{\partial x\,\partial y^2}=-\cos\theta x\sin\theta y$$

and
$$\frac{\partial^3 f}{\partial x\,\partial y}=\frac{\partial^3 f}{\partial y^3}=-\sin\theta x\cos\theta y$$

$$\therefore \sin x\sin y=0+0+\frac{1}{2!}(0+2xy)(1)+0$$
$$+\frac{1}{3!}\{x^3\cdot(-\cos\theta x\sin\theta y)+3x^2y(-\sin\theta x\cos\theta y)$$
$$+3xy^2(-\cos\theta x\sin\theta y)+y^3(-\sin\theta x,\cos\theta y)$$

or
$$\sin x\sin y=xy-\frac{1}{6}\{(x^3+3xy^2)\cos\theta x\sin\theta y)$$
$$+(y^3+3x^2y)\sin\theta x\cos\theta y\},\ \text{where, } 0<\theta<1.$$

Example 7. *Show that*

$$e^{ax}\sin by=by+abxy+\frac{1}{3!}e^{a\theta x}[(a^3x^3-3ab^2xy^2)\sin b\theta y$$
$$+(3a^2bx^2y-b^2y^3)\cos b\theta y],$$

where, $0<\theta<1$.

Solution : By Taylor's theorem for $n = 3$, the expansion of the function $f(x, y)$ about $(0, 0)$ is

$$f(x, y) = f(0,0) + \left\{x\frac{\partial}{\partial x} + y\frac{\partial}{\partial y}\right\} f(0,0)$$

$$+ \frac{1}{2!}\left(x^2\frac{\partial^2}{\partial x^2} + 2xy\frac{\partial^2}{\partial x\,\partial y} + y^2\frac{\partial^2}{\partial y^2}\right) f(0,0)$$

$$+ \frac{1}{3!}\left(x^3\frac{\partial^3}{\partial x^3} + 3x^2y\frac{\partial^3}{\partial x^2\,\partial y} + 3xy^2\frac{\partial^3}{\partial x\,\partial y^2} + y^3\frac{\partial^3}{\partial y^3}\right) f(\theta x, \theta y),$$

where, $0 < \theta < 1$.

Here $f(x, y) = e^{ax}\sin by$, $\therefore\ f(0,0) = 0$

Now, $\dfrac{\partial f}{\partial x} = ae^{ax}\sin by$, $\dfrac{\partial f}{\partial y} = be^{ax}\cos by$

$$\frac{\partial^2 f}{\partial x^2} = a^2e^{ax}\sin by,\ \frac{\partial^2 f}{\partial x\,\partial y} = abe^{ax}\cos by$$

$$\frac{\partial^2 f}{\partial y^2} = -b^2e^{ax}\sin by,\ \frac{\partial^3 f}{\partial x^3} = a^3e^{ax}\sin by$$

$$\frac{\partial^3 f}{\partial x^2\,\partial y} = a^2be^{ax}\cos by$$

$$\frac{\partial^3 f}{\partial x\,\partial y^2} = ab^2e^{ax}\sin by,\ \frac{\partial^3 f}{\partial y^3} = -b^3e^{ax}\cos by$$

Thus, at the point $(0, 0)$

$$f_x = 0, f_y = b,\ f_{xx} = f_{yy} = 0,\ f_{xy} = ab$$

Also, at the point $(\theta x, \theta y)$

$$\frac{\partial^3 f}{\partial x^3} = a^3e^{a\theta x}\sin b\theta y,\ \frac{\partial^3 f}{\partial x^2\,\partial y} = a^2b\,e^{a\theta x}\cos b\theta y$$

$$\frac{\partial^3 f}{\partial x\,\partial y^2} = -ab^2e^{a\theta x}\sin b\theta y,\ \frac{\partial^3 f}{\partial y^3} = -b^3e^{a\theta x}\cos b\theta y$$

$$\therefore e^{ax}\sin bx = 0 + \{0 + by\} + \frac{1}{2!}\{0 + 2xy(ab) + 0\}$$

$$+ \frac{1}{3!}e^{a\theta x}\{(a^3x^3 - 3ab^2xy^2)\sin b\theta y + (3a^2bx^2y - b^3y^3)\cos b\theta y\}$$

or $$e^{ax}\sin bx = by + abxy + \frac{1}{3!}e^{a\theta x}\{(a^3x^3 - 3ab^2xy^2)\times\sin b\theta y$$

$$+ (3a^2bx^2y - b^3y^3)\cos b\theta y\}.$$

Example 8. *Prove that the error in the area A of an ellipse due to small errors in the lenghts of the semi-axes a, b is given by*

$$\frac{\delta A}{A} = \frac{\delta a}{a} + \frac{\delta b}{b}.$$

Solution : We have

$$A = \pi a b$$

$\therefore$ $$\frac{\partial A}{\partial a} = \pi b \text{ and } \frac{\partial A}{\partial b} = \pi a.$$

Since $$\delta A = \frac{\partial A}{\partial a}\,\delta a + \frac{\partial A}{\partial b}\,\delta b$$

$$= \pi b\,\delta a + \pi a\,\delta b$$

$\therefore$ $$\frac{\delta A}{A} = \frac{\pi b\,\delta a + \pi a\,\delta b}{\pi a b}.$$

$\Rightarrow$ $$\frac{\delta A}{A} = \frac{\delta a}{a} + \frac{\delta b}{b}.$$

Example 9. *The side a and the opposite angle A of a triangle ABC remain constant; show that when the other sides and angles are slightly varied,*

$$\frac{\delta b}{\cos B} + \frac{\delta c}{\cos C} = 0.$$

Solution : By sine formula in any triangle, we have

$$\frac{a}{\sin A} = \frac{b}{\sin B} = \frac{c}{\sin C}$$

$\therefore$ $$a \sin B = b \sin A \quad \text{... (i)}$$

and $$a \sin C = c \sin A. \quad \text{... (ii)}$$

Since a and A are constants, we have from equations (i) and (ii) on taking differentials.

$$a \cos B\,\delta B = \sin A\,\delta b$$

and $$a \cos C\,\delta C = \sin A\,\delta c$$

$\therefore$ $$\frac{\delta b}{\cos B} = \frac{a\,\delta B}{\sin A}$$

and $$\frac{\delta c}{\cos C} = \frac{a\,\delta C}{\sin A}$$

$\therefore$ $$\frac{\delta b}{\cos B} + \frac{\delta c}{\cos C} = \frac{a}{\sin A}(\delta B + \delta C) \quad \text{... (iii)}$$

But $$A + B + C = 180^\circ \text{ and } A \text{ is constant.}$$

$$\delta B + \delta C = 0.$$

Hence, from equation (iii), we get

$$\frac{\delta b}{\cos B} + \frac{\delta c}{\cos C} = 0.$$

Example 10. *If the sides and the angles of a plane triangle ABC vary in such a way that its circum-radius remains constant, prove that*

$$\frac{\delta a}{\cos A} + \frac{\delta b}{\cos B} + \frac{\delta c}{\cos C} = 0,$$

where, $\delta a, \delta b, \delta c$ denote small increments in the sides a, b, c respectively.

Solution : If R is the circum-radius, then

$$R = \frac{a}{2\sin A} = \frac{b}{2\sin B} = \frac{c}{2\sin C}$$

$$\Rightarrow \quad a = 2R\sin A, \quad b = 2R\sin B, c = 2R\sin C$$

$$\Rightarrow \quad \delta a = 2R\cos A\,\delta A$$

$$\delta b = 2R\cos B\,\delta B$$

$$\delta c = 2R\cos C\,\delta C$$

$$\Rightarrow \quad \frac{\delta a}{\cos A} + \frac{\delta b}{\cos B} + \frac{\delta c}{\cos C} = 2R(\delta A + \delta B + \delta C) \qquad \text{... (i)}$$

But $\quad A + B + C = \pi \qquad$ [for a triangle ABC]

$\therefore \quad \delta A + \delta B + \delta C = 0$

$\therefore$ Equation (i) gives

$$\frac{\delta a}{\cos A} + \frac{\delta b}{\cos B} + \frac{\delta c}{\cos C} = 0.$$

EXERCISE 11

1. Expand $x^2 + xy - y^2$ in powers of $x - 1$ and $y + 2$.
2. Expand $\sin xy$ in powers of $x - 1$ and $y - \frac{\pi}{2}$ upto and including second degree terms.
3. Expand $e^x \sin y$ in powers of x and y as far as terms of third degree.
4. Expand $x^3 + 3x^2y + 4xy^2 + y^3$ by Taylor's theorem in powers of $(x - 1)$ and $(y - 1)$.
5. Expand $f(x, y) = xy^2 + \cos xy$ in powers of $(x - 1)$ and $\left(y - \frac{\pi}{2}\right)$.

6. About the point (1, 1), expand $\tan^{-1}(y/x)$ as far as terms of third degree.

7. About the point (0, 0), expand $f(x, y) = e^x \cos y$, as far as term of the third degree.

8. Expand $e^x \tan^{-1} y$ about (1, 1) upto second degree in $(x-1)$ and $(y-1)$.

9. About the point (0, 0), expand $f(x, y) = \cos(x+y)$ as far as terms of third degree.

10. In a triangle, if $\sin^2 A + \sin^2 B + \sin^2 C = \text{constant}$, prove that

$$\frac{dA}{dB} = \frac{\tan C - \tan B}{\tan A - \tan C}.$$

11. If Δ, the area of a triangle ABC, be a function of b, c, A then prove that

$$\frac{\delta\Delta}{\Delta} = \frac{\delta b}{b} + \frac{\delta c}{c} + \cot A \cdot \delta A.$$

12. In a plane triangle, if the angles and sides receive small variations, show that

(i) $\cos C\,\delta a + \cos A\,\delta c = 0$, b, B being constants

(ii) $c\delta A + a\cos B\,\delta c = 0$; a, b being constants.

[**Hint :** (i) Use $b^2 = a^2 + c^2 - 2ac\cos B$, $a = b\cos C + c\cos B$ etc.

(ii) use $\dfrac{\sin A}{a} = \dfrac{\sin B}{b} = \dfrac{\sin(A+C)}{b}$

and $b\cos A + a\cos B = c$.]

ANSWERS

1. $-5 + 5(y+2) + (x-1)^2 + (x-1)(y+2) - (y+2)^2$

2. $1 - \frac{\pi^2}{8}(x-1)^2 - \frac{\pi}{2}(x-1)\left(y - \frac{\pi}{2}\right) - \frac{1}{2}\left(y - \frac{\pi}{2}\right)^2$

3. $y - \frac{y^3}{6} + xy + \frac{x^2 y}{2} + \ldots$

4. $9 + 13(x-1) + 14(y-1) + 6(x-1)^2 + 14(x-1)(y-6)$
$+ 7(y-1)^2 - (n-1)^3 + (x-1)^2(y-1) + 4(x-1)(y-1)^2 + (y-1)^2$

5. $\frac{\pi^2}{4} + \left(\frac{\pi^2}{4} - \frac{\pi}{2}\right)(x-1) + (x-1)\left(y - \frac{\pi}{2}\right)$

$$+ \frac{1}{2}\left\{2(\pi - 1)(x - 1)\left(y - \frac{\pi}{2}\right) + 2\left(y - \frac{\pi}{2}\right)^2\right\}$$

$$+ \frac{1}{3!}\left\{\frac{\pi^3}{8}(n - 1)^3 + \frac{3\pi^2}{4}(x - 1)^2\left(y - \frac{\pi}{2}\right)\right.$$

$$\left. + 3\left(2 + \frac{\pi}{2}\right) + \left(y - \frac{\pi}{2}\right)^2\right\}$$

6. $\frac{\pi}{4} - \frac{1}{2}(n - 1) + \frac{1}{2}(y - 1) + \frac{1}{2!}\left\{\frac{1}{2}(x - 1)^2 - \frac{1}{2}(y - 1)^2\right\} + \ldots$

7. $1 + x + \frac{1}{2!}(x^2 - y^2) + \frac{1}{3!}(x^3 - 3xy^2) + \ldots.$

8. $1 - \frac{1}{2!}(x + y)^2 + \frac{1}{3!}(x + y)^3 \sin\theta(x + y) + \ldots.$

12

Jacobians

12.1. Jacobian of Functions of Two Independent Variables

Definition : *If u and v be the differentiable functions of two independent variables x and y, then the determinant*

$$\begin{vmatrix} \dfrac{\partial u}{\partial x} & \dfrac{\partial u}{\partial y} \\ \dfrac{\partial v}{\partial x} & \dfrac{\partial v}{\partial y} \end{vmatrix}$$

is called the Jacobian of u and v with respect to x and y. It is denoted by $\dfrac{\partial(u, v)}{\partial(x, y)}$ *or* $J(u, v)$.

12.2. Jacobian of Functions of Three Independent Variables

Definition : *If u, v and w be the differentiable functions of three independent variables x, y and z, then the determinant*

$$\begin{vmatrix} \dfrac{\partial u}{\partial x} & \dfrac{\partial u}{\partial y} & \dfrac{\partial u}{\partial z} \\ \dfrac{\partial v}{\partial x} & \dfrac{\partial v}{\partial y} & \dfrac{\partial v}{\partial z} \\ \dfrac{\partial w}{\partial x} & \dfrac{\partial w}{\partial y} & \dfrac{\partial w}{\partial z} \end{vmatrix}$$

is called the ***Jacobian*** *of u, v and w with respect to x, y and z. It is denoted by* $\dfrac{\partial (u, v, w)}{\partial (x, y, z)}$ *or* $J(u, v, w)$.

12.3. Jacobian of Functions of *n* Independent Variables

Definition : *If* $u_1, u_2, u_3, ..., u_n$ *be the differentiable functions of n independent variables* $x_1, x_2, .., x_n$, *then the determinant*

$$\begin{vmatrix} \frac{\partial u_1}{\partial x_1} & \frac{\partial u_1}{\partial x_2} & \frac{\partial u_1}{\partial x_3} & \cdots & \frac{\partial u_1}{\partial x_n} \\ \frac{\partial u_2}{\partial x_1} & \frac{\partial u_2}{\partial x_2} & \frac{\partial u_2}{\partial x_3} & \cdots & \frac{\partial u_2}{\partial x_n} \\ \cdots & \cdots & \cdots & \cdots & \cdots \\ \cdots & \cdots & \cdots & \cdots & \cdots \\ \frac{\partial u_n}{\partial x_1} & \frac{\partial u_n}{\partial x_2} & \frac{\partial u_n}{\partial x_3} & \cdots & \frac{\partial u_n}{\partial x_n} \end{vmatrix}$$

*is called **Jacobian** of $u_1, u_2, ..., u_n$ with respect to the variables $x_1, x_2, x_3, \ldots x_n$. It is denoted by*

$$\frac{\partial (u_1, u_2, ..., u_n)}{\partial (x_1, x_2, ..., x_n)} \text{ or } J(u_1, u_2, ..., u_n).$$

12.4. Some Theorems on Jacobians

Theorem 1. ***If $u_1, u_2, ..., u_n$ are functions of $x_1, x_2, ..., x_n$ are functions of $x_1, x_2, \ldots x_n$ of the form***

$$u_1 = f_1(x_1)$$
$$u_2 = f_2(x_1, x_2)$$
$$\cdots \quad \cdots$$
$$u_n = f_n(x_1, x_2, ..., x_n)$$

then $$\frac{\partial (u_1, u_2, ..., u_n)}{\partial (x_1, x_2, ..., x_n)} = \frac{\partial u_1}{\partial x_1} \cdot \frac{\partial u_2}{\partial x_2} \cdot \frac{\partial u_3}{\partial x_3} \cdots \frac{\partial u_n}{\partial x_n}.$$

Proof : We know that

$$\frac{\partial (u_1, u_2, ..., u_n)}{\partial (x_1, x_2, ..., x_n)} = \begin{vmatrix} \frac{\partial u_1}{\partial x_1} & \frac{\partial u_1}{\partial x_2} & \frac{\partial u_1}{\partial x_3} & \cdots & \frac{\partial u_1}{\partial x_n} \\ \frac{\partial u_2}{\partial x_1} & \frac{\partial u_2}{\partial x_2} & \frac{\partial u_2}{\partial x_3} & \cdots & \frac{\partial u_2}{\partial x_n} \\ \cdots & \cdots & \cdots & \cdots & \cdots \\ \cdots & \cdots & \cdots & \cdots & \cdots \\ \frac{\partial u_n}{\partial x_1} & \frac{\partial u_n}{\partial x_2} & \frac{\partial u_n}{\partial x_3} & \cdots & \frac{\partial u_n}{\partial x_n} \end{vmatrix} \quad ..(1)$$

Since u_1 is a function of x_1 only, therefore,

$$\frac{\partial u_1}{\partial x_2} = 0, \ \frac{\partial u_1}{\partial x_3} = 0, \ \ldots \ \frac{\partial u_1}{\partial x_n} = 0.$$

Also u_2 is a function of x_1 and x_2, hence only $\frac{\partial u_2}{\partial x_1}$ and $\frac{\partial u_2}{\partial x_2}$ will exist.

The rest will be zero,

i.e., $$\frac{\partial u_2}{\partial x_3} = 0, \frac{\partial u_2}{\partial x_4} = 0, \ldots, \frac{\partial u_2}{\partial x_n} = 0.$$

Again, since u_3 is a function of x_1, x_2 and x_3, therefore $\frac{\partial u_3}{\partial x_1}, \frac{\partial u_3}{\partial x_2}, \frac{\partial u_3}{\partial x_3}$ will exist and the rest will be zero.

Similarly, u_n is a function of $x_1, x_2, \ldots, x_n$ therefore, all $\frac{\partial u_n}{\partial x_1}, \frac{\partial u_n}{\partial x_2}, \ldots, \frac{\partial u_n}{\partial x_n}$ will exist.

Putting these values in (1), we get

$$\frac{\partial (u_1, u_2, u_3, \ldots, u_n)}{\partial (x_1, x_2, \ldots, x_n)} = \begin{vmatrix} \frac{\partial u_1}{\partial x_1} & 0 & 0 & \ldots & 0 \\ \frac{\partial u_2}{\partial x_1} & \frac{\partial u_2}{\partial x_2} & 0 & \ldots & 0 \\ \frac{\partial u_3}{\partial x_1} & \frac{\partial u_3}{\partial x_2} & \frac{\partial u_3}{\partial x_3} & \ldots & 0 \\ \ldots & \ldots & \ldots & \ldots & \ldots \\ \ldots & \ldots & \ldots & \ldots & \ldots \\ \frac{\partial u_n}{\partial x_1} & \frac{\partial u_n}{\partial x_2} & \frac{\partial u_n}{\partial x_3} & \ldots & \frac{\partial u_n}{\partial x_n} \end{vmatrix}$$

Expanding the determinant in terms of first row, we get

$$\frac{\partial (u_1, u_2, \ldots, u_n)}{\partial (x_1, x_2, \ldots, x_n)} = \frac{\partial u_1}{\partial x_1} \cdot \frac{\partial u_2}{\partial x_2} \cdot \frac{\partial u_3}{\partial x_3} \ldots \frac{\partial u_n}{\partial x_n}.$$

12.5. Jacobian of Composite Functions

Theorem 2. ***If*** u_1, u_2, u_3 ***are functions of*** y_1, y_2, y_3 ***and*** y_1, y_2, y_3 ***are functions of*** x_1, x_2, x_3 ***then***

$$\frac{\partial (u_1, u_2, u_3)}{\partial (x_1, x_2, x_3)} = \frac{\partial (u_1, u_2, u_3)}{\partial (y_1, y_2, y_3)} \times \frac{\partial (y_1, y_2, y_3)}{\partial (x_1, x_2, x_3)}.$$

Proof : u_1, u_2 and u_3 are functions of y_1, y_2 and y_3. Also y_1, y_2, y_3 are functions of x_1, x_2 and x_3.

Hence we get

$$\frac{\partial u_1}{\partial x_1} = \frac{\partial u_1}{\partial y_1} \cdot \frac{\partial y_1}{\partial x_1} + \frac{\partial u_1}{\partial y_2} \cdot \frac{\partial y_2}{\partial x_1} + \frac{\partial u_1}{\partial y_3} \cdot \frac{\partial y_3}{\partial x_1}$$

$$= \sum_{i=1}^{3} \frac{\partial u_1}{\partial y_i} \cdot \frac{\partial y_i}{\partial x_1}$$

$$\frac{\partial u_1}{\partial x_2} = \frac{\partial u_1}{\partial y_1} \cdot \frac{\partial y_1}{\partial x_2} + \frac{\partial u_1}{\partial y_2} \cdot \frac{\partial y_2}{\partial x_2} + \frac{\partial u_1}{\partial y_3} \cdot \frac{\partial y_3}{\partial x_3}$$

$$= \sum \frac{\partial u_1}{\partial y_i} \cdot \frac{\partial y_i}{\partial x_2}.$$

Similarly, $$\frac{\partial u_1}{\partial x_3} = \sum \frac{\partial u_1}{\partial y_i} \cdot \frac{\partial y_i}{\partial x_3}$$

$$\frac{\partial u_2}{\partial x_1} = \sum \frac{\partial u_2}{\partial y_i} \cdot \frac{\partial y_i}{\partial x_1}$$

$$\frac{\partial u_2}{\partial x_2} = \sum \frac{\partial u_2}{\partial y_i} \cdot \frac{\partial y_i}{\partial x_2}$$

$$\frac{\partial u_2}{\partial x_3} = \sum \frac{\partial u_2}{\partial y_i} \cdot \frac{\partial y_i}{\partial x_3}$$

$$\frac{\partial u_3}{\partial x_1} = \sum \frac{\partial u_3}{\partial y_i} \cdot \frac{\partial y_i}{\partial x_1}$$

$$\frac{\partial u_3}{\partial x_2} = \sum \frac{\partial u_3}{\partial y_i} \cdot \frac{\partial y_i}{\partial x_2}$$

and $$\frac{\partial u_3}{\partial x_3} = \sum \frac{\partial u_3}{\partial y_3} \cdot \frac{\partial y_3}{\partial x_3}.$$

Now consider

$$\frac{\partial(u_1, u_2, u_3)}{\partial(y_1, y_2, y_3)} \times \frac{\partial(y_1, y_2, y_3)}{\partial(x_1, x_2, x_3)}$$

$$= \begin{vmatrix} \frac{\partial u_1}{\partial y_1} & \frac{\partial u_1}{\partial y_2} & \frac{\partial u_1}{\partial y_3} \\ \frac{\partial u_2}{\partial y_1} & \frac{\partial u_2}{\partial y_2} & \frac{\partial u_2}{\partial y_3} \\ \frac{\partial u_3}{\partial y_1} & \frac{\partial u_3}{\partial y_2} & \frac{\partial u_3}{\partial y_3} \end{vmatrix} \times \begin{vmatrix} \frac{\partial y_1}{\partial x_1} & \frac{\partial y_1}{\partial x_2} & \frac{\partial y_1}{\partial x_3} \\ \frac{\partial y_2}{\partial x_1} & \frac{\partial y_2}{\partial x_2} & \frac{\partial y_2}{\partial x_3} \\ \frac{\partial y_3}{\partial x_1} & \frac{\partial y_3}{\partial x_2} & \frac{\partial y_3}{\partial x_3} \end{vmatrix}$$

$$= \begin{vmatrix} \sum \frac{\partial u_1}{\partial y_i} \cdot \frac{\partial y_i}{\partial x_1} & \sum \frac{\partial u_1}{\partial y_i} \cdot \frac{\partial y_i}{\partial x_2} & \sum \frac{\partial u_1}{\partial y_i} \cdot \frac{\partial y_i}{\partial x_3} \\ \sum \frac{\partial u_2}{\partial y_i} \cdot \frac{\partial y_i}{\partial x_1} & \sum \frac{\partial u_2}{\partial y_i} \cdot \frac{\partial y_i}{\partial x_2} & \sum \frac{\partial u_2}{\partial y_i} \cdot \frac{\partial y_i}{\partial x_3} \\ \sum \frac{\partial u_3}{\partial y_i} \cdot \frac{\partial y_i}{\partial x_1} & \sum \frac{\partial u_3}{\partial y_i} \cdot \frac{\partial y_i}{\partial x_2} & \sum \frac{\partial u_3}{\partial y_i} \cdot \frac{\partial y_i}{\partial x_3} \end{vmatrix}$$

Putting the values of each element of the determinant from the above relations, we get

$$= \begin{vmatrix} \frac{\partial u_1}{\partial x_1} & \frac{\partial u_1}{\partial x_2} & \frac{\partial u_1}{\partial x_3} \\ \frac{\partial u_2}{\partial x_1} & \frac{\partial u_2}{\partial x_2} & \frac{\partial u_2}{\partial x_3} \\ \frac{\partial u_3}{\partial x_1} & \frac{\partial u_3}{\partial x_2} & \frac{\partial u_3}{\partial x_3} \end{vmatrix}$$

$$= \frac{\partial (u_1, u_2, u_3)}{\partial (x_1, x_2, x_3)}.$$

$$\therefore \qquad \frac{\partial (u_1, u_2, u_3)}{\partial (y_1, y_2, y_3)} \cdot \frac{\partial (y_1, y_2, y_3)}{\partial (x_1, x_2, x_3)} = \frac{\partial (u_1, u_2, u_3)}{\partial (x_1, x_2, x_3)}.$$

Generalization :

Theorem 3. ***If*** $u_1, u_2, u_3, \ldots, u_n$ ***are functions of*** $y_1, y_2, y_3, \ldots, y_n$ ***and*** $y_1, y_2, y_3, \ldots, y_n$ ***are functions of*** $x_1, x_2, x_3, \ldots, x_n$ ***then***

$$\frac{\partial (u_1, u_2, \ldots, u_n)}{\partial (x_1, x_2, \ldots, x_n)} = \frac{\partial (u_1, u_2, u_3, \ldots, u_n)}{\partial (y_1, y_2, \ldots, y_n)} \frac{\partial (y_1, y_2, \ldots, y_n)}{\partial (x_1, x_2, \ldots, x_n)}.$$

Proof : By hypothesis, variables $u_1, u_2, \ldots, u_n$ are functions of variables $y_1, y_2, \ldots, y_n$ which are themselves functions of variables $x_1, x_2, \ldots, x_n$.

Therefore, $$\frac{\partial u_j}{\partial x_i} = \frac{\partial u_j}{\partial y_1}\frac{\partial y_1}{\partial x_i} + \frac{\partial u_j}{\partial y_2}\frac{\partial y_2}{\partial x_i} + \ldots + \frac{\partial u_j}{\partial y_n}\frac{\partial y_n}{\partial x_i}$$

$$= \sum_{r=1}^{n} \frac{\partial u_j}{\partial y_r}\frac{\partial y_r}{\partial x_i}. \qquad \ldots (1)$$

Thus, $$\frac{\partial u_1}{\partial x_i} = \sum_{r=1}^{n} \frac{\partial u_1}{\partial y_r}\frac{\partial y_r}{\partial x_i}$$

$$\frac{\partial u_2}{\partial x_i} = \sum_{r=1}^{n} \frac{\partial u_2}{\partial y_r}\frac{\partial y_r}{\partial x_i} \quad \ldots \text{ and so on.}$$

Now, by 'row by column rule' for multiplication of determinants,

$$\frac{\partial (u_1, u_2, \ldots u_n)}{\partial (y_1, y_2, \ldots, y_n)} \cdot \frac{\partial (y_1, y_2, \ldots, y_n)}{\partial (x_1, x_2, \ldots, x_n)}$$

$$= \begin{vmatrix} \frac{\partial u_1}{\partial y_1} & \frac{\partial u_1}{\partial y_2} & \cdots & \frac{\partial u_1}{\partial y_n} \\ \frac{\partial u_2}{\partial y_1} & \frac{\partial u_2}{\partial y_2} & \cdots & \frac{\partial u_2}{\partial y_n} \\ \cdots & \cdots & \cdots & \cdots \\ \frac{\partial u_n}{\partial y_1} & \frac{\partial u_n}{\partial y_2} & \cdots & \frac{\partial u_n}{\partial y_n} \end{vmatrix} \begin{vmatrix} \frac{\partial y_1}{\partial x_1} & \frac{\partial y_1}{\partial x_2} & \cdots & \frac{\partial y_1}{\partial x_n} \\ \frac{\partial y_2}{\partial x_1} & \frac{\partial y_2}{\partial x_2} & \cdots & \frac{\partial y_2}{\partial x_n} \\ \cdots & \cdots & \cdots & \cdots \\ \frac{\partial y_n}{\partial x_1} & \frac{\partial y_n}{\partial x_2} & \cdots & \frac{\partial y_n}{\partial x_n} \end{vmatrix}$$

$$= \begin{vmatrix} \sum \frac{\partial u_1}{\partial y_r} \frac{\partial y_r}{\partial x_1} & \sum \frac{\partial u_1}{\partial y_r} \frac{\partial y_r}{\partial x_2} & \cdots & \sum \frac{\partial u_1}{\partial y_r} \frac{\partial y_r}{\partial x_n} \\ \sum \frac{\partial u_2}{\partial y_r} \frac{\partial y_r}{\partial x_1} & \sum \frac{\partial u_2}{\partial y_r} \frac{\partial y_r}{\partial x_2} & \cdots & \sum \frac{\partial u_2}{\partial y_r} \frac{\partial y_r}{\partial x_n} \\ \cdots & \cdots & \cdots & \cdots \\ \sum \frac{\partial u_n}{\partial y_r} \frac{\partial y_r}{\partial x_1} & \sum \frac{\partial u_n}{\partial y_r} \frac{\partial y_r}{\partial x_2} & \cdots & \sum \frac{\partial u_n}{\partial y_r} \frac{\partial y_r}{\partial x_n} \end{vmatrix}$$

$$= \begin{vmatrix} \frac{\partial u_1}{\partial x_1} & \frac{\partial u_1}{\partial x_2} & \cdots & \frac{\partial u_1}{\partial x_n} \\ \frac{\partial u_2}{\partial x_1} & \frac{\partial u_2}{\partial x_2} & \cdots & \frac{\partial u_2}{\partial x_n} \\ \cdots & \cdots & \cdots & \cdots \\ \frac{\partial u_n}{\partial x_1} & \frac{\partial u_n}{\partial x_2} & \cdots & \frac{\partial u_n}{\partial x_n} \end{vmatrix}$$

$$= \frac{\partial (u_1, u_2, \ldots, u_n)}{\partial (x_1, x_2, \ldots, x_n)}.$$

12.6. Jacobian of Implicit Function

Theorem 4. ***If*** u_1, u_2 ***and*** u_3 ***be the functions of independent variables*** x_1, x_2 ***and*** x_3 ***given by the implicit relations***

$$F_1 (u_1, u_2, u_3, x_1, x_2, x_3) = 0$$
$$F_2 (u_1, u_2, u_3, x_1, x_2, x_3) = 0$$
$$F_3 (u_1, u_2, u_3, x_1, x_2, x_3) = 0$$

then
$$\frac{\partial (u_1, u_2, u_3)}{\partial (x_1, x_2, x_3)} = (-1)^3 \frac{\partial (F_1, F_2, F_3)}{\partial (x_1, x_2, x_3)} \Big/ \frac{\partial (F_1, F_2, F_3)}{\partial (u_1, u_2, u_3)}.$$

Proof : Differentiating F_1, F_2, F_3 with respect to x_1, x_2, x_3, we get

$$\frac{\partial F_1}{\partial x_1} + \frac{\partial F_1}{\partial u_1} \cdot \frac{\partial u_1}{\partial x_1} + \frac{\partial F_1}{\partial u_2} \cdot \frac{\partial u_2}{\partial x_1} + \frac{\partial F_1}{\partial u_3} \cdot \frac{\partial u_3}{\partial x_1} = 0$$

$$\Rightarrow \qquad \sum_{r=1}^{3} \frac{\partial F_1}{\partial u_r} \cdot \frac{\partial u_r}{\partial x_1} = -\frac{\partial F_1}{\partial x_1}$$

$$\frac{\partial F_1}{\partial x_2} + \frac{\partial F_1}{\partial u_1} \cdot \frac{\partial u_1}{\partial x_2} + \frac{\partial F_1}{\partial u_2} \cdot \frac{\partial u_2}{\partial x_2} + \frac{\partial F_1}{\partial u_3} \cdot \frac{\partial u_3}{\partial x_3} = 0$$

$$\Rightarrow \qquad \sum_{r=1}^{3} \frac{\partial F_1}{\partial u_r} \cdot \frac{\partial u_r}{\partial x_2} = -\frac{\partial F_1}{\partial x_2}$$

$$\frac{\partial F_1}{\partial x_3} + \frac{\partial F_1}{\partial u_1} \cdot \frac{\partial u_1}{\partial x_3} + \frac{\partial F_1}{\partial u_2} \cdot \frac{\partial u_2}{\partial x_3} + \frac{\partial F_1}{\partial u_3} \cdot \frac{\partial u_3}{\partial x_3} = 0$$

$$\Rightarrow \qquad \sum_{r=1}^{3} \frac{\partial F_1}{\partial u_r} \cdot \frac{\partial u_r}{\partial x_3} = -\frac{\partial F_1}{\partial x_3}$$

Similarly,

$$\sum_{r=1}^{3} \frac{\partial F_2}{\partial u_r} \cdot \frac{\partial u_r}{\partial x_1} = -\frac{\partial F_2}{\partial x_1},$$

$$\sum_{r=1}^{3} \frac{\partial F_2}{\partial u_r} \cdot \frac{\partial u_r}{\partial x_2} = -\frac{\partial F_2}{\partial x_2},$$

$$\sum_{r=1}^{3} \frac{\partial F_2}{\partial u_r} \cdot \frac{\partial u_r}{\partial x_3} = -\frac{\partial F_2}{\partial x_3},$$

$$\sum_{r=1}^{3} \frac{\partial F_3}{\partial u_r} \cdot \frac{\partial u_r}{\partial x_1} = -\frac{\partial F_3}{\partial x_1}$$

$$\sum_{r=1}^{3} \frac{\partial F_3}{\partial u_r} \cdot \frac{\partial u_r}{\partial x_2} = -\frac{\partial F_3}{\partial x_2}$$

and

$$\sum_{r=1}^{3} \frac{\partial F_3}{\partial u_r} \cdot \frac{\partial u_r}{\partial x_3} = -\frac{\partial F_3}{\partial x_3}.$$

Now consider,

$$\frac{\partial (F_1, F_2, F_3)}{\partial (u_1, u_2, u_3)} \times \frac{\partial (u_1, u_2, u_3)}{\partial (x_1, x_2, x_3)} = \begin{vmatrix} \frac{\partial F_1}{\partial u_1} & \frac{\partial F_1}{\partial u_2} & \frac{\partial F_1}{\partial u_3} \\ \frac{\partial F_2}{\partial u_1} & \frac{\partial F_2}{\partial u_2} & \frac{\partial F_2}{\partial u_3} \\ \frac{\partial F_3}{\partial u_1} & \frac{\partial F_3}{\partial u_2} & \frac{\partial F_3}{\partial u_3} \end{vmatrix} \times \begin{vmatrix} \frac{\partial u_1}{\partial x_1} & \frac{\partial u_1}{\partial x_2} & \frac{\partial u_1}{\partial x_3} \\ \frac{\partial u_2}{\partial x_1} & \frac{\partial u_2}{\partial x_2} & \frac{\partial u_2}{\partial x_3} \\ \frac{\partial u_3}{\partial x_1} & \frac{\partial u_3}{\partial x_2} & \frac{\partial u_3}{\partial x_3} \end{vmatrix}$$

$$= \begin{vmatrix} \sum \frac{\partial F_1}{\partial u_r} \cdot \frac{\partial u_r}{\partial x_1} & \sum \frac{\partial F_1}{\partial u_r} \cdot \frac{\partial u_r}{\partial x_2} & \sum \frac{\partial F_1}{\partial u_r} \cdot \frac{\partial u_r}{\partial x_3} \\ \sum \frac{\partial F_2}{\partial u_r} \cdot \frac{\partial u_r}{\partial x_1} & \sum \frac{\partial F_2}{\partial u_r} \cdot \frac{\partial u_r}{\partial x_2} & \sum \frac{\partial F_2}{\partial u_r} \cdot \frac{\partial u_r}{\partial x_3} \\ \sum \frac{\partial F_3}{\partial u_r} \cdot \frac{\partial u_r}{\partial x_1} & \sum \frac{\partial F_3}{\partial u_r} \cdot \frac{\partial u_r}{\partial x_2} & \sum \frac{\partial F_3}{\partial u_r} \cdot \frac{\partial u_r}{\partial x_3} \end{vmatrix}$$

Putting the values of each summation, we get

$$= \begin{vmatrix} -\dfrac{\partial F_1}{\partial x_1} & -\dfrac{\partial F_1}{\partial x_2} & -\dfrac{\partial F_1}{\partial x_3} \\ -\dfrac{\partial F_2}{\partial x_1} & -\dfrac{\partial F_2}{\partial x_2} & -\dfrac{\partial F_2}{\partial x_3} \\ -\dfrac{\partial F_3}{\partial x_1} & -\dfrac{\partial F_3}{\partial x_2} & -\dfrac{\partial F_3}{\partial x_3} \end{vmatrix}$$

$$= (-1)^3 \begin{vmatrix} \dfrac{\partial F_1}{\partial x_1} & \dfrac{\partial F_1}{\partial x_2} & \dfrac{\partial F_1}{\partial x_3} \\ \dfrac{\partial F_2}{\partial x_1} & \dfrac{\partial F_2}{\partial x_2} & \dfrac{\partial F_2}{\partial x_3} \\ \dfrac{\partial F_3}{\partial x_1} & \dfrac{\partial F_3}{\partial x_2} & \dfrac{\partial F_3}{\partial x_3} \end{vmatrix}$$

$$= (-1)^3 \frac{\partial (F_1, F_2, F_3)}{\partial (x_1, x_2, x_3)}.$$

$$\therefore \qquad \frac{\partial (u_1, u_2, u_3)}{\partial (x_1, x_2, x_3)} = (-1)^3 \frac{\partial (F_1, F_2, F_3)}{\partial (x_1, x_2, x_3)} \Big/ \frac{\partial (F_1, F_2, F_3)}{\partial (u_1, u_2, u_3)}.$$

12.7. Generalizations

If $u_1, u_2, ..., u_n$ be the functions of n independent variables $x_1, x_2, ..., x_n$, given by the implicit relations

$$F_1 (u_1, u_2, ..., u_n, x_1, x_2, ..., x_n) = 0$$

$$F_2 (u_1, u_2, ..., u_n, x_1, x_2, ..., x_n) = 0$$

$$\ldots \qquad \ldots \qquad \ldots$$

$$F_n (u_1, u_2, ..., u_n, x_1, x_2, ..., x_n) = 0,$$

then
$$\frac{\partial (u_1, u_2, ..., u_n)}{\partial (x_1, x_2, ..., x_n)} = (-1)^n \frac{\dfrac{\partial (F_1, F_2, ..., F_n)}{\partial (x_1, x_2, ..., x_n)}}{\dfrac{\partial (F_1, F_2, ..., F_n)}{\partial (u_1, u_2, ..., u_n)}}.$$

12.8. Corollary

If $u_1, u_2, ..., u_n$ be the functions of n independent variables $x_1, x_2, ..., x_n$, given by implicit relations of the form

$$F_1 (u_1, x_1, x_2, ..., x_n) = 0$$

$$F_2 (u_1, u_2, x_2, x_3, ..., x_n) = 0$$

$$F_3(u_1, u_2, u_3, x_3, x_4, ..., x_n) = 0$$

$$\cdots \qquad \cdots \qquad \cdots$$

$$F_n(u_1, u_2, ..., u_n, x_n) = 0,$$

then
$$\frac{\partial(u_1, u_2, ..., u_n)}{\partial(x_1, x_2, ..., x_n)} = (-1)^n \frac{\dfrac{\partial F_1}{\partial x_1} \cdot \dfrac{\partial F_2}{\partial x_2} \cdots \dfrac{\partial F_n}{\partial x_n}}{\dfrac{\partial F_1}{\partial u_1} \cdot \dfrac{\partial F_2}{\partial x_2} \cdots \dfrac{\partial F_n}{\partial u_n}}.$$

Proof : Since, by hypothesis
$$\frac{\partial(F_1, F_2, ..., F_n)}{\partial(x_1, x_2, ..., x_n)} = \frac{\partial F_1}{\partial x_1} \cdot \frac{\partial F_2}{\partial x_2} \cdots \frac{\partial F_n}{\partial x_n}$$

and
$$\frac{\partial(F_1, F_2, ..., F_n)}{\partial(u_1, u_2, ..., u_n)} = \frac{\partial F_1}{\partial u_1} \cdot \frac{\partial F_2}{\partial u_2} \cdots \frac{\partial F_n}{\partial u_n}$$

$\therefore$
$$\frac{\partial(u_1, u_2, ..., u_n)}{\partial(x_1, x_2, ..., x_n)} = (-1)^n \frac{\dfrac{\partial(F_1, F_2, ..., F_n)}{\partial(x_1, x_2, ..., x_n)}}{\dfrac{\partial(F_1, F_2, ..., F_n)}{\partial(u_1, u_2, ..., u_n)}}$$

$$= (-1)^n \frac{\dfrac{\partial F_1}{\partial x_1} \cdot \dfrac{\partial F_2}{\partial x_2} \cdots \dfrac{\partial F_n}{\partial x_n}}{\dfrac{\partial F_1}{\partial u_1} \cdot \dfrac{\partial F_2}{\partial u_2} \cdots \dfrac{\partial F_n}{\partial u_n}}$$

12.9. Functional Relationship

Theorem 5. ***If functions u, v, w of three independent variables x, y, z are not independent then the Jacobian of u, v, w with respect to x, y, z vanishes.***

Proof. It is given that u, v and w are not independent variables, then there will be a relation $F(u, v, w) = 0$, which will connect these independent variables.

Differentiating this relation with respect to x, y and z, we get

$$\frac{\partial F}{\partial u} \cdot \frac{\partial u}{\partial x} + \frac{\partial F}{\partial v} \cdot \frac{\partial v}{\partial x} + \frac{\partial F}{\partial w} \cdot \frac{\partial w}{\partial x} = 0 \qquad ...(1)$$

$$\frac{\partial F}{\partial u} \cdot \frac{\partial u}{\partial y} + \frac{\partial F}{\partial v} \cdot \frac{\partial v}{\partial y} + \frac{\partial F}{\partial w} \cdot \frac{\partial w}{\partial y} = 0 \qquad ...(2)$$

$$\frac{\partial F}{\partial u} \cdot \frac{\partial u}{\partial z} + \frac{\partial F}{\partial v} \cdot \frac{\partial v}{\partial z} + \frac{\partial F}{\partial w} \cdot \frac{\partial w}{\partial z} = 0 \qquad ...(3)$$

Eliminating $\dfrac{\partial F}{\partial u}, \dfrac{\partial F}{\partial v}$ and $\dfrac{\partial F}{\partial w}$ from (1), (2) and (3), we get

$$\begin{vmatrix} \frac{\partial u}{\partial x} & \frac{\partial v}{\partial x} & \frac{\partial w}{\partial x} \\ \frac{\partial u}{\partial y} & \frac{\partial v}{\partial y} & \frac{\partial w}{\partial y} \\ \frac{\partial u}{\partial z} & \frac{\partial v}{\partial z} & \frac{\partial w}{\partial z} \end{vmatrix} = 0$$

or
$$\frac{\partial (u, v, w)}{\partial (x, y, z)} = 0.$$

Theorem 6. ***Let*** $u_1, u_2, u_3, ..., u_n$ ***be the functions of*** n ***independent variables*** $x_1, x_2, ..., x_n$. ***Then a necessary and sufficient condition for the existence of a relation of the form*** $F(u_1, u_2, ..., u_n) = 0$ ***is that the Jacobian*** $\frac{\partial (u_1, u_2, ..., u_n)}{\partial (x_1, x_2, ..., x_n)}$ ***should vanish identically.***

Proof : Necessary condition.

Let there exists a relation

$$F(u_1, u_2, ..., u_n) = 0 \qquad ...(1)$$

i.e., $u_1, u_2, ..., u_n$ are not independent then we are to show that

$$\frac{\partial (u_1, u_2, ..., u_n)}{\partial (x_1, x_2, ..., x_n)} = 0.$$

Now, differentiating (1) partially w.r.t. $x_1, x_2, ..., x_n$ respectively, we get

$$\frac{\partial F}{\partial u_1} \cdot \frac{\partial u_1}{\partial x_1} + \frac{\partial F}{\partial u_2} \cdot \frac{\partial u_2}{\partial x_1} + ... + \frac{\partial F}{\partial u_n} \cdot \frac{\partial u_n}{\partial x_1} = 0$$

$$\frac{\partial F}{\partial u_1} \cdot \frac{\partial u_1}{\partial x_2} + \frac{\partial F}{\partial u_2} \cdot \frac{\partial u_2}{\partial x_2} + ... + \frac{\partial F}{\partial u_n} \cdot \frac{\partial u_n}{\partial x_n} = 0$$

$$\cdots \qquad \cdots \qquad \cdots$$

$$\frac{\partial F}{\partial u_1} \cdot \frac{\partial u_1}{\partial x_n} + \frac{\partial F}{\partial u_2} \cdot \frac{\partial u_2}{\partial x_n} + ... + \frac{\partial F}{\partial u_n} \cdot \frac{\partial u_n}{\partial x_n} = 0.$$

Since, we cannot have at the same time

$$\frac{\partial F}{\partial u_1} = \frac{\partial F}{\partial u_2} = \frac{\partial F}{\partial u_3} = ... = \frac{\partial F}{\partial u_n} = 0$$

otherwise relation (1) will reduce to a trivial identity, hence eliminating, $\frac{\partial F}{\partial u_1}, \frac{\partial F}{\partial u_2}, ..., \frac{\partial F}{\partial u_n}$ from these above equations, we get

$$\begin{vmatrix} \frac{\partial u_1}{\partial x_1} & \frac{\partial u_2}{\partial x_1} & \frac{\partial u_3}{\partial x_1} & \cdots & \frac{\partial u_n}{\partial x_1} \\ \frac{\partial u_1}{\partial x_2} & \frac{\partial u_2}{\partial x_2} & \frac{\partial u_3}{\partial x_2} & \cdots & \frac{\partial u_n}{\partial x_2} \\ \frac{\partial u_1}{\partial x_3} & \frac{\partial u_2}{\partial x_3} & \frac{\partial u_3}{\partial x_3} & \cdots & \frac{\partial u_n}{\partial x_3} \\ \cdots & \cdots & \cdots & \cdots & \cdots \\ \cdots & \cdots & \cdots & \cdots & \cdots \\ \frac{\partial u_1}{\partial x_n} & \frac{\partial u_2}{\partial x_n} & \frac{\partial u_3}{\partial x_n} & \cdots & \frac{\partial u_n}{\partial x_n} \end{vmatrix} = 0$$

Now interchanging rows and columns, we get

$$\begin{vmatrix} \frac{\partial u_1}{\partial x_1} & \frac{\partial u_1}{\partial x_2} & \cdots & \frac{\partial u_1}{\partial x_n} \\ \frac{\partial u_2}{\partial x_1} & \frac{\partial u_2}{\partial x_2} & \cdots & \frac{\partial u_2}{\partial x_n} \\ \cdots & \cdots & \cdots & \cdots \\ \frac{\partial u_n}{\partial x_1} & \frac{\partial u_n}{\partial x_2} & \cdots & \frac{\partial u_n}{\partial x_n} \end{vmatrix} = 0$$

or $$\frac{\partial (u_1, u_2, ..., u_n)}{\partial (x_1, x_2, ..., x_n)} = 0.$$

The Condition is Sufficient :

If $\frac{\partial (u_1, u_2, ..., u_n)}{\partial (x_1, x_2, ..., x_n)} = 0$, then we have to prove that $u_1, u_2, ..., u_n$ are not independent, *i.e.*, there must exist a relation between $u_1, u_2, ..., u_n$.

Let the functions $u_1, u_2, ..., u_n$ and the independent variables $x_1, x_2, ..., x_n$ are connected by following set of equations,

$$\begin{aligned} F_1 (u_1, x_1, x_2, ..., x_n) &= 0 \\ F (u_1, u_2, x_2, x_3, ..., x_n) &= 0 \\ \cdots \quad \cdots \quad &\cdots \\ F_r (u_1, u_2, ..., u_r, x_r, x_{r+1}, ..., x_n) &= 0 \\ \cdots \quad \cdots \quad &\cdots \\ F_n (u_1, u_2, ..., u_n, x_n) &= 0. \end{aligned}$$

Now, $$\frac{\partial (u_1, u_2, u_3, ..., u_n)}{\partial (x_1, x_2, ..., x_n)} = (-1)^n \frac{\frac{\partial F_1}{\partial x_1} \cdot \frac{\partial F_2}{\partial x_2} \cdots \frac{\partial F_n}{\partial x_n}}{\frac{\partial F_1}{\partial u_1} \cdot \frac{\partial F_2}{\partial u_2} \cdots \frac{\partial F_n}{\partial u_n}}$$

Since $$\frac{\partial (u_1, u_2, ..., u_n)}{\partial (x_1, x_2, ..., x_n)} = 0 \qquad ...(2)$$

Therefore from (2)

$$\frac{\partial F_1}{\partial x_1} \cdot \frac{\partial F_2}{\partial x_2} \cdots \frac{\partial F_n}{\partial x_n} = 0.$$

Hence, $\frac{\partial F_r}{\partial x_r} = 0$, for some value of r between 1 and n. Thus for particular value of r, F_r must be free from x_r. Consequently F_r is of the form,

$$F_r (u_1, u_2, ..., u_r, x_{r+1}, x_{r+2}, ..., x_n) = 0$$

which contains only the independent variables $x_{r+1}, x_{r+2}, ..., x_n$ in addition to the u's. We can now eliminate the $(n - r)$ independent variables $x_{r+1}, x_{r+2}, ..., x_n$ from the $(n - r + 1)$ equations given by $F_r = F_{r+1} = ... = F_n = 0$ and we can find a functional relation, say

$$F (u_1, u_2, ..., u_n) = 0$$

between the u-variables.

Theorem 7. $\frac{\partial (u, v, w)}{\partial (x, y, z)} \times \frac{\partial (x, y, z)}{\partial (u, v, w)} = 1.$

Proof : Let $u = f_1 (x, y, z), v = f_2 (x, y, z)$ and $w = f_3 (x, y, z)$ then we may write these equations as

$$x = \phi_1 (u, v, w), \; y = \phi_2 (u, v, w) \text{ and } z = \phi_3 (u, v, w)$$

Now differentiating $u = f_1 (x, y, z)$ partially with respect to u, v and w, we get

$$\frac{\partial u}{\partial u} = \frac{\partial u}{\partial x} \cdot \frac{\partial x}{\partial u} + \frac{\partial u}{\partial y} \cdot \frac{\partial y}{\partial u} + \frac{\partial u}{\partial z} \cdot \frac{\partial z}{\partial u}$$

$$1 = \frac{\partial u}{\partial x} \cdot \frac{\partial x}{\partial u} + \frac{\partial u}{\partial y} \cdot \frac{\partial y}{\partial u} + \frac{\partial u}{\partial z} \cdot \frac{\partial z}{\partial u} \qquad ...(1)$$

$$\frac{\partial u}{\partial v} = \frac{\partial u}{\partial x} \cdot \frac{\partial x}{\partial v} + \frac{\partial u}{\partial y} \cdot \frac{\partial y}{\partial v} + \frac{\partial u}{\partial z} \cdot \frac{\partial z}{\partial v}$$

$$0 = \frac{\partial u}{\partial x} \cdot \frac{\partial x}{\partial v} + \frac{\partial u}{\partial y} \cdot \frac{\partial y}{\partial v} + \frac{\partial u}{\partial z} \cdot \frac{\partial z}{\partial v} \qquad ...(2)$$

Similarly,

$$0 = \frac{\partial u}{\partial x} \cdot \frac{\partial x}{\partial w} + \frac{\partial u}{\partial y} \cdot \frac{\partial y}{\partial w} + \frac{\partial u}{\partial z} \cdot \frac{\partial z}{\partial w} \qquad ...(3)$$

Now differentiating $v = f_2 (x, y, z)$ and $w = f_3 (x, y, z)$ partially with respect to u, v and w, we get

$$\left.\begin{aligned} 0 &= \frac{\partial v}{\partial x}\cdot\frac{\partial x}{\partial u} + \frac{\partial v}{\partial y}\cdot\frac{\partial y}{\partial u} + \frac{\partial v}{\partial z}\cdot\frac{\partial z}{\partial u} \\ 1 &= \frac{\partial v}{\partial x}\cdot\frac{\partial x}{\partial v} + \frac{\partial v}{\partial y}\cdot\frac{\partial y}{\partial v} + \frac{\partial v}{\partial z}\cdot\frac{\partial z}{\partial v} \\ 0 &= \frac{\partial v}{\partial x}\cdot\frac{\partial x}{\partial w} + \frac{\partial v}{\partial y}\cdot\frac{\partial y}{\partial w} + \frac{\partial v}{\partial z}\cdot\frac{\partial z}{\partial w} \end{aligned}\right\} \qquad ...(4)$$

and

$$\left.\begin{aligned} 0 &= \frac{\partial w}{\partial x}\cdot\frac{\partial x}{\partial u} + \frac{\partial w}{\partial y}\cdot\frac{\partial y}{\partial u} + \frac{\partial w}{\partial z}\cdot\frac{\partial z}{\partial u} \\ 0 &= \frac{\partial w}{\partial x}\cdot\frac{\partial x}{\partial v} + \frac{\partial w}{\partial y}\cdot\frac{\partial y}{\partial v} + \frac{\partial w}{\partial z}\cdot\frac{\partial z}{\partial v} \\ 1 &= \frac{\partial w}{\partial x}\cdot\frac{\partial x}{\partial w} + \frac{\partial w}{\partial y}\cdot\frac{\partial y}{\partial w} + \frac{\partial w}{\partial z}\cdot\frac{\partial z}{\partial w} \end{aligned}\right\} \qquad ...(5)$$

Now consider

$$\frac{\partial (u, v, w)}{\partial (x, y, z)} \times \frac{\partial (x, y, z)}{\partial (u, v, w)}$$

$$= \begin{vmatrix} \frac{\partial u}{\partial x} & \frac{\partial u}{\partial y} & \frac{\partial u}{\partial z} \\ \frac{\partial v}{\partial x} & \frac{\partial v}{\partial y} & \frac{\partial v}{\partial z} \\ \frac{\partial w}{\partial x} & \frac{\partial w}{\partial y} & \frac{\partial w}{\partial z} \end{vmatrix} \times \begin{vmatrix} \frac{\partial x}{\partial u} & \frac{\partial x}{\partial v} & \frac{\partial x}{\partial w} \\ \frac{\partial y}{\partial u} & \frac{\partial y}{\partial v} & \frac{\partial y}{\partial w} \\ \frac{\partial z}{\partial u} & \frac{\partial z}{\partial v} & \frac{\partial z}{\partial w} \end{vmatrix}$$

$$= \begin{vmatrix} \frac{\partial u}{\partial x}\cdot\frac{\partial x}{\partial u} + \frac{\partial u}{\partial y}\cdot\frac{\partial y}{\partial u} + \frac{\partial u}{\partial z}\cdot\frac{\partial z}{\partial u} & \sum \frac{\partial u}{\partial x}\cdot\frac{\partial x}{\partial v} & \sum \frac{\partial u}{\partial x}\cdot\frac{\partial x}{\partial w} \\ \frac{\partial v}{\partial x}\cdot\frac{\partial x}{\partial u} + \frac{\partial v}{\partial y}\cdot\frac{\partial y}{\partial u} + \frac{\partial v}{\partial z}\cdot\frac{\partial z}{\partial u} & \sum \frac{\partial v}{\partial x}\cdot\frac{\partial x}{\partial v} & \sum \frac{\partial v}{\partial x}\cdot\frac{\partial x}{\partial w} \\ \frac{\partial w}{\partial x}\cdot\frac{\partial x}{\partial u} + \frac{\partial w}{\partial y}\cdot\frac{\partial y}{\partial u} + \frac{\partial w}{\partial z}\cdot\frac{\partial z}{\partial u} & \sum \frac{\partial w}{\partial x}\cdot\frac{\partial x}{\partial v} & \sum \frac{\partial w}{\partial x}\cdot\frac{\partial x}{\partial w} \end{vmatrix}$$

Using the above relation, we get

$$= \begin{vmatrix} 1 & 0 & 0 \\ 0 & 1 & 0 \\ 0 & 0 & 1 \end{vmatrix} = 1.$$

Hence $\dfrac{\partial (u, v, w)}{\partial (x, y, z)} \times \dfrac{\partial (x, y, z)}{\partial (u, v, w)} = 1.$

ILLUSTRATIVE EXAMPLES

Example 1. *If $u^3 + v^3 = x + y$ and $u^2 + v^2 = x^3 + y^3$, then prove that*

$$\frac{\partial (u, v)}{\partial (x, y)} = \frac{y^2 - x^2}{2uv\,(u - v)}.$$

Solution : Here $f_1 \equiv u^3 + v^3 - (x+y) = 0$

and $f_2 \equiv u^2 + v^2 - (x^3 + y^3) = 0$

$$\therefore \quad \frac{\partial (u, v)}{\partial (x, y)} = (-1)^2 \frac{\dfrac{\partial (f_1, f_2)}{\partial (x, y)}}{\dfrac{\partial (f_1, f_2)}{\partial (u, v)}}$$

$$= \begin{vmatrix} \dfrac{\partial f_1}{\partial x} & \dfrac{\partial f_1}{\partial y} \\ \dfrac{\partial f_2}{\partial x} & \dfrac{\partial f_2}{\partial y} \end{vmatrix} \Bigg/ \begin{vmatrix} \dfrac{\partial f_1}{\partial u} & \dfrac{\partial f_1}{\partial v} \\ \dfrac{\partial f_2}{\partial u} & \dfrac{\partial f_2}{\partial v} \end{vmatrix}$$

$$= \begin{vmatrix} -1 & -1 \\ -3x^2 & -3y^2 \end{vmatrix} \Bigg/ \begin{vmatrix} 3u^2 & 3v^2 \\ 2u & 2v \end{vmatrix}$$

$$= \frac{3(y^2 - x^2)}{6uv(u - v)} = \frac{y^2 - x^2}{2uv(u - v)}.$$

Example 2. *If* $x + y + z = u$, $y + z = uv$ *and* $z = uvw$, *then prove that*

$$\frac{\partial (x, y, z)}{\partial (u, v, w)} = u^2 v.$$

Solution : Let $f_1 \equiv x + y + z - u = 0$

$f_2 \equiv y + z - uv = 0$

and $f_3 \equiv z - uvw = 0$

$$\therefore \quad \frac{\partial (x, y, z)}{\partial (u, v, w)} = (-1)^3 \frac{\dfrac{\partial (f_1, f_2, f_3)}{\partial (u, v, w)}}{\dfrac{\partial (f_1, f_2, f_3)}{\partial (x, y, z)}} = -\frac{\dfrac{\partial f_1}{\partial u} \cdot \dfrac{\partial f_2}{\partial v} \cdot \dfrac{\partial f_3}{\partial w}}{\dfrac{\partial f_1}{\partial x} \cdot \dfrac{\partial f_2}{\partial y} \cdot \dfrac{\partial f_3}{\partial z}}$$

$$= -\left| \frac{(-1) \cdot (-u) \cdot (-uv)}{1 \cdot 1 \cdot 1} \right|$$

$$= u^2 v.$$

Example 3. *If* $x = r \sin \theta \cos \phi$, $y = r \sin \theta \sin \phi$, $z = r \cos \theta$ *then show that*

$$\frac{\partial (x, y, z)}{\partial (r, \theta, \phi)} = r^2 \sin \theta$$

Solution : Here given that

$$x = r \sin \theta \cos \phi, \; y = r \sin \theta \sin \phi, \; z = r \cos \theta$$

$$\therefore \quad \frac{\partial x}{\partial r} = \sin\theta\cos\phi$$

$$\frac{\partial x}{\partial \theta} = r\cos\theta\cos\phi, \quad \frac{\partial x}{\partial \phi} = -r\sin\theta\sin\phi, \quad \frac{\partial y}{\partial r} = \sin\theta\sin\phi$$

$$\frac{\partial y}{\partial \theta} = -r\cos\theta\sin\phi, \quad \frac{\partial y}{\partial \phi} = r\sin\theta\cos\phi$$

and
$$\frac{\partial z}{\partial r} = \cos\theta, \quad \frac{\partial z}{\partial \theta} = -r\sin\theta, \quad \frac{\partial z}{\partial \phi} = 0.$$

$$\therefore \quad \frac{\partial(x, y, z)}{\partial(r, \theta, \phi)} = \begin{vmatrix} \frac{\partial x}{\partial r} & \frac{\partial x}{\partial \theta} & \frac{\partial x}{\partial \phi} \\ \frac{\partial y}{\partial r} & \frac{\partial y}{\partial \theta} & \frac{\partial y}{\partial \phi} \\ \frac{\partial z}{\partial r} & \frac{\partial z}{\partial \theta} & \frac{\partial z}{\partial \phi} \end{vmatrix}$$

$$= \begin{vmatrix} \sin\theta\cos\phi & r\cos\theta\cos\phi & -r\sin\theta\sin\phi \\ \sin\theta\sin\phi & r\cos\theta\sin\phi & r\sin\theta\cos\phi \\ \cos\theta & -r\sin\theta & 0 \end{vmatrix}$$

$$= r^2[\sin^3\theta\cos^2\phi + \sin^3\theta\sin^2\phi + \cos^2\theta\sin\theta\cos^2\phi + \cos^2\theta\sin\theta\sin^2\phi]$$

$$= r^2[\sin^3\theta + \cos^2\theta\sin\theta]$$

$$= r^2\sin\theta(\sin^2\theta + \cos^2\theta)$$

$$= r^2\sin\theta.$$

Example 4. *If* $x = a\cosh\alpha\cos\beta$, $y = a\sinh\alpha\sin\beta$

then show that $\dfrac{\partial(x, y)}{\partial(\alpha, \beta)} = \dfrac{a^2}{2}[\cosh 2\alpha - \cos 2\beta]$.

Solution :
$$\frac{\partial(x, y)}{\partial(\alpha, \beta)} = \begin{vmatrix} \frac{\partial x}{\partial \alpha} & \frac{\partial x}{\partial \beta} \\ \frac{\partial y}{\partial \alpha} & \frac{\partial y}{\partial \beta} \end{vmatrix} \quad \text{... (i)}$$

Differentiate partially with respect to α and β from the equations

$$x = a\cosh\alpha\cos\beta \quad \text{... (ii)}$$

and
$$y = a\sinh\alpha\sin\beta, \quad \text{... (iii)}$$

we get

$$\frac{\partial x}{\partial \alpha} = a\sinh\alpha\cos\beta, \quad \frac{\partial x}{\partial \beta} = -a\cosh\alpha\cdot\sin\beta$$

$$\frac{\partial y}{\partial \alpha} = a \cosh \alpha \sin \beta \text{ and } \frac{\partial y}{\partial \beta} = a \sinh \alpha \cos \beta$$

Putting in (i)

$$\frac{\partial (x, y)}{\partial (\alpha, \beta)} = \begin{vmatrix} a \sinh \alpha \cos \beta & -a \cosh \alpha \sin \beta \\ a \cosh \alpha \sin \beta & a \sinh \alpha \cos \beta \end{vmatrix}$$

$$= a^2 (\sinh^2 \alpha \cos^2 \beta + \cosh^2 \alpha \sin^2 \beta)$$

$$= a^2 [(\cosh^2 \alpha - 1) \cos^2 \beta + \cosh^2 \alpha (1 - \cos^2 \beta)]$$

$$= a^2 [\cosh^2 \alpha \cos^2 \beta - \cos^2 \beta + \cosh^2 \alpha - \cosh^2 \alpha \cos^2 \beta]$$

$$= a^2 [\cosh^2 \alpha - \cos^2 \beta]$$

$$= \frac{a^2}{2} [\cosh 2\alpha + 1 - 1 - \cos 2\beta]$$

$$= \frac{a^2}{2} [\cosh 2\alpha - \cos 2\beta].$$

Example 5. *If* $y_1 = \cos x_1, y_2 = \sin x_1 \cos x_2$ *and* $y_3 = \sin x_1 \sin x_2 \cos x_3$ *then show that*

$$\frac{\partial (y_1, y_2, y_3)}{\partial (x_1, x_2, x_3)} = -\sin^3 x_1 \sin^2 x_2 \sin x_3.$$

Solution : Differentiating y_1, y_2, y_3 partially with respect to x_1, x_2, x_3, we get

$$\frac{\partial y_1}{\partial x_1} = -\sin x_1, \frac{\partial y_1}{\partial x_2} = 0, \frac{\partial y_1}{\partial x_3} = 0$$

$$\frac{\partial y_2}{\partial x_1} = \cos x_1 \cos x_2, \frac{\partial y_2}{\partial x_2} = -\sin x_1 \sin x_2, \frac{\partial y_2}{\partial x_3} = 0$$

$$\frac{\partial y_3}{\partial x_1} = \cos x_1 \sin x_2 \cos x_3, \frac{\partial y_3}{\partial x_3} = \sin x_1 \cos x_2 \cos x_3$$

$$\frac{\partial y_3}{\partial x_3} = -\sin x_1 \sin x_2 \sin x_3.$$

Putting these values in

$$\frac{\partial (y_1, y_2, y_3)}{\partial (x_1, x_2, x_3)} = \begin{vmatrix} \frac{\partial y_1}{\partial x_1} & \frac{\partial y_1}{\partial x_2} & \frac{\partial y_1}{\partial x_3} \\ \frac{\partial y_2}{\partial x_1} & \frac{\partial y_2}{\partial x_2} & \frac{\partial y_2}{\partial x_3} \\ \frac{\partial y_3}{\partial x_1} & \frac{\partial y_3}{\partial x_2} & \frac{\partial y_3}{\partial x_3} \end{vmatrix}$$

$$= \begin{vmatrix} -\sin x_1 & 0 & 0 \\ \cos x_1 \cos x_2 & -\sin x_1 \sin x_2 & 0 \\ \cos x_1 \sin x_2 \cos x_3 & \sin x_1 \cos x_2 \cos x_3 & -\sin x_1 \sin x_2 \sin x_3 \end{vmatrix}$$

Expanding in terms of R_3, we get

$$= -\sin x_1 \sin x_2 \sin x_3 \, [\sin^2 x_1 \sin x_2]$$

$$= -\sin^3 x_1 \sin^2 x_2 \sin x_3.$$

Example 6. *Show that* $ax^2 + 2hxy + by^2$ *and* $Ax^2 + 2Hxy + By^2$ *are independent or*

$$\frac{a}{A} = \frac{b}{B} = \frac{h}{H}.$$

Solution : Let $\quad u = ax^2 + 2hxy + by^2 \quad$ ... (i)

and $\quad v = Ax^2 + 2Hxy + By^2 \quad$ ... (ii)

Differentiating equation (i) and (ii) w.r.t. x and y, we get

$$\frac{\partial u}{\partial x} = 2ax + 2hy = 2(ax + hy)$$

$$\frac{\partial u}{\partial y} = 2hx + 2by = 2(hx + by)$$

$$\frac{\partial v}{\partial x} = 2Ax + 2Hy = 2(Ax + Hy)$$

and

$$\frac{\partial v}{\partial y} = 2Hx + 2By = 2(Hx + By)$$

Again let if u and v are not independent, then

$$J(u, v) = \frac{\partial (u, v)}{\partial (x, y)} = 0$$

or

$$\begin{vmatrix} 2(ax + hy) & 2(hx + by) \\ 2(Ax + Hy) & 2(Hx + By) \end{vmatrix} = 0$$

or

$$(ax + hy)(Hx + By) - (hx + by)(Ax + Hy) = 0$$

or

$$(aH - Ah)x^2 + (aB - Ab)xy + (Bh - bH)y^2 = 0$$

$\because$ x and y are independent.

$\therefore$ In the above equation the coefficients of x^2 and y^2 are must be separately zero, *i.e.*,

$$aH - Ah = 0 \text{ and } Bh - bH = 0$$

Hence

$$\frac{a}{A} = \frac{h}{H} = \frac{b}{B}.$$

Example 7. *If* $u^3 + v^3 + w^3 = x + y + z$, $u^2 + v^2 + w^2 = x^3 + y^3 + z^3$ *and* $u + v + w = x^2 + y^2 + z^2$ *then show that*

$$\frac{\partial (u, v, w)}{\partial (x, y, z)} = \frac{(x - y)(y - z)(z - x)}{(u - v)(v - w)(w - u)}.$$

Solution : Let

$$F_1 \equiv (u^3 + v^3 + w^3 - x - y - z),$$

$$F_2 \equiv (u^2 + v^2 + w^2 - x^3 - y^3 - z^3)$$

and

$$F_3 \equiv (u + v + w - x^2 - y^2 - z^2)$$

$$\frac{\partial (u, v, w)}{\partial (x, y, z)} = (-1)^3 \frac{\dfrac{\partial (F_1, F_2, F_3)}{\partial (x, y, z)}}{\dfrac{\partial (F_1, F_2, F_3)}{\partial (u, v, w)}} \quad \text{... (i)}$$

$$\frac{\partial (F_1, F_2, F_3)}{\partial (x, y, z)} = \begin{vmatrix} \dfrac{\partial F_1}{\partial x} & \dfrac{\partial F_1}{\partial y} & \dfrac{\partial F_1}{\partial z} \\ \dfrac{\partial F_2}{\partial x} & \dfrac{\partial F_2}{\partial y} & \dfrac{\partial F_2}{\partial z} \\ \dfrac{\partial F_3}{\partial x} & \dfrac{\partial F_3}{\partial y} & \dfrac{\partial F_3}{\partial z} \end{vmatrix} \quad \text{... (ii)}$$

From F_1, F_2 and F_3, we get

$$\frac{\partial F_1}{\partial x} = -1, \; \frac{\partial F_1}{\partial y} = -1, \; \frac{\partial F_1}{\partial z} = -1$$

$$\frac{\partial F_2}{\partial x} = -3x^2, \; \frac{\partial F_2}{\partial y} = -3y^2, \; \frac{\partial F_2}{\partial z} = -3z^2$$

$$\frac{\partial F_3}{\partial x} = -2x, \; \frac{\partial F_3}{\partial y} = -2y, \; \frac{\partial F_3}{\partial z} = -2z.$$

Putting these values in (ii), we get

$$\frac{\partial (F_1, F_2, F_3)}{\partial (x, y, z)} = \begin{vmatrix} -1 & -1 & -1 \\ -3x^2 & -3y^2 & -3z^2 \\ -2x & -2y & -2z \end{vmatrix}$$

$C_1 \to C_1 - C_3$ and $C_2 \to C_2 - C_3$, we get

$$= \begin{vmatrix} 0 & 0 & -1 \\ 3(z^2 - x^2) & 3(z^2 - y^2) & -3z^2 \\ 2(z - x) & 2(z - y) & -2z \end{vmatrix}$$

$$= 6 \begin{vmatrix} 0 & 0 & -1 \\ (z-x)(z+x) & (z-y)(z+y) & -3z^2 \\ (z-x) & z-y & -2z \end{vmatrix}$$

$$= 6(z-x)(z-y) \begin{vmatrix} 0 & 0 & -1 \\ z+x & z+y & -3z^2 \\ 1 & 1 & -2z \end{vmatrix}$$

$$= 6(z-x)(z-y)(-1) \begin{vmatrix} z+x & z+y \\ 1 & 1 \end{vmatrix}$$

$$= -6(z-x)\ (z-y)\ (x-y)$$

$$= 6(x-y)\ (y-z)\ (z-x).$$

Now consider

$$\frac{\partial(F_1, F_2, F_3)}{\partial(u, v, w)} = \begin{vmatrix} \frac{\partial F_1}{\partial u} & \frac{\partial F_1}{\partial v} & \frac{\partial F_1}{\partial w} \\ \frac{\partial F_2}{\partial u} & \frac{\partial F_2}{\partial v} & \frac{\partial F_2}{\partial w} \\ \frac{\partial F_3}{\partial u} & \frac{\partial F_3}{\partial v} & \frac{\partial F_3}{\partial w} \end{vmatrix}.$$

Again from F_1, F_2 and F_3, we get

$$\frac{\partial F_1}{\partial u} = 3u^2,\ \frac{\partial F_1}{\partial v} = 3v^2,\ \frac{\partial F_1}{\partial w} = 3w^2$$

$$\frac{\partial F_2}{\partial u} = 2u,\ \frac{\partial F_2}{\partial v} = 2v,\ \frac{\partial F_2}{\partial w} = 2w$$

$$\frac{\partial F_3}{\partial u} = 1,\ \frac{\partial F_3}{\partial v} = 1,\ \frac{\partial F_3}{\partial w} = 1$$

Putting in (iii),

$$\frac{\partial(F_1, F_2, F_3)}{\partial(u, v, w)} = \begin{vmatrix} 3u^2 & 3v^2 & 3w^2 \\ 2u & 2v & 2w \\ 1 & 1 & 1 \end{vmatrix}$$

$C_1 \to C_1 - C_3$, $C_2 \to C_2 - C_3$, gives us

$$= 6 \begin{vmatrix} u^2-w^2 & v^2-w^2 & 3w^2 \\ u-w & v-w & 2w \\ 0 & 0 & 1 \end{vmatrix}$$

$$= 6(u-w)(v-w) \begin{vmatrix} u+w & v+w & 3w^2 \\ 1 & 1 & 2w \\ 0 & 0 & 1 \end{vmatrix}$$

$$= 6(u-w)(v-w)\begin{vmatrix} u+w & v+w \\ 1 & 1 \end{vmatrix}$$

$$= 6(u-w)(v-w)(u-v)$$

$$= -6(u-v)(v-w)(w-u)$$

$$\therefore \qquad \frac{\partial(u, v, w)}{\partial(x, y, z)} = (-1)^3 \frac{6(x-y)(y-z)(z-x)}{6(u-v)(v-w)(w-u)}$$

$$= (-1)\frac{(x-y)(y-z)(z-x)}{(u-v)(v-w)(w-u)}.$$

Example 8. *If* $y_1 = 1 - x_1, y_2 = x_1(1-x_2), y_3 = x_1 x_2 (1-x_3) \ldots$
$y_n = x_1 x_2 x_3 \ldots x_{n-1}(1-x_n)$

then show that

$$\frac{\partial(y_1, y_2, \ldots, y_n)}{\partial(x_1, x_2, \ldots, x_n)} = (-1)^n x_1^{n-1} x_2^{n-2} \ldots x_{n-1}.$$

Solution : We know that

$$\frac{\partial(y_1, y_2, \ldots, y_n)}{\partial(x_1, x_2, \ldots, x_n)} = \frac{\partial y_1}{\partial x_1} \cdot \frac{\partial y_2}{\partial x_2} \cdots \frac{\partial y_n}{\partial x_n}$$

because, by partial differentiation of $y_1, y_2, \ldots, y_n$ with respect to $x_1, x_2, \ldots, x_n$

we get $\quad \frac{\partial y_1}{\partial x_1} = -1, \frac{\partial y_1}{\partial x_2} = \frac{\partial y_1}{\partial x_3} = \ldots = \frac{\partial y_1}{\partial x_n} = 0$

$$\frac{\partial y_2}{\partial x_1} = (1-x_2), \frac{\partial y_2}{\partial x_2} = -x_1, \frac{\partial y_2}{\partial x_3} = \ldots = \frac{\partial y_n}{\partial x_n} = 0$$

$$\frac{\partial y_3}{\partial x_1} = x_2(1-x_3), \frac{\partial y_3}{\partial x_2} = x_1(1-x_3), \frac{\partial y_3}{\partial x_3} = -x_1 x_2$$

$$\frac{\partial y_3}{\partial x_4} = \ldots = \frac{\partial y_3}{\partial x_n} = 0$$

Hence, we get

$$\frac{\partial(y_1, \ldots, y_n)}{\partial(x_1, \ldots, x_n)} = \begin{vmatrix} \frac{\partial y_1}{\partial x_1} & \frac{\partial y_1}{\partial x_2} & \frac{\partial y_1}{\partial x_3} & \ldots & \frac{\partial y_1}{\partial x_n} \\ \frac{\partial y_2}{\partial x_1} & \frac{\partial y_2}{\partial x_2} & \frac{\partial y_2}{\partial x_3} & \ldots & \frac{\partial y_2}{\partial x_n} \\ \ldots & \ldots & \ldots & \ldots & \ldots \\ \ldots & \ldots & \ldots & & \ldots \\ \frac{\partial y_n}{\partial x_1} & \frac{\partial y_n}{\partial x_2} & \frac{\partial y_n}{\partial x_3} & \ldots & \frac{\partial y_n}{\partial x_n} \end{vmatrix}$$

$$= \begin{vmatrix} -1 & 0 & 0 & .... & 0 \\ (1-x_2) & -x_1 & 0 & .... & 0 \\ x_2(1-x_3) & x_1(1-x_3) & -x_1x_2 & .... & 0 \\ .... & .... & .... & .... & .... \\ .... & .... & -x_1 & .... & x_{n-1} \\ x_2 - x_{n-1}(1-x_n) & .... & -x_1 & .... & x_{n-1} \end{vmatrix}$$

$$= (-1)(-x_1)(-x_1x_2)(-x_1x_2x_3) \; .... \; (-x_1x_2 \ldots x_{n-1})$$

$$= (-1)^n x_1^{n-1} x_2^{n-2} \; .... \; x_{n-2}^2 \cdot x_{n-1}.$$

Example 9. *If* $y_1 = \dfrac{x_2x_3}{x_1}$, $y_2 = \dfrac{x_1x_3}{x_2}$, $y_3 = \dfrac{x_1x_2}{x_3}$ *then show that*

$$\frac{\partial(y_1, y_2, y_3)}{\partial(x_1, x_2, x_3)} = 4.$$

Solution :

$$y_1 = \frac{x_2x_3}{x_1}. \text{ Then}$$

$$\frac{\partial y_1}{\partial x_1} = -\frac{x_2x_3}{x_1^2}, \; \frac{\partial y_1}{\partial x_2} = \frac{x_3}{x_1}, \; \frac{\partial y_1}{\partial x_3} = \frac{x_2}{x_1}$$

$$y_2 = \frac{x_1x_3}{x_2}.$$

Then

$$\frac{\partial y_2}{\partial x_1} = \frac{x_3}{x_2}, \; \frac{\partial y_2}{\partial x_2} = \frac{-x_1x_3}{x_2^2}, \; \frac{\partial y_2}{\partial x_3} = \frac{x_1}{x_2}$$

$$y_3 = \frac{x_1x_2}{x_3}.$$

Then

$$\frac{\partial y_3}{\partial x_1} = \frac{x_2}{x_3}, \; \frac{\partial y_3}{\partial x_2} = \frac{x_1}{x_3}, \; \frac{\partial y_3}{\partial x_3} = -\frac{x_1x_2}{x_3}.$$

$\therefore$

$$\frac{\partial(y_1, y_2, y_3)}{\partial(x_1, x_2, x_3)} = \begin{vmatrix} \dfrac{\partial y_1}{\partial x_1} & \dfrac{\partial y_1}{\partial x_2} & \dfrac{\partial y_1}{\partial x_3} \\ \dfrac{\partial y_2}{\partial x_1} & \dfrac{\partial y_2}{\partial x_2} & \dfrac{\partial y_2}{\partial x_3} \\ \dfrac{\partial y_3}{\partial x_1} & \dfrac{\partial y_3}{\partial x_2} & \dfrac{\partial y_3}{\partial x_3} \end{vmatrix}$$

$$= \begin{vmatrix} \dfrac{-x_2x_3}{x_1^2} & \dfrac{x_3}{x_1} & \dfrac{x_2}{x_1} \\ \dfrac{x_3}{x_2} & \dfrac{-x_2x_3}{x_2^2} & \dfrac{x_1}{x_2} \\ \dfrac{x_2}{x_3} & \dfrac{x_1}{x_3} & \dfrac{-x_1x_2}{x_3^2} \end{vmatrix}$$

$$= \frac{1}{x_1^2\, x_2^2\, x_2^2} \begin{vmatrix} -x_2 x_3 & x_1 x_3 & x_1 x_2 \\ x_2 x_3 & -x_1 x_3 & x_1 x_2 \\ x_3 x_3 & x_1 x_3 & -x_1 x_2 \end{vmatrix}$$

$R_3 \to R_3 - (2R_1 + R_2)$

$$= \frac{1}{x_1^2 x_2^2 x_3^2} \begin{vmatrix} -x_2 x_3 & x_1 x_3 & x_1 x_2 \\ x_2 x_3 & -x_1 x_3 & x_1 x_2 \\ 2x_2 x_3 & 0 & 0 \end{vmatrix}$$

$$= \frac{1}{x_1^2 x_2^2 x_3^2} \cdot 2x_2 x_3 \begin{vmatrix} +x_1 x_3 & x_1 x_3 \\ -x_1 x_3 & +x_1 x_2 \end{vmatrix}$$

$$= \frac{1}{x_1^2 x_2^2 x_3^2} \times 2x_2 x_3 \, [x_1^2 x_2 x_3 + x_1^2 x_2 x_3]$$

$$= \frac{4x_1^2 x_2^2 x_3^2}{x_1^2 x_2^2 x_3^2} = 4.$$

Example 10. *If* $u = x(1 - r^2)^{-1/2}$, $v = y(1 - r^2)^{-1/2}$, $w = z(1 - r^2)^{-1/2}$ *where* $r^2 = x^2 + y^2 + z^2$, *then show that*

$$\frac{\partial (u, v, w)}{\partial (x, y, z)} = (1 - r^2)^{-5/2}.$$

Solution : $u = x(1 - r^2)^{-1/2}$, putting the value of r

$$= x[1 - x^2 - y^2 - z^2]^{-1/2}$$

$$x^2 = u^2 [1 - x^2 - y^2 - z^2]$$

$$y^2 = v^2 [1 - x^2 - y^2 - z^2]$$

and $$z^2 = w^2 [1 - x^2 - y^2 - z^2].$$

Let $$f_1 \equiv x^2 - u^2 [1 - x^2 - y^2 - z^2] = 0 \quad \text{... (i)}$$

$$f_2 \equiv y^2 - v^2 [1 - x^2 - y^2 - z^2] = 0 \quad \text{... (ii)}$$

$$f_3 \equiv z^2 - w^2 [1 - x^2 - y^2 - z^2] = 0 \quad \text{.... (iii)}$$

We know that,

$$\frac{\partial (u, v, w)}{\partial (x, y, z)} = (-1)^3 \frac{\dfrac{\partial (f_1, f_2, f_3)}{\partial (x, y, z)}}{\dfrac{\partial (f_1, f_2, f_3)}{\partial (u, v, w)}} \quad \text{... (iv)}$$

$$\frac{\partial(f_1, f_2, f_3)}{\partial(x, y, z)} = \begin{vmatrix} \frac{\partial f_1}{\partial x} & \frac{\partial f_1}{\partial y} & \frac{\partial f_1}{\partial z} \\ \frac{\partial f_2}{\partial x} & \frac{\partial f_2}{\partial y} & \frac{\partial f_2}{\partial z} \\ \frac{\partial f_3}{\partial x} & \frac{\partial f_3}{\partial y} & \frac{\partial f_3}{\partial z} \end{vmatrix}$$

By (i), (ii) and (iii),

$$= \begin{vmatrix} 2x(1+u^2) & 2yu^2 & 2zu^2 \\ 2xv^2 & 2y(1+v^2) & 2zv^2 \\ 2xw^2 & 2yw^2 & 2z(1+w^2) \end{vmatrix}$$

$$= 8xyz \begin{vmatrix} 1+u^2 & u^2 & u^2 \\ v^2 & 1+v^2 & v^2 \\ w^2 & w^2 & 1+w^2 \end{vmatrix}$$

$C_1 \to C_1 + (C_2 + C_3)$

$$= 8xyz \begin{vmatrix} 1+u^2+v^2+w^2 & 1+v^2+u^2+w^2 & 1+u^2+v^2+w^2 \\ v^2 & 1+v^2 & v^2 \\ w^2 & w^2 & 1+w^2 \end{vmatrix}$$

$$= 8xyz(1+u^2+v^2+w^2) \begin{vmatrix} 1 & 1 & 1 \\ v^2 & 1+v^2 & v^2 \\ w^2 & w^2 & 1+w^2 \end{vmatrix}$$

$C_1 \to C_2 - C_1, C_3 \to C_3 - C_1$ $= 8xyz(1+u^2+v^2+w^2) \begin{vmatrix} 1 & 0 & 0 \\ v^2 & 1 & 0 \\ w^2 & 0 & 1 \end{vmatrix}$

$$= 8xyz(1+u^2+v^2+w^2) \qquad \text{... (v)}$$

Now consider

$$\frac{\partial(f_1, f_2, f_3)}{\partial(u, v, w)} = \begin{vmatrix} \frac{\partial f_1}{\partial u} & \frac{\partial f_1}{\partial v} & \frac{\partial f_1}{\partial w} \\ \frac{\partial f_2}{\partial u} & \frac{\partial f_2}{\partial v} & \frac{\partial f_2}{\partial w} \\ \frac{\partial f_3}{\partial u} & \frac{\partial f_3}{\partial v} & \frac{\partial f_3}{\partial w} \end{vmatrix}$$

By (i), (ii) and (iii)

$$= \begin{vmatrix} -2u(1-x^2-y^2-z^2) & 0 & 0 \\ 0 & -2v(1-x^2-y^2-z^2) & 0 \\ 0 & 0 & -2w(1-x^2-y^2-z^2) \end{vmatrix}$$

$$= -8uvw(1-x^2-y^2-z^2)^3$$

$$= -8uvw(1-r^2)^3. \qquad \text{... (vi)}$$

Using (v) and (vi) in (iv), we get

$$\frac{\partial(u,v,w)}{\partial(x,y,z)} = (-1)^3 \frac{8(1+u^2+v^2+w^2)\,xyz}{-8uvw(1-r^2)^3}$$

$$= \frac{\left[1+\frac{x^2}{1-r^2}+\frac{y^2}{1-r^2}+\frac{z^2}{1-r^2}\right]xyz}{xyz(1-r^2)^{-3/2}\cdot(1-r^2)^3}$$

$$= \frac{\left[1+\frac{r^2}{1-r^2}\right]}{(1-r^2)^{3/2}} = \frac{1}{(1-r^2)(1-r^2)^{3/2}}$$

$$= \frac{1}{(1-r^2)^{5/2}}.$$

Example 11. *If* $u = x + 2y + z,\ v = x - 2y + 3z$

and $w = 2xy - xz + 4yz - 2z^3,$

show that they are not independent. Find the relation among u, v and w.

Solution : u, v and w will not be independent if

$$\frac{\partial(u,v,w)}{\partial(x,y,z)} = 0.$$

Hence,

$$\frac{\partial(u,v,w)}{\partial(x,y,z)} = \begin{vmatrix} \frac{\partial u}{\partial x} & \frac{\partial u}{\partial y} & \frac{\partial u}{\partial z} \\ \frac{\partial v}{\partial x} & \frac{\partial v}{\partial y} & \frac{\partial v}{\partial z} \\ \frac{\partial w}{\partial x} & \frac{\partial w}{\partial y} & \frac{\partial w}{\partial z} \end{vmatrix}$$

Differentiating partially u, v and w with respect to x, y and z, we get

$$\frac{\partial u}{\partial x} = 1,\ \frac{\partial u}{\partial y} = 2,\ \frac{\partial u}{\partial z} = 1$$

$$\frac{\partial v}{\partial x} = 1,\ \frac{\partial v}{\partial y} = -2,\ \frac{\partial v}{\partial z} = 3$$

$$\frac{\partial w}{\partial x} = 2y - z,\ \frac{\partial w}{\partial y} = 2x + 4z,\ \frac{\partial w}{\partial z} = -x + 4y - 4z$$

$$\frac{\partial (u, v, w)}{\partial (x, y, z)} = \begin{vmatrix} 1 & 2 & 1 \\ 1 & -2 & 3 \\ 2y-z & 2x+4z & -x+4y-4z \end{vmatrix}$$

$C_2 \to C_2 - 2C_1,\ C_3 \to C_3 - C_1,$

$$= \begin{vmatrix} 1 & 0 & 0 \\ 1 & -4 & 2 \\ 2y-z & 2x-4y+6z & -x+2y-3z \end{vmatrix}$$

$$= \begin{vmatrix} -4 & 2 \\ 2x-4y+6z & -x+2y-3z \end{vmatrix}$$

$$= -4(-x + 2y - 3z) - 2(2x - 4y + 6z)$$

$$= 4x - 8y + 12z - 4x + 8y - 12z = 0$$

Hence, u, v and w are not independent.

Now,
$$u + v = 2x + 4z$$
$$u - v = 4y - 2z$$

Multiplying these we get,

$$(u+v)(u-v) = (2x + 4z)(4y - 2z)$$
$$u^2 - v^2 = 4(2xy + 4yz - zx - 2z^2)$$
$$u^2 - v^2 = 4w.$$

Example 12. *If u, v, w are the roots of the cubic*

$$(\lambda - x)^3 + (\lambda - y)^3 + (\lambda - z)^3 = 0$$

in λ, find $\partial (u, v, w)/\partial (x, y, z)$.

Solution : If u, v, w be the roots of the equation

$$(\lambda - x)^3 + (\lambda - y)^3 + (\lambda - z)^3 = 0$$

$$3\lambda^3 - 3\lambda^2 (x+y+z) + 3\lambda (x^2+y^2+z^2) - (x^3+y^3+z^3) = 0$$

then,
$$S_1 = u+v+w = \frac{3(x+y+z)}{3}$$

$$S_2 = uv+vw+wu = \frac{3(x^2+y^2+z^2)}{3}$$

$$S_3 = uvw = \frac{x^2+y^2+z^2}{3}.$$

Let,

$$F_1 \equiv u + v + w - x - y - z = 0$$

$$F_2 \equiv uv + vw + wu - (x^2 + y^2 + z^2) = 0$$

$$F_3 \equiv uvw - \frac{1}{3}(x^3 + y^3 + z^3) = 0$$

$$\therefore \qquad \frac{\partial(u, v, w)}{\partial(x, y, z)} = (-1)^3 \frac{\dfrac{\partial(F_1, F_2, F_3)}{\partial(x, y, z)}}{\dfrac{\partial(F_1, F_2, F_3)}{\partial(u, v, w)}} \qquad \text{... (i)}$$

$$\frac{\partial(F_1, F_2, F_3)}{\partial(x, y, z)} = \begin{vmatrix} \dfrac{\partial F_1}{\partial x} & \dfrac{\partial F_1}{\partial y} & \dfrac{\partial F_1}{\partial z} \\ \dfrac{\partial F_2}{\partial x} & \dfrac{\partial F_2}{\partial y} & \dfrac{\partial F_2}{\partial z} \\ \dfrac{\partial F_3}{\partial x} & \dfrac{\partial F_3}{\partial y} & \dfrac{\partial F_3}{\partial z} \end{vmatrix}$$

Putting the values, we get

$$\frac{\partial(F_1, F_2, F_3)}{\partial(x, y, z)} = \begin{vmatrix} -1 & -1 & -1 \\ -2x & -2y & -2z \\ -x^2 & -y^2 & -z^2 \end{vmatrix}$$

$C_2 \to C_2 - C_1,\ C_3 \to C_3 - C_1$

$$= (-2) \begin{vmatrix} 1 & 0 & 0 \\ x & y-x & z-x \\ x^2 & y^2-x^2 & z^2-x^2 \end{vmatrix}$$

$$= (-2) \begin{vmatrix} y-x & z-x \\ (y-x)(y+x) & (z-x)(z+x) \end{vmatrix}$$

$$= -2(y-x)(z-x)[z+x-y-x]$$

$$= -2(x-y)(y-z)(z-x)$$

Now, $$\frac{\partial(F_1, F_2, F_3)}{\partial(u, v, w)} = \begin{vmatrix} \dfrac{\partial F_1}{\partial u} & \dfrac{\partial F_1}{\partial v} & \dfrac{\partial F_1}{\partial w} \\ \dfrac{\partial F_2}{\partial u} & \dfrac{\partial F_2}{\partial v} & \dfrac{\partial F_2}{\partial w} \\ \dfrac{\partial F_3}{\partial u} & \dfrac{\partial F_3}{\partial v} & \dfrac{\partial F_3}{\partial w} \end{vmatrix}$$

$$= \begin{vmatrix} 1 & 1 & 1 \\ v+w & w+u & u+v \\ vw & uw & uv \end{vmatrix}$$

$C_2 \to C_2 - C_1,\ C_3 \to C_3 - C_1,\ C_2 \to C_2 - C_1,\ C_3 \to C_3 - C_1$

$$= \begin{vmatrix} 1 & 0 & 0 \\ v+w & u-v & u-w \\ vw & w(u-v) & v(u-w) \end{vmatrix}$$

$$= \begin{vmatrix} u-v & u-w \\ w(u-v) & v(u-w) \end{vmatrix}$$

$$= (u-v)\ (u-w) \begin{vmatrix} 1 & 1 \\ w & v \end{vmatrix}$$

$$= (u-v)\ (u-w)\ (v-w) = -(u-v)\ (v-w)\ (w-u)$$

Hence,

$$\frac{\partial(u, v, w)}{\partial(x, y, z)} = \frac{-2(x-y)(y-z)(z-x)}{-(u-v)\ (v-w)\ (w-v)}(-1)^3$$

$$J(u, v, w) = \frac{-2(x-y)(y-z)(z-x)}{(u-v)\ (v-w)\ (w-v)}$$

Example 13. *If the Jacobians of* u_1 and u_2 *w.r.t.* x_1, x_2 *is* $J = J_1$ *and Jacobians of* x_1 *and* x_2 *w.r.t.,* u_1, u_2 *is* $J = J_2$*, then prove that*

$$J_1 \cdot J_2 = 1$$

Solution : $J_1 \times J_2 = \dfrac{\partial(u_1, u_2)}{\partial(x_1, x_2)} \times \dfrac{\partial(x_1, x_2)}{\partial(u_1, u_2)}$

$$= \begin{vmatrix} \frac{\partial u_1}{\partial x_1} & \frac{\partial u_1}{\partial x_2} \\ \frac{\partial u_2}{\partial x_1} & \frac{\partial u_2}{\partial x_2} \end{vmatrix} \begin{vmatrix} \frac{\partial x_1}{\partial u_1} & \frac{\partial x_1}{\partial u_2} \\ \frac{\partial x_2}{\partial u_1} & \frac{\partial x_2}{\partial u_2} \end{vmatrix}$$

$$= \begin{vmatrix} \frac{\partial u_1}{\partial x_1}\cdot\frac{\partial x_1}{\partial u_1} + \frac{\partial u_1}{\partial x_2}\cdot\frac{\partial x_2}{\partial u_1} & \frac{\partial u_1}{\partial x_1}\cdot\frac{\partial x_1}{\partial u_2} + \frac{\partial u_1}{\partial x_2}\cdot\frac{\partial x_2}{\partial u_2} \\ \frac{\partial u_2}{\partial x_1}\cdot\frac{\partial x_1}{\partial u_1} + \frac{\partial u_2}{\partial x_2}\cdot\frac{\partial x_2}{\partial u_1} & \frac{\partial u_2}{\partial x_1}\cdot\frac{\partial x_1}{\partial u_2} + \frac{\partial u_2}{\partial x_2}\cdot\frac{\partial x_2}{\partial u_2} \end{vmatrix}$$

$$= \begin{vmatrix} \frac{\partial u_1}{\partial u_1} & \frac{\partial u_1}{\partial u_2} \\ \frac{\partial u_2}{\partial u_1} & \frac{\partial u_2}{\partial u_2} \end{vmatrix} = \begin{vmatrix} 1 & 0 \\ 0 & 1 \end{vmatrix} = 1.$$

Example 14. *Prove that* $\dfrac{\partial(x_1, x_2 \ldots., x_n)}{\partial(y_1, y_2, \ldots., y_n)} \cdot \dfrac{\partial(y_1, y_2, \ldots., y_n)}{\partial(x_1, x_2 \ldots., x_n)} = 1.$

Solution : Here $\dfrac{\partial(x_1, x_2 \ldots., x_n)}{\partial(y_1, y_2, \ldots., y_n)} \cdot \dfrac{\partial(y_1, y_2, \ldots., y_n)}{\partial(x_1, x_2 \ldots., x_n)}$

$$= \begin{vmatrix} \dfrac{\partial x_1}{\partial y_1} & \dfrac{\partial x_1}{\partial y_2} & \dots & \dfrac{\partial x_1}{\partial y_n} \\ \dfrac{\partial x_2}{\partial y_1} & \dfrac{\partial x_2}{\partial y_2} & \dots & \dfrac{\partial x_2}{\partial y_n} \\ \dots & \dots & \dots & \dots \\ \dots & \dots & \dots & \dots \\ \dfrac{\partial x_n}{\partial y_1} & \dfrac{\partial x_n}{\partial y_2} & \dots & \dfrac{\partial x_n}{\partial y_n} \end{vmatrix} \times \begin{vmatrix} \dfrac{\partial y_1}{\partial x_1} & \dfrac{\partial y_1}{\partial x_2} & \dots & \dfrac{\partial y_1}{\partial x_n} \\ \dfrac{\partial y_2}{\partial x_1} & \dfrac{\partial y_2}{\partial x_2} & \dots & \dfrac{\partial y_2}{\partial n_n} \\ \dots & \dots & \dots & \dots \\ \dots & \dots & \dots & \dots \\ \dfrac{\partial y_n}{\partial x_1} & \dfrac{\partial y_n}{\partial x_2} & \dots & \dfrac{\partial y_n}{\partial x_n} \end{vmatrix}$$

$$= \begin{vmatrix} \sum_{i=1}^{n} \dfrac{\partial x_i}{\partial y_i} \cdot \dfrac{\partial y_1}{\partial x_1} & \sum_{i=1}^{n} \dfrac{\partial x_1}{\partial y_i} \cdot \dfrac{\partial y_i}{\partial x_2} & \dots & \sum_{i=1}^{n} \dfrac{\partial x_1}{\partial y_i} \cdot \dfrac{\partial y_i}{\partial x_n} \\ \sum_{i=1}^{n} \dfrac{\partial x_2}{\partial y_i} \cdot \dfrac{\partial y_i}{\partial x_1} & \sum_{i=1}^{n} \dfrac{\partial x_2}{\partial y_i} \cdot \dfrac{\partial y_i}{\partial x_2} & \dots & \sum_{i=1}^{n} \dfrac{\partial x_2}{\partial y_i} \cdot \dfrac{\partial y_i}{\partial x_n} \\ \dots & \dots & \dots & \dots \\ \dots & \dots & \dots & \dots \\ \sum_{i=1}^{n} \dfrac{\partial x_n}{\partial y_i} \cdot \dfrac{\partial y_i}{\partial x_1} & \sum_{i=1}^{n} \dfrac{\partial x_n}{\partial y_i} \cdot \dfrac{\partial y_i}{\partial x_2} & \dots & \sum_{i=1}^{n} \dfrac{\partial x_n}{\partial y_i} \cdot \dfrac{\partial y_i}{\partial x_n} \end{vmatrix}$$

$$= \begin{vmatrix} \dfrac{\partial x_1}{\partial x_1} & \dfrac{\partial x_1}{\partial x_2} & \dots & \dfrac{\partial x_1}{\partial x_n} \\ \dfrac{\partial x_2}{\partial x_1} & \dfrac{\partial x_2}{\partial x_2} & \dots & \dfrac{\partial x_2}{\partial x_n} \\ \dots & \dots & \dots & \dots \\ \dots & \dots & \dots & \dots \\ \dfrac{\partial x_n}{\partial x_1} & \dfrac{\partial x_n}{\partial x_2} & \dots & \dfrac{\partial x_n}{\partial x_n} \end{vmatrix}$$

$$= \begin{vmatrix} 1 & 0 & \dots & 0 \\ 0 & 1 & \dots & 0 \\ \dots & \dots & \dots & \dots \\ \dots & \dots & \dots & \dots \\ 0 & 0 & \dots & 1 \end{vmatrix} = 1.$$

EXERCISE 12 (A)

1. Find the Jacobians :

   $u = ax + by, \; v = cx + dy.$

2. If $x = u\,(1 + v)$ and $y = v\,(1 + u)$, then find $\dfrac{\partial\,(x, y)}{\partial\,(u, v)}$.

3. If $u = x + y + z + t$, $v = x + y - z - t$, $w = xy - 3t$ and $r = x^2 + y^2 - z^2 - t^2$ then prove that :

$$\frac{\partial (u, v, w, t)}{\partial (x, y, z, t)} = 0.$$

4. If $u^3 + v + w = x + y^2 + z^2$, $u + v^3 + w = x^2 + y + z^2$, $u + v + w^3 = x^2 + y^2 + z$, then prove that :

$$\frac{\partial (u, v, w)}{\partial (x, y, z)} = \frac{1 - 4(xy + yz + zx) + 16xyz}{2 - 3(u^2 + v^2 + w^2) + 27u^2 v^2 w^2}.$$

5. Prove that by Jacobians, following functions are independent :

$u = x + y - z$, $v = x - y + z$ and $w = x^2 + y^2 + z^2 - 2yz$.

Find also the relation.

6. If x, y, z are functionally related to $f(x, y, z) = 0$ then prove that :

$$\frac{\partial (y, z)}{\partial (x, y)} = \left(\frac{\partial y}{\partial x}\right) z = \text{constant}.$$

7. Prove that $\dfrac{\partial (u, v)}{\partial (x, y)} - \dfrac{\partial (x, y)}{\partial (u, v)} = 1.$

8. If the root of K is λ in $\dfrac{x}{a+k} + \dfrac{y}{b+k} + \dfrac{z}{c+k} = 1$, then prove that

$$\frac{\partial (x, y, z)}{\partial (\lambda, \mu, \gamma)} = -\frac{(\gamma - \lambda)(\lambda - \mu)(\mu - \gamma)}{(a - b)(b - c)(c - a)}.$$

9. If $x = r \cos \theta$ and $y = r \sin \theta$, then prove that

$$\frac{\partial (x, y)}{\partial (r, \theta)} = r.$$

10. If

$$x_1 + x_2 + x_3 + \dots + x_n = y_1$$
$$x_2 + x_3 + \dots + x_n = y_2 y_1$$
$$x_3 + x_4 + \dots + x_n = y_3 y_2 y_1$$
$$x_n = y_1 y_2 y_3 \dots y_n,$$

then prove that : $\dfrac{\partial (x_1, x_2, \dots, x_n)}{\partial (y_1, y_2, \dots, y_n)} = y^{n-1} \cdot y_2^{n-2} \cdot \dots \cdot y_{n-2}^2 \cdot y_{n-1}.$

11. If $x = \sin \theta \sqrt{1 - c^2 \sin^2 \phi}$, $y = \cos \theta \cos \phi$, then prove that :

$$\frac{\partial (x, y)}{\partial (\theta, \phi)} = -\sin \phi \frac{(1 - c^2) \cos^2 \theta + c^2 \phi}{\sqrt{1 - c^2 \sin^2 \phi}}.$$

12. If $u = x^2 + y^2 + z^2$, $v = x + y + z$, $w = xy + yz + zx$, then show that Jacobians $\dfrac{\partial (u, v, w)}{\partial (x, y, z)}$ exist and find the relation in u, v, w.

13. If $u = \dfrac{x+y}{z}$, $v = \dfrac{y+z}{x}$, $w = \dfrac{y(x+y+z)}{xz}$, then show that u, v and w are not independent. Find also the relation.

14. If $u^3 = xyz$, $\dfrac{1}{v} = \dfrac{1}{x} + \dfrac{1}{y} + \dfrac{1}{z}$ and $w^2 = x^2 + y^2 + z^2$ then prove that :

$$\frac{\partial (u, v, w)}{\partial (x, y, z)} = \frac{v(x-y)(y-z)(z-x)(x+y+z)}{3u^2 w(xy+yz+zx)}.$$

15. If $u = y\sqrt{(1+x^2)} + x\sqrt{(1-y^2)}$ and $v = \sqrt{(1-x^2)(1-y^2)} - xy$, then prove that u and v are not independent and find also the relation.

16. If $f(0) = 0$, $f'(x) = \dfrac{1}{1+x^2}$, $f'(y) = \dfrac{1}{1+y^2}$, without using the method of integration, prove that :

$$f(x) + f(y) = f\left(\frac{x+y}{1-xy}\right).$$

ANSWERS

1. $ad - bc$ 2. $1 + u + v$
5. $v^2 + u^2 + 2w^2 = 2uv$ 12. $v^2 = u + 2w$
13. $uv = w + 1$ 15. $u^2 + v^2 = 1$

13

Maxima and Minima

13.1. Increasing and Decreasing Functions

(i) If $y = f(x)$ is a function of x such that y increases as x increases in a certain interval, then y is called an *increasing function* of x in that interval. Thus, $y = f(x)$ is an increasing function of x over a sub-set of its domain if $x_1 < x_2 \Leftrightarrow f(x_1) < f(x_2)$.

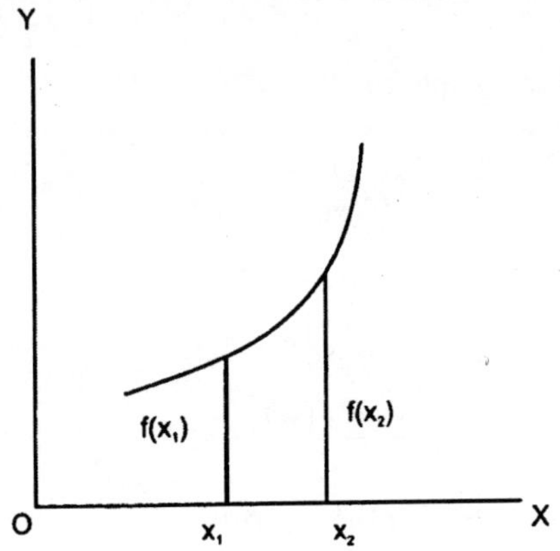

Fig. 13.1 Increasing Function

(ii) If $y = f(x)$ is a function of x such that y decreases as x increases in a certain interval then y is called a *decreasing function* of x in that interval.

Thus, $y = f(x)$ is a decreasing function of x over a sub-set of its domain if $x_1 < x_2 \Leftrightarrow f(x_1) > f(x_2)$.

13.2. Test for Increasing and Decreasing Functions

By Lagrange's Mean Value Theorem, we know that if $f(x)$ is differentiable in $[a, b]$ and if x_1, x_2 are any two points of the interval $[a, b]$ such that $x_1 < x_2$ then there exists a number ξ between x_1 and x_2 such that

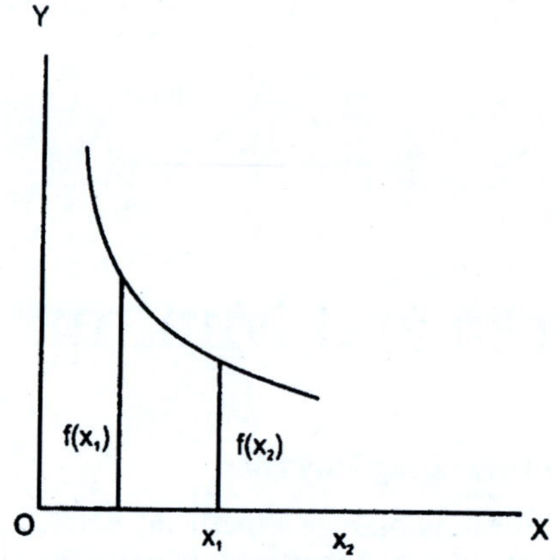

Fig. 13.2 Decreasing Function

$$\frac{f(x_2)-f(x_1)}{x_2-x_1} = f'(\xi)$$

(i) **Let $f(x)$ be an increasing function of x**

Then by definition

$$f(x_2) > f(x_1) \qquad \because\ x_2 > x_1$$

$$\therefore \quad f(x_2)-f(x_1) > 0$$

Also $$x_2 - x_1 > 0 \qquad \because\ x_2 > x_1$$

$$\therefore \quad \frac{f(x_2) - f(x_1)}{x_2-x_1} > 0$$

$$\therefore \quad f'(\xi) > 0$$

since x_1 and x_2 are arbitrary in $[a, b]$

$$\therefore \quad f'(x) > 0 \text{ in } [a, b]$$

or, $$\frac{dy}{dx} > 0 \text{ in } [a, b]$$

(ii) **Let $f(x)$ be a decreasing function of x**

Then by definition

$$f(x_2) < f(x_1) \qquad \because\ x_2 > x_1$$

$$\therefore \quad f(x_2)-f(x_1) < 0$$

Also $$x_2 - x_1 > 0 \qquad \because\ x_2 > x_1$$

$$\therefore \qquad \frac{f(x_2) - f(x_1)}{x_2 - x_1} < 0$$

$$\therefore \qquad f'(\xi) < 0$$

since x_1 and x_2 are arbitrary in $[a, b]$

$$\therefore \qquad f'(x) < 0 \quad \text{in } [a, b]$$

or

$$\frac{dy}{dx} < 0 \quad \text{in } [a, b]$$

Corollary. If $y = f(x)$ is a constant function in the interval $[a, b]$ then $\frac{dy}{dx} = 0$ in $[a, b]$.

Since $f(x)$ is a constant function

$$\therefore \qquad f(x_1) = f(x_2) \qquad \text{where } x_2 > x_1$$

$$\therefore \qquad \frac{f(x_2) - f(x_1)}{x_2 - x_1} = 0 \qquad \because x_2 - x_1 \neq 0$$

$$\therefore \qquad f'(\xi) = 0$$

since x_1 and x_2 are arbitrary in $[a, b]$

$$\therefore \qquad f'(\xi) = 0 \quad \text{in } [a, b]$$

or

$$\frac{dy}{dx} = 0 \quad \text{in } [a, b]$$

Conclusions :

(i) $\frac{dy}{dx}$ is positive for an increasing function.

(ii) $\frac{dy}{dx}$ is negative for a decreasing function.

(iii) $\frac{dy}{dx}$ is zero for a constant function.

13.3. Maximum and Minimum values of $y = f(x)$

Definition :

(i) A function $f(x)$ is said to have a maximum value at $x = a$, if $f(a)$ is greater than any other value that the function can have in a small neighbourhood of $x = a$.

Or

A function $f(x)$ is said to have a maximum value at $x = a$, if we can find a positive number δ, however small, such that $f(x) < f(a)$ for all values of x, other than a, lying in the interval $[a - \delta,\ a + \delta]$.

Or

A function $f(x)$ is said to have a maximum value at $x = a$ if $f(x)$ ceases to increase at $x = a$ and begins to decrease as x increases beyond a.

(ii) A function $f(x)$ is said to have a minimum value at $x = a$, if $f(a)$ is less than any other value that the function can have in a small neighbourhood of $x = a$.

Or

A function $f(x)$ is said to have a minimum value at $x = a$, if we can find a positive number δ, however small, such that $f(x) > f(a)$ for all values of x, other than a, lying in the interval $[a - \delta,\ a + \delta]$.

Or

A function $f(x)$ is said to have a minimum value at $x = a$ if $f(x)$ ceases to decrease at $x = a$ and begins to increase as x increases beyond a.

Notes:

1. Maximum and minimum values are also called *Extreme values* or *Turning values* or *Stationary values.*
2. The points where a function has a maximum or a minimum values are called *Extreme Points* or *Turning Points* or *Stationary Points.*

13.4. Geometrical Meaning

In the figure given below, the graph of the function $y = f(x)$ in the interval $[a, b]$ has been shown.

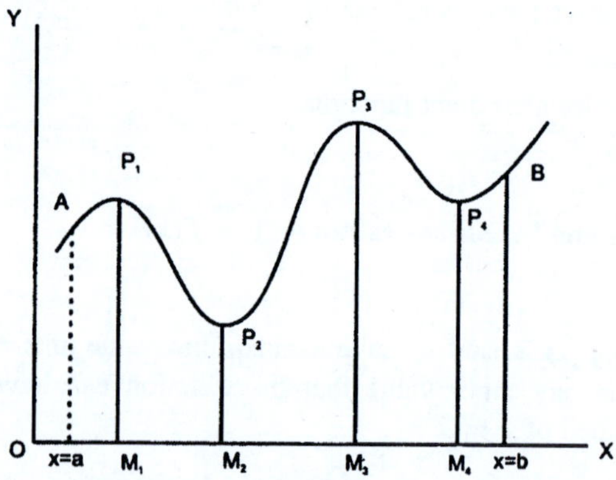

Fig. 13.3

We see that the function has a maximum value at the point P_1, since the value of the function at P_1 is greater than the value of the function for every other value of x in a small neighbourhood of P_1. Similarly the function has a maximum value at the point P_3.

It is interesting to note here that the value of the function at P_1 is less than the value of the function at P_3, although the function has a maximum value at P_1 as well as at P_3. The reason is that the requirement for a function to have a maximum at P_1 is that the value of the function at P_1 is greater than all other values of the function in a *small neighbourhood* of P_1.

Similarly we can discuss the minimum values of the function at P_2 and P_4.

The value of the function at P_3 is known as the Absolute Maximum value in the interval $[a, b]$.

The value of the function at P_2 is known as the Absolute Minimum value in the interval $[a, b]$.

Thus a function may have several maximum and minimum values in any given interval.

Also a minimum value at some point may even be greater than a maximum value at some other point. For example, the minimum value at P_4 is greater than the maximum value at P_1.

Again, it is important to note that a maximum (or a minimum) value at a point say $x = a$ is not necessarily the greatest (or the least) value of the function in that interval, but it is solely a greatest (or a least) value is an interval sufficiently small and containing the value $x = a$. That is why we sometimes say that '*$f(x)$ has a local or relative maxima (or minima) at $x = a$*'.

13.5. Greatest and Least values of a function in an Interval

The greatest and the least values of a function $f(x)$ is an interval $[a, b]$ are the greatest and the smallest values respectively of $f(x)$ out of the values of $f(x)$ at $x = a$, $x = b$ and at points where $f(x)$ is maximum or minimum.

13.6. Conditions for Maxima and Minima

Let $f(x)$ be a function x which is capable of being expanded by Taylor's Theorem in the neighbourhood of $x = a$.

At $x = a$, the value of $f(x)$ is $f(a)$. Consider two values of x *viz.* $a + h$ and $a - h$ in the small neighbourhood of a on either side of $x = a$, where h is very small and positive. Let the values of $f(x)$ at $a + h$ and $a - h$ be $f(a + h)$ and $f(a - h)$ respectively. Then if there is a maximum at $x = a$, then from definition

$$f(a) > f(a + h)$$

and $$f(a) > f(a-h)$$

i.e., $$f(a+h) - f(a) \text{ is } -\text{ve}$$

and $$f(a-h) - f(a) \text{ is } -\text{ve}$$

Similarly, if there is a minimum at $x = a$, then from definition

$$f(a) < f(a+h)$$

and $$f(a) < f(a-h)$$

i.e., $$f(a+h) - f(a) \text{ is } +\text{ve}$$

and $$f(a-h) - f(a) \text{ is } +\text{ve}$$

Thus, we see that for a maxima or minima at $x = a$, $f(a+h) - f(a)$ and $f(a-h) - f(a)$ both have the same sign.

Now by Taylor's theorem, we have

$$f(a+h) = f(a) + hf'(a) + \frac{h^2}{\underline{|2}} f''(a) + \frac{h^3}{\underline{|3}} f'''(a) + \ldots\ldots$$

and $$f(a-h) = f(a) - hf'(a) + \frac{h^2}{\underline{|2}} f''(a) - \frac{h^3}{\underline{|3}} f'''(a) + \ldots\ldots$$

$$\therefore \quad f(a+h) - f(a) = hf'(a) + \frac{h^2}{\underline{|2}} f''(a) + \frac{h^3}{\underline{|3}} f'''(a) + \ldots\ldots \qquad \ldots(1)$$

and $$f(a-h) - f(a) = -hf'(a) + \frac{h^2}{\underline{|2}} f''(a) - \frac{h^3}{\underline{|3}} f'''(a) + \ldots\ldots \qquad \ldots(2)$$

Since h is very small, we can neglect powers of h higher than the first and so from (1) and (2), we get

$$f(a+h) - f(a) = hf'(a) \qquad \ldots(3)$$

and $$f(a-h) - f(a) = -hf'(a) \qquad \ldots(4)$$

But we have seen above that for maxima or minima, both $f(a+h) - f(a)$ and $f(a-h) - f(a)$ should be of the same sign. So from (3) and (4), we conclude that $f'(a)$ should be zero otherwise they will have different signs, since h is + ve.

Hence, the necessary condition that $f(x)$ has a maxima or minima at $x = a$ is

$$f'(a) = 0.$$

Now if $f'(a) = 0$, we have from (1) and (2)

$$f(a+h) - f(a) = \frac{h^2}{\underline{|2}} f''(a) \qquad \ldots(5)$$

and $$f(a-h) - f(a) = \frac{h^2}{\underline{|2}} f''(a) \qquad \ldots(6)$$

(neglecting powers of h higher than 2)

Now we find that $f(a+h)-f(a)$ and $f(a-h)-f(a)$, both, are of the same sign.

Now two cases arise.

CASE I. *When $f''(a)$ is positive :*

In this case

$$f(a+h)-f(a) \text{ and } f(a-h)-f(a) \text{ are both positive.}$$

Hence there is a minimum at $x = a$.

CASE II. *When $f''(a)$ is negative :*

In this case

$$f(a+h)-f(a) \text{ and } f(a-h)-f(a) \text{ are both negative.}$$

Hence there is a maximum at $x = a$.

Hence we conclude that

(i) *the function has a maximum at $x = a$ if $f'(a) = 0$ and $f''(a)$ is negative, and*

(ii) *the function $f(x)$ has a minimum at $x = a$ if $f'(a) = 0$ and $f''(a)$ is positive.*

Generalisation

If $f'(a) = 0$ and $f''(a) = 0$, then from (1) and (2) we have

$$f(a+h)-f(a) = \frac{h^3}{\lfloor 3} f'''(a) \quad \text{...(7)}$$

and $$f(a-h)-f(a) = \frac{h^3}{\lfloor 3} f'''(a) \quad \text{...(8)}$$

(neglecting powers of h higher than the third)

For maximum or minimum value of $f(x)$ at $x = a$, we know that $f(a+h)-f(a)$ and $f(a-h)-f(a)$ must have the same signs and for that $f'''(a)$ should be zero otherwise they will have opposite signs, since h is +ve, as is evident from (7) and (8).

Hence if $f'''(a)$ is also zero, then we have from (1) and (2),

$$f(a+h)-f(a) = \frac{h^4}{\lfloor 4} f''''(a) \quad \text{...(9)}$$

and $$f(a-h)-f(a) = \frac{h^4}{\lfloor 4} f''''(a) \quad \text{...(10)}$$

(neglecting powers of h higher than fourth)

and these are of the same signs.

$\because$ If $f(x)$ is maximum at $x = a$, we know that $f(a+h) - f(a)$ and $f(a-h) - f(a)$ are both negative and hence $f''''(a)$ must be negative.

i.e., If $f'(a) = 0$, $f''(a) = 0$, *then function* $f(x)$ *is maximum at* $x = a$, *provided* $f'''(a) = 0$ *and* $f''''(a)$ *are negative.*

Similarly, *if* $f'(a) = 0, f''(a) = 0$ *then function* $f(x)$ *is minimum at* $x = a$ *provided* $f'''(a) = 0$ *and* $f''''(a)$ *is positive.*

Proceeding in this manner, we find that in general if $f'(a) = 0 = f''(a) = f'''(a) = \ldots = f^{n-1}(a)$ and $f^n(a) \neq 0$, then for a maximum or a minimum, n must be even. Also for a maximum $f^n(a)$ must be negative and for a minimum $f^n(a)$ must be positive.

Working Rule for finding the maximum and minimum values of $y = f(x)$.

First Method :

(i) Find $\dfrac{dy}{dx}$ and equate to zero. Solve this equation for real values of x. Let these values be a, b, c, ...

(ii) Find $\dfrac{d^2y}{dx^2}$. Put $x = a, b, c$,..... in it turn by turn.

If $\dfrac{d^2y}{dx^2}$ is –ve when $x = a$, then $f(x)$ is maximum at $x = a$ and the corresponding maximum value of $f(x)$ is $f(a)$.

If $\dfrac{d^2y}{dx^2}$ is +ve when $x = a$, then $f(x)$ is minimum at $x = a$ and the corresponding minimum value of $f(x)$ is $f(a)$.

Similarly for the points $x = b, c$, ... etc.

(iii) If $\dfrac{d^2y}{dx^2} = 0$ at $x = a$ but $\dfrac{d^3y}{dx^3} \neq 0$ at $x = a$, then there is neither a maximum nor a minimum at $x = a$.

But if $f'''(a) = 0$, then we find $f''''(x)$ and put $x = a$ in it.

If $f''''(a)$ is negative, we have a maximum at $x = a$ and if $f''''(a)$ is positive, we have a minimum at $x = a$.

If $f''''(a) = 0$, then we must find $f''''(x)$ and so on.

Second Method :

Sometimes the process of finding $\dfrac{d^2y}{dx^2}$ becomes tedious. In such cases, we follow this method preferably. It is called *First Derivative Test.*

(i) Find $\frac{dy}{dx}$ and equate it to zero. Solve this equation for real values of x. Let these values be $a, b, c, ...$

(ii) Consider the value $x = a$. Study the sign of $\frac{dy}{dx}$ for values of x slightly less than a and slightly greater than a.

(iii) If $\frac{dy}{dx}$ changes sign from +ve to −ve, then $f(x)$ is maximum at $x = a$.

If $\frac{dy}{dx}$ changes sign from −ve to +ve, then $f(x)$ is minimum at $x = a$.

If $\frac{dy}{dx}$ does not change sign, then $x = a$ is neither a point of maxima nor a point of minima.

13.7. Points of Inflexion Stationary Values

For the function $y = f(x)$ to have a maximum or minimum value at $x = a$, $\frac{dy}{dx} = 0$ at $x = a$. But if $\frac{dy}{dx} = 0$ at $x = 0$, it is not essential that the function may have a maximum or minimum value at $x = a$. For it may happen that $\frac{dy}{dx}$ does not change sign from +ve to −ve or from −ve to +ve as x passes through the value $x = a$ and consequently the function may go on increasing or decreasing.

Thus the condition that $\frac{dy}{dx} = 0$ *for a maximum or a minimum value of the function* $y = f(x)$, *is only the necessary condition but not the sufficient.*

Definition :

A point on the curve $y = f(x)$ at which $\frac{dy}{dx} = 0$ but $\frac{dy}{dx}$ does not change sign as x passes through the point is called a *point of inflexion*.

Or

A curve ordinarily does not cross its tangent, but if at some point, it crosses its tangent, the point is called a *point of inflexion*.

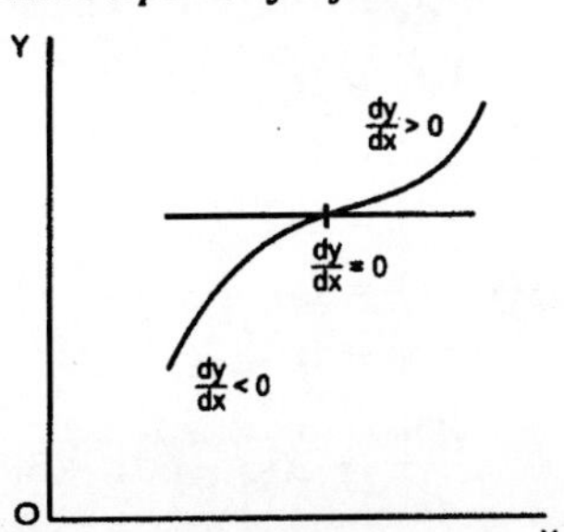

Fig. 13.4

Definition :

The values of $y = f(x)$ at the points where $\frac{dy}{dx} = 0$ are called stationary values of the function.

Notes:

1. Maximum and minimum values are all stationary values, but the converse is not true *i.e.*, a stationary value of (a) need not be a maximum or a minimum value, because the curve may have a point of inflexion at $x = a$.
2. $\frac{dy}{dx} = 0$ at $x = a$ implies that the tangent to the curve $y = f(x)$ at $x = a$ is parallel to the axis of x. Hence at stationary points, tangent is parallel to x-axis.

13.8. Application to Problems

In problems of geometry and algebra, sometimes we have

$$u = f(x, y) \quad \text{...(1)}$$

where

$$\phi(x, y) = 0 \quad \text{...(2)}$$

and we are required to discuss the maxima and minima of u. For this purpose, we eliminate x or y with the help of (2) and express u as a function of x or y (*i.e.*, single variable) and apply the usual methods.

ILLUSTRATIVE EXAMPLES

Example 1. *Show that $x^5 - 5x^4 + 5x^3 - 1$ has a maximum at $x = 1$, a minimum at $x = 3$ and neither when $x = 0$.*

Solution : Let

$$y = x^5 - 5x^4 + 5x^3 - 1$$

$\therefore$

$$\frac{dy}{dx} = 5x^4 - 20x^3 + 15x^2$$

$$= 5(x^4 - 4x^3 + 3x^2)$$

For a maxima or minima, we have

$$\frac{dy}{dx} = 0$$

or,
$$5(x^4 - 4x^3 + 3x^2) = 0$$

or,
$$x^4 - 4x^3 + 3x^2 = 0$$

or,
$$x^2(x^2 - 4x + 3) = 0$$

or,
$$x^2(x - 1)(x - 3) = 0$$

or,
$$x = 0, 1, 3$$

Now,
$$\frac{d^2y}{dx^2} = 20x^3 - 60x^2 + 30x$$

$$= -10 \text{ at } x = 1 \textit{ i.e., } \text{negative.}$$

Hence $x = 1$ gives a maximum.

Again $\frac{d^2y}{dx^2} = 90$ at $x = 3$ *i.e.,* + ve.

Hence $x = 3$ gives a minimum.

Also $\frac{d^2y}{dx^2} = 0$ at $x = 0$

and $\frac{d^3y}{dx^2} = 60x^2 - 120x + 30$

$= 30$ at $x = 0$ *i.e.,* $\neq 0$.

Hence there is neither a maxima nor a minima $x = 0$.

Example 2. *Show that the function* $\sin x\,(1 + \cos x)$ *is maximum when* $x = \frac{\pi}{3}$.

Solution : Let $y = \sin x\,(1 + \cos x)$

$$\therefore \quad \frac{dy}{dx} = \cos x\,(1 + \cos x) + \sin x\,(-\sin x)$$

$$= \cos x + \cos^2 x - \sin^2 x$$

$$= \cos x + \cos^2 x - (1 - \cos^2 x)$$

$$= 2\cos^2 x + \cos x - 1$$

For a maxima or minima we have

$$\frac{dy}{dx} = 0$$

or, $2\cos^2 x + \cos x - 1 = 0$

or, $\cos x = \frac{-1 \pm \sqrt{1+8}}{4}$

or, $\cos x = \frac{1}{2}, -1$

or, $x = 2n\pi \pm \frac{\pi}{3},\ 2n\pi \pm \pi$ where $n \in 1$.

Now, $\frac{d^2y}{dx^2} = 4\cos x\,(-\sin x) - \sin x.$

$$= -\sin x - 2\sin 2x$$

when $x = 2n\pi \pm \pi,\ \frac{d^2y}{dx^2} = 0$

So, $\frac{d^3y}{dx^3} = -\cos x - 4\cos 2x$

When $$x = 2n\pi \pm \pi$$

$$\frac{d^3y}{dx^3} = -3\ (\neq 0)$$

Hence, the function has neither a maximum nor a minimum at $x = 2n\pi \pm \pi$.

When $x = 2n\pi + \frac{\pi}{3}$, $\frac{d^2y}{dx^2} = -\frac{\sqrt{3}}{2} - 2\frac{\sqrt{3}}{2} = -\frac{3\sqrt{3}}{2}$ which is – ve

When $x = 2n\pi - \frac{\pi}{3}$, $\frac{d^2y}{dx^2} = \frac{\sqrt{3}}{2} + \frac{2\sqrt{3}}{2} = \frac{3\sqrt{3}}{2}$ which is + ve

Hence the given function has a maximum at $x = 2n\pi + \frac{\pi}{3}$ and a minimum at $x = 2n\pi - \frac{\pi}{3}$.

In particular, when $n = 0$, the given function has a maximum at $x = \frac{\pi}{3}$.

Example 3. *Find the largest and smallest values of $x^3 - 18x^2 + 96x$ in the interval [0, 9].*

Solution. Let $$y = f(x) = x^3 - 18x^2 + 96x$$

then $$\frac{dy}{dx} = 3x^2 - 36x + 96$$

$$= 3\,(x^2 - 12x + 32)$$

For maxima and minima,

$$\frac{dy}{dx} = 0$$

$$\therefore \quad 3\,(x^2 - 12x + 32) = 0$$

or, $$x^2 - 12x + 32 = 0$$

or, $$(x-4)\,(x-8) = 0$$

or, $$x = 4, 8$$

i.e., there can be maximum or minimum values at $x = 4$ or 8 which lie in the interval [0, 9].

Now, $$\frac{d^2y}{dx^2} = 6x - 36$$

$$= -12 \text{ at } x = 4$$

and $$12 \text{ at } x = 8$$

Hence, there is a maximum at $x = 4$ and a minimum at $x = 8$.

To find the largest and smallest values of $f(x)$

$$f(4) = 160 \qquad \text{maximum value}$$

$$f(8) = 128 \qquad \text{minimum value}$$

$$\left.\begin{aligned} f(0) &= 0 \\ f(9) &= 135 \end{aligned}\right\} \quad \text{values of the function at the end points of the interval}$$

The greatest and smallest values of the function are the greatest and smallest values of $f(4)$, $f(8)$, $f(0)$ and $f(9)$.

Thus, greatest value $= 160$

and smallest value $= 0$.

Example 4. *Prove that* $\sin^p \theta \cos^q \theta$ *could be maximum when*

$$\theta = \tan^{-1}\left(\sqrt{\frac{p}{q}}\right).$$

Solution : Let $y = \sin^p \theta \cos^q \theta$

Then
$$\frac{dy}{d\theta} = p \sin^{p-1} \theta \cos \theta \cos^q \theta + \sin^p \theta \cdot q \cos^{q-1} \theta \, (-\sin \theta$$

$$= p \sin^{p-1} \theta \cos^{q+1} \theta - q \sin^{p+1} \theta \cos^{q-1} \theta$$

$$= \sin^{p-1} \theta \cos^{q-1} \theta \, (p \cos^2 \theta - q \sin^2 \theta)$$

$$= \sin^{p-1} \cos^{q+1} \theta \, (p - q \tan^2 \theta)$$

For maxima or minima,

$$\frac{dy}{d\theta} = 0$$

$\therefore$
$$p - q \tan^2 \theta \qquad [\because \sin \theta \neq 0; \cos \theta \neq 0 \text{ For, in that case } y = 0]$$

or,
$$\tan \theta = \sqrt{\frac{p}{q}}$$

Now,
$$\frac{dy}{d\theta} = \sin^{p-1} \cos^{q+1} \theta \, (p - q \tan^2 \theta)$$

$$= \sin^p \theta \cos^q \theta \, (p \cot \theta - q \tan \theta)$$

$$= y \, (p \cot \theta - q \tan \theta)$$

$\therefore$
$$\frac{d^2y}{d\theta^2} = \frac{dy}{d\theta} (p \cot \theta - q \tan \theta) + y \, (-p \operatorname{cosec}^2 \theta - q \sec^2 \theta)$$

$\therefore$ For maxima or minima,

$$\frac{dy}{d\theta} = 0$$

$$\therefore \qquad \frac{d^2y}{d\theta^2} = -y\,(p \ \text{cosec}^2\,\theta + q\sec^2\theta)$$

$$= \text{negative when } \tan\theta = \frac{p}{q} \ \ i.e., \theta = \tan^{-1}\frac{p}{q}$$

Hence y is maximum when

$$\theta = \tan^{-1}\frac{p}{q}.$$

Example 5. *Examine the function* $(x-3)^5\,(x+1)^2$ *for maximum and minimum.*

Solution. Let $\qquad y = (x-3)^5\,(x+1)^2$

Then $\qquad \dfrac{dy}{dx} = 5\,(x-3)^4\,(x+1)^2 + (x-3)^5\,2\,(x+1)$

$$= (x+1)\,(x-3)^4\,[5\,(x+1) + 2\,(x-3)]$$

$$= (x+1)\,(x-3)^4\,(7x-1)$$

For maxima or minima

$$\frac{dy}{dx} = 0$$

$$\therefore \qquad (x+1)\,(x-3)^4\,(7x-1) = 0$$

$$\therefore \qquad x = -1, 3, \frac{1}{7}.$$

Now let us test these values one by one.

(i) $x = -1$

When x is slightly < -1

$$\frac{dy}{dx} = (-)\,(+)\,(-) = +\text{ve}$$

When x is slightly > -1

$$\frac{dy}{dx} = (+)\,(+)\,(-) = -\text{ve}$$

$\therefore \qquad \dfrac{dy}{dx}$ changes sign from + ve to – ve.

$\therefore$ The function has a maximum value at $x = -1$

and maximum value $= (-1-3)^5\,(-1+1)^2 = 0.$

(ii) $x = 3$

When x is slightly < 3

$$\frac{dy}{dx} = (+)\,(+)\,(+) = +\text{ve}$$

When x is slightly > 3

$$\frac{dy}{dx} = (+)\ (+)\ (+) = +\text{ve}$$

Thus $\frac{dy}{dx}$ does not changes sign in the immediate neighbourhood of $x = 3$.

∴ the function has a point of inflexion at $x = 3$.

(iii) $x = \frac{1}{7}$

Whwn x is slightly < $\frac{1}{7}$

$$\frac{dy}{dx} = (+)\ (+)\ (-) = -\text{ve}$$

When x is slightly > $\frac{1}{7}$

$$\frac{dy}{dx} = (+)\ (+)\ (+) = +\text{ve}$$

Thus $\frac{dy}{dx}$ change sign from – ve to + ve.

Hence the function has a minimum value at $x = \frac{1}{7}$ and the minimum

$$\text{value} = \left(\frac{1}{7} - 3\right)^5 \left(\frac{1}{7} + 1\right)^2 = -\frac{5^5\, 2^{16}}{7^7}.$$

Example 6. *Show that the maximum and minimum values of $x^2 + y^2$ where $ax^2 + 2hxy + by^2 = 1$ are given by the roots of the quadratic*

$$\left(a - \frac{1}{r^2}\right)\left(b - \frac{1}{r^2}\right) = h^2.$$

Solution : Given $\qquad ax^2 + 2hxy + by^2 = 1 \qquad ...(1)$

also $\qquad x^2 + y^2 = r^2 \qquad$ (changing to polar form)

So in this problem we are to find the maximum and minimum value of r^2.

From (1), changing to polar form by putting $x = r \cos \theta$ and $y = r \sin \theta$, we get

$$r^2 (a \cos^2 \theta + 2h \cos \theta \sin \theta + b \sin^2 \theta) = 1$$

$$\Rightarrow \qquad \frac{1}{r^2} = a\cos^2\theta + 2h\cos\theta\sin\theta + b\sin^2\theta \qquad ...(2)$$

where $\quad R = \dfrac{1}{r^2}$

Now, $$\frac{dR}{d\theta} = -2a\cos\theta\sin\theta + 2h\cos^2\theta - 2h\sin^2\theta + 2b\sin\theta\cos\theta$$
$$= 2h\cos 2\theta - (a-b)\sin 2\theta$$

For maximum and minimum values of r^2 *i.e.*, minimum and maximum values of R, we must have $\dfrac{dR}{d\theta} = 0$.

$\therefore$ equating $\dfrac{dR}{d\theta}$ to zero, we get

$$2h\cos 2\theta - (a-b)\sin 2\theta = 0$$

$$\Rightarrow \qquad \tan 2\theta = \frac{2h}{a-b}$$

$$\therefore \qquad \sin 2\theta = \frac{2h}{\sqrt{(a-b)^2+4h^2}}$$

and $$\cos 2\theta = \frac{a-b}{\sqrt{(a-b)^2+4h^2}}$$

From (2),

$$\frac{1}{r^2} = R = a\cos^2\theta + 2h\cos\theta\sin\theta + b\sin^2\theta$$

or, $$\frac{1}{r^2} = \frac{a(1+\cos 2\theta)}{2} + h\sin 2\theta + \frac{b(1-\cos 2\theta)}{2}$$

$$= \frac{a}{2}\left[\frac{\sqrt{(a-b)^2+4h^2}+a-b}{\sqrt{(a-b)^2+4h^2}}\right] + \frac{2h^2}{\sqrt{(a-b)^2+4h^2}}$$
$$+ \frac{b}{2}\left[\frac{\sqrt{(a-b)^2+4h^2}-(a-b)}{\sqrt{(a-b)^2+4h^2}}\right]$$

or, $$\sqrt{(a-b)^2+4h^2}\left[\frac{1}{r^2}-\frac{a}{2}-\frac{b}{2}\right] = \frac{a}{2}(a-b) + 2h^2 - \frac{b}{2}(a-b)$$
$$= \frac{1}{2}[(a-b)^2+4h^2]$$

or, $$\frac{2}{r^2} - (a+b) = \sqrt{(a-b)^2+4h^2}$$

or, $$\frac{4}{r^4} - \frac{4(a+b)}{r^2} + (a+b)^2 = (a-b)^2 + 4h^2$$

(Squaring both sides)

or, $$\frac{1}{r^4} - \frac{a+b}{r^2} = h^2 - ab$$

or, $$\frac{1}{r^4} - \frac{a}{r^2} - \frac{b}{r^2} + ab = h^2$$

or, $$\left(a - \frac{1}{r^2}\right)\left(b - \frac{1}{r^2}\right) = h^2.$$

Example 7. *If* $\frac{dy}{dx} = (x-a)^{2n} (x-b)^{2p+1}$ *when* n *and* p *are positive integers, show that* $x = a$ *gives neither a maximum nor a minimum value of* y *but* $x = b$ *gives a minimum.*

Solution. Given

$$\frac{dy}{dx} = (x-a)^{2n} (x-b)^{2p+1} \quad \text{... (1)}$$

For maxima or minima

$$\frac{dy}{dx} = 0$$

$\therefore$ $$(x-a)^{2n} (x-b)^{2p+1} = 0$$

$\therefore$ $$x = a, b$$

Differentiating (1) $2n$ times by Leibnitz's Theorem, we get

$$\frac{d^{2n+1}y}{dx^{2n+1}} = \lfloor 2n \; (x-b)^{2p+1} + \text{terms containing powers of } (x-a)$$

$\therefore$ at $x = a$,

$$\frac{d^{2n+1}y}{dx^{2n+1}} = \lfloor 2n \; (a-b)^{2p+1} \neq 0$$

Also $\frac{d^{2n+1}y}{dx^{2n+1}}$ is an odd order derivative at y, since $(2n+1)$ is odd. And we can see that $(2n)$th derivative of y which is of even order will be zero when $x = a$.

Hence y has neither a maxima nor a minima at $x = a$.

Again differentiating (1), $(2p+1)$ times we get

$$\frac{d^{2p+2}y}{dx^{2p+2}} = (x-a)^{2n} \lfloor 2p+1 + \text{terms containing powers of } (x-b)$$

$\therefore$ at $x = b$,

$$\frac{d^{2p+2}y}{dx^{2p+2}} = (b-a)^{2n} \lfloor 2n+1 \text{ which is +ve.}$$

Also $\dfrac{d^{2p+2}y}{dx^{2p+2}}$ is an even order derivative of y.

Moreover $\dfrac{d^{2p+1}y}{dx^{2p+1}}$ contains $(x-b)$ in each term and so at $x = b$, $\dfrac{d^{2p+1}y}{dx^{2p+1}} = 0$ and all other derivatives, of order less than $(2p+1)$ are zero at $x = b$.

Hence y is minimum at $x = b$.

Example 8. *Assuming that the petrol burnt (per hour) in driving a motor boat varies as the cube of its velocity, show that the most economical speed, when going against a current of c miles per hour is $\frac{3}{2}c$ miles per hour.*

Solution : Let the velocity of the motor boat be v miles/hour. Then the velocity of the boat relative to the current when it is going against the current $= (v-c)$ m.p.h.

If d is the distance (constant) covered up, then

$$\text{the time of run} = \frac{d}{v-c} \text{ hours}$$

Also the petrol burnt per hour $= kv^3$, where k is a constant.

If y denote the total amount of petrol burnt, then

$$y = kv^3 \cdot \frac{d}{v-c} = dk\,\frac{v^3}{v-c}$$

For maxima or minima, $\dfrac{dy}{dv} = 0$

$$\therefore \quad dk \cdot \frac{(v-c)\ 3v^2 - v^3}{(v-c)^2} = 0$$

$$\Rightarrow \quad v^2\,[3v - 3c - v] = 0$$

$$\Rightarrow \quad 2v - 3c = 0 \qquad [\because\ v \neq 0]$$

$$\Rightarrow \quad v = \frac{3c}{2}$$

Clearly $\dfrac{dy}{dv}$ changes sign from – ve to +ve as v passes through $\frac{3}{2}c$. Hence y is minimum when $v = \frac{3}{2}c$.

Example 9. *One corner of a long rectangular sheet of paper of width foot is folded so as to reach the opposite edge of the sheet. Find the minimum length of the crease.*

Solution : Let $ABCD$ be the rectangular sheet of paper of which the corner B is folded along EF so as to reach the opposite edge (*i.e.*, AD) at B'. Let l be the length of the crease EF.

Let $\angle BFE = \theta$

B′

Then $\angle EFB' = \theta$

and $\angle B'FA = \pi - 2\theta$

Now, $FB = l\cos\theta = B'F$

and
$$AF = B'F \cos B'FA$$
$$= l\cos\theta\cos(\pi - 2\theta)$$
$$= -l\cos\theta\cos 2\theta$$

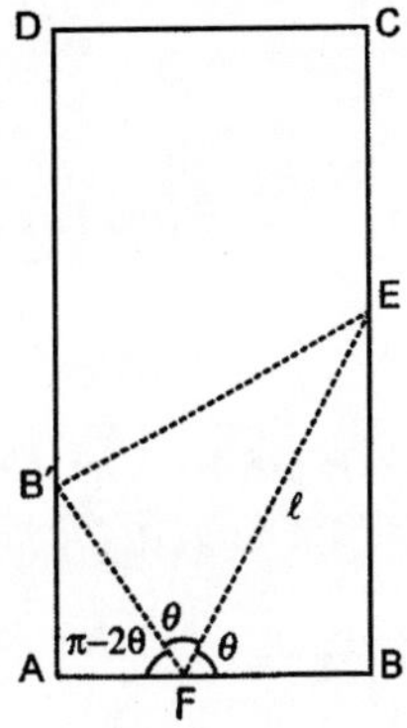

Fig. 13.5

The width of the sheet is 1 foot.

$$\therefore \quad 1 = AB = AF + FB$$
$$= l\cos\theta - l\cos\theta\cos 2\theta$$

i.e.,
$$\frac{1}{l} = \cos\theta(1 - \cos 2\theta) = z \text{ (say)} \qquad \text{...(1)}$$

Now clearly l will be minimum when z is maximum.

Now
$$\frac{dz}{d\theta} = -\sin\theta(1 - \cos 2\theta) + \cos\theta(2\sin 2\theta)$$
$$= 0 \text{ for } z \text{ to be maximum or minimum}$$

$$\Rightarrow \quad 2\sin\theta(-\sin^2\theta + 2\cos^2\theta) = 0$$

$\Rightarrow \quad \tan\theta = \sqrt{2}$ as $\sin\theta = 0$ is not possible if there is to be folding

Again,
$$\frac{d^2z}{d\theta^2} = 4\cos^3\theta - 14\sin^2\theta\cos\theta$$
$$= 2\cos\theta(2\cos^2\theta - 7\sin^2\theta)$$
$$= 2\cdot\frac{1}{\sqrt{3}}\left(2\cdot\frac{1}{3} - 7\cdot\frac{2}{3}\right) \quad \left[\text{When } \tan\theta = \sqrt{2}, \text{ then } \sin\theta = \sqrt{\frac{2}{3}}, \text{ and } \cos\theta = \frac{1}{\sqrt{3}}\right]$$
$$= -\text{ve}$$

Hence when $\tan\theta = \sqrt{2}$, z is maximum *i.e.*, l is minimum.

From (1)

$$l = \frac{1}{\cos\theta\,(1-\cos 2\theta)}$$

$$= \frac{1}{2\sin^2\theta\cos\theta}$$

$$= \frac{1}{2\cdot\frac{2}{3}\frac{1}{\sqrt{3}}} \qquad \left[\because \text{ When } \tan\theta = \sqrt{2},\right.$$

$$\left.\text{then } \sin\theta = \sqrt{\frac{2}{3}} \text{ and } \cos\theta = \frac{1}{\sqrt{3}}\right]$$

$$= \frac{3\sqrt{3}}{4} \text{ ft.}$$

Example 10. *The perimeter of a ractangle is* 100 cm. *For maximum area, find the sides of the rectangle.*

Solution : Let the sides of the rectangle be x cm and y cm. Then Perimeter = $2(x+y)$ cm.

But perimeter = 100 cm (given)

$\therefore \quad 2(x+y) = 100$

$\therefore \quad x + y = 50$ ...(1)

Let A be the area of the rectangle. Then

$A = xy$ ...(2)

From (1),

$y = 50 - x$

$\therefore \quad A = x(50-x)$

$\Rightarrow \quad A = 50x - x^2$ ...(3)

In order that A is maximum or minimum, we have

$$\frac{dA}{dx} = 0$$

i.e., $\quad 50 - x = 0$

i.e., $\quad x = 25$

When x is slightly < 5, $\frac{dA}{dx}$ is + ve.

When x is slightly > 25, $\frac{dA}{dx}$ is − ve

Hence $\frac{dA}{dx}$ changes sign from + ve to − ve.

$\therefore$ Area is maximum when $x = 25$

Hence the sides of the rectangle are 25 cm each *i.e.*, it is a square.

Example 11. *Show that the rectangle of maximum area inscribed in a circle is a square.*

Solution. Let $ABCD$ be a rectangle inscribed in the circle $x^2 + y^2 = a^2$

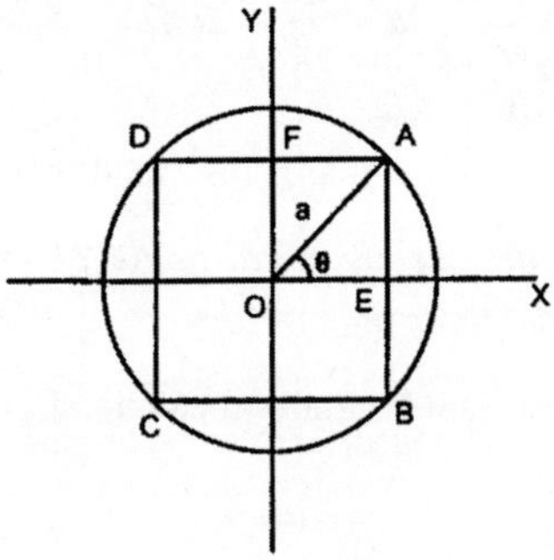

Fig. 13.6

Let the coordinates of A be $(a \cos \theta, a \sin \theta)$

Then, $AD = 2 \; AF = 2 \; OE = 2a \cos \theta$

$AB = 2 \; AE = 2a \sin \theta$

Let S denote the area of the rectangle $ABCD$. Then,

$$S = AD \times AB$$
$$= 4a^2 \sin \theta \cos \theta$$
$$= 2a^2 \sin 2\theta$$
$$\frac{dS}{d\theta} = 4a^2 \cos 2\theta$$

For maxima or minima,

$$\frac{dS}{d\theta} = 0$$
$$\Rightarrow 4a^2 \cos 2\theta = 0$$
$$\Rightarrow \quad \cos 2\theta = 0 = \cos \frac{\pi}{2}$$
$$\Rightarrow \quad \theta = \frac{\pi}{4}$$
$$\frac{d^2S}{d\theta^2} = -8a^2 \sin 2\theta$$

$$= -8a^2 \sin \frac{\pi}{2} \quad \text{at } \theta = \frac{\pi}{4}$$

$$= 8a^2 \text{ which is } -\text{ve.}$$

$\therefore$ S is maximum when $\theta = \frac{\pi}{4}$

Also, $AD = 2a \cos \frac{\pi}{4} = 2a. \frac{1}{\sqrt{2}} = a\sqrt{2}$

$AB = 2a \sin \frac{\pi}{4} = 2a. \frac{1}{\sqrt{2}} = a\sqrt{2}$

$\therefore$ $AB = AD$

Hence, the rectangles of maximum area inscribed in a circle is a square.

Example 12. *Show that the semivertical angle of the cone of maximum volume and of given slant height is* $\tan^{-1}\sqrt{2}$.

Solution : Let l be the given slant height of the cone. Let θ be its semivertical angle. Then,

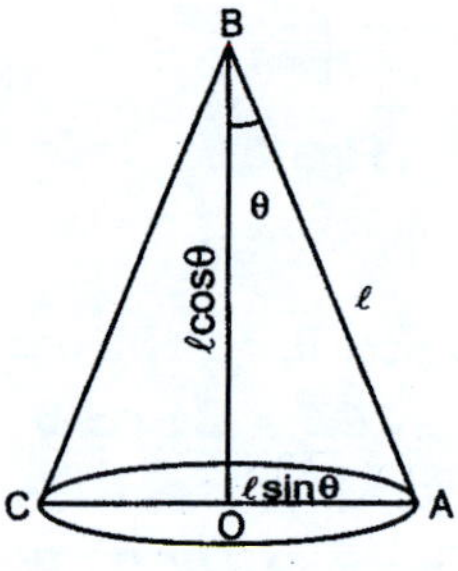

Fig 13.7

Height of cone $= l \cos \theta$
and radius of cone $= l \sin \theta$
Let V denote the volume of the cone,

then $V = \frac{1}{3}\pi (l \sin \theta)^2 l \cos \theta$ $\quad [\because \text{Volume of cone} = \frac{1}{3}\pi r^2 h]$

$$= \frac{1}{3}\pi l^3 (\sin^2 \theta \cos \theta)$$

$$\frac{dV}{d\theta} = \frac{1}{3}\pi l^3 [2 \sin \theta \cos^2 \theta - \sin^3 \theta]$$

For maxima or minima,

$$\frac{dV}{d\theta} = 0$$

$$\therefore \frac{1}{3}\pi l^3 (2 \sin \theta \cos^2 \theta - \sin^3 \theta) = 0$$

$\therefore\ \sin\theta\,(2\cos^2\theta - \sin^2\theta) = 0$

$\therefore$ Either $\sin\theta = 0$ *i.e.*, $\theta = 0$ which is not possible.

or, $\quad 2\cos^2\theta - \sin^2\theta = 0$

i.e., $\quad \tan^2\theta = 2$

i.e., $\quad \tan\theta = \sqrt{2}$ ($\because$ θ is acute)

$$\frac{d^2V}{d\theta^2} = \frac{1}{3}\pi l^3\,[2\cos^3\theta - 4\sin^2\theta\cos\theta - 3\sin^2\theta\cos\theta]$$

$$= \frac{1}{3}\pi l^3\,[2\cos^3\theta - 7\sin^2\theta\cos\theta]$$

$$= \frac{1}{3}\pi l^3\cos^3\theta\,[2 - 7\tan^2\theta]$$

which is –ve for $\tan\theta = \sqrt{2}$

$\therefore$ V is maximum when $\tan\theta = \sqrt{2}$ *i.e.*, when $\theta = \tan^{-1}\sqrt{2}$

Example 13. *Tangents are drawn to the ellipse* $\frac{x^2}{a^2} + \frac{y^2}{b^2} = 1$ *and the circle* $x^2 + y^2 = a^2$ *at the points where a common ordinate cuts them. Show that if* θ *be the greater inclination of these tangents then*

$$\tan\theta = \frac{a-b}{2\sqrt{ab}}.$$

Solution. Let the common ordinate cut the ellipse and circle at Q and P respectively. Let the eccentric angle of P be ϕ. Then,

$\angle POM = \phi$

$\therefore\quad Q \to (a\cos\phi,\ b\sin\phi)$

and $\quad P \to (a\cos\phi,\ a\sin\phi)$

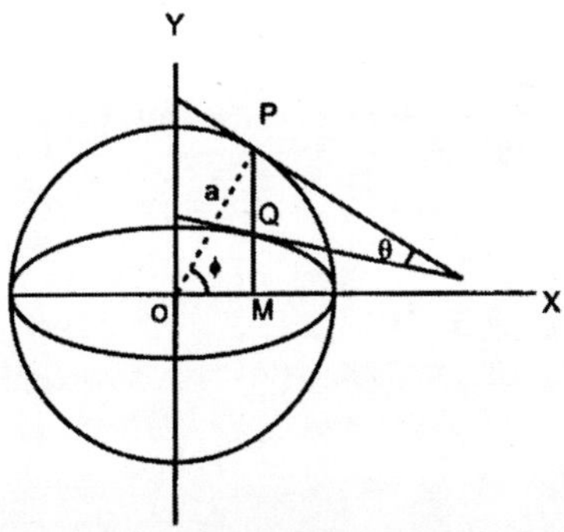

Fig. 13.8

Tangent to the given ellipse at Q is, $\frac{x}{a}\cos\phi+\frac{y}{b}\sin\phi=1$ gradient of this tangent $=-\frac{b}{a}\cot\phi=m_1$ (say)

Tangent to the given circle at P is, $x\cos\phi+y\sin\phi=a$

gradient of this tangent $=-\cot\phi=m_2$ (say)

Let θ be the angle between these tangents. Then we have

$$\tan\theta=\frac{-\frac{b}{a}\cot\phi+\cot\phi}{1+\frac{b}{a}\cot^2\phi}$$

$$=\frac{a-b}{a\tan\phi+b\cot\phi} \qquad ...(1)$$

Now θ will be maximum when $\tan\theta$ is maximum *i.e.*, when $a\tan\phi+b\cot\phi$ is minimum.

Let $$z=a\tan\phi+b\cot\phi$$

then $$\frac{dz}{d\phi}=a\sec^2\phi-b\operatorname{cosec}^2\phi$$

and $$\frac{d^2z}{d\phi^2}=2a\sec^2\phi\tan\phi+2b\operatorname{cosec}^2\phi\cot\phi$$

For z to be maximum or minimum, $\frac{dz}{d\phi}=0$

i.e., $a\sec^2\phi-b\operatorname{cosec}^2\phi=0$

i.e., $$\tan^2\phi=\frac{b}{a}$$

i.e., $$\tan\phi=\pm\sqrt{\frac{b}{a}}$$

When $\tan\phi=\sqrt{\frac{b}{a}}$, $\frac{d^2z}{d\phi^2}=+$ ve because for $\tan\phi$, all trigonometrical ratios are +ve.

When $\tan\phi=-\sqrt{\frac{b}{a}}$, $\frac{d^2z}{d\phi^2}=-$ ve.

Hence θ is maximum when

$$\tan\phi=+\sqrt{\frac{b}{a}}$$

Hence from (1), we have

$$\tan\theta = \frac{a-b}{+a\sqrt{\frac{b}{a}} + b\sqrt{\frac{a}{b}}}$$

$$= \frac{a-b}{+2\sqrt{ab}}$$

$$= \frac{-b+a}{2\sqrt{ab}}$$

Example 14. *A given quantity of metal is to be cast into a half cylinder with a rectangular base and semi circular ends. Show that in order that the total surface area may be minimum, the ratio of the length of the cylinder to the diameter of its semi-circular ends is* $\pi : \pi + 2$.

Solution. Let r be the radius of the half cylinder and l, its length.

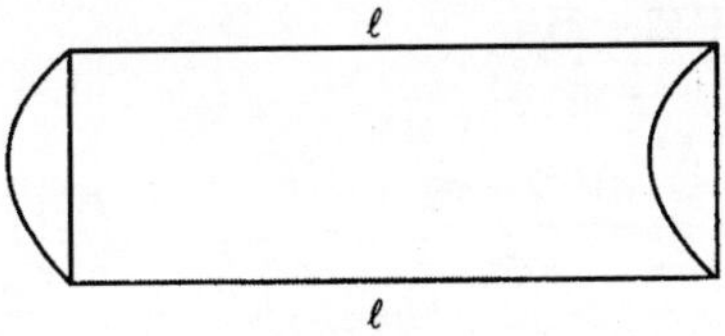

Fig 13.9

Area of Base (which is a rectangle)

$$= l \times 2r$$

$$= 2rl$$

Area of curved surface

$$= \frac{1}{2} \cdot 2\pi rl$$

$$= \pi rl$$

Area of two semi circular ends

$$= 2 \cdot \frac{1}{2}\pi r^2 = \pi r^2$$

Let S be the total surface area. Then,

$$S = 2rl + \pi rl + \pi r^2 \quad \text{...(1)}$$

Volume $$V = \frac{1}{2}\pi r^2 l \quad \text{...(2)}$$

It is required to find $\frac{l}{2r}$ when S is minimum.

$$S = \pi r^2 + (2+\pi)\, rl$$

$$= \pi r^2 + (2+\pi)\cdot r\,\frac{2V}{\pi r^2} \text{ from (2)}$$

$$= \pi r^2 + \frac{2V(2+\pi)}{\pi r}$$

$$\frac{dS}{dr} = 2\pi r - \frac{2V(2+\pi)}{\pi r^2}$$

For S to be maximum or minimum,

$$\frac{dS}{dr} = 0$$

$$\therefore \quad 2\pi r - \frac{2V(2+\pi)}{\pi r^2} = 0$$

$$\Rightarrow \quad \pi^2 r^3 = V(2+\pi)$$

$$\Rightarrow \quad r^3 = V\frac{V(2+\pi)}{\pi^2}$$

$$\frac{d^2s}{dr^2} = 2\pi + \frac{4V(2+\pi)}{\pi r^3}$$

$$= 2\pi + \frac{4V(2+\pi)}{\pi}\;\frac{\pi^2}{V(2+\pi)} \text{ when } r^3 = \frac{V(2+\pi)}{\pi^2}$$

$$= 2\pi + 4\pi$$

$$= 6\pi \text{ which is } +\text{ve}.$$

$\therefore$ S is minimum when

$$r^3 = \frac{V(2+\pi)}{\pi^2}$$

i.e., when $\pi^2 r^3 = \frac{1}{2}\pi r^2 l\,(2+\pi)$

i.e., when $2\pi r = l\,(2+\pi)$

i.e., when $\frac{l}{2r} = \frac{\pi}{\pi+2}$

Example 15. *A cone is circumscribed about a sphere of radius r. Show that when the volume of the cone is a minimum, its altitude is 4a and its semi vertical angle is* $\sin^{-1}\left(\frac{1}{3}\right)$.

Solution. Let A be the vertex of the cone and O, the centre of sphere.

Let θ be the semi vertical angle and

$$AO = x.$$

$$AP = AO + OP = x + a$$

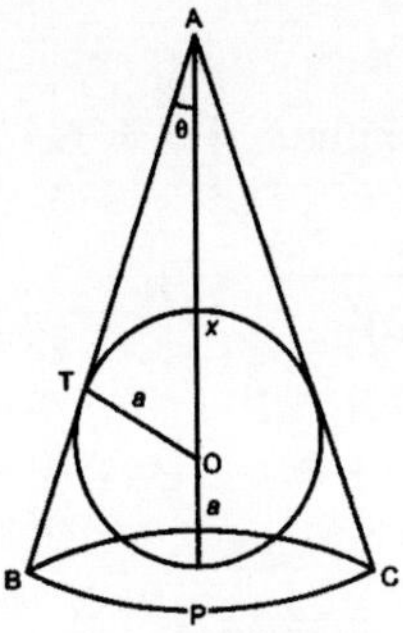

Fig. 13.10

From ΔATO,

$$\sin\theta = \frac{a}{x}$$

$$\tan\theta = \frac{a}{\sqrt{x^2 - a^2}}$$

Also from ΔABP,

$$\tan\theta = \frac{BP}{AP}$$

$$\therefore \quad BP = AP\tan\theta$$

$$= (x+a)\frac{a}{\sqrt{x^2 - a^2}}$$

If V denotes the volume of the cone, then

$$V = \frac{1}{3}\pi r^2 h$$

$$= \frac{1}{3}\pi\, BP^2 \cdot AP$$

$$= \frac{1}{3}\pi\frac{a^2(x+a)^2}{x^2 - a^2}(x+a)$$

$$= \frac{1}{3}\pi a^2\frac{(x+a)^2}{x-a}$$

$$\therefore \quad \frac{dV}{dx} = \frac{1}{3}\pi a^2\frac{(x-a)\cdot 2(x+a) - (x+a)^2\cdot 1}{(x-a)^2}$$

$$= \frac{1}{3}\pi a^2\frac{(x+a)[2x - 2a - x - a]}{(x-a)^2}$$

$$= \frac{1}{3}\pi a^2\frac{(x+a)(x-3a)}{(x-a)^2}$$

For V to be maximum or minimum, $\frac{dV}{dx} = 0$

$$\therefore \quad \frac{1}{3}\pi a^2 \frac{(x+a)(x-3a)}{(x-a)^2} = 0$$

$$\therefore \quad (x+a)(x-3a) = 0$$

$$\therefore \quad x = -a, 3a$$

But x cannot be –ve

$$\therefore \quad x = 3a$$

When x is slightly $< 3a$, $\frac{dV}{dx}$ is negative.

When x is slightly $> 3a$, $\frac{dV}{dx}$ is +ve.

$\therefore \frac{dV}{dx}$ changes origin from – ve to + ve at $x = 3a$

$\therefore V$ is minimum for $x = 3a$

Then altitude $= AP$

$= x + a$

$= 3a + a$

$= 4a$

and $\sin\theta = \frac{a}{x}$

$= \frac{a}{3a}$

$= \frac{1}{3}$

$$\therefore \quad \theta = \sin^{-1}\left(\frac{1}{3}\right)$$

EXERCISE 13

1. Find the maximum and minimum value of $x^3 - 2x^2 + x + 6$.
2. Examine the function $(x-2)^6 (x-3)^5$ for maxima and minima.
3. Find the maximum value of $(x-1)(x-2)(x-3)$.
4. Find the largest and smallest values of $3x^4 - 2x^3 - 6x^2 + 6x + 1$ in the interval $[0, 2]$.

5. If $y = a \log x + bx^2 + x$ has its extremum values at $x = -1$ and $x = 2$, then prove that $a = 2$ and $b = -\frac{1}{2}$.

6. Discuss the maxima and minima of the function $x + \sin 2x, 0 < x < 2\pi$.

7. Show that the maximum value of $\left(\frac{1}{x}\right)^x$ is $e^{1/e}$.

8. Find the maximum value of $\frac{\log_e x}{x}$ in the interval $0 < x < \infty$.

9. Prove that $\frac{x}{1 + x \tan x}$ is maximum when $x = \cos x$.

10. Investigate the maxima and minima of $\sin nx \sin^n x$, where n is a positive integer.

11. Prove that the minimum radius vector of the curve $\frac{a^2}{x^2} + \frac{b^2}{y^2} = 1$ is of length $a + b$.

12. Find the maximum and minimum radii vectors of the curve

$$\frac{c^4}{r^2} = \frac{a^2}{\sin^2 \theta} + \frac{b^2}{\cos^2 \theta}$$

13. The velocity of waves of wavelength λ on deep water is proportional to $\sqrt{\frac{\lambda}{a} + \frac{a}{\lambda}}$, where a is a certain linear magnitude. Prove that the velocity is minimum when $\lambda = a$.

14. A rectangular sheet of metal has four equal square portions removed at the corners, and the sides are then turned up so as to form an open rectangular box. Show that when the volume contained in the box is a maximum, the depth will be

$$\frac{1}{6}\ \{ (a + b) - (a^2 - ab + b^2)^{1/2} \}$$

15. The lower corner of a leaf in a book is folded over so as just to reach the inner edge of the page. Show that the fraction of the width folded over when the area of the folded part is minimum is $\frac{2}{3}$.

16. A window is in the form of a rectangle, surmounted by a semi-circle. If the perimeter be 30 meters, find the dimensions so that the greatest possible amount of light may be admitted.

17. An open rectangular tank, with a square base and vertical sides; is to be constructed of sheet metal to hold a given quantity of water. Show that the cost of the material will be least when the depth is half the width.

18. The strength of a beam varies as the product of its breadth and the square of its depth. Find the dimensions of the strongest beam which can be cut from a circular log of radius a.

19. Prove that the least perimeter of an isoscales triangle which can be circumscribed to a circle of radius r is $6\sqrt{3}r$.

20. Show that the right circular cylinder of given surface (including the ends) is such that its height is equal to the diameter of the base.

21. Show that the radius of the right circular cylinder of greatest curved surface which can be inscribed in a given cone is half that of the cone.

22. Show that the volume of the greatest cylinder which can be inscribed in a cone of height h and semi vertical angle α is

$$\frac{4}{27}\pi h^3 \tan^2 \alpha.$$

23. An open cylindrical can of given capacity is to be made from a metal sheet of uniform thickness. If no allowance is to be made for waste of material, what will be the most economical ratio of the radius the height of the can ?

24. A thin closed rectangular box is to have one edge n times the length of another edge, and the volume of the box is given to be v. Prove that the least surface S is given by

$$nS^3 = 54\ (n+1)^2 v^2.$$

25. A person being in a boat a km from the nearest point to the beach, wishes to reach as quickly as possible a point b km from that point along the shore. The ratio of his rate of walking to his rate of rowing is $\sec \alpha$. Prove that he should land at a distance $b - a \cot \alpha$ from the place to be reached.

26. A tree trunk l metre long is in the shape of a frustum of a cone; the radii of its ends being a and b $(a > b)$. It is required to cut from it a beam of uniform square section. Prove that the beam of greatest volume that can be cut is $\dfrac{al}{3(a-b)}$ metre long.

27. From a fixed point A on the circumference of a circle of radius a, the perpendicular AY is let fall on the tangent at P. Prove that the greatest area APY can have is $\dfrac{3\sqrt{3}}{8}a^2$.

28. A normal is drawn at a variable point P of the ellipse $\frac{x^2}{a^2}+\frac{y^2}{b^2}=1$; find the maximum distance of the normal from the centre of the ellipse.

29. A grocer requires cylindrical vessels of thin metal with lids, each to contain exactly a given volume V. Show that if he wishes to be as economical as possible in metal, the radius r of the base is given by $2\pi r^3 = V$. If, for other reasons, it is impracticable to use vessels in which the diameter exceeds three-fourth of the height, what should be the radius of the base of each vessel ?

30. The amount of the fuel consumed per hour by a certain steamer varies as the cube of its speed. When the speed is 15 m.p.h., the fuel consumed is $4\frac{1}{2}$ tons of coal per hour at Rs. 4 per ton. The other expenses total Rs. 100 per hour. Find the moist economical speed and the cost of a voyage of 1980 miles.

ANSWERS

1. $\frac{166}{27}, 9$

2. $x = 2$ gives maximum value $0, x=\frac{28}{11}$ gives minimum value $-\frac{5^5\,6^8}{11^{11}}$, $x = 3$ is a point of inflexion.

3. $\frac{2}{3\sqrt{3}}$

4. Largest value $= 21$,
Smallest value $= 1$

6. The function has a maxima at $x = \frac{\pi}{3}$ and maximum value is $\frac{\pi}{3}+\frac{\sqrt{3}}{2}$.

8. $\frac{1}{e}$.

10. Maxima or Minima at $x=\frac{k\pi}{n+1}$ according as k is odd or even, neither maxima nor minima at, $x = m\pi$; $m = 0, 1, 2, ...$

12. Maximum $= \frac{c^2}{a+b}$ when $\tan\theta = \pm\sqrt{\frac{a}{b}}$

16. Radius of the semicircle = Height of rectangle

$= \dfrac{30}{\pi + 4}$ metres.

18. Depth = $\sqrt{\dfrac{2}{3}}.\ 2a$

Breadth = $\sqrt{\dfrac{1}{3}}.\ 2a$

23. Height of the cylinder = Radius of the base.

28. $a - b$.

29. $\left(\dfrac{3V}{8\pi}\right)^{\frac{1}{3}}$.

30. $5\,(75)^{1/3}$ m.p.h., Rs. 3,960 $(45)^{1/3}$.

14
Maxima and Minima of Functions of Two Variables

14.1. Extreme Values (Maximum or Minimum)

Let $f(x, y)$ be a function of two independent variables x and y. Let us suppose that $f(x, y)$ is continuous and finite for all values of x and y in small neighbourhood of (a, b). Then $f(a, b)$ is said to be an extreme value of f at (a, b) if the difference

$$f(x, y) - f(a, b) \qquad ...(1)$$

keeps the same sign for every point (x, y) in a small neighbourhood of (a, b). The point (a, b) is called extremum point or critical point.

14.2. Maximum

The extreme value $f(a, b)$ is said to be a maximum value at (a, b) if the difference (1) is negative, *i.e.*, if

$$f(a + h, b + k) < f(a, b)$$

for all sufficiently small independent values of h and k.

14.3. Minimum

The extreme value $f(a, b)$ is said to be a minimum at (a, b) if the difference (1) is positive, *i.e.*, if

$$f(a + h, b + k) > f(a, b),$$

for all sufficiently small independent values of h and k.

14.4.Necessary Condition for the Existence of Maxima or Minima at a Point

Theorem : ***A necessary condition for $f(x, y)$ to have an extreme value (maximum or minimum) at (a, b) is that $f_x(a, b) = 0$ and $f_y(a, b) = 0$, provided these partial derivatives exist.***

Proof : To determine maxima or minima of a function $f(x, y)$ at a point (a, b), we investigate the sign of

$$f(a + h, b + k) - f(a, b) \qquad ...(1)$$

which has invariable sign for all sufficiently small and finite values of h and k, positive or negative.

By Taylor's theorem,

$$f(a + h, b + k) = f(a, b) + \left(h\frac{\partial f}{\partial x} + k\frac{\partial f}{\partial y}\right)_{\substack{x=a\\y=b}}$$

$$+ \frac{1}{2!}\left(h^2\frac{\partial^2 f}{\partial x^2} + 2hk\frac{\partial^2 f}{\partial x\,\partial y} + k^2\frac{\partial^2 f}{\partial y^2}\right)_{\begin{pmatrix}x=a\\y=b\end{pmatrix}} + \ldots \qquad \ldots(2)$$

Now, for small values of h and k the second and higher order terms are much smaller numerically than the first order terms, and may be disregarded when determining the sign of (1). Thus, the sign of (1) depends upon the first degree expression

$$\left(h\frac{\partial f}{\partial x} + k\frac{\partial f}{\partial y}\right)_{(a,\, b)}$$

But this expression change their sign when the sign of h and k are changed. Hence, in order that (1) may preserve a fixed sign it is necessary that

$$\left(h\frac{\partial f}{\partial x} + k\frac{\partial f}{\partial y}\right)_{\begin{pmatrix}x=a\\y=b\end{pmatrix}} = 0 \qquad \ldots(3)$$

Since h and k are independent and non-zero. Therefore, (3) implies that

$$\left(\frac{\partial f}{\partial x}\right)_{(a,\, b)} = f_x(a, 0) = 0$$

and

$$\left(\frac{\partial f}{\partial y}\right)_{(a,\, b)} = f_y(a, 0) = 0$$

Hence, the necessary conditions that $f(x, y)$ should have a maximum or minimum at (a, b) are that

$$f_x(a, b) = 0,\ f_y(a, b) = 0.$$

Note : The function $f(x, y) = |x| + |y|$ has an extreme value (minimum) at (0, 0) even though the partial derivatives f_x and f_y do not exist at (0, 0).

14.5. Sufficient Conditions for $f(x, y)$ to have an Extreme Value (Maxima & Minima) at (a, b) (Lagrange's Condition)

Let $f(x, y)$ possesses continuous second order partial derivatives in a certain neighbourhood of (a, b) and these derivatives at (a, b), *viz.*, $f_{xy}(a, b)$,

$f_{xx}(a, b)$, $f_{yy}(a, b)$ are not all zero. Then by Taylor's theorem, we have

$$f(a + h, b + k) = f(a, b) + \left(h \frac{\partial f}{\partial x} + k \frac{\partial f}{\partial y}\right)_{(a, b)}$$

$$+ \frac{1}{2!}\left(h^2 \frac{\partial^2 f}{\partial x^2} + 2hk \frac{\partial^2 f}{\partial x \partial y} + k^2 \frac{\partial^2 f}{\partial y^2}\right)_{(a, b)} + R_3$$

where, R_3 consists of terms of third and higher order in h and k.

But necessary conditions for maximum or minimum at (a, b) are satisfied, *i.e.*,

$$f_x(a, b) = 0 = f_y(a, b).$$

Therefore,

$$f(a + h, b + k) - f(a, b) = \frac{1}{2!}(rh^2 + 2hks + tk^2) + R_3 \qquad ...(1)$$

where, $r = f_{xx}(a, b)$, $s = f_{xy}(a, b)$ $t = f_{yy}(a, b)$

Now, by taking h and k sufficiently small, the second degree terms in R.H.S. of (1) may be made to govern the sign of right hand side and therefore, of the L.H.S. also.

Thus, we can write

$$rh^2 + 2hks + tk^2 = \frac{1}{r}[r^2h^2 + 2hkrs + rtk^2]$$

$$= \frac{1}{r}[r^2h^2 + 2hkrs + s^2k^2 + rtk^2 - s^2k^2]$$

$$= \frac{1}{r}[(rh + sk)^2 + k^2(rt - s^2)] \qquad ...(2)$$

The first term inside the brackets is positive. The second will also be positive if $rt - s^2 > 0$. Thus, the expression (2) will have the same sign as r, for all values of h and k. This sign is determined by the sign of r. Now, we consider following cases.

CASE I. If $rt - s^2 > 0$ and $r > 0$, then

$$f(a + h, b + k) - f(a, b) > 0,$$

for all small values of h and k. Hence, $f(x, y)$ has a maxima at (a, b).

CASE II. If $rt - s^2 > 0$ and $r < 0$, then

$$f(a + h, b + k) - f(a, b) = 0,$$

Hence, $f(x, y)$ has a minima at (a, b).

CASE III. If $rt - s^2 = 0$, then further investigation is needed to determine whether $f(x, y)$ is a maximum or minimum at $x = a,\ y = b$ or not.

CASE IV. Saddle Points : If $rt - s^2 < 0$, then $f(x, y)$ has neither a maxima nor minima at (a, b). In this case $f(a+h, b+k) - f(a, b)$ is not of invariable sign, *i.e.*, it has one sign for some values of h, k while it has another sign for other values of h, k. Such point is called a saddle point.

14.6. Definitions : Stationary Points and Stationary Values

A point (a, b) is called a stationary point if first order partial derivatives of the function $f(x, y)$ vanish at that point. Thus, if $f(x, y)$ is a function of two independent variables, then

$$df = \frac{\partial f}{\partial x}\,dx + \frac{\partial f}{\partial y}\,dy.$$

Now at the stationary points, $\frac{\partial f}{\partial x} = 0 = \frac{\partial f}{\partial y}$. Therefore,

$$df = 0.$$

Hence, stationary points can be obtained by solving following equations simultaneously :

$$\left(\frac{\partial f}{\partial x}\right) = 0$$

and

$$\left(\frac{\partial f}{\partial y}\right) = 0$$

the value of the function obtained at stationary points are called stationary values.

Note : Extreme points, the points of "ridge" of maximum (or minimum) saddle point are classified as stationary points.

14.7. Ridge of Maximum (or Minimum)

If the surface falls (or rises) in all directions except that of the ridge where it remains stationary is called 'ridge' of maximum (or minimum). Thus, we can explain it geometrically as,

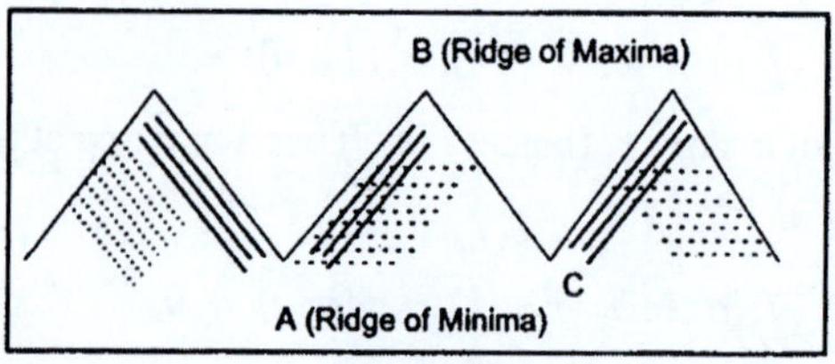

Fig. 14.1

14.8. Working Rule for Determining the Maxima and Minima of Function of Two Variables

(1) Find $\frac{\partial f}{\partial x}$ and $\frac{\partial f}{\partial y}$, and equate them to zero.

(2) Solve these simultaneous equations $\frac{\partial f}{\partial x} = 0$, $\frac{\partial f}{\partial y} = 0$ for x and y.

Then pairs of values of x and y, thus obtained will give stationary values of $f(x, y)$. Let (a, b) be one of the pair of roots.

(3) Find $\frac{\partial^2 f}{\partial x^2}, \frac{\partial^2 f}{\partial x \, \partial y}, \frac{\partial^2 f}{\partial y^2}$ and substitute the point (a, b) in these. Calculate $rt - s^2$ for the point (a, b).

(4) (i) If $(rt - s^2) > 0$ and $r < 0$, then $f(x, y)$ has a maximum at (a, b) and this point is called point of maxima.

(ii) If $(rt - s^2) > 0$ and $r > 0$, then $f(x, y)$ has minimum at (a, b) and this point is called point of minima.

(iii) If $rt - s^2 < 0$, then $f(x, y)$ has neither maximum nor minimum at (a, b) and therefore function has saddle point there.

(iv) If $rt - s^2 = 0$, the case is undecided and further investigation is necessary to decide it.

ILLUSTRATIVE EXAMPLES

Example 1. *Find the maximum or minimum values of the function*

$$x^3 y^2 (1 - x - y).$$

Solution : Let $u = x^3 y^2 (1 - x - y)$.

For maxima or minima, we have

$$\frac{\partial u}{\partial x} = 3x^2 y^2 (1 - x - y) - x^3 y^2 = 0 \quad \text{...(i)}$$

$$\frac{\partial u}{\partial y} = 2x^3 y (1 - x - y) - x^3 y^2 = 0. \quad \text{...(ii)}$$

Subtracting equation (ii) from (i), we get

$$x^2 y (1 - x - y)(3y - 2x) = 0$$

$$\therefore \quad y = \frac{2}{3} x$$

Putting the value of y in equation (i), we get

$$3x^2 \frac{4}{9}x^2\left(1 - x - \frac{2}{3}x\right) - x^3\frac{4}{9}x^2 = 0$$

or
$$\frac{4}{9}x^4\ [3 - 5x - x] = 0$$

or
$$\frac{4}{9}x^4(3 - 6x) = 0$$

$\therefore$
$$x = \frac{1}{2}$$

and putting the value of x in $y = \frac{2}{3}x$

$\therefore$
$$y = \frac{1}{3}$$

so there can be a maxima or minima at $\left(\frac{1}{2}, \frac{1}{3}\right)$.

Now,
$$r = \frac{\partial^2 u}{\partial x^2} = 6xy^2 - 12x^2y^2 - 6xy^3$$

The value of r at the point $\left(\frac{1}{2}, \frac{1}{3}\right)$

$$= -\frac{1}{9}$$

$$t = \frac{\partial^2 u}{\partial y^2} = 2x^3 - 2x^4 - 6x^2 y$$

The value of t at the point $\left(\frac{1}{2}, \frac{1}{3}\right)$

$$= -\frac{1}{8}$$

and
$$s = \frac{\partial^2 u}{\partial x\,\partial y} = -6x^2y - 8x^3y - 9x^2y^2$$

The value of s at the point $\left(\frac{1}{2}, \frac{1}{3}\right)$

$$= -\frac{1}{12}$$

so that
$$rt - s^2 = \left(-\frac{1}{9}\right)\left(-\frac{1}{8}\right) - \left(\frac{1}{12}\right)^2$$

$$= \text{positive.}$$

Also r is negative. Hence the function has a maximum at $x = \frac{1}{2}, y = \frac{1}{2}$.

The maximum value of

$$u = \left(\frac{1}{2}\right)^3 \left(\frac{1}{3}\right)^2 \left(1 - \frac{1}{2} - \frac{1}{3}\right)$$

$$= \frac{1}{432}.$$

Example 2. *Find maxima and minima of the following function :*

$$\sin x + \sin y + \sin (x + y)$$

Solution : Let $\quad u = \sin x + \sin y + \sin (x + y).$

For maximum or minimum of u, we have

$$\frac{\partial u}{\partial x} = \cos x + \cos (x + y) = 0 \qquad \text{...(i)}$$

and $$\frac{\partial u}{\partial y} = \cos y + \cos (x + y) = 0 \qquad \text{...(ii)}$$

From equation (i) and (ii), it is clear that

$$\cos x = \cos y$$

$\therefore$ $$x = y$$

From equation (i),

$$\cos x + \cos (x + x) = 0$$

or $$\cos x + \cos 2x = 0$$

or $$\cos x + 2 \cos^2 x - 1 = 0$$

or $$2 \cos^2 x + \cos x - 1 = 0$$

or $$2 \cos^2 x + 2 \cos x - \cos x - 1 = 0$$

or $$2 \cos x (\cos x + 1) - 1 (\cos x + 1) = 0$$

or $$(2 \cos x - 1) (\cos x + 1) = 0.$$

When $$2 \cos x - 1 = 0,$$

then $$\cos x = \frac{1}{2} \cos \pi/3$$

$\therefore$ $$x = 2 n\pi \pm \pi/3.$$

When $$\cos x + 1 = 0,$$

then $$\cos x = -1 = \cos \pi$$

$\therefore$ $$x = 2 n\pi \pm \pi.$$

Now to discuss maxima or minima, we have

$$r = \frac{\partial^2 u}{\partial x^2} = -\sin x - \sin(x+y)$$

$$s = \frac{\partial^2 u}{\partial x\,\partial y} = -\sin(x+y)$$

and
$$t = \frac{\partial^2 u}{\partial y^2} = -\sin y - \sin(x+y)$$

When $x = \pi/3 = y$

$$r = -\sin\frac{\pi}{3} - \sin\left(\frac{2\pi}{3}\right)$$

$$= -\frac{\sqrt{3}}{2} - \frac{\sqrt{3}}{2} = -\sqrt{3},$$

$$s = -\sin\left(\frac{\pi}{3} + \frac{\pi}{3}\right) = -\frac{\sqrt{3}}{2},$$

and
$$t = -\sin\frac{\pi}{3} - \sin\left(\frac{2\pi}{3}\right) = -\sqrt{3}$$

so that
$$rt - s^2 = (-\sqrt{3})\,(-\sqrt{3}) - \left(-\frac{\sqrt{3}}{2}\right)^2$$

$$= 3 - \frac{3}{4} = \frac{9}{4} = +\text{ve}$$

Since $rt - s^2$ is +ve and r and t are both −ve, hence we have a maximum at

$$x = y = 2n\pi \pm \pi/3$$

When $x = \pi = y$

$$r = -\sin\pi - \sin(\pi + \pi) = 0$$

$$s = 0$$

$$t = 0$$

so that $rt - s^2 = 0$

Hence, the function has neither a maximum nor a minimum at

$$x = y = 2n\pi \pm \pi/3.$$

Example 3. *Discuss the maximum or minimum values of the function*

$$u = xy + \frac{a^3}{x} + \frac{a^3}{y}.$$

Solution : We have

$$u = xy + \frac{a^3}{x} + \frac{a^3}{y}.$$

For maximum or minimum, we have

$$\frac{\partial u}{\partial x} = y - \frac{a^3}{x^2} = 0 \qquad \text{or} \qquad a^3 = x^2 y$$

$$\frac{\partial u}{\partial y} = x - \frac{a^3}{y^2} = 0 \qquad \text{or} \qquad a^3 = x y^2$$

so that $\qquad a^3 = x^2 y = x y^2$

or $\qquad x = y = a.$

Thus, there can be a maximum or minimum at $x = y = a$.

Now, $\qquad r = \dfrac{\partial^2 u}{\partial x^2} = \dfrac{2a^3}{x^3} = 2$ at (a, a)

$$s = \frac{\partial^2 u}{\partial x \, \partial y} = 1$$

and $\qquad t = \dfrac{\partial^2 u}{\partial y^2} = \dfrac{2a^3}{y^3} = 2$ at (a, a)

$\therefore \qquad rt - s^2 = (2)(2) - (1)^2$

$= 3$ which is positive.

Also, $\qquad r = 2$ which is positive.

Hence, there exists a minimum at $x = a,\ y = a$.

Putting $x = a,\ y = a$ in u, we see that

$$u_{min.} = a^2 + a^2 + a^2 = 3a^2.$$

Example 4. *Find the maximum value of u where*

$$u = \sin x \sin y \sin (x + y).$$

Solution : We have

$$u = \sin x \sin y \sin (x + y).$$

For maximum or minimum of u, we have

$$\frac{\partial u}{\partial x} = \sin y \ [\cos x \sin (x + y) + \sin x \cos (x + y)] = 0$$

which gives $\qquad \tan (x + y) = -\tan x \qquad \text{... (i)}$

and $\qquad \dfrac{\partial u}{\partial y} = \sin x \, [\cos y \sin (x + y) + \sin y \cos (x + y)] = 0$

which gives $\tan(x + y) = -\tan y$ ...(ii)

From equation (i) and (ii), we see that

$$\tan x = \tan y$$

or $$x = y$$

∴ From equation (i) and (ii),

$$\tan 2x = -\tan x = \tan(\pi - x)$$

or $$2x = \pi - x$$

or $$3x = \pi$$

or $$x = \pi/3 = y.$$

Moreover,

$$\frac{\partial u}{\partial x} = 0 \text{ will also give}$$

$$\sin y = 0 \quad \Rightarrow \quad y = 0$$

and $$\frac{\partial u}{\partial y} = 0 \text{ will also give}$$

$$\sin x = 0 \quad \Rightarrow \quad x = 0$$

We see that when $x = 0, y = 0$; $u = 0$. These values of x and y clearly correspond to a minimum value of u which is not required in the problem.

Now in order to discuss maximum or minimum at $x = \pi/3 = y$, we have

$$\frac{\partial u}{\partial x} = \sin y \sin(2x + y)$$

and $$\frac{\partial u}{\partial y} = \sin x \sin(x + 2y)$$

∴ $$r = \frac{\partial^2 u}{\partial x \partial y} = 2 \sin y \cos(2x + y)$$

$$= 2 \sin \pi/3 \cos \pi \text{ at } x = y = \pi/3$$

$$= -\sqrt{3}$$

$$s = \frac{\partial^2 u}{\partial x \partial y} = \sin y \cos(2x + y) + \cos y \sin(2x + y)$$

$$= \sin(2x + 2y)$$

$$= \sin \frac{4\pi}{3} \text{ at } x = y = \pi/3$$

$$= -\frac{\sqrt{3}}{2}$$

and $$t = \frac{\partial^2 u}{\partial y^2} = 2 \sin x \cos(x + 2y)$$

$$= 2 \sin \pi/3 \cos \pi$$
$$= -\sqrt{3}$$

so that $$rt - s^2 = 3 - \frac{3}{4} = \frac{9}{4} = \text{positive}$$

and $$r = -\sqrt{3} = \text{negative.}$$

Hence, there is a maximum at $x = y = \pi/3$.

Putting $x = y = \pi/3$ in u,

$$u_{max.} = \sin\frac{\pi}{2} \sin\frac{\pi}{3} \sin\frac{2\pi}{3}$$
$$= \frac{3\sqrt{3}}{8}.$$

Example 5. *Find a point within a triangle such that the sum of the square of its distances from the three angular points is a minimum.*

Solution : Let the vertices of the triangle be (x_1, y_1), (x_2, y_2) and (x_3, y_3). Let (x, y) be a point inside the triangle.

Let u denote the sum of the squares of the distances of (x, y) from three vertices, then

$$u = [(x-x_1)^2 + (y-y_1)^2] + [(x-x_2)^2 + (y-y_2)^2] + [(x-x_3)^2 + (y-y_3)^2].$$

For maximum or minimum of u, we have

$$\frac{\partial u}{\partial x} = 2(x-x_1) + 2(x-x_2) + 2(x-x_3) = 0$$

or $$3x = x_1 + x_2 + x_3$$

or $$x = \frac{x_1 + x_2 + x_3}{3}$$

and $$\frac{\partial u}{\partial y} = 2(y-y_1) + 2(y-y_2) + 2(y-y_3) = 0$$

or $$3y = y_1 + y_2 + y_3$$

or $$y = \frac{y_1 + y_2 + y_3}{3}$$

Thus, the critical point is

$$\left(\frac{x_1 + x_2 + x_3}{3}, \frac{y_1 + y_2 + y_3}{3}\right)$$

which is the centroid of the triangle.

Again, $$r = \frac{\partial^2 u}{\partial x^2} = 6; \; s = \frac{\partial^2 u}{\partial x \, \partial y} = 0; \; t = \frac{\partial^2 u}{\partial y^2} = 6$$

so that, $$rt - s^2 = 36 = +\text{ve}$$

Also, $r = 6$ which is +ve.

Hence, u is minimum.

Example 6. *Find the maximum and minimum of the function*

$$u = x^3 + y^3 - 63\,(x+y) + 12xy.$$

Solution : $$u = x^3 + y^3 - 63\,(x+y) + 12xy \qquad \text{...(i)}$$

$$\frac{\partial u}{\partial x} = 3x^2 - 63 + 12y \text{ and } \frac{\partial u}{\partial y} = 3y^2 - 63 + 12x$$

For maximum and minimum $\frac{\partial u}{\partial x} = 0$ and $\frac{\partial u}{\partial y} = 0$

$$3x^2 - 63 + 12y = 0 \;\Rightarrow\; x^2 + 4y = 21 \qquad \text{...(ii)}$$

and $$3y^2 - 63 + 12x = 0 \;\Rightarrow\; y^2 + 4x = 21 \qquad \text{...(iii)}$$

Subtracting equation (ii) and (iii),

$$x^2 - y^2 + 4\,(y - x) = 0$$

$$(x - y)\,(x + y - 4) = 0$$

$$x = y \text{ or } x + y = 4$$

Putting them in equation (ii)

$$y^2 + 4y - 21 = 0$$

$$(y + 7)\,(y - 3) = 0$$

$$y = -7 \text{ or } 3.$$

But $$x = y \Rightarrow x = -7 \text{ or } 3.$$

The maximum or minimum is possible at $(-7, -7)$ and $(3, 3)$.

Now, if $x + y = 4 \Rightarrow x = 4 - y$, using in (ii)

$$y^2 - 8y + 16 + 4y - 21 = 0$$

$$y^2 - 4y - 5 = 0$$

$$(y - 5)\,(y + 1) = 0$$

$$y = 5 \text{ or } -1.$$

If $y = 5$ then $\quad x = 4 - y = 4 - 5 = -1$

$y = -1$ then $\quad x = 4 + 1 = 5.$

The maximum or minimum is possible at $(-1, 5)$ and $(5, -1)$.

The maximum or minimum of the function is possible at $(-7, -7)$, $(3, 3)$, $(-1, 5)$ and $(5, -1)$.

$$r = \frac{\partial^2 u}{\partial x^2} = 6x, \; s = \frac{\partial^2 u}{\partial x \, \partial y} = 12, \; t = \frac{\partial^2 u}{\partial y^2} = 6y$$

At $(-7, -7)$, $r = -42 < 0$, $s = 12$, $t = -42 < 0$.

Now, $rt - s^2 = (-42)(-42) - 144 > 0$, therefore the function has maxima at $(-7, -7)$.

At $(3, 3)$, $r = 18 > 0$, $s = 12$ and $t = 18 > 0$.

Now, $\quad rt - s^2 = 18 \times 18 - 144 > 0$

Therefore, function has minima at $(3, 3)$.

At $(-1, 5)$, $r = -6 < 0$, $s = 12 > 0$ and $t = 30 > 0$

$$rt - s^2 = (-6)(30) - 144 < 0$$

$$rt - s^2 < 0.$$

$\Rightarrow$ There is neither maxima nor minima at $(-1, 5)$.

Similarly, we can see that there is neither maximum nor minimum of function at $(5, -1)$.

Example 7. *Find the maxima and minima value of* $u = \sin x \sin y \sin z$, *where x, y and z are the angle vertex of a triangle.*

Solution : $\quad u = \sin x \sin y \sin z \qquad$ ...(i)

where, $\quad z = \pi - (x + y).$

$$u = \sin x \sin y \sin (\pi - \overline{x + y})$$

$$= \sin x \sin y \sin (x + y)$$

$$\frac{\partial u}{\partial x} = \cos x \sin y \sin (x + y) + \sin x \sin y \cos (x + y)$$

$$= \sin y \sin (2xh + y) = 0 \qquad \text{...(ii)}$$

Similarly,

$$\frac{\partial u}{\partial y} = \sin x \sin (2y + y) = 0 \qquad \text{...(iii)}$$

From equation (ii),

$$\sin y = 0 \text{ or } \sin (y + 2x) = 0$$

$$y = 0$$

or $$\sin(x + x + y) = 0$$

$$\sin x \cos(x + y) + \cos x \sin(x + y) = 0$$

$$\tan(x + y) = -\tan x$$

$$\tan(x + y) = \tan(\pi - x) \quad \text{...(iv)}$$

$$x + y = \pi - x$$

$$2x + y = \pi \quad \text{...(v)}$$

Similarly, from equation (iii),

$$y = 0 \text{ or } \tan(x + y) = -\tan y \quad \text{...(vi)}$$

By equation (iv) and (vi),

$$\tan x = \tan y \quad \Rightarrow \quad x = y$$

By equation (v), $3y = \pi \quad \Rightarrow \quad y = \frac{\pi}{3}$

Similarly, $x = \frac{\pi}{3}$

There can be maxima or minima at $\left(\frac{\pi}{3}, \frac{\pi}{3}\right)$ and $(0, 0)$.

At (0, 0), $u = 0$

Therefore, for $\left(\frac{\pi}{3}, \frac{\pi}{3}\right)$

$$r = \frac{\partial^2 u}{\partial x^2} = 2 \sin y \cos(2x + y)$$

$$= 2 \sin \frac{\pi}{3} \cos\left(\frac{2\pi}{3} + \frac{\pi}{3}\right) = -\sqrt{3} < 0$$

$$s = \frac{\partial^2 u}{\partial x \partial y} = \sin(2x + 2y) = \sin\left(\frac{2\pi}{3} + \frac{2\pi}{3}\right)$$

$$= \sin\left(\frac{4\pi}{3}\right) = -\frac{\sqrt{3}}{2} < 0$$

$$t = \frac{\partial^2 u}{\partial y^2} = 2 \sin x \cos(x + 2y)$$

$$= 2 \sin \frac{\pi}{3} \cos \pi = -\sqrt{3} < 0$$

Now, $$rt - s^2 = (-\sqrt{3})(-\sqrt{3}) - \left(-\frac{\sqrt{3}}{2}\right)^2 = \frac{3}{2} > 0$$

Thus, $$rt - s^2 > 0 \text{ and } r < 0.$$

There will be maxima of the function at $\left(\frac{\pi}{3}, \frac{\pi}{3}, \frac{\pi}{3}\right)$.

Example 8. *Find a point inside a triangle such that the sum of square of the distance of it from vertices is minimum.*

Solution : Let $A(x_1, y_1, z_1)$; $B(x_2, y_2, z_2)$ and $C(x_3, y_3, z_3)$ be the vertices of a triangle ABC.

Let $P(x, y, z)$ be any other point inside the triangle.

$$u = PA^2 + PB^2 + PC^2$$

$$= (x-x_1)^2 + (y-y_1)^2 + (z-z_1)^2 + (x-x_2)^2 + (y-y_2)^2 + (z-z_2)^2 + (x-x_3)^2 + (y-y_3)^2 + (z-z_3)^2$$

$$\frac{\partial u}{\partial x} = 2[x - x_1 + y - y_1 + z - z_1] = 0 \Rightarrow x = \frac{x_1 + x_2 + x_3}{3}$$

$$\frac{\partial u}{\partial y} = 2[y - y_1 + y - y_2 + y - y_3] = 0 \Rightarrow y = \frac{y_1 + y_2 + y_3}{3}$$

$$\frac{\partial u}{\partial z} = 2[z - z_1 + z - z_2 + z - z_3] = 0 \Rightarrow z = \frac{z_1 + z_2 + z_3}{3}$$

$$P = \left(\frac{x_1 + x_2 + x_3}{3}, \frac{y_1 + y_2 + y_3}{3}, \frac{z_1 + z_2 + z_3}{3}\right)$$

For maxima and minima,

$$r = \frac{\partial^2 u}{\partial x^2} = 6, \quad s = \frac{\partial^2 u}{\partial x\, \partial y} = 0 \quad \text{and } t = \frac{\partial^2 u}{\partial y^2} = 6$$

$$rt - s^2 = 36 - 0 = 36 > 0.$$

r and $rt - s^2$ are both positive. There will be minima at the

$$\left(\frac{x_1 + x_2 + x_3}{3}, \frac{y_1 + y_2 + y_3}{3}, \frac{z_1 + z_2 + z_3}{3}\right)$$

which is centroid of the triangle.

Example 9. *A rectangular box is placed on x-y plane. The one end of the box is at the origin. If the vertex opposite to origin be on the plane* $6x + 4y + 3z = 24$, *then find the maximum value of this box.*

Solution : The one vertex of the box is on the plane

$$6x + 4y + 3z = 24 \qquad \text{...(i)}$$

Let v be the volume of the box then

$$v = xyz \qquad \text{...(ii)}$$

By equation (i), $z = \frac{1}{3}[24 - 4y - 6x]$, we get

$$v = \frac{1}{3}xy\ [24 - 4y - 6x]$$

$$= 8xy - 2x^2y - \frac{4}{3}xy^2$$

For maxima and minima, value of v,

$$\frac{\partial v}{\partial x} = 8y - 4xy - \frac{4}{3}y^2 = \frac{4y^2}{3}[6 - 3x - y] = 0$$

$$y[6 - 3x - y] = 0. \qquad \text{...(iii)}$$

Similarly, $$\frac{\partial x}{\partial y} = 8x - 2x^2 - \frac{8}{3}xy = \frac{2x}{3}[12 - 3x^2 - 4y] = 0$$

$$x[12 - 3x - 4y] = 0. \qquad \text{...(iv)}$$

Solving equation (iii) and (iv),

$$x = 0,\ y = 0;\ x = \frac{4}{3},\ y = 2.$$

Point (0, 0) can not have maxima or minima because $v = 0$, at this point.

Now, when $x = \frac{4}{3}$ and $y = 2$, we get

$$r = \frac{\partial^2 u}{\partial x^2} = -4y = -8 < 0,\ s = \frac{\partial^2 u}{\partial x\, \partial y} = 8 - 4x - \frac{8y}{3} = -\frac{8}{3} < 0.$$

$$t = \frac{\partial^2 y}{\partial y^2} = \frac{-32}{3} < 0.$$

$$rt - s^2 = (-8)\left(-\frac{8}{3}\right) - \left(\frac{32}{3}\right)^2 > 0$$

where, $y = 2$ and $x = \frac{4}{3}$ then value of v will be maximum.

Maximum value of $v = \frac{1}{3}\frac{4}{3}(2)(24 - 8 - 8) = \frac{64}{9}$ cube unit.

Example 10. *Find the maximum and minimum value of the following function :*

(i) $f(x, y) = xy(a - x - y)$

(ii) $f(x, y) = x^4 + x^2y + y^2$

Solution : For the maximum and minimum value of the function,

$$\frac{\partial f}{\partial x} = ay - 2xy - y^2 = 0 \qquad ...(i)$$

and
$$\frac{\partial f}{\partial y} = ax - x^2 - 2xy = 0 \qquad ...(ii)$$

Subtracting equation (ii) from (i), we get

$$a(y - x)(y^2 - x^2) = 0$$

or
$$(y - x)(a - y - x) = 0$$

$$\therefore \quad y = x \text{ or } y + x = a.$$

Substituting $y = x$ in equation (i), we have

$$ya - 2y^2 - y^2 = 0$$

or
$$3y^2 - ya = 0$$

or
$$y = 0,\ y = a/3.$$

When $y = a - x$, then from equation (ii),

$$xa - x^2 - 2x(a - x) = 0$$

or
$$ax - x^2 - 2ax + 2x^2 = 0$$

or
$$x^2 - ax = 0$$

$$\therefore \quad x = 0,\ x = a.$$

$\therefore$ f can be maximum or minimum at the following points :

$$(0, 0);\ (a/3,\ a/3);\ (0, a);\ (a, 0).$$

$$\therefore \quad r = \frac{\partial^2 f}{\partial x^2} = -2y = 0,\ -\frac{2a}{3},\ -2a, 0$$

$$s = \frac{\partial^2 f}{\partial x\, \partial y} = a - 2x - 2y = a, -\frac{a}{2},\ -a,\ -a$$

and
$$t = \frac{\partial^2 f}{\partial y^2} = -2x = 0,\ -\frac{2a}{3},\ 0,\ -2a$$

$$\therefore \quad rt - s^2 = -\text{ve},\ +\text{ve},\ -\text{ve},\ -\text{ve}$$

Hence, f is maximum at $\left(\frac{a}{3}, \frac{a}{3}\right)$.

$\therefore$ The maximum value of $f = \frac{a^2}{9}\left(a - \frac{a}{3} - \frac{a}{3}\right) = \frac{a^3}{27}.$

(ii) The given function is

$$f(x, y) = x^4 + x^2y + y^2 \qquad ...(i)$$

For the maximum or minimum values,

$$\frac{\partial f}{\partial x} = 4x^3 + 2xy = 0 \qquad \text{...(ii)}$$

and
$$\frac{\partial f}{\partial y} = x^2 + 2y = 0 \qquad \text{...(iii)}$$

Solving the equation (ii) and (iii), we get

$$x = 0, \quad y = 0$$

$\therefore$
$$\frac{\partial^2 f}{\partial x^2} = 12x^2 + 2y,$$

$$\frac{\partial^2 f}{\partial x \, \partial y} = 2x,$$

and
$$\frac{\partial^2 f}{\partial y^2} = 2.$$

$\therefore$ At (0, 0)

$$r = \frac{\partial^2 f}{\partial x^2} = 0, \ s = \frac{\partial^2 f}{\partial x \, \partial y} = 0, \ t = \frac{\partial^2 f}{\partial y^2} = 2$$

Hence, $rt - s^2 = 0$

$\therefore f(0, 0) = 0$ also with for $x \neq 0, \ y \neq 0$

$$f(x, y) = x^4 + x^2 y + y^2$$
$$= \left(x^2 + \frac{1}{2}y\right)^2 + \frac{3}{4}y^2$$

which is also positive.

Hence, from the definition of maximum and minimum for the given function is minimum at (0, 0) and the minimum value of f is $= f(0, 0) = 0$.

EXERCISE 14 (A)

Discuss the maximum or minimum of u in the following cases :

1. $u = x^3 + y^3 - 3axy.$
2. $u = x^2 y^2 - 5x^2 - 8xy - 5y^2.$
3. $u = ax^3 y^2 - x^4 y^2 - x^3 y^3.$
4. $u = x^4 + 2x^2 y - x^2 + 3y^2.$
5. $u = 2(x - y)^2 - x^4 - y^4.$

6. $u = x^2 + y^2 + (ax + by + c)^2$.

7. $u = xy^2 (3x + 6y - 2)$.

8. $u = y^2 + 4xy + 3x^2 + x^3$.

9. $u = 2a^2xy - 3ax^2y - ay^3 + x^3y + xy^3$.

10. $u = x^2 + y^2 + 6x + 12$.

11. $u = x^2 + y^2 + 2/x + 2/y$.

12. $u = 2 \sin \dfrac{x + y}{2} \cos \dfrac{x - y}{2} + \cos (x + y)$.

13. $u = \dfrac{x + y}{x^2 + 2y^2 + 6}$.

14. $u = \dfrac{x + y - 1}{x^2 + y^2}$.

15. $u = (x - 1)(y - 1)(x^2 + y^2 - 4)$.

16. In a plane triangle ABC, find the maximum value of $\cos A \cos B \cos C$.

17. Find a point within a triangle such that the sum of the distances from the angular points may be a minimum.

18. $ABCD$ is a quadrilateral having no re-entrant angle, and P is a point in its plane. Find the position of P for which the sum of the distances from the vertices is a minimum.

ANSWERS

1. $x = y = a$ gives a maximum or minimum according as

$$a < 0 \text{ or } > 0.$$

2. $x = y = 0$ gives a maximum.

3. $x = a/2,\ y = a/3$ give a maximum.

4. $x = \pm \dfrac{\sqrt{3}}{2},\ y = -\dfrac{1}{4}$ give a minimum.

5. Neither maximum nor minimum exists at (0, 0);

$$x = \pm \sqrt{2},\ y = \pm \sqrt{2} \text{ give maximums.}$$

6. $\dfrac{x}{a} = \dfrac{y}{b} = -\dfrac{c}{a^2 + b^2 + 1}$ give a minimum.

7. Minimum at $\left(\frac{1}{6}, \frac{1}{6}\right)$.

8. Minimum at $\left(\frac{2}{3}, -\frac{4}{3}\right)$.

9. Maximum at $(a/2,\ a/2)$ and $(3\,a/2,\ -\,a/2)$;
Minimum at $(a/2,\ -a/2)$ and $(3\,a/2,\ a/2)$.

10. Minimum at $(-3, 0)$.

11. Minimum at $(1, 1)$.

12. When $x = \pi/2 = y \rightarrow$ neither a maxima nor a minima;
When $x = y = n\,\pi + (-\,1)\,\pi/6$, maxima;
When $x = y = 3\,\pi/2$, minima.

13. Maxima at $(2, 1)$, Minima at $(-\,2, -\,1)$.

14. Maximum at $(1, 1)$.

15. Minima at $x = y = \dfrac{1 \pm \sqrt{17}}{4}$;
Maximum at $\left(-\frac{1}{2}, \frac{3}{2}\right)$ and $\left(\frac{3}{2}, -\frac{1}{2}\right)$.

16. Maximum value is $\frac{1}{8}$ at $A = B = C = \pi/3$.

17. The point is such that each side subtends an angle of 120° at that point.

18. P is the point of intersection of the diagonals of the quadrilateral.

14.9. Lagrange's Method of Undetermined Multipliers

In this method we shall discuss the determination of stationary points from a modified point of view.

Let
$$u = \phi\,(x_1,\ x_2,\ \ldots..,\ x_n) \qquad \ldots(1)$$

be a function of n variables $x_1, x_2, x_3, \ldots.., x_n$ which are connected by m equations

$$\left.\begin{aligned} f_1\,(x_1, x_2, \ldots.., x_n) &= 0 \\ f_2\,(x_1, x_2, \ldots, x_n) &= 0 \\ \ldots \quad \ldots & \quad \ldots \\ f_m\,(x_1, x_2, \ldots, x_n) &= 0 \end{aligned}\right\} \qquad \ldots(2)$$

so that only $n - m$ of the variables are independent.

Since we can eliminate m variables from (1) with the help of the equations (2), the values of the remaining $n - m$ independent variables giving maxima and minima can be found using the method studied in earlier articles. But the elimination is tedious. To avoid the absolute elimination, we may with advantage make use of the undetermined multipliers as follows :

For maxima and minima of u, we have

$$du = 0$$

or $$\frac{\partial u}{\partial x_1} dx_1 + \frac{\partial u}{\partial x_2} dx_2 + \frac{\partial u}{\partial x_3} dx_3 + \ldots + \frac{\partial u}{\partial x_n} dx_n = 0. \quad \ldots(3)$$

Also equations (2) on differentiation give

$$\left.\begin{aligned} df_1 &= \frac{\partial f_1}{\partial x_1} dx_1 + \frac{\partial f_1}{\partial x_2} dx_2 + \frac{\partial f_1}{\partial x_3} dx_3 + \ldots + \frac{\partial f_1}{\partial x_n} dx_n = 0 \\ df_2 &= \frac{\partial f_2}{\partial x_1} dx_1 + \frac{\partial f_2}{\partial x_2} dx_2 + \frac{\partial f_2}{\partial x_3} dx_3 + \ldots + \frac{\partial f_2}{\partial x_n} dx_n = 0 \\ &\ldots \quad \ldots \quad \ldots \quad \ldots \quad \ldots \\ df_m &= \frac{\partial f_m}{\partial x_1} dx_1 + \frac{\partial f_m}{\partial x_2} dx_2 + \frac{\partial f_m}{\partial x_3} dx_3 + \ldots + \frac{\partial f_m}{\partial x_n} dx_n = 0 \end{aligned}\right\} \quad \ldots(4)$$

Multiplying relations (4) by m undetermined multipliers $\lambda_1, \lambda_2, \ldots, \lambda_m$ and adding to (3), we get a result which may be written as

$$P_1\, dx_1 + P_2\, dx_2 + P_3\, dx_3 + \ldots + P_n\, dx_n = 0 \quad \ldots(5)$$

where, $$P_r = \frac{\partial u}{\partial x_r} + \lambda_1 \frac{\partial f_1}{\partial x_r} + \lambda_2 \frac{\partial f_2}{\partial x_r} + \ldots + \lambda_m \frac{\partial f_m}{\partial x_r};$$

$$r = 1, 2, 3, \ldots, n$$

Since $\lambda_1, \lambda_2, \ldots, \lambda_m$ are arbitrary, we can choose them so as to satisfy the following m linear equations :

$$P_1 = P_2 = P_3 = \ldots = P_m = 0. \quad \ldots(6)$$

Hence, equation (5) reduces to

$$P_{m+1}\, dx_{m+1} + P_{m+2}\, dx_{m+2} + \ldots + P_n\, dx_n = 0. \quad \ldots(7)$$

Since equation (2) can be solved to give any of the m variables say $x_1, x_2, \ldots, x_m$, in terms of the remaining $(n-m)$ variables $x_{m+1}, x_{m+2}, \ldots, x_n$ without any loss of generality, hence the $(n-m)$ quantities $dx_{m+1}, dx_{m+2}, \ldots, dx_n$ are all independent, therefore, their coefficients must be separately zero.

As a result, we get the following additional $(n-m)$ equations :

$$P_{m+1} = P_{m+2} = \ldots = P_n = 0. \quad \ldots(8)$$

Hence, the $(m + n)$ equations given by

$$f_1 = f_2 = f_3 = \ldots = f_m = 0$$

and $$P_1 = P_2 = P_3 = \ldots = P_n = 0$$

determine the m undetermined multipliers $\lambda_1, \lambda_2, \ldots, \lambda_m$ and values of the n variables $x_1, x_2, x_3, \ldots, x_n$ for which maximum and minimum values of u are possible.

14.10. Application of the Method of Undetermined Multipliers

Although the method explained in Art. 14.9 can be applied to determine the extreme (maximum or minimum) values of the given function, however, it is more convenient to find out the extreme values of V with the help of a new function F defined as

$$F = \phi + \lambda_1 f_1 + \lambda_2 f_2 + \ldots + \lambda_m f_m$$

and by following the method given below. This method has the advantage over the method given in Art. 8.6 in that it enables us to decide whether the values are maximum or minimum. Though the method is quite general, however, we give below the method for four variables x, y, u, v connected by two relations.

Let $V = \phi(x, y, u, v)$ be subjected to the conditions

$$f_1(x, y, u, v) = 0 \qquad \ldots(1)$$

and $$f_2(x, y, u, v) = 0 \qquad \ldots(2)$$

For maxima and minima of V, we have

$$dV = \frac{\partial \phi}{\partial x} dx + \frac{\partial \phi}{\partial y} dy + \frac{\partial \phi}{\partial u} du + \frac{\partial \phi}{\partial v} dv = 0 \qquad \ldots(3)$$

Also, from equation (1) and (2), we have

$$df_1 = \frac{\partial f_1}{\partial x} dx + \frac{\partial f_1}{\partial y} dy + \frac{\partial f_1}{\partial u} du + \frac{\partial f_1}{\partial v} dv = 0 \qquad \ldots(4)$$

and $$df_2 = \frac{\partial f_2}{\partial x} dx + \frac{\partial f_2}{\partial y} dy + \frac{\partial f_2}{\partial u} du + \frac{\partial f_2}{\partial v} dv = 0 \qquad \ldots(5)$$

Multiplying (4) by λ_1, (5) λ_2 and adding their sum to (3), we get

$$\left(\frac{\partial \phi}{\partial x} + \lambda_1 \frac{\partial f_1}{\partial x} + \lambda_2 \frac{\partial f_2}{\partial x}\right) dx + \left(\frac{\partial \phi}{\partial y} + \lambda_1 \frac{\partial f_1}{\partial y} + \lambda_2 \frac{\partial f_2}{\partial y}\right) dy + \left(\frac{\partial \phi}{\partial u} + \lambda_1 \frac{\partial f_1}{\partial u} + \lambda_2 \frac{\partial f_2}{\partial u}\right) du + \left(\frac{\partial \phi}{\partial v} + \lambda_1 \frac{\partial f_1}{\partial v} + \lambda_2 \frac{\partial f_2}{\partial v}\right) dv = 0 \qquad \ldots(6)$$

Since λ_1 and λ_2 are arbitrary, we can choose them to satisfy the two linear equations :

$$\frac{\partial \phi}{\partial x} + \lambda_1 \frac{\partial f_1}{\partial x} + \lambda_2 \frac{\partial f_2}{\partial x} = 0 \qquad ...(7)$$

and $$\frac{\partial \phi}{\partial y} + \lambda_1 \frac{\partial f_1}{\partial y} + \lambda_2 \frac{\partial f_2}{\partial y} = 0 \qquad ...(8)$$

Using (7) and (8), equation (6) now reduces to

$$\left(\frac{\partial \phi}{\partial u} + \lambda_1 \frac{\partial f_1}{\partial u} + \lambda_2 \frac{\partial f_2}{\partial u}\right) du + \left(\frac{\partial \phi}{\partial v} + \lambda_1 \frac{\partial f_1}{\partial v} + \lambda_2 \frac{\partial f_2}{\partial v}\right) dv = 0.$$

Now since the given function contains four variables and we are given two equations of condition, so therefore, only two of the variables are independent and it is immaterial which two of the four variables are regarded as independent. Let them be u and v, then since du and dv are independent, their coefficients must separately be zero. Thus

$$\frac{\partial \phi}{\partial u} + \lambda_1 \frac{\partial f_1}{\partial u} + \lambda_2 \frac{\partial f_2}{\partial u} = 0 \qquad ...(9)$$

and $$\frac{\partial \phi}{\partial v} + \lambda_1 \frac{\partial f_1}{\partial v} + \lambda_2 \frac{\partial f_2}{\partial v} = 0. \qquad ...(10)$$

We have now six equations namely (1), (2), (7), (8), (9) and (10) to determine the two multipliers λ_1, λ_2 and values of the four variables x, y, u, v for which maximum and minimum values of v are possible.

Now we define a new function $F(x, y, u, v)$ where

$$F(x, y, u, v) = \phi\ (x, y, u, v) + \lambda_1 f_1 (x, y, u, v) + \lambda_2 f_2 (x, y, u, v)$$

assuming that x, y, u, v are now all *independent variables*. Hence, for maxima and minima of F, we must have

$$\frac{\partial F}{\partial x} = \frac{\partial \phi}{\partial x} + \lambda_1 \frac{\partial f_1}{\partial x} + \lambda_2 \frac{\partial f_2}{\partial x} = 0 \qquad ...(i)$$

$$\frac{\partial F}{\partial y} = \frac{\partial \phi}{\partial y} + \lambda_1 \frac{\partial f_1}{\partial y} + \lambda_2 \frac{\partial f_2}{\partial y} = 0 \qquad ...(ii)$$

$$\frac{\partial F}{\partial u} = \frac{\partial \phi}{\partial u} + \lambda_1 \frac{\partial f_1}{\partial u} + \lambda_2 \frac{\partial f_2}{\partial u} = 0 \qquad ...(iii)$$

and $$\frac{\partial F}{\partial v} = \frac{\partial \phi}{\partial v} + \lambda_1 \frac{\partial f_1}{\partial v} + \lambda_2 \frac{\partial f_2}{\partial v} = 0 \qquad ...(iv)$$

Equations (i), (ii), (iii) and (iv) are exactly the same as the equations (7), (8), (9) and (10) obtained above.

Hence, the maxima and minima of $F(x, y, u, v)$ are same as those of $V(x, y, u, v)$ assuming that in $F(x, y, u, v)$ the variables x, y, u, v are now all independent.

We now proceed to find whether the values of V obtained with the help of the above equations are maximum or minimum. For this purpose we adopt the following procedure :

From (3), we get

$$d^2V = \left(dx\frac{\partial}{\partial x} + dy\frac{\partial}{\partial y} + du\frac{\partial}{\partial u} + dv\frac{\partial}{\partial v}\right)^2 \phi + \frac{\partial \phi}{\partial x}d^2x + \frac{\partial \phi}{\partial y}d^2y + \frac{\partial \phi}{\partial u}d^2u + \frac{\partial \phi}{\partial v}d^2v \quad ...(11)$$

Also,

$$d^2f_1 = \left(dx\frac{\partial}{\partial x} + dy\frac{\partial}{\partial y} + du\frac{\partial}{\partial u} + dv\frac{\partial}{\partial v}\right)^2 f_1 + \frac{\partial f_1}{\partial x}d^2x + \frac{\partial f_1}{\partial y}d^2y + \frac{\partial f_1}{\partial u}d^2u + \frac{\partial f_1}{\partial v}d^2v = 0 \quad ...(12)$$

and

$$d^2f_2 = \left(dx\frac{\partial}{\partial x} + dy\frac{\partial}{\partial y} + du\frac{\partial}{\partial u} + dv\frac{\partial}{\partial v}\right)^2 f_2 + \frac{\partial f_2}{\partial x}d^2x + \frac{\partial f_2}{\partial y}d^2y + \frac{\partial f_2}{\partial u}d^2u + \frac{\partial f_2}{\partial v}d^2v = 0 \quad ...(13)$$

Multiplying equation (12) by λ_1, (13) by λ_2 and adding their sum to (11) and using the results (i), (ii), (iii) and (iv), we have

$$d^2V = \left(dx\frac{\partial}{\partial x} + dy\frac{\partial}{\partial y} + du\frac{\partial}{\partial u} + dv\frac{\partial}{\partial v}\right)^2 (\phi + \lambda_1 f_1 + \lambda_2 f_2)$$

$$= \left(dx\frac{\partial}{\partial x} + dy\frac{\partial}{\partial y} + du\frac{\partial}{\partial u} + dv\frac{\partial}{\partial v}\right)^2 F$$

$$= d^2F$$

Hence, d^2V is equal to d^2F, where d^2F, is obtained by assuming all the variables x, y, u, v as independent. Thus, from above it is clear that d^2V and d^2F have the same signs. Hence, V will be maximum or minimum according as F is maximum or minimum.

ILLUSTRATIVE EXAMPLES

Example 1. *Find the maxima and minima of* $u = x^2 + y^2 + z^2$, *where,*

$$ax^2 + by^2 + cz^2 = 1$$

Solution : Define a function F, where,

$$F = x^2 + y^2 + z^2 + \lambda\,(ax^2 + by^2 + cz^2 - 1)$$

For maxima or minima of F, we have

$$\frac{\partial F}{\partial x} = 2x + 2\lambda\, ax = 0 \qquad \text{...(i)}$$

$$\frac{\partial F}{\partial y} = 2y + 2\lambda\, by = 0 \qquad \text{...(ii)}$$

$$\frac{\partial F}{\partial z} = 2z + 2\lambda\, cz = 0. \qquad \text{...(iii)}$$

Multiplying equation (i) by x, (ii) by y and (iii) by z and adding, we have

$$2(x^2 + y^2 + z^2) + 2\lambda(ax^2 + by^2 + cz^2) = 0$$

or $$2u + 2\lambda = 0$$

or $$\lambda = -u.$$

putting $\lambda = -u$ in equation (i), we get

or $$2x - 2uax = 0$$

or $$x(1 - ua) = 0$$

or $$1 - ua = 0$$

or $$1 = ua$$

or $$\frac{1}{a} - u = 0$$

Similarly, $$\frac{1}{b} - u = 0$$

and $$\frac{1}{c} - u = 0.$$

Hence, the maximum and minimum values of u are the roots of the equation

$$\left(\frac{1}{a} - u\right)\left(\frac{1}{b} - u\right)\left(\frac{1}{c} - u\right) = 0.$$

Example 2. *Find the minimum value of $x^2 + y^2 + z^2$ having given that*

$$ax + by + cz = p.$$

Solution : Let $$u = x^2 + y^2 + z^2 \qquad \text{...(i)}$$

and $$ax + by + cz - p = 0 \qquad \text{...(ii)}$$

For maximum or minimum of u, we have

$$du = 2x\,dx + 2y\,dy + 2z\,dz = 0$$

$$\Rightarrow \quad x\,dx + y\,dy + z\,dz = 0 \qquad \text{...(iii)}$$

Also, differentiating (ii), we get

$$a\,dx + b\,dy + c\,dz = 0. \qquad \text{...(iv)}$$

Multiplying equation (iii) by 1, (iv) by λ and adding, we get

$$(x + \lambda a)\,dx + (y + \lambda b)\,dy + (z + \lambda c)\,dz = 0.$$

Now, equating to zero the coefficients of dx, dy and dz, we get

$$x + \lambda a = 0 \qquad \text{...(v)}$$

$$y + \lambda b = 0 \qquad \text{...(vi)}$$

$$z + \lambda c = 0 \qquad \text{...(vii)}$$

Multiplying (v), (vi), (vii) by x, y, z respectively and adding, we get

$$x^2 + y^2 + z^2 + \lambda\,(ax + by + cz) = 0$$

or $$u + \lambda p = 0$$

or $$\lambda = -\frac{u}{p}$$

$\therefore$ From equation (v),

$$x - \frac{u}{p}a = 0 \quad \text{or} \quad x = \frac{au}{p}$$

Similarly, $$y = \frac{bu}{p} \quad \text{and} \quad z = \frac{cu}{p}$$

$\therefore$ $$u = x^2 + y^2 + z^2$$

$$= \frac{a^2u^2 + b^2u^2 + c^2u^2}{p^2}$$

or $$u = \frac{p^2}{a^2 + b^2 + c^2}. \qquad \text{...(viii)}$$

Discrimination : Now to discuss maxima or minima, we see that the three variables x, y and z are connected by one relation, so that there is only one dependent variable say z and two independent variables say x and y.

Now, differentiating equation (ii) w.r.t. x partially, we get

$$a + c\frac{\partial z}{\partial x} = 0$$

or $$\frac{\partial z}{\partial x} = -\frac{a}{c}.$$

Similarly, $$\frac{\partial z}{\partial y} = -\frac{b}{c}.$$

Also $$\frac{\partial u}{\partial x} = 2x + 2z\frac{\partial z}{\partial x} = 2x + 2z\left(-\frac{a}{c}\right)$$

$$= 2x - 2z\left(\frac{a}{c}\right)$$

$$r = \frac{\partial^2 u}{\partial x^2} = 2 - \frac{2a}{c}\frac{\partial z}{\partial x}$$

$$= 2 - \frac{2a}{c}\left(-\frac{a}{c}\right)$$

$$= 2 + \frac{2a^2}{c^2}$$

$$= \frac{2}{c^2}(a^2 + c^2) = +\text{ve}$$

$$s = \frac{\partial^2 u}{\partial x\,\partial y} = -\frac{2a}{c}\frac{\partial z}{\partial y}$$

$$= -\frac{2a}{c}\cdot\left(-\frac{b}{c}\right)$$

$$= \frac{2ab}{c^2}$$

$$t = \frac{\partial^2 u}{\partial y^2} = \frac{2}{c^2}(b^2 + c^2)$$

$$\therefore \quad rt - s^2 = \frac{2}{c^2}(a^2 + c^2)\frac{2}{c^2}(b^2 + c^2) - \frac{4a^2b^2}{c^4}$$

$$= \frac{4}{c^4}[(a^2 + c^2)(b^2 + c^2) - a^2b^2]$$

$$= \frac{4}{c^2}(a^2 + b^2 + c^2) = +\text{ ve.}$$

Since r and $rt - s^2$ are both positive, hence, the value of u is given by (8) is a minimum.

Example 3. *In any triangle ABC, find the maximum value of* $\cos A \cos B \cos C$ *by Lagrange's method.*

Solution : Since A, B, C are angles of triangle,

$$\therefore \quad A + B + C = \pi. \quad \text{...(i)}$$

Let $\quad u = \cos A \cos B \cos C.$

Differentiating u logarithmically, we get for maxima or minima,

$$\frac{1}{u}du = -[\tan A\,dA + \tan B\,dB + \tan C\,dC] = 0 \qquad [\because du = 0]$$

or $\tan A\,dA + \tan B\,dB + \tan C\,dC = 0$ ..(ii)

and from equation (i), $dA + dB + dC = 0.$ ...(iii)

Multiplying equation (ii), (iii) by 1, λ respectively, adding and then equating to zero the coefficients of dA, dB, dC, we get

$$\tan A + \lambda = 0$$

$$\tan B + \lambda = 0$$

and $$\tan C + \lambda = 0$$

$\therefore$ $\tan A = \tan B = \tan C$

or $A = B = C = \pi/3$ $[\because$ from equation (i) $A + B + C = \pi]$

Discrimination : Now we shall show that u is maximum when

$$A = B = C = \pi/3.$$

Differentiating equation (1) w.r.t. A partially, treating C as the dependent variable, we have

$$1 + \frac{\partial C}{\partial A} = 0 \quad \text{or} \quad \frac{\partial C}{\partial A} = -1.$$

Also, $$\frac{\partial C}{\partial B} = -1.$$

Now, $$\frac{\partial u}{\partial A} = \cos B\left[-\sin A\cos C - \cos A\sin C\frac{\partial C}{\partial A}\right]$$

$$= \cos B\,[-\sin A\cos C + \cos A\sin C] \qquad \left[\because \frac{\partial C}{\partial A} = -1\right]$$

$$= \cos B\sin(C - A)$$

$\therefore$ $$r = \frac{\partial^2 u}{\partial A^2} = \cos B\cos(C - A)\left(\frac{\partial C}{\partial A} - 1\right)$$

$$= -2\cos B\cos(C - A)$$

$$s = \frac{\partial^2 u}{\partial A\,\partial B} = \frac{\partial}{\partial B}[\cos B\sin(C - A)]$$

$$= -\sin B\sin(C - A) + \cos(C - A)\frac{\partial C}{\partial B}\cos B$$

$$= -\sin B\sin(C - A) - \cos B\cos(C - A)$$

$$= -\cos(B + C + A).$$

Similarly,

$$t = -2\cos A\cos(C - B)$$

When $A = B = C = \pi/3$, then

$$r = -2 \cdot \frac{1}{2} \cdot 1 = -1 = -\text{ve}$$

$$s = -\frac{1}{2}$$

$$t = -1$$

$\therefore$ $$rt - s^2 = 1 - \frac{1}{4} = \frac{3}{4} = +\text{ve}.$$

Since $rt - s^2$ is +ve and r is −ve, hence, at $A = B = C = \pi/3$, u is maximum and the maximum value of $u = \frac{1}{2} \cdot \frac{1}{2} \cdot \frac{1}{2} = \frac{1}{8}$.

Example 4. *Find the maximum or minimum values of $u = a^2 x^2 + b^2 y^2 + c^2 z^2$ subject to the conditions $x^2 + y^2 + z^2 = 1$ and $lx + my + nz = 0$.*

Solution : Let

$$F = a^2 x^2 + b^2 y^2 + c^2 z^2 + \lambda_1 (x^2 + y^2 + z^2 - 1) + \lambda_2 (lx + my + nz).$$

For maxima and minima of F, we have

$$\frac{\partial F}{\partial x} = 2a^2 x + 2\lambda_1 x + l\lambda_2 = 0 \quad \text{...(i)}$$

$$\frac{\partial F}{\partial y} = 2b^2 y + 2\lambda_1 y + m\lambda_2 = 0 \quad \text{...(ii)}$$

and $$\frac{\partial F}{\partial z} = 2c^2 z + 2\lambda_1 z + n\lambda_2 = 0. \quad \text{...(iii)}$$

Multiplying equations (i), (ii) and (iii) by x, y and z respectively and adding, we get

$$2u + 2\lambda_1 \cdot 1 + \lambda_2 \cdot 0 = 0$$

or $$\lambda_1 = -u$$

Substituting this in equation (i), we get

$$2a^2 x - 2ux + l\lambda_2 = 0$$

or $$x = \frac{l\lambda_2}{2(u - a^2)}$$

Similarly, $$y = \frac{m\lambda_2}{2(u - b^2)}$$

and $$z = \frac{n\lambda_2}{2(u - c^2)}$$

Substituting these values of x, y and z in $lx + my + nz = 0$, we get

$$l \cdot \frac{l\lambda_2}{2(u-a^2)} + m \cdot \frac{m\lambda_2}{2(u-b^2)} + n \cdot \frac{n\lambda_2}{2(u-c^2)} = 0$$

or
$$\frac{l^2}{u-a^2} + \frac{m^2}{u-b^2} + \frac{n^2}{u-c^2} = 0$$

[cancelling $\lambda^2/2$ throughout]

which gives the required maximum and minimum values of u.

Example 5. *Find the maxima and minima of* $u = x^2 + y^2 + z^2$ *subject to the conditions*

$$ax^2 + by^2 + cz^2 = 1 \quad \textit{and} \quad lx + my + nz = 0.$$

Interpret the result geometrically.

Solution : Let us define a function F such that

$$F = x^2 + y^2 + z^2 + \lambda_1 (ax^2 + by^2 + cz^2 - 1) + \lambda_2 (lx + my + nz).$$

For maxima and minima of F, we have

$$\frac{\partial F}{\partial x} = 2x + 2ax\lambda_1 + l\lambda_2 = 0 \quad \text{...(i)}$$

$$\frac{\partial F}{\partial y} = 2y + 2by\lambda_1 + m\lambda_2 = 0 \quad \text{...(ii)}$$

and
$$\frac{\partial F}{\partial z} = 2z + 2cz\lambda_1 + n\lambda_2 = 0. \quad \text{...(iii)}$$

Multiplying equations (i), (ii), (iii) by x, y, z respectively and adding, we obtain,

$$2(x^2 + y^2 + z^2) + 2\lambda_1 (ax^2 + by^2 + cz^2) + \lambda_2 (lx + my + nz) = 0$$

$$\therefore \quad 2u + \lambda_1 \cdot 1 + \lambda_2 \cdot 0 = 0$$

$$\therefore \quad \lambda_1 = -u$$

Putting $\lambda_1 = -u$ in equation (i), we get

$$2x - 2axu + l\lambda_2 = 0$$

or
$$x = \frac{l\lambda_2}{2(au-1)}$$

Similarly,
$$y = \frac{m\lambda_2}{2(bu-1)}$$

and
$$z = \frac{n\lambda_2}{2(cu-1)}$$

Substituting these values of x, y, z in $lx + my + nz = 0$, we obtain

$$l \cdot \frac{l\lambda_2}{2(au-1)} + m \cdot \frac{m\lambda_2}{2(bu-1)} + n \cdot \frac{n\lambda_2}{2(cu-1)} = 0$$

or
$$\frac{l^2}{au-1} + \frac{m^2}{bu-1} + \frac{n^2}{cu-1} = 0$$

[cancelling $\lambda_2/2$ throughout]

which gives the required maximum and minimum values of u.

Geometrical Interpretation : $x^2 + y^2 + z^2$ is the square of the distance of any point (x, y, z) from the origin (0, 0, 0). Hence, in this problem, we have found the maximum and minimum values of the square of distance of the origin from the point of intersection of the central conicoid $ax^2 + by^2 + cz^2 = 1$ by the central plane $lx + my + nz = 0$.

Example 6. *Show that the maximum and minimum radii vectors of the section of the surface* $(x^2 + y^2 + z^2)^2 = \frac{x^2}{a^2} + \frac{y^2}{b^2} + \frac{z^2}{c^2}$

by the plane $lx + my + nz = 0$ *are given by*

$$\frac{a^2 l^2}{1 - a^2 r^2} + \frac{b^2 m^2}{1 - b^2 r^2} + \frac{c^2 n^2}{1 - c^2 r^2} = 0.$$

Solution : Here $u = x^2 + y^2 + z^2$, where, $u = r^2$

$$f_1 = \frac{x^2}{a^2} + \frac{y^2}{b^2} + \frac{z^2}{c^2} = u^2$$

$$f_2 = lx + my + nz = 0$$

$\therefore$
$$du = 2x\,dx + 2y\,dy + 2z\,dz = 0 \quad \text{...(i)}$$

[$\because$ for max. or mini. of u, $du = 0$]

$$df_1 = \frac{2x}{a^2}dx + \frac{2y}{b^2}dy + \frac{2z}{c^2}dz = 0 \quad \text{...(ii)}$$

$$df_2 = l\,dx + m\,dy + n\,dz = 0. \quad \text{...(iii)}$$

Multiplying equation (i), (ii), (iii) by $\frac{1}{2}, \frac{1}{2}\lambda_1, \lambda_2$ respectively and adding and equating to zero the coefficients of dx, dy, dz, we get

$$x + \lambda_1 \frac{x}{a^2} + \lambda_2 l = 0 \quad \text{...(iv)}$$

$$y + \lambda_1 \frac{y}{b^2} + \lambda_2 m = 0 \quad \text{...(v)}$$

$$z + \lambda_1 \frac{z}{c^2} + \lambda_2 n = 0. \qquad \text{...(vi)}$$

Multiplying equation (iv), (v), (vi) by x, y, z respectively and adding we get

$$x^2 + y^2 + z^2 + \lambda_1 \left(\frac{x^2}{a^2} + \frac{y^2}{b^2} + \frac{z^2}{c^2}\right) + \lambda_1 (l x + m y + n z) = 0$$

or $$u + \lambda_1 u^2 + \lambda_2 \cdot 0 = 0$$

or $$\lambda_1 = -\frac{1}{u}$$

Putting this value of λ_1 in equation (iv), we get

$$x - \frac{x}{u a^2} + \lambda_2 l = 0$$

or $$x (l - u a^2) = \lambda_2 l u a^2$$

or $$x = \frac{\lambda_2 l u a^2}{1 - u a^2}$$

Similarly, $$y = \frac{\lambda_2 m u b^2}{1 - u b^2}$$

and $$z = \frac{\lambda_2 n u c^2}{1 - u c^2}$$

Substituting these values of x, y, z in $l x + m y + n z = 0$, we get

$$l \frac{\lambda_2 l u a^2}{1 - u a^2} + m \frac{\lambda_2 m u b^2}{1 - u b^2} + n \frac{\lambda_2 n u c^2}{1 - u c^2} = 0$$

or $$\frac{l^2 a^2}{1 - u a^2} + \frac{m^2 b^2}{1 - u b^2} + \frac{n^2 c^2}{1 - u c^2} = 0$$

or $$\frac{l^2 a^2}{1 - a^2 r^2} + \frac{m^2 b^2}{1 - b^2 r^2} + \frac{n^2 c^2}{1 - c^2 r^2} = 0. \qquad [\text{Putting } u = r^2]$$

Example 7. *Find the maximum and minimum values of*

$$\frac{x^2}{a^4} + \frac{y^2}{b^4} + \frac{z^2}{c^4}$$

when $l x + m y + n z = 0$ *and* $\frac{x^2}{a^2} + \frac{y^2}{b^2} + \frac{z^2}{c^2} = 1.$

Interpret the result geometrically.

Solution : Let $$u = \frac{x^2}{a^4} + \frac{y^2}{b^4} + \frac{z^2}{c^4}.$$

Hence, for maximum or minimum of u, we have

$$du = \frac{2x}{a^4}\,dx + \frac{2y}{b^4}\,dy + \frac{2z}{c^4}\,dz = 0$$

or $$\frac{x}{a^4}\,dx + \frac{y}{b^4}\,dy + \frac{z}{c^4}\,dz = 0 \qquad \text{...(i)}$$

Also, $$l\,dx + m\,dy + n\,dz = 0 \qquad \text{...(ii)}$$

and $$\frac{x}{a^2}\,dx + \frac{y}{b^2}\,dy + \frac{z}{c^2}\,dz = 0. \qquad \text{...(iii)}$$

Multiplying equation (i), (ii) and (iii) by 1, λ_1, and λ_2 respectively, and adding, and then equating to zero, the coefficients of dx, dy and dz, we get

$$\frac{x}{a^4} + \lambda_1 l + \lambda_2 \frac{x}{a^2} = 0 \qquad \text{...(iv)}$$

$$\frac{y}{b^4} + \lambda_1 m + \lambda_2 \frac{y}{b^2} = 0 \qquad \text{...(v)}$$

and $$\frac{z}{c^4} + \lambda_1 n + \lambda_2 \frac{z}{c^2} = 0. \qquad \text{...(vi)}$$

Multiplying equation (iv), (v), (vi) by x, y, z respectively and adding we get

$$\left(\frac{x^2}{a^4} + \frac{y^2}{b^4} + \frac{z^2}{c^4}\right) + \lambda_1 (l\,x + m\,y + n\,z) + \lambda_2 \left(\frac{x^2}{a^2} + \frac{y^2}{b^2} + \frac{z^2}{c^2}\right) = 0$$

or $$u + \lambda_1 \cdot 0 + \lambda_2 \cdot 1 = 0$$

$\therefore$ $$\lambda_2 = -u.$$

Putting this value of λ_2 in equation (iv), we get

$$\frac{x}{a^4} + \lambda_1 l - \frac{u\,x}{a^2} = 0$$

or $$\frac{x}{a^4}(-1 + u\,a^2) = \lambda_1 l$$

or $$x = \frac{\lambda_1 l\, a^4}{u\,a^2 - 1}$$

Similarly, $$y = \frac{\lambda_1 m\, b^4}{u\,b^2 - 1}$$

and $$z = \frac{\lambda_1 n\, c^4}{u\,c^2 - 1}$$

Substituting these values of x, y, z in the relation

$$l x + m y + n z = 0,$$

we get
$$l \cdot \frac{\lambda_1 l a^4}{u a^2 - 1} + m \cdot \frac{\lambda_1 m b^4}{u b^2 - 1} + n \cdot \frac{\lambda_1 n c^4}{u c^2 - 1} = 0$$

or
$$\frac{l^2 a^4}{u a^2 - 1} + \frac{m^2 b^4}{u b^2 - 1} + \frac{n^2 c^4}{u c^2 - 1}$$

[cancelling λ_1 throughout]

which gives the required maximum and minimum values of u.

Geometrical Interpretation : $\frac{x^2}{a^2} + \frac{y^2}{b^2} + \frac{z^2}{c^2} = 1$ represents an ellipsoid. The equation of tangent plane at any point (x, y, z) of it is $\frac{Xx}{a^2} + \frac{Yy}{b^2} + \frac{Zz}{c^2} = 1$. Let p be the length of the perpendicular from origin $O(0, 0, 0)$ upon this plane. Then

$$p^2 = \frac{1}{\frac{x^2}{a^4} + \frac{y^2}{b^4} + \frac{z^2}{c^4}}$$

or
$$\frac{1}{p^2} = \frac{x^2}{a^4} + \frac{y^2}{b^4} + \frac{z^2}{c^4}.$$

If the point (x, y, z) also lies on $l x + m y + n z = 0$, then it satisfies both the given conditions namely,

$$l x + m y + n z = 0 \quad \text{and} \quad \frac{x^2}{a^2} + \frac{y^2}{b^2} + \frac{z^2}{c^2} = 1.$$

Thus, in the present problem we have discussed maximum and minimum values of the perpendicular distance from the origin to the tangent planes of the ellipsoid $\frac{x^2}{a^2} + \frac{y^2}{b^2} + \frac{z^2}{c^2} = 1$ at the points which also lie on the plane $l x + m y + n z = 0$.

Example 8. *Find the maximum value of $x^p y^q z^r$ when the variables are subject to the conditions $a x + b y + c z = p + q + r$.*

Solution : Let $u = x^p y^q z^r$.

Taking logarithm, we get

$$\log u = p \log x + q \log y + r \log z.$$

Differentiating, we get

$$\frac{1}{u}\,du = \frac{p}{x}\,dx + \frac{q}{y}\,dy + \frac{r}{z}\,dz.$$

For u to be maximum or minimum, we have $du = 0$

Hence, we get

$$(p/x)\ dx + (q/y)\ dy + (r/z)\ dz = 0. \qquad \text{...(i)}$$

Again from the condition

$$ax + by + cz = p + q + r,$$

on differentiation, we get

$$a\ dx + b\ dy + c\ dz = 0. \qquad \text{...(ii)}$$

Multiplying equation (i) and (ii) by 1 and λ respectively, adding and then equating to zero, the coefficients of dx, dy, dz, we get

$$p/x + \lambda a = 0 \qquad \text{...(iii)}$$

$$q/y + \lambda b = 0 \qquad \text{...(iv)}$$

$$r/z + \lambda c = 0. \qquad \text{...(v)}$$

Multiplying equation (iii), (iv), (v) by x, y, z respectively and adding, we get

$$p + q + r + \lambda\,(ax + by + cz) = 0$$

or $$p + q + r + \lambda\,(p + q + r) = 0$$

or $$\lambda = -1.$$

Hence, from equation (iii), we get

$$p/x - a = 0 \quad \text{or} \quad x = p/a$$

Similarly, we get

$$y = q/b$$

and $$z = r/c$$

$\therefore$ $$u = x^p y^q z^r$$

$$= \left(\frac{p}{a}\right)^p \left(\frac{q}{b}\right)^q \left(\frac{r}{c}\right)^r. \qquad \text{...(vi)}$$

Discrimination : Now to discuss its maxima or minima, differentiating

$$ax + by + cz = p + q + r$$

partially with respect to x, we get

$$a + c\,\frac{\partial z}{\partial x} = 0$$

or
$$\frac{\partial z}{\partial x} = -\frac{a}{c}$$

also
$$\frac{\partial z}{\partial y} = -\frac{b}{c}.$$

Now, differentiating u logarithmically, w.r.t. x, partially, we get

$$\frac{1}{u}\frac{\partial u}{\partial x} = \frac{p}{x} + \frac{r}{z}\frac{\partial z}{\partial x} = \frac{p}{x} + \frac{r}{z}\left(-\frac{a}{c}\right)$$

or
$$\frac{1}{u}\frac{\partial u}{\partial x} = \frac{p}{x} - \frac{r\,a}{z\,c}.$$

Differentiating again partially with respect to x, we get

$$\frac{1}{u}\frac{\partial^2 u}{\partial x^2} - \frac{1}{u^2}\left(\frac{\partial u}{\partial x}\right)^2 = -\frac{p}{x^2} + \frac{r}{z^2}\frac{a}{c}\frac{\partial z}{\partial x}$$

$$= -\frac{p}{x^2} + \frac{r}{z^2}\frac{a}{c}\left(-\frac{a}{c}\right)$$

$$= -\frac{p}{x^2} - \frac{r\,a^2}{z^2\,c^2}$$

$$= -\frac{c^2 z^2 p + r a^2 x^2}{c^2 z^2 x^2}.$$

Since $\frac{\partial u}{\partial x} = 0$ for maximum or minimum

$\therefore$
$$\frac{1}{u}\frac{\partial^2 u}{\partial x^2} = -\frac{c^2 z^2 p + r a^2 x^2}{c^2 z^2 x^2}$$

or
$$\frac{\partial^2 u}{\partial x^2} = r = -\frac{c^2 z^2 p + r a^2 x^2}{c^2 z^2 x^2}\,u$$

$$= -\text{ve quantity.}$$

It is easy to show that $rt - s^2$ is +ve.

Since r is − ve, hence, there is a maximum value of u given by equation (vi).

Example 9. *Prove that of all rectangular parallelopiped of the same volume, the cube has the least surface.*

Solution : Let x, y, z be the lengths of the three coterminous edges of the rectangular parallelopiped. Then volume is given by

$$x\,y\,z = V \text{ (constant)} \qquad \text{...(i)}$$

If S is the surface, then

$$S = 2\,(y\,z + z\,x + x\,y). \qquad \text{...(ii)}$$

Thus, we are to find the minimum value of S under the condition (i).

For S to be maximum or minimum we have $$dS = 0$$

or $$2\,[y\,dz + z\,dy + z\,dx + x\,dz + x\,dy + y\,dx] = 0$$

or $$(y + z)\,dx + (z + x)\,dy + (x + y)\,dz = 0. \quad \text{...(iii)}$$

Also, from equation (i), we have

$$yz\,dx + zx\,dy + xy\,dz = 0. \quad \text{...(iv)}$$

Multiplying equation (iii), (iv) by 1, λ respectively, adding and then equating to zero, the coefficients of dx, dy, dz, we get

$$y + z + \lambda yz = 0$$

$$z + x + \lambda zx = 0$$

and $$x + y + \lambda xy = 0.$$

So that $$-\lambda = \frac{y+z}{yz} = \frac{z+x}{zx} = \frac{x+y}{xy}$$

or $$\frac{1}{y} + \frac{1}{z} = \frac{1}{z} + \frac{1}{x} = \frac{1}{x} + \frac{1}{y}.$$

From first two, we get

$$\frac{1}{y} + \frac{1}{z} = \frac{1}{z} + \frac{1}{x}$$

or $$\frac{1}{y} = \frac{1}{x}$$

or $$y = x.$$

From last two, we get

$$\frac{1}{z} + \frac{1}{x} = \frac{1}{x} + \frac{1}{y}$$

or $$\frac{1}{z} = \frac{1}{y}$$

or $$z = y$$

Thus, $$x = y = z = (xyz)^{1/3} = V^{1/3}$$

Thus, all the edges are equal when S is maximum or minimum.

Discrimination : Now to discuss maximum or minimum, we have on differentiating equation (i) w.r.t. x partially regarding z as a function of x and y, we get

$$yz + xy\frac{\partial z}{\partial x} = 0$$

or $$\frac{\partial z}{\partial x} = -\frac{z}{x}$$

Similarly, $$\frac{\partial z}{\partial y} = -\frac{z}{y}$$

Now, $$\frac{\partial S}{\partial x} = 2\left[y\frac{\partial z}{\partial x} + x\frac{\partial z}{\partial x} + z + y\right]$$

$$= 2\left[y\cdot\left(-\frac{z}{x}\right) + x\left(-\frac{z}{x}\right) + z + y\right]$$

$$= 2\left[y - \frac{yz}{x}\right]$$

$$r = \frac{\partial^2 S}{\partial x^2} = 2\frac{yz}{x^2} - 2\frac{v}{x}\frac{\partial z}{\partial x}$$

$$= 2\frac{yz}{x^2} - \frac{2y}{x}\left(-\frac{z}{x}\right) = \frac{4yz}{x^2}$$

$$= 4 \text{ when } x = y = z$$

$$s = \frac{\partial^2 S}{\partial y\,\partial x} = 2\left[1 - \frac{1}{x}\left\{z + y\frac{\partial z}{\partial y}\right\}\right]$$

$$= 2\left[1 - \frac{1}{x}\left\{z + y\left(-\frac{z}{y}\right)\right\}\right]$$

$$= 2.$$

Similarly, $$t = \frac{\partial^2 S}{\partial y^2} = 4$$

$\therefore$ $$rt - s^2 = 4\cdot 4 - (2)^2 = 12\ (+\text{ ve})$$

and $$r = 4\ (+\text{ ve}).$$

Hence, when $x = y = z$, *i.e.*, when the rectangular parallelopiped is a cube, its surface is minimum (least).

Example 10. *Divide a number n into three parts x, y, z such that $a\,y\,z + b\,z\,x + c\,x\,y$ shall have maximum or minimum and determine which it is.*

Solution : The number n has been divided into three parts x, y, z.

$\therefore$ $$x + y + z = n \qquad \text{...(i)}$$

Let $$u = a\,y\,z + b\,z\,x + c\,x\,y. \qquad \text{...(ii)}$$

For u to be maximum or minimum, we have

$$du = 0$$

or $$(b\,z + c\,y)\,dx + (c\,x + a\,z)\,dy + (a\,y + b\,x)\,dz = 0. \qquad \text{...(iii)}$$

Also, from equation (i) on differentiation, we get

$$dx + dy + dz = 0. \qquad \text{...(iv)}$$

Multiplying equation (iii) and (iv) by 1 and λ respectively, adding and equating to zero, the coefficients of dx, dy, dz, we get

$$bz + cy + \lambda = 0 \quad \text{...(v)}$$

$$cx + az + \lambda = 0 \quad \text{...(vi)}$$

$$ay + bx + \lambda = 0. \quad \text{...(vii)}$$

Multiplying equation (v), (vi), (vii) by x, y, z respectively and adding, we get

$$2u + n\lambda = 0$$

or
$$\lambda = -\frac{2u}{n}$$

$\therefore$ Equation (v), (vi) and (vii) can be written as

$$0 \cdot x + c \cdot y + b \cdot z - \frac{2u}{n} = 0$$

$$c \cdot x + 0 \cdot y + a \cdot z - \frac{2u}{n} = 0$$

and
$$b \cdot x + a \cdot y + 0 \cdot z - \frac{2u}{n} = 0$$

Also, equation (i) is $\quad x + y + z - n = 0$

Eliminating $x, y, z, -1$ from above equations determinentically, we get

$$\begin{vmatrix} 0 & c & b & \frac{2u}{n} \\ c & 0 & a & \frac{2u}{n} \\ b & a & 0 & \frac{2u}{n} \\ 1 & 1 & 1 & n \end{vmatrix} = 0 \quad \text{...(viii)}$$

which gives the maximum or minimum values of u.

Discrimination : Now to discuss maxima or minima, differentiating equation (i) w.r.t. x partially treating z as a function of x and y, we get

$$1 + \frac{\partial z}{\partial x} = 0$$

or
$$\frac{\partial z}{\partial x} = -1$$

Similarly,
$$\frac{\partial z}{\partial y} = -1$$

Now,
$$\frac{\partial u}{\partial x} = bz + cy + (bx + ay)\frac{\partial z}{\partial x}$$

$$= bz + cy - (bx + ay)$$

$$r = \frac{\partial^2 u}{\partial x^2} = b \frac{\partial z}{\partial x} - b = -2b$$

$$s = \frac{\partial^2 u}{\partial x \, \partial y} = b \frac{\partial z}{\partial y} + c - a$$

$$= -b + c - a = c - a - b$$

$$t = \frac{\partial^2 u}{\partial y^2} = -2a \qquad \text{[Thus } r \text{ and } t \text{ have the same sign]}$$

$$rt - s^2 = (-2b)(-2a) - (c - a - b)^2$$

$$= 2ab + 2bc + 2ca - a^2 - b^2 - c^2$$

$$= a(b + c - a) + b(c + a - b) + c(a + b - c)$$

If we assume that a, b, c form the three sides of a triangle, then $b + c - a$, $c + a - b$, $a + b - c$ are +ve.

$\therefore \qquad rt - s^2 = +\text{ve}$

and $\qquad r = -2b = -\text{ve}.$

Hence, the values of u given by (viii) are maximum under the assumption that a, b, c form the sides of a triangle.

Example 11. *Find the rectangular parallelopiped of maximum volume that can be inscribed in the ellipsoid*

$$\frac{x^2}{a^2} + \frac{y^2}{b^2} + \frac{z^2}{c^2} = 1.$$

Solution : Let (x, y, z) be the coordinates of an angular point of the rectangular parallelopiped (that lies in the positive octant) then the lengths of the edges of inscribed rectangular parallelopiped are $2x$, $2y$, $2z$. Therefore, the volume V is given by

$$V = 2x \cdot 2y \cdot 2z = 8xyz \qquad \text{...(i)}$$

Since the point (x, y, z) lies on the ellipsoid, hence x, y, z admit the relation

$$\frac{x^2}{a^2} + \frac{y^2}{b^2} + \frac{z^2}{c^2} = 1. \qquad \text{...(ii)}$$

Thus, we are to find the maximum value of V given by equation (i) under the condition (ii).

Differentiating equation (i) logarithmically, we get

$$\frac{dV}{V} = \frac{dx}{x} + \frac{dy}{y} + \frac{dz}{z}$$

But for V to be maximum or minimum, $dV = 0$,

$$\therefore \quad \frac{dx}{x} + \frac{dy}{y} + \frac{dz}{z} = 0. \qquad \text{...(iii)}$$

Again from equation (ii) on differentiation, we get

$$\frac{2x}{a^2}\,dx + \frac{2y}{b^2}\,dy + \frac{2z}{c^2}\,dz = 0$$

or

$$\frac{x}{a^2}\,dx + \frac{y}{b^2}\,dy + \frac{z}{c^2}\,dz = 0. \qquad \text{...(iv)}$$

Multiplying equation (iii), (iv) by 1, λ respectively, adding and then equating to zero the coefficients of dx, dy, dz, we get

$$\frac{1}{x} + \lambda\frac{x}{a^2} = 0 \qquad \text{...(v)}$$

$$\frac{1}{y} + \lambda\frac{y}{b^2} = 0 \qquad \text{...(vi)}$$

$$\frac{1}{z} + \lambda\frac{z}{c^2} = 0. \qquad \text{...(vii)}$$

Multiplying equation (v), (vi), (vii) by $x,\ y,\ z$ respectively and adding, we get

$$3 + \lambda\left(\frac{x^2}{a^2} + \frac{y^2}{b^2} + \frac{z^2}{c^2}\right) = 0$$

or

$$3 + \lambda\, 1 = 0$$

or

$$\lambda = -3$$

$\therefore$

$$\frac{1}{x} + \frac{\lambda x}{a^2} = 0 \text{ gives}$$

$$\frac{1}{x} - \frac{3x}{a^2} = 0$$

or

$$x^2 = \frac{a^2}{3}$$

or

$$x = \frac{a}{\sqrt{3}}$$

Similarly,

$$y = \frac{b}{\sqrt{3}}$$

and

$$z = \frac{c}{\sqrt{3}}.$$

Discrimination : Now we shall show that these values of x, y, z correspond to a maximum value of V.

Since $x,\ y,\ z$ are connected by a relation (ii), there are two independent variables ($x,\ y$ say) and one dependent variable (z here).

Differentiating equation (ii) with respect to x partially, regarding z as dependent variable, we have

$$\frac{2x}{a^2} + \frac{2z}{c^2}\frac{\partial z}{\partial x} = 0$$

$$\Rightarrow \qquad \frac{\partial z}{\partial x} = -\frac{x c^2}{z a^2}.$$

Now,
$$\frac{\partial V}{\partial x} = 8yz + 8xy\frac{\partial z}{\partial x} = 8yz + 8xy\left(-\frac{x c^2}{z a^2}\right)$$

$$r = \frac{\partial^2 V}{\partial x^2} = 8y\frac{\partial z}{\partial x} - \frac{8yc^2}{za^2}\cdot 2x - \frac{8x^2yc^2}{a^2}\left(-\frac{1}{z^2}\frac{\partial z}{\partial x}\right)$$

$$= 8y\left(-\frac{xc^2}{za^2}\right) - \frac{16c^2xy}{za^2} + \frac{8x^2yc^2}{a^2z^2}\left(-\frac{xc^2}{za^2}\right)$$

$$= -\frac{8yc^2x}{a^2z} - \frac{16c^2xy}{a^2z} - \frac{8x^3yc^4}{a^4z^3}$$

$$= -\text{ve clearly when } x = \frac{a}{\sqrt{3}},\ y = \frac{b}{\sqrt{3}},\ z = \frac{c}{\sqrt{3}}.$$

Hence, when $x = \frac{a}{\sqrt{3}},\ y = \frac{b}{\sqrt{3}},\ z = \frac{c}{\sqrt{3}}$; V is maximum.

Putting these values of $x,\ y,\ z$ in (i), we get

$$V_{maximum} = 8\,\frac{a}{\sqrt{3}}\,\frac{b}{\sqrt{3}}\,\frac{c}{\sqrt{3}} = \frac{8abc}{3\sqrt{3}}.$$

Example 12. *If* $u = a^3x^2 + b^3y^2 + c^3z^2$ *where* $1/x + 1/y + 1/z = 1$, *show that an extreme value is given by* $ax = by = cz$ *and this gives a true maximum or minimum if* $abc(a + b + c)$ *is positive.*

Solution : If u is maximum or minimum, then

$$du = 0$$

or
$$2a^3x\,dx + 2b^3y\,dy + 2c^3z\,dz = 0$$

or
$$a^3x\,dx + b^3y\,dy + c^3z\,dz = 0. \qquad \text{...(i)}$$

Again, from $\frac{1}{x} + \frac{1}{y} + \frac{1}{z} = 1$, on differentiation, we get

$$\frac{dx}{x^2} + \frac{dy}{y^2} + \frac{dz}{z^2} = 0 \qquad \text{...(ii)}$$

Multiplying equation (i), (ii) by 1, λ respectively, adding and then equating to zero the coefficients of dx, dy, dz, we get

$$a^3 x + \frac{\lambda}{x^2} = 0$$

$$b^3 y + \frac{\lambda}{y^2} = 0$$

$$c^3 z + \frac{\lambda}{z^2} = 0$$

or
$$a^3 x^3 = b^3 y^3 = c^3 z^3 = -\lambda$$

or
$$a x = b y = c z.$$

Discrimination : Now to discuss maximum or minimum, we have from

$$1/x + 1/y + 1/z = 1,$$

$$\frac{1}{x^2} + \frac{1}{z^2}\frac{\partial z}{\partial x} = 0,$$

or
$$\frac{\partial z}{\partial x} = -\frac{z^2}{x^2}$$

Also
$$\frac{\partial z}{\partial y} = -\frac{z^2}{y^2}$$

Now,
$$\frac{\partial u}{\partial x} = 2a^3 x + 2c^3 z \frac{\partial z}{\partial x}$$

$$= 2a^3 x - 2c^3 z \left(\frac{z^2}{x^2}\right)$$

$$= 2a^3 x \quad \frac{2c^3 z^3}{x^2}$$

$\therefore$
$$\frac{\partial^2 u}{\partial x^2} = 2a^3 - 2c^3 \left[\frac{3z^2}{x^2}\frac{\partial z}{\partial x} - \frac{2z^3}{x^3}\right]$$

$$= 2a^3 + 2c^3 \left[\frac{3z^4}{x^4} + \frac{2z^3}{x^3}\right]$$

$$= 2a^3 + 2c^3 \left[\frac{3a^4}{c^4} + \frac{2a^3}{c^3}\right] \qquad \left[\begin{array}{l} \because ax = cz \\ \text{or } z/x = a/c \end{array}\right]$$

i.e.,
$$r = \frac{\partial^2 u}{\partial x^2} = 6a^3 + \frac{6a^4}{c}.$$

Similarly, $$t = \frac{\partial^2 u}{\partial y^2} = 6\,b^3 + \frac{6\,b^4}{c}$$

and $$s = \frac{\partial^2 u}{\partial x\,\partial y} = \frac{\partial}{\partial y}\left(\frac{\partial u}{\partial x}\right)$$

$$= -\,2\,c^3\,\frac{3\,z^2}{x^2}\,\frac{\partial z}{\partial y}$$

$$= \frac{6\,c^3\,a^4}{x^2\,y^2} = 6\,c^3\left(\frac{z^2}{x^2}\right)\left(\frac{z^2}{y^2}\right)$$

$$= 6\,c^3\left(\frac{a^2}{c^2}\right)\left(\frac{b^2}{c^2}\right) = \frac{6\,a^2\,b^2}{c}.$$

Now, for a true maximum or minimum,

$$rt - s^2 > 0$$

or $$\left(6\,a^3 + \frac{6\,a^4}{c}\right)\left(6\,a^3 + \frac{6\,b^4}{c}\right) - \left(\frac{6\,a^2\,b^2}{c}\right)^2 > 0$$

or $$\frac{36\,a^3\,b^3}{c^2}\,[c^2 + a\,c + b\,c] > 0$$

or $$\left(\frac{6\,a\,b}{c}\right)^2 a\,b\,c\,\,(a+b+c) > 0$$

or $$a\,b\,c\,\,(a + b + c) > 0.$$

$$\left[\text{Since } \left(6\,\frac{a\,b}{c}\right)^2 \text{ is essentially +ve}\right]$$

Example 13. *If A, B and C be the angles of a triangle ABC, then find the maxima and minima of* $u = \sin A \sin B \sin C$.

Solution : It is given $A + B + C = \pi$ ...(i)

and $$u = \sin A \sin B \sin C \qquad ...(ii)$$

$$F \equiv u + \lambda\,(A + B + C - \pi)$$

$$= \sin A \sin B \sin C + \lambda\,(A + B + C - \pi) \qquad ...(iii)$$

For extreme value of F

$$\frac{\partial F}{\partial A} = \cos A \sin B \sin C + \lambda = 0.$$

$$\lambda = -\,\cos\,A\,\,\sin\,B\,\,\sin\,C \qquad ...(iv)$$

$$\frac{\partial F}{\partial B} = \sin A \cos B \sin C + \lambda = 0$$

$$\lambda = -\sin A \cos B \sin C \qquad \text{...(v)}$$

and $$\frac{\partial F}{\partial C} = \sin A \sin B \cos C + \lambda = 0$$

$$\lambda = -\sin A \sin B \cos C. \qquad \text{...(vi)}$$

By equation (iv) and equation (v)

$$\cos A \sin B \sin C = \sin A \cos B \sin C$$

$$\cot A = \cot B$$

By equation (v) and equation (vi)

$$\cot B = \cot C$$

or $$\cot A = \cot B = \cot C$$

or $$A = B = C.$$

Using in equation (i) $A = B = C = \dfrac{\pi}{3}$

For maximum value of u, we take from equation (i) and (ii)

$$u = \sin A \sin B \sin (\pi - \overline{A + B})$$

$$= \sin A \sin B \sin (A + B)$$

$$\frac{\partial u}{\partial A} = \cos A \sin B \sin (A + B) + \sin A \sin B \cos (A + B)$$

$$= \sin B \sin (2A + B)$$

$$s = \frac{\partial^2 u}{\partial A\, \partial B} = \cos B \sin (2A + B) + \sin B \cos (2A + B)$$

$$= \sin (2A + 2B).$$

$$r = \frac{\partial^2 u}{\partial A^2} = 2 \cos (2A + B) \sin B$$

Similarly, $$t = \frac{\partial^2 u}{\partial B^2} = \cos (2B + A) \sin A.$$

If $A = B = C = \dfrac{\pi}{3}$, then

$$r = 2 \cos \left(\frac{3\pi}{3}\right) \cdot \sin \frac{\pi}{3} = -\sqrt{3}$$

$$s = \cos \frac{\pi}{3} \sin \pi + \cos \pi \sin \frac{\pi}{3} = -\frac{\sqrt{3}}{2}$$

$$t = -\sqrt{3}$$

$$rt - s^2 = (-\sqrt{3})(-\sqrt{3}) - \left(-\frac{\sqrt{3}}{2}\right)^2 = \frac{9}{4} > 0$$

$r < 0$ and $rt - s^2 > 0$.

Therefore, the value of $\sin A \sin B \sin C$ will be maximum when $A = B = C = \frac{\pi}{3}$.

Example 14. *Find the maximum value of rectangular parallelopiped in the side the ellipsoid* $\frac{x^2}{a^2} + \frac{y^2}{b^2} + \frac{z^2}{c^2} = 1$.

Solution : Let the co-ordinates of a vertex of parallelopiped which is in positive octant be (x, y, z).

The sides of the parallelopiped will be $2x$, $2y$ and $2z$.

$$\text{Volume } V = 8xyz. \qquad \text{...(i)}$$

The point (x, y, z) is on,

$$\frac{x^2}{a^2} + \frac{y^2}{b^2} + \frac{z^2}{c^2} = 2. \qquad \text{...(ii)}$$

Differentiate equation (i) by taking log of both sides

$$\log V = \log 8 + \log x + \log y + \log z$$

$$\frac{1}{v} dv = \frac{1}{x} dx + \frac{1}{y} dy + \frac{1}{z} dz \qquad \text{...(iii)}$$

Differentiating equation (i),

$$\frac{x}{a^2} dx + \frac{y}{b^2} dy + \frac{z}{c^2} dz = 0. \qquad \text{...(iv)}$$

Multiplying equation (iv) by λ and adding to equation (iii), we get

$$\left(\frac{1}{x} + \frac{\lambda x}{a^2}\right) dx + \left(\frac{1}{y} + \frac{\lambda y}{b^2}\right) dy + \left(\frac{1}{z} + \frac{\lambda^z}{c^2}\right) dz = 0.$$

Equating the coefficients of dx, dy and dz to zero

$$\frac{1}{x} + \frac{\lambda x}{a^2} = 0 \qquad \text{...(v)}$$

$$\frac{1}{y} + \frac{\lambda y}{b^2} = 0 \qquad \text{...(vi)}$$

$$\frac{1}{z} + \frac{\lambda z}{c^2} = 0 \qquad \text{...(vii)}$$

After multiplying equations (v), (vi) and (vii) by x, y and z respectively, add them

$$3 + \lambda\left(\frac{x^2}{a^2} + \frac{y^2}{b^2} + \frac{z^2}{c^2}\right) = 0$$

Using (ii)

$$3 + \lambda = 0 \Rightarrow \lambda = -3.$$

Putting the value of λ in equation (v), (vi) and (vii) ·

$$x = \frac{a}{\sqrt{3}},\ y = \frac{b}{\sqrt{3}} \text{ and } z = \frac{c}{\sqrt{3}}$$

Taking z as dependent variable and x, y as independent variables, differentiate it with respect to x and y partially.

$$\frac{2x}{a^2} + \frac{2z}{c^2}\frac{dz}{dx} = 0 \text{ or } \frac{dz}{dx} = -\frac{x}{a^2}\cdot\frac{c^2}{z}$$

Similarly,

$$\frac{dz}{dy} = -\frac{y}{b^2}\cdot\frac{c^2}{z}$$

Differentiating equation (i) w.r.t. x,

$$\frac{\partial v}{\partial x} = 8\left(yz + xy\frac{\partial z}{\partial x}\right) = 8\left[yz - \frac{yx^2c^2}{za^2}\right]$$

$$r = \frac{\partial^2 v}{\partial x^2} = 8\left[y\frac{\partial z}{\partial x} - \frac{2xy}{z}\frac{c^2}{a^2} + \frac{yx^2}{z^2}\frac{c^2}{a^2}\frac{\partial z}{\partial x}\right]$$

$$= 8\left[\frac{-yxc^2}{a^2z} - \frac{2xyc^2}{za^2} - \frac{yx^2c^2}{z^2a^2} \times \frac{xc^2}{a^2z}\right]$$

x, y, z are points therefore,

$$r < 0.$$

Similarly, we can show that $rt - s^2 > 0$.

The value of V will be maximum when

$$x = \frac{a}{\sqrt{3}},\ y = \frac{b}{\sqrt{3}} \text{ and } z = \frac{c}{\sqrt{3}}.$$

The maximum value of V

$$V = 8\cdot\frac{a}{\sqrt{3}}\cdot\frac{b}{\sqrt{3}}\cdot\frac{c}{\sqrt{3}} = \frac{8abc}{3\sqrt{3}}$$

Example 15. *If V be the value of the given cuboid then show that its whole surface will be minimum.*

Solution : Let x, y and z be the sides of the cuboid.

$$\text{Volume } V = xyz$$

and $\quad$ whole surface $s = 2(xy + yz + zx)$

For maximum or minimum of s,

$$ds = 2[(y + z)\,dx + (z + x)dy + (x + y)\,dz] = 0$$

$$(y + z)\,dx + (z + x)\,dy + (x + y)\,dz = 0 \qquad \text{...(i)}$$

Volume V is constant, $\quad yzdx + zxdy + yxdz = 0 \qquad$ ...(ii)

After multiplying equation (ii) by λ add to equation (i)

$$(y + z + \lambda yz)\,dx + (z + x + \lambda xz)\,dy + (x + y + \lambda xy)\,dz = 0$$

Equating the coefficients of dx, dy and dz to zero

$$y + z + \lambda yz = 0 \qquad \text{...(iii)}$$

$$x + z + \lambda xz = 0 \qquad \text{...(iv)}$$

$$x + y + \lambda xy = 0 \qquad \text{...(v)}$$

From there $\quad \dfrac{y + z}{yz} = \dfrac{x + z}{xz} = \dfrac{x + y}{xy} = -\lambda$

or $\quad \dfrac{1}{z} + \dfrac{1}{y} = \dfrac{1}{x} + \dfrac{1}{z} = \dfrac{1}{x} + \dfrac{1}{y} = -\lambda$

or $\quad \dfrac{1}{z} + \dfrac{1}{y} = \dfrac{1}{x} + \dfrac{1}{z}$

$$y = x$$

$$y = z$$

Thus, we get $\quad x = y = z$

Putting in $\quad V = xyz$

$$V = x^3 \text{ or } x = V^{1/3} = y = z$$

V is constant, therefore,

$$0 = yz + xy\frac{\partial z}{\partial x}$$

$$\frac{\partial z}{\partial x} = \frac{-z}{x}$$

and $\quad \dfrac{\partial z}{\partial y} = \dfrac{-z}{y}$

Differentiating s, w.r.t. x and y partially

$$\frac{\partial s}{\partial x} = 2\left[y + y\frac{\partial z}{\partial x} + z + x\frac{\partial z}{\partial x}\right]$$

$$= 2\left[y + z + (x + y)\frac{\partial z}{\partial x}\right]$$

$$= 2\left[y + z + (x + y)\left(\frac{-z}{x}\right)\right]$$

$$= 2\left[y - \frac{yz}{x}\right]$$

$$r_1 = \frac{\partial^2 s}{\partial x^2} = 2\left[\frac{yz}{x^2} - \frac{y}{x}\cdot\frac{\partial z}{\partial x}\right] = 2\left[\frac{yz}{x^2} + \frac{yz}{x^2}\right] = 4\frac{yz}{x^2}$$

$= 4$, because $x = y = z$.

$r_1 > 0$.

$$\frac{\partial s}{\partial y} = 2\left[x + z + y\frac{\partial z}{\partial y} + x\frac{\partial z}{\partial y}\right]$$

$$= 2\left[(x + z) + (y + x)\frac{\partial z}{\partial y}\right]$$

$$= 2\left[(x + z) - (x + y)\frac{z}{y}\right] = 2\left[x - \frac{xz}{y}\right]$$

$$t_1 = \frac{\partial^2 s}{\partial y^2} = 2\left[\frac{xz}{y^2} - \frac{x}{y}\frac{\partial z}{\partial y}\right] = 2\left[\frac{x^2}{y^2} + \frac{x^2}{y^2}\right] = 4\frac{x^2}{y^2}$$

$= 4$, because $x = y = z$.

$$s_1 = \frac{\partial^2 s}{\partial x\,\partial y} = 2\left[1 - \frac{z}{y} - \frac{x}{y}\frac{\partial z}{\partial x}\right] = 2\left[1 - \frac{z}{y} - \frac{x}{y}\left(\frac{-z}{x}\right)\right]$$

$$= 2$$

$$r_1 t_1 - s_1^2 = 16 - 4 = 12 > 0.$$

$r_1 > 0$ and $r_1 t_1 - s_1^2 > 0 \Rightarrow$ The whole surface of cuboid is minimum if volume is given.

EXERCISE 14 (B)

1. Show that the maximum value of u when $u = x^2 y^3 z^4$ and $2x + 3y + 4z = a$ is $(a/9)^9$.
2. Find the maximum or minimum values of $x^4 y z^2$ subject to the condition $a^2 x^2 - 2by^3 + z^4 = c^4$.
3. If $f = \dfrac{5xyz}{x + 2y + 4z}$ and $xyz = 8$, find x, y, z for which f is a maximum.
4. Discuss the maxima or minima of the function.

$$u = \sin x \sin y \sin z, \text{ where, } x + y + z = 180°.$$

5. Find the maximum or minimum value of $x^4 + y^4 + z^4$, where,

$$xyz = c^3.$$

6. Find the minimum value of $x + y + z$ under the condition

$$a/x + b/y + c/z = 1.$$

7. Find the minimum value of $x^2 + y^2 + z^2 + w^2 + ...$ under the condition $ax + by + cz + aw + ... = k$.

8. Given $x + y + z = a$, find the maximum value of xyz.

9. If $u = x^2 + y^2 + z^2$, where,

$$ax^2 + by^2 + cz^2 + 2hxy + 2fyz + 2gzx + 1,$$

find the maximum and minimum value of u.

10. If $(x^2 + y^2 + z^2) = a^2x^2 + b^2y^2 + c^2z^2$ and $lx + my + nz = 0$, show that the maximum and minimum values of $r^2 = x^2 + y^2 + z^2$ are given by the equation $\dfrac{l^2}{r^2 - a^2} + \dfrac{m^2}{r^2 - b^2} + \dfrac{n^2}{r^2 - c^2} = 0.$

11. Find the maxima and minima of $x^2 + y^2 + z^2$ subject to the conditions

$$ax + by + cz = 1$$

$$a'x + b'y + c'z = 1.$$

12. Find the maxima and minima of $u = x^2 + y^2 + z^2$ subject to the conditions $ax^2 + by^2 + cz^2 + 2hxy + 2fyz + 2gzx = 0$,
and $lx + my + nz = 0$.

13. Find the maximum value of $x^m y^n z^p$ with the condition $x + y + z = a$.

14. Find the minima of $u = x^2 + y^2 + z^2$ when $xy + yz + zx = 3a^2$.

15. Find a plane triangle such that $u = \sin^m A \cdot \sin^n B \cdot \sin^p C$ has a maximum value.

16. Find the maximum and minimum values of $\dfrac{x^2}{a^4} + \dfrac{y^2}{b^4} + \dfrac{c^2}{c^4}$, where,

$lx + my + nz = 0$ and $\dfrac{x^2}{a^2} + \dfrac{y^2}{b^2} + \dfrac{z^2}{c^2} = 1$. Interpret the result geometrically.

17. Find the maxima and minima of $x^2 + y^2 + z^2$ subject to following conditions :

$$ax + by + cz = 1, \quad a'x + b'y + c'z = 1.$$

ANSWERS

2. $a^2 x^2 = \dfrac{12\,c^4}{17},\; b\,y^3 = \dfrac{c^4}{17},\; z^4 = \dfrac{3\,c^4}{17}$ give the maximum value of u.

3. $x = 4,\; y = 2,\; z = 1.$

4. u is maximum at $x = y = z = \pi/3$.

5. Minimum when $x = y = z = c$.

6. $\dfrac{x}{\sqrt{a}} = \dfrac{y}{\sqrt{b}} = \dfrac{z}{\sqrt{c}} = \sqrt{a} + \sqrt{b} + \sqrt{c}$ give a minimum.

7. $k^2 (a^2 + b^2 + c^2 + \ldots)$

8. $a^3/27$.

9. Maximum or Minimum values of u are given by

$$\begin{vmatrix} a - 1/u & h & g \\ h & b - 1/u & f \\ g & f & c - 1/u \end{vmatrix} = 0.$$

11. Maximum or minimum values of u are given by

$$\begin{vmatrix} u & 1 & 1 \\ 1 & a^2 + b^2 + c^2 & a\,a' + b\,b' + c\,c' \\ 1 & a\,a' + b\,b' + c\,c' & a'^2 + b'^2 + c'^2 \end{vmatrix} = 0.$$

12. $$\begin{vmatrix} a - \frac{1}{4} & h & g & l \\ h & b - \frac{1}{4} & f & m \\ g & f & c - \frac{1}{4} & n \\ l & m & n & o \end{vmatrix} = 0.$$

13. Maximum value of $x^m y^n z^p$ corresponds to

$$x = \frac{a\,m}{m + n + p},\quad y = \frac{a\,n}{m + n + p},\quad z = \frac{a\,p}{m + n + p}.$$

14. u has a minimum at the points (a, a, a) and $(-a, -a, -a)$.

15. $\dfrac{\tan A}{m} = \dfrac{\tan B}{n} = \dfrac{\tan C}{p}$.

16. $\dfrac{l^2 a^4}{1 - a^2 u} + \dfrac{m^2 b^4}{1 - b^2 u} + \dfrac{u^2 c^4}{1 - c^2 u} = 0.$

17. $$\begin{vmatrix} u & 1 & 1 \\ 1 & a^2 + b^2 + c^2 & a\,a' + b\,b' + c\,c' \\ 1 & a\,a' + b\,b' + c\,c' & a'^2 + b'^2 + c'^2 \end{vmatrix} = 0.$$

15

Asymptotes

15.1. Asymptote

Some curves are limited in extent, for example the circle; others extend to infinity, for example the parabola or the hyperbola. In the latter case it is interesting to enquire what happens to the tangent when the point at which the tangent is drawn are three possibilities, corresponding to three possibilities in the cases of series, viz., that a series may be divergent, oscillatory or convergent. Thus, it is possible that the tangent may go further and further away from the origin, or it may keep oscillating or it may tend to a definite straight line. In the last case the straight line to which the tangent tends is called the *asymptote.* The formal definition is as follows—

A straight line at a finite distance from origin, which cuts a curve in two points at infinite distance from origin and yet is not itself wholly at infinity, is called an asymptote to the curve.

15.2 Condition for the Existence of the Asymptote

Let the equation of the curve be

$$y = f(x) \qquad ...(1)$$

Then the tangent to the curve at (x, y) is,

$$Y - y = \frac{dy}{dx}(X - x)$$

$$\Rightarrow \qquad Y = X\frac{dy}{dx} + y - x\frac{dy}{dx}. \qquad ...(2)$$

The tangent given by (2) will be an asymptote (excluding the case of asymptote parallel to Y-axis) if

$$\lim_{x \to \infty} \frac{dy}{dx} \text{ and } \lim_{x \to \infty} \left(y - x\frac{dy}{dx}\right)$$

both respectively tend to finite value say m and c. If these conditions are satisfied, then the asymptote will be

$$y = mx + c \qquad ...(3)$$

15.3. If $y = mx + c$ is an asymptote to a curve, then

$$m = \lim_{x \to \infty} \frac{y}{x}.$$

Proof: By article 15.2, we know that if $y = mx + c$ is an asymptote to a curve, then

$$\lim_{x \to \infty} \left(y - x \frac{dy}{dx} \right) = c$$

$$\Rightarrow \qquad \lim_{x \to \infty} \left(\frac{y}{x} - \frac{dy}{dx} \right) = \lim_{x \to 0} \frac{c}{x} = 0$$

$$\lim_{x \to \infty} \frac{y}{x} = \lim_{x \to \infty} \frac{dy}{dx}$$

$$\lim_{x \to \infty} \frac{y}{x} = m$$

15.4. Asymptotes of Algebraic Curves

Let a rational integral algebraic equation of nth degree be

$$f(x, y) = 0 \qquad ...(1)$$

Let $y = mx + c$ be an asymptote of (1).

Putting $mx + c$ for y in (1), let the resulting equation be of the form

$$a_0 x^n + a_1 x^{n-1} + a_2 x^{n-2} + \ldots + a_{n-1} x + a_n = 0. \qquad ...(2)$$

The coefficients a_0, a_1, a_2, ..., a_{n-1}, a_n are functions of m and c.

Equation (2) when solved, gives the points of intersection of (1) with $y = mx + c$. Since $y = mx + c$ is an asymptote of (1), therefore it must meet (1) in at least two points at infinity. Thus, (2) must have two infinite roots. Therefore we must have

$$a_0 = 0 \text{ and } a_1 = 0.$$

Solving these two equations we can find m and c. Hence, the asymptotes of the curve (1) can be found.

Working rule : Suppose $y = mx + c$ is an asymptote of the curve. Put $mx + c$ for y in the equation of the curve and arrange it in descending powers of x. Equate to zero the coefficients of two highest degree terms. Solve these equations to find the values of m and c. For every pair of values of m and c, there is one asymptote $y = mx + c$ of the curve.

ILLUSTRATIVE EXAMPLES

Example 1. *Find the asymptote of the curve*

$$x^3 + 2x^2 y - xy^2 - 2y^3 + 4y^2 + 2xy + y - 1 = 0$$

Solution : Let $y = mx + c$ be an asymptote of the given curve. Then putting $mx + c$ for y in the equation of the curve, we have

$$x^3 + 2x^2 (mx + c) - x(mx + c)^2 - 2 (mx + c)^3 + 4 (mx + c)^2 + 2x (mx + c) + (mx + c) - 1 = 0$$

$$\Rightarrow (1 + 2m - m^2 - 2m^3) x^3 + 2 (c - mc - 3m^2 c + 2m^2 + m) x^2 + \dots + \dots + \dots = 0$$

Equating to zero the coefficients of two highest degree terms in x, we get

$$1 + 2m - m^2 - 2m^3 = 0 \quad \text{...(1)}$$

and $$c - mc - 3m^2 c + 2m^2 + m = 0$$

i.e., $$c (1 - m - 3m^2) + 2m^2 + m = 0. \quad \text{...(2)}$$

Now, from (1), we have on factorization

$$(1 + m) (1 + 2m) (1 - m) = 0$$

i.e., $$m = -1, 1, -\frac{1}{2}$$

From (2),

$$c = -\frac{2m^2 + m}{1 - m - 3m^2}$$

Putting in above $m = -1, 1, -\frac{1}{2}$, we get

$$c = 1, 1, 0 \text{ respectively.}$$

Hence, the asymptotes are,

$$y = -x + 1 \quad \text{i.e., } x + y - 1 = 0$$

$$y = x + 1 \quad \text{i.e., } x - y + 1 = 0$$

$$y = -\frac{1}{2}x + 0 \quad \text{i.e., } x + 2y = 0$$

Example 2. *Find the asymptotes of the curve*

$$x^3 + 2x^2 y - xy^2 - 2y^3 + xy - y^2 = 1$$

Solution : The given curve is,

$$x^3 + 2x^2 y - xy^2 - 2y^3 + xy - y^2 = 1 \quad \text{...(1)}$$

Let $y = mx + c$ be an asymptote of (1). Then putting $mx + c$ for y in (1), we get

$$x^3 + 2x^2 (mx + c) - x (mx + c)^2 - 2 (mx + c)^3 + x (mx + c)$$

$$\Rightarrow \quad x^3 + 2mx^3 + 2cx^2 - m^2x^3 - c^2x - 2mcx^2 - 2m^3x^3 - 2c^3$$
$$- 6m^2cx^2 - 6c^2mx + mx^2 + cx - m^2x^2 - c^2 - 2mcx - 1 = 0$$
$$\Rightarrow x^3(1 + 2m - m^2 - 2m^3) + x^2(2c - 2mc - 6m^2c + m - m^2)$$
$$+ x(-c^2 - 6c^2m + c - 2mc) - 2c^3 - c^2 - 1 = 0$$

Equating to zero the coefficients of two highest degree terms in x, we get

$$1 + 2m - m^2 - 2m^3 = 0 \qquad ...(2)$$

and
$$2c - 2mc - 6m^2c + m - m^2 = 0$$

i.e.,
$$2c(1 - m - 3m^2) = m^2 - m$$

i.e.,
$$c = \frac{m^2 - m}{2(1 - m - 3m^2)} \qquad ...(3)$$

Now, from (1), we have on factorization

$$(1 + m)(1 - m)(1 + 2m) = 0$$

$$\therefore \qquad m = 1, 1, -\frac{1}{2}$$

Putting $m = -1, 1, -\frac{1}{2}$ in (iii), we get $c = -1, 0, \frac{1}{2}$ respectively.

Hence, the asymptotes are,

$$y = -x - 1 \quad \text{i.e.,} \quad x + y + 1 = 0$$
$$y = x + 0 \quad \text{i.e.,} \quad x - y = 0$$
$$y = -\frac{1}{2}x + \frac{1}{2} \quad \text{i.e.,} \quad 2x + y - 1 = 0$$

15.5. Another Method for Finding Asymptotes of General Algebraic Curves

Let the general algebraic equation of the curve of nth degree be

$$(a_0x^n + a_1x^{n-1}y + a_2x^{n-2}y^2 + \ldots + a_ny^n) + (b_0x^{n-1} + b_1x^{n-2}y$$
$$+ b_2x^{n-3}y^2 + \ldots + b_{n-1}y^{n-1}) + (c_0x^{n-2} + c_1x^{n-3}y + \ldots +$$
$$c_{n-3}y^{n-2}) + \ldots + (k_0x + k_1y) + k = 0. \qquad ...(1)$$

The above equation can be written in the form

$$x^n\phi_n\left(\frac{y}{x}\right) + x^{n-1}\phi_{n-1}\left(\frac{y}{x}\right) + x^{n-2}\phi_{n-2}\left(\frac{y}{x}\right) + \ldots$$

$$\ldots + x\,\phi_1\left(\frac{y}{x}\right) + \phi_0\left(\frac{y}{x}\right) = 0 \qquad \ldots(2)$$

where, $\phi_r\left(\frac{y}{x}\right)$ is an expression of rth degree in $\frac{y}{x}$.

Let $y = mx + c$ be an asymptote of the curve, where m and c are finite.

To find *m* : Divide both sides of (2) by x^n, we get

$$\phi_n\left(\frac{y}{x}\right) + \frac{1}{x}\,\phi_{n-1}\left(\frac{y}{x}\right) + \frac{1}{x^2}\,\phi_{n-2}\left(\frac{y}{x}\right) + \ldots$$

$$\ldots + \frac{1}{x^{n-1}}\,\phi_1\left(\frac{y}{x}\right) + \frac{1}{x^n}\,\phi_0\left(\frac{y}{x}\right) = 0 \qquad \ldots(3)$$

Proceeding to limit as $x \to \infty$ (excluding the case of asymptotes parallel to Y-axis), so that $\lim_{x \to \infty} \frac{y}{x} = m$, we get

$$\phi_n(m) = 0. \qquad \ldots(4)$$

This gives the slopes (gradients) of the asymptotes.

Equation (4) is an equation of nth degree in m. Hence, it will give n values of m. Hence, (1) will have n asymptotes in general. We are concerned with only real roots and therefore reject imaginary roots if there are any. Differentiating (5) w.r.t. x, we have

$$\left[\phi_n'\left(\frac{y}{x}\right) + \frac{1}{x}\,\phi_{n-1}'\left(\frac{y}{x}\right) + \ldots\right]\frac{\left\{x\frac{dy}{dx} - y\right\}}{x^2}$$

$$-\frac{1}{x^2}\,\phi_{n-1}\left(\frac{y}{x}\right) - \frac{2}{x^3}\,\phi_{n-2}\left(\frac{y}{x}\right) + \ldots = 0. \qquad \ldots(5)$$

Now, multiplying (5) throughout by x^2 and taking limit as $x \to \infty$, we have

$$c\,\phi_n'(m) + \phi_{n-1}(m) = 0$$

$$\left[\because \quad c = \lim_{x \to \infty}\left(y - x\frac{dy}{dx}\right)\right]$$

$$\Rightarrow \qquad c = \frac{\phi_{n-1}(m)}{\phi_n'(m)} \qquad \ldots(6)$$

Equation (6) determines the value of c corresponding to every value of m found earlier from (4).

Hence, for every pair of values of m and c, there exists an asymptote of the form $y = mx + c$.

Working rule : (i) In the highest degree terms of the given equation of the curve put $x = 1$ and $y = m$. Thus, we get $\phi_n (m)$.

(ii) Equate $\phi_n (m)$ to zero and solve for m. Let its roots be $m_1, m_2, ..., m_n$.

(iii) Find $\phi_{n-1} (m)$ by putting $x = 1$, $y = m$ in the next higher degree terms of the equation.

(iv) Find the values $c_1, c_2, c_3, ...$ of c corresponding to the values $m_1, m_2, m_3, ...$ of m using the equation

$$c = -\frac{\phi_{n-1} (m)}{\phi_n' (m)},$$

where, $\phi'_n (m)$ denotes the expression obtained on differentiating $\phi_n (m)$ w.r.t. m.

(v) Then the required asymptotes are,

$$y_1 = m_1 x + c_1,$$

$$y_2 = m_2 x + c_2,$$

$$\dots \quad \dots \quad \dots$$

$$\dots \quad \dots \quad \dots$$

(vi) If any value of m comes out to be imaginary, we get an imaginary asymptote corresponding to that value of m. In this case we say that the asymptote does not exist.

(vii) If $\phi_n' (m) = 0$ for some value of m and $\phi_{n-1} (m) \neq 0$ corresponding to that value, then there exists no asymptote corresponding to that value of m.

ILLUSTRATIVE EXAMPLES

Example 1. *Find the asymptotes of following curve—*

$$x^3 + 2x^2 y - xy^2 - 2y^3 + xy - y^2 = 1.$$

Solution : Here the highest degree is 3. So putting $x = 1$ and $y = m$ in third degree terms, we see that

$$\phi_n (m) = 1 + 2m - m^2 - 2m^3$$

Now $\phi_n (m) = 0$ gives

$$\Rightarrow \quad 1 + 2m - m^2 - 2m^3 = 0$$

$$\Rightarrow \quad 1 - m^2 + 2m - 2m^3 = 0$$

$$\Rightarrow \quad 1 - m^2 + 2m (1 - m^2) = 0$$

$$\Rightarrow \quad (1 - m^2) (1 + 2m) = 0$$

$$\Rightarrow \quad (1-m)(1+m)(1+2m) = 0$$

$$\Rightarrow \quad m = 1, -1, -\frac{1}{2}$$

Now, putting $x = 1$ and $y = m$ in second degree terms, we see that

$$\phi_{n-1}(m) = m - m^2$$

Now,
$$c = -\frac{\phi_{n-1}(m)}{\phi_n'(m)}$$

$$= \frac{m^2 - m}{2 - 2m - 6m^2}$$

When $m = 1, \quad c = 0$

When $m = -1, \quad c = -1$

When $m = -\frac{1}{2}, \quad c = \frac{1}{2}$

Hence, the required asymptotes are—

$y = 1 \cdot x + 0, \; y = (-1)\,x - 1$

and $y = -\frac{1}{2}x + \frac{1}{2}$

$y = x, \; y = -x - 1$

and $2y = -x + 1$

Example 2. *Find the asymptotes of the curve*

$$x^3 + 2x^2y - xy^2 - 2y^3 + 3xy + 3y^2 + x + 1 = 0$$

Solution : Here highest degree is 3, i.e., $n = 3$

Putting $x = 1$ and $y = m$ in third degree terms, we get

$$\phi_n(m) = 1 + 2m - m^2 - 2m^3$$

Now, $\phi_n(m) = 0$ gives

$$1 + 2m - m^2 - 2m^3 = 0$$

$$\Rightarrow \quad (m-1)(m+1)(2m+1) = 0$$

$$\Rightarrow \quad m = 1, -1, -\frac{1}{2}$$

Again put $x = 1$ and $y = m$ in second degree terms, we get

$$\phi_{n-1}(m) = 3m + 3m^2$$

Differentiating $\phi_n(m)$ w.r.t. m, we get

$$\phi_n'(m) = 2 - 2m - 6m^2$$

Now, $$c = -\frac{\phi_{n-1}(m)}{\phi_n'(m)}$$

$$= -\frac{3m + 3m^2}{2 - 2m - 6m^2}$$

$$= \frac{3m + 3m^2}{2m + 6m^2 - 2}$$

When $m = 1$, $c = 1$.

When $m = -1$, $c = 0$.

When $m = -\frac{1}{2}$, $c = \frac{1}{2}$.

Hence, the asymptotes are—

$y = 1 \cdot x + 1$ i.e., $y - x - 1 = 0$

$y = -1 \cdot x + 0$ i.e., $y + x = 0$

$y = -\frac{1}{2}x + \frac{1}{2}$ i.e., $2y + x - 1 = 0$

Example 3. *Find the asymptotes of the curve*

$$y^3 - x^2y - 2xy^2 + 2x^3 - 7xy + 3y^2 + 2x^2 + 2x + 2y + 1 = 0$$

Solution : Here highest degree is 3, i.e., $n = 3$.

Putting $x = 1$ and $y = m$ in third degree terms, we get

$$\phi_n(m) = m^3 - m - 2m^2 + 2$$

Now, $\phi_n(m) = 0$ gives

$$m^3 - m - 2m^2 + 2 = 0$$

$$\Rightarrow \quad m^2(m-2) - 1(m-2) = 0$$

$$\Rightarrow \quad (m-2)(m^2-1) = 0$$

$$\Rightarrow \quad (m-2)(m-1)(m+1) = 0$$

$$\Rightarrow \quad m = 1, -1, 2$$

Again, putting $x = 1$ and $y = m$ in the second degree terms, we get

$$\phi_{n-1}(m) = -7m + 3m^2 + 2$$

Differentiating $\phi_n(m)$ w.r.t. m, we get

$$\phi_n'(m) = 3m^2 - 1 - 4m$$

Now, $$c = -\frac{\phi_{n-1}(m)}{\phi_n'(m)}$$

$$= -\frac{-7m + 3m^2 + 2}{3m^2 - 1 - 4m}$$

$$= \frac{7m - 3m^2 - 2}{3m^2 - 1 - 4m}$$

When $m = 1$, $c = -1$.

When $m = -1$, $c = -2$.

When $m = 2$, $c = 0$.

Hence, the asymptotes are—

$$y = -1,\ y = -x - 2 \text{ and } y = 2x$$

Example 4. *Find the asymptotes of the curve*

$$x^3 + y^3 = 3axy$$

Solution : Here the highest degree is 3, i.e., $n = 3$.

Putting $x = 1$ and $y = m$ in third degree terms, we get

$$\phi_n(m) = 1 + m^3$$

Now, $\phi_n(m) = 0$ gives

$$1 + m^3 = 0$$

$$\Rightarrow \quad (1 + m)(1 - m + m^2) = 0$$

$$\Rightarrow \quad m = -1, \frac{1}{2} \pm \frac{\sqrt{3}}{2} i.$$

Discarding imaginary values of m, we get $m = -1$.

Again, putting $x = 1$ and $y = m$ in second degree terms, we get

$$\phi_{n-1}(m) = -3am$$

Differentiating $\phi_n(m)$ w.r.t. m, we get

$$\phi_n'(m) = 3m^2$$

$$\therefore \quad c = -\frac{\phi_{n-1}(m)}{\phi_n'(m)} = -\left(\frac{-3am}{3m^2}\right) = \frac{a}{m}$$

Putting $m = -1$, we get $c = \frac{a}{-1} = -a$

Hence, the asymptote is $y = -1(x) - a$, i.e., $y + x + a = 0$.

EXERCISE 15 (A)

Find the asymptotes to the following curves :

1. $y^3 - x^2 y + 2y^2 + 4y + x = 0$

2. $x^3 + 3x^2y - xy^2 - 3y^3 + x^2 - 2xy + 3y^2 + 4x + 7 = 0$

3. $2y^3 - 2x^2y - 4xy^2 + 4x^3 - 14xy + 6y^2 + 4x^2 - 6y + 4x + 7 = 0$

4. $x^3 - 2y^3 + 2x^2y - xy^2 + xy - y + 1 = 0$

5. $3x^3 + 2x^2y - 7xy^2 + 2y^3 - 14xy + 7y^2 + 4x + 5y = 0$

6. $y^3 - 3x^2y + xy^2 - 3x^3 + 2y^2 + 2xy + 4x + 5y + 6 = 0$

7. $2x^3 - x^2y - 2xy^2 + y^3 - 4x^2 + 8xy + 4x + 1 = 0$

8. $4x^2(y - x) + y(y - 2)(x - y) = 4x + 4y - 7$

ANSWERS

1. $y = 0,\ y = x - 1,\ y = -x - 1.$
2. $4x - 4y + 1 = 0,\ 2x + 2y - 3 = 0,\ 4x + 12y + 9 = 0$
3. $2x - y = 0,\ x - y - 1 = 0,\ x + y + 2 = 0$
4. $y = x + \frac{1}{6},\ y = -x - \frac{1}{2},\ y = -\frac{1}{2}x + \frac{1}{3}$
5. $6x - 6y - 7 = 0,\ 6x - 2y - 3 = 0,\ 3x + 6y - 5 = 0$
6. $x + y = 0,\ \sqrt{3x} - y - 1 = 0,\ \sqrt{3x} + y + 1 = 0$
7. $x - y + 2 = 0,\ x + y - 2 = 0,\ 2x - y - 4 = 0$
8. $x - y = 0,\ 2x - y + 1 = 0,\ 2x + y - 1 = 0$

15.6. Asymptotes Might Not Exist

If one or more values of m, found from $\phi_n(m) = 0$, make $\phi_n(m)$ zero, but do not make $\phi_{n-1}(m)$ zero, the equation for determining the corresponding value of c becomes

$$0 \cdot c + \phi_{n-1}(m) = 0$$

This means that the equation in c was $Fc + G = 0$, where

$$\lim_{x \to \infty} F = 0 \text{ and } \lim_{x \to \infty} G = \phi_{n-1}(m)$$

Hence, $\lim\limits_{x \to \infty} c = +\infty$ or $-\infty$, and this corresponds to the case when the tangent goes further and further away as $x \to \infty$.

ILLUSTRATIVE EXAMPLES

Example 1. *Show that the parabola* $y^2 = 4ax$ *has no asymptotes.*

Solution : Here the highest degree is 2, i.e., $n = 2$.

Putting $x = 1$ and $y = m$ in second degree terms, we get

$$\phi_n(m) = m^2.$$

Now, $\phi_n(m) = 0$ gives

$$m^2 = 0$$

$\therefore$ $m = 0, 0.$

Again, putting $x = 1$ and $y = m$ in first degree terms, we get

$$\phi_{n-1}(m) = -4a$$

Differentiating $\phi_n(m)$ w.r.t. m, we get

$$\phi_n'(m) = 2m$$

$\therefore$
$$c = -\frac{\phi_{n-1}(m)}{\phi_n'(m)} = +\frac{4a}{2m} = +\frac{2a}{m}$$

Putting $m = 0$, we get $c = \dfrac{2a}{0,}$ *i.e.*, $c \to \infty$

Hence, the given curve has no asymptotes.

Example 2. *Find the asymptotes of the curve* $y^2 = x$.

Solution : Putting $y = mx$ in $y^2 = x$, we get

$$(mx + c)^2 = x$$

$$\Rightarrow \quad m^2x^2 + x(2mc - 1) + c^2 = 0$$

Equating to zero the coefficients of x^2 and x, we have

$$m^2 = 0 \text{ and } 2mc - 1 = 0$$

The first gives $m = 0$. Then the second reduces to $-1 = 0$, which is impossible. Hence $y^2 = x$ has no asymptotes.

15.7. Two Parallel Asymptotes

The curve is said to have two parallel asymptotes if any two values of m obtained from $\phi_n(m) = 0$ are equal. This repeated value of m will make $\phi_n'(m)$ and $\phi_{n-1}(m)$, both, zero. Hence, the equation

$$c = -\frac{\phi_{n-1}(m)}{\phi_n'(m)}$$

takes the form of an identity $c \cdot 0 + 0 = 0$.

In such case, the values of c corresponding to repeated value of m are obtained by the equation

$$\frac{c^2}{2!}\phi_n''(m) + c\,\phi_{n-1}'(m) + \phi_{n-2}(m) = 0.$$

This equation is quadratic in c. Hence, it will give two values of c corresponding to two equal values of m. Thus, we have two parallel asymptotes for two equal values of m.

15.8. Three Parallel Asymptotes

The curve is said to have three parallel asymptotes, if any three values of m obtained from $\phi_n(m) = 0$ are equal. In such cases, the values of c corresponding to three equal values of m are determined from the following cubic in c.

$$\frac{1}{3!} c^3 \phi_n'''(m) + \frac{c^2}{2!} \phi_{n-1}''(m) + c\,\phi_{n-2}'(m) + \phi_{n-3}(m) = 0$$

This method can in general be applied to any number of parallel asymptotes.

ILLUSTRATIVE EXAMPLES

Example 1. *Find the asymptotes of the curve*

$$x^3 + 3x^2y - 4y^3 - x + y + 3 = 0$$

Solution : Here given equation is

$$x^3 + 3x^2y - 4y^3 - x + y + 3 = 0 \qquad \ldots(1)$$

Here the highest degree is 3, i.e., $n = 3$.

Putting $x = 1$ and $y = 3$ in third degree terms, we get

$$\phi_n(m) = 1 + 3m - 4m^3$$

Now, $\phi_n(m) = 0$, gives

$$1 + 3m - 4m^3 = 0$$

$$\Rightarrow \quad 1 + 4m - m - 4m^3 = 0$$

$$\Rightarrow \quad (1 - m) + 4m - 4m^3 = 0$$

$$\Rightarrow \quad (1 - m) + 4m(1 - m^2) = 0$$

$$\Rightarrow \quad (1 - m)[1 + 4m(1 + m)] = 0$$

$$\Rightarrow \quad (1 - m)(2m + 1)^2 = 0$$

$$\Rightarrow \quad m = 1 \text{ and } m = -\tfrac{1}{2}, -\tfrac{1}{2}$$

Put $x = 1$ and $y = m$ in second degree terms, we get

$$\phi_{n-1}(m) = 0.$$

Put $x = 1$ and $y = m$ in first degree terms, we get

$$\phi_{n-2}(m) = -1 + m.$$

The value of c corresponding to the non-repeated value of m viz., 1 is given by

$$c = \frac{-\phi_{n-1}(m)}{\phi'_n(m)} = \frac{0}{3 - 12m^2} = 0$$

The corresponding asymptote is

$$y = x + 0$$

Again the value of c corresponding to two repeated values of m $\left(\text{viz., } -\frac{1}{2}, -\frac{1}{2}\right)$ are given by

$$\frac{c^2}{2!}\phi''_n(m) + c\phi'_n(m) + \phi_{n-2}(m) = 0$$

$$\Rightarrow \qquad \frac{c^2}{2}(-24m) + c \times 0 + (-1 + m) = 0$$

$$\Rightarrow \qquad \frac{c^2}{2}\left(-24 \times -\frac{1}{2}\right) + \left(-1 - \frac{1}{2}\right) = 0$$

$$\Rightarrow \qquad 6c^2 = \frac{3}{2}$$

$$\Rightarrow \qquad c^2 = \frac{1}{4}$$

$$\Rightarrow \qquad c = \pm\frac{1}{2}$$

Hence, the corresponding asymptotes are

$$y = -\frac{1}{2}x \pm \frac{1}{2}$$

Example 2. *Find all the asymptotes of the curve*

$$y^3 - 5xy^2 + 8x^2y - 4x^3 - 3y^2 + 9xy - 6x^2 + 2y - 2x + 1 = 0$$

Solution : The curve is of degree 3, i.e., $n = 3$.

Putting $x = 1$ and $y = m$ in the highest, i.e., third degree terms, we have

$$\phi_n(m) = m^3 - 5m^2 + 8m - 4.$$

Now, $\phi_n(m) = 0$ gives

$$m^3 - 5m^2 + 8m - 4 = 0$$

$$\Rightarrow \qquad (m - 1)(m^2 - 4m + 4) = 0$$

$$\Rightarrow \qquad (m - 1)(m - 2)^2 = 0$$

$\therefore \quad m = 1, 2, 2$

Again, putting $x = 1, y = m$ in the second degree terms, we have

$$\phi_{n-1}(m) = -3m^2 + 9m - 6$$

Differentiating $\phi_n(m)$ w.r.t. m, we get

$$\phi_n'(m) = 3m^2 - 10m + 8.$$

Now,
$$c = -\frac{\phi_{n-1}(m)}{\phi_n'(m)}$$

$$= -\frac{-3m^2 + 9m - 6}{3m^2 - 10m + 8}$$

$$= \frac{3m^2 - 9m + 6}{3m^2 - 10m + 8}$$

When $m = 1$,
$$c = \frac{3 - 9 + 6}{3 - 10 + 8} = \frac{0}{1} = 0$$

When $m = 2$,
$$c = \frac{12 - 18 + 6}{12 - 20 + 8} = \frac{0}{0} \text{ which is indeterminate.}$$

$\therefore$ In this case c is given by

$$\frac{c^2}{2!}\phi_n''(m) + c\phi_{n-1}'(m) + \phi_{n-2}(m) = 0 \qquad \text{...(1)}$$

Differentiating $\phi_n'(m)$ and $\phi_{n-1}(m)$ w.r.t. m, we get

$$\phi_n''(m) = 6m - 10$$

$$\phi_{n-1}'(m) = -6m + 9$$

Again, putting $x = 1, y = m$ in first degree terms, we have

$$\phi_{n-2}(m) = 2m - 2$$

Now, (1) gives

$$\frac{c^2}{2} \cdot (6m - 10) + c(-6m + 9) + 2m - 2 = 0$$

$$c^2(3m - 5) + c(-6m + 9) + 2m - 2 = 0$$

Putting $m = 2$, we have

$$c^2 - 3c + 2 = 0$$

$$\Rightarrow \quad (c - 1)(c - 2) = 0$$

$$\Rightarrow \quad c = 1, 2.$$

Putting the pairs of values of m and c in the equation $y = mx + c$, the required asymptotes are,

$$y = 1 \cdot x + 0 \quad \text{i.e.,} \quad x - y = 0$$

$$y = 2 \cdot x + 1 \quad \text{i.e.,} \quad 2x - y + 1 = 0$$

$$y = 2 \cdot x + 2 \quad \text{i.e.,} \quad 2x - y + 2 = 0$$

Example 3. *Find all the asymptotes of the curve*

$$y^3 + x^2 y + 2xy^2 - y + 1 = 0$$

Solution : Here the given equation is

$$y^3 + x^2 y + 2xy^2 - y + 1 = 0 \qquad ...(1)$$

Putting $x = 1$ and $y = m$ in third degree terms, we get

$$\phi_n(m) = m^3 + m + 2m^2$$

Now, $$\phi_n(m) = 0$$

$\Rightarrow$ $$m^3 + m + 2m^2 = 0$$

$\Rightarrow$ $$m(m+1)^2 = 0$$

$\Rightarrow$ $$m = 0, m = -1, -1$$

Again, putting $x = 1$ and $y = m$ in second degree terms, we have

$$\phi_{n-1}(m) = 0.$$

Again putting $x = 1$ and $y = m$ in first degree terms, we have

$$\phi_{n-2}(m) = -m$$

Differentiating $\phi_n(m)$ w.r.t. m, we get

$$\phi'_n(m) = 3m^2 + 4m + 1$$

and $$\phi''_n(m) = 6m + 4$$

The value of c corresponding to the non-repeated value of m viz., 0 is given by

$$c = \frac{-\phi_{n-1}(m)}{\phi'_n(m)}$$

$$= -\frac{0}{1}$$

$\therefore$ The corresponding asymptote is

$$y = 0.$$

Again the value of c corresponding to two repeated values of m (viz., $-1, -1$) are

$$\frac{c^2}{2!} \phi''_n(m) + c\, \phi'_n(m) + \phi_{n-2}(m) = 0$$

$$\Rightarrow \qquad \frac{c^2}{2}(6m+4) + c \cdot 0 + (-m) = 0$$

$$\Rightarrow \qquad \frac{c^2}{2}(-6+4) + 1 = 0 \qquad \text{[Since } m = -1]$$

$$\Rightarrow \qquad c^2 = 1$$

$$\Rightarrow \qquad c = \pm 1.$$

Hence, the corresponding asymptotes are

$$y = -x \pm 1.$$

Example 4. *Find all the asymptotes of the curve*

$$(x+y)^2 (x+2y+2) = x + 9y - 2$$

Solution: Here the highest degree is 3, i.e., $n = 3$. The given equation can be written as

$$(x+y)^2 (x+2y) + 2(x+y)^2 - (x+9y) + 2 = 0 \qquad ...(1)$$

Putting $x = 1$ and $y = m$ in third degree terms, we get

$$\phi_n(m) = (1+m)^2 (1+2m)$$

Now, $\qquad \phi_n(m) = 0$

$$\Rightarrow \qquad m = -1, -1 \text{ and } m = -½$$

Again putting $x = 1$ and $y = m$ in second degree terms, we get

$$\phi_{n-1}(m) = 2(1+m)^2$$

Again by putting $x = 1$ and $y = m$ in first degree terms, we get

$$\phi_{n-2}(m) = -(1+9m)$$

Now,
$$\phi_n'(m) = (1+m)^2 \cdot 2 + 2(1+m)(1+2m)$$
$$= (1+m)[2m + 2 + 4m + 2]$$
$$= (1+m)(6m+4)$$

$$\phi_n''(m) = (1+m)\,6 + (6m+4)$$
$$= 12m + 10$$

and
$$\phi_{n-1}'(m) = 4(1+m).$$

The value of c corresponding to the non-repeated value of m viz., $-\frac{1}{2}$ is given by

$$c = -\frac{\phi_{n-1}(m)}{\phi_n'(m)}$$

$$= \frac{-2(1+m)^2}{(1+m)(6m+4)}$$

$$= \frac{-2\left(1-\frac{1}{2}\right)^2}{\left(1-\frac{1}{2}\right)\left(6\times-\frac{1}{2}+4\right)}$$

$$= \frac{-2\,\frac{1}{2}}{(-3+4)}$$

$$= -1$$

$\therefore$ The corresponding asymptote is

$$y = -\frac{1}{2}x - 1$$

Again the value of c corresponding to two repeated values of m (viz., $-1, -1$) are

$$\frac{c^2}{2!}\phi_n''(m) + c\,\phi_{n-1}'(m) + \phi_{n-2}(m) = 0$$

$$\Rightarrow \quad \frac{c^2}{2!}(12m+10) + c\cdot 4(1+m) - (1+9m) = 0$$

$$\Rightarrow \quad \frac{c^2}{2}(-12+10) + c\cdot 4(1-1) - (1-9) = 0$$

$$\Rightarrow \quad \frac{c^2}{2}\times -2 + 8 = 0$$

$$\Rightarrow \quad c^2 = 8$$

$$\Rightarrow \quad c = \pm 2\sqrt{2}$$

Hence, the corresponding asymptotes are

$$y = -x \pm 2\sqrt{2}$$

i.e., $$y + x \pm \sqrt{2} = 0.$$

EXERCISE 15 (B)

Find the asymptotes of the following curves—

1. $x^3 + x^2y - xy^2 - y^3 - 3x - y - 1 = 0$
2. $x^3 + x^2y - xy^2 - y^3 + x^2 - y^2 = 2$
3. $4x^3 - 3xy^2 - y^3 + 2x^2 - xy - y^2 = 1$
4. $x^3 - 5x^2y + 8xy^2 - 4y^3 + x^2 - 3xy + 2y^2 - 1 = 0$
5. $y^3 - xy^2 - x^2y + x^3 + x^2 - y^2 - 1 = 0$
6. $3x^3 + 17x^2 y + 21xy^2 - 9y^3 - 2ax^2 - 12axy - 18ay^2 - 3a^2x + a^2 y = 0$
7. $x^3 - x^2 y - xy^2 + y^3 + 2x^2 - 4y^2 + 2xy + x + y + 1 = 0$

ANSWERS

1. $y = x,\ y = -x - 1,\ y = -x + 1.$
2. $y = x,\ y = -x,\ y = -x - 1.$
3. $y = x,\ y = -x + 1,\ 2y = -x - 1.$
4. $y = x,\ 2y = x,\ 2y = x + 1.$
5. $y + x = 0,\ y = x,\ y = x + 1.$
6. $3x - y + 2a = 0,\ x + 3y - a = 0,\ x + 3y + a = 0.$
7. $x + y - 1 = 0;\ 2x - 2y + 3\sqrt{5} = 0;\ 2x - 2y + 3 - \sqrt{5} = 0$

15.9. Asymptotes Parallel to the Axes

Let the equation of the curve be

$$(a_0 x^n + a_1 x^{n-1} y + a_2 x^{n-2} y^2 + \dots + a_{n-1} xy^{n-1} + a_n y^n) + (b_1 x^{n-1} + b_2 x^{n-2} y + \dots + b_n y^{n-1}) + (c_2 x^{n-2} + c_3 x^{n-3} y + \dots + c_n y^{n-2}) + \dots + \dots + \dots = 0. \quad \dots(1)$$

Arranging in descending powers of x, we get

$$a_0 x^n + (a_1 y + b_1)\, x^{n-1} + (a_2 y^2 + b_2 y + c_2)\, x^{n-2} + \dots = 0 \quad \dots(2)$$

Now, if $a_0 = 0$ and y be so chosen that $a_2 y + b_2 = 0$, [i.e., the coefficient of two highest powers {nth and $(n-1)^{th}$} of x in equation (2) vanish], then two roots of (2) are infinite. Hence, the line $a_1 y + b_1 = 0$ will be an asymptote. This line is of the form $y = k$, a line parallel to X-axis.

Again, if $a_0 = 0$, $a_1 = 0$ and y be so chosen that $a_2 y^2 + b_2 y + c^2 = 0$ [i.e., the coefficients of x^n, x^{n-1} are zero], then three roots of (2) are infinite.

Hence, the equation $a_2 y^2 + b_2 y + c_2 = 0$ will give two asymptotes which are parallel to X-axis.

Hence, *"The asymptotes parallel to the axis of X can be obtained by equating to zero the coefficient of the highest degree terms in x, provided that it is not merely a constant."*

Similarly, *"The asymptotes parallel to the axis of Y can be obtained by equating to zero the coefficient of the highest power of x, provided that this is not merely a constant."*

ILLUSTRATIVE EXAMPLES

Example 1. *Find the asymptotes parallel to X-axis for the curve*

$$y^4 + x^2y^2 + 2xy^2 - 4x^2 - y + 1 = 0$$

Solution : The highest power of x is 2. The coefficient of x^2 is $y^2 - 4$. Therefore, equations to asymptote parallel to X-axis are—

$$\Rightarrow \quad y^2 - 4 = 0\, y^2 = 4 \Rightarrow y = \pm 2$$

There are two asymptotes parallel to X-axis. They are—

$$y = 2 \quad \text{and} \quad y = -2$$

Example 2. *Find the asymptotes parallel to Y-axis for the curve*

$$x^4 + x^2y^2 - a^2 (x^2 + y^2) = 0$$

Solution : The highest power of y is 2. The coefficient of y^2 is $x^2 - a^2$. The asymptotes parallel to Y-axis are—

$$x^2 - a^2 = 0 \quad \Rightarrow \quad x = \pm a$$

The asymptotes are $x = a$ and $x = -a$.

Example 3. *Find the asymptotes, parallel to the axes, for the curve*

$$x^2y^2 = a^2 (x^2 + y^2)$$

Solution : The given curve is

$$x^2 y^2 = a^2 (x^2 + y^2)$$

$$\Rightarrow \quad x^2 (y^2 - a^2) - a^2 y^2 = 0 \qquad ...(1)$$

Equating to zero the coefficient of highest power of x, i.e., x^2 in (1), we get

$$y^2 - a^2 = 0 \quad \Rightarrow \quad y = \pm a$$

which are the asymptotes parallel to X-axis.

Again, the equation of the given curve can be written as

$$y^2 (x^2 - a^2) - a^2 x^2 = 0$$

Equating to zero the coefficient of highest power of y, i.e., y^2 in the above equation, we get

$$x^2 - a^2 = 0 \Rightarrow x = \pm a$$

which are the asymptotes parallel to the Y-axis.

Hence, the asymptotes of the given curve parallel to axes are—

$$x = \pm a; \ y = \pm a$$

Example 4. *Find the asymptotes for the curve* $\dfrac{a^2}{x^2} - \dfrac{b^2}{y^2} = 1.$

Solution : The equation to the curve is,

$$\frac{a^2}{x^2} - \frac{b^2}{y^2} = 1$$

$$\Rightarrow \quad a^2 y^2 - x^2 b^2 = x^2 y^2$$

$$\Rightarrow \quad x^2 y^2 + x^2 b^2 - a^2 y^2 = 0$$

The coefficient of highest power of x that is coefficient of x^2 is $y^2 + b^2$.

Equation to asymptotes parallel to X-axis are :

$$y^2 + b^2 = 0 \Rightarrow y = \pm ib$$

which are imaginary.

Now, coefficient of highest power of y that is coefficient of y^2 is $x^2 - a^2$.

Equation to asymptotes parallel to Y-axis are—

$$x^2 - a^2 = 0 \Rightarrow x = \pm a$$

The real asymptotes parallel to Y-axis are—

$$x = a \quad \text{and} \quad x = -a$$

Example 5. *Show that the asymptotes of the curve*

$$x^2 y^2 - a^2 (x^2 + y^2) - a^3 (x + y) + a^4 = 0$$

form a square, through two of whose angular points, the curve passes.

Solution : The equation of the given curve can be written as

$$x^2 (y^2 - a^2) - a^2 y^2 - a^3 (x + y) + a^4 = 0$$

The asymptotes parallel to X-axis are obtained by equating to zero, the coefficient of highest power of x, i.e., $y^2 - a^2$ in the equation of the curve.

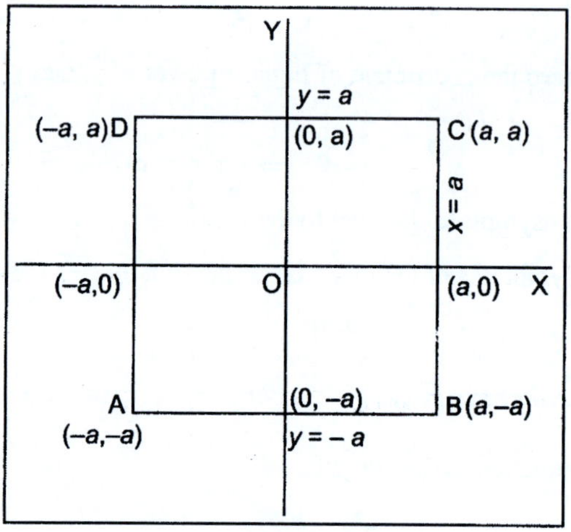

Fig. 15.1

Hence, they are—

$$y^2 - a^2 = 0$$

$\Rightarrow$ $$y = \pm a$$

The asymptotes parallel to Y-axis are obtained by equating to zero, the coefficient of highest power of y in the equation of the curve which can be written as

$$y^2 (x^2 - a^2) - a^2x^2 - a^3 (x + y) + a^4 = 0$$

Hence, they are—

$$x^2 - a^2 = 0$$

i.e., $$x = \pm a$$

Since the equation of the given curve is of 4th degree, hence no more asymptotes are possible.

Clearly, the asymptotes form a square $ABCD$ whose angular points are $A(-a, -a)$, $B(a, -a)$, $C(a, a)$ and $D(-a, a)$. It is easy to see that the points $B(a, -a)$ and $D(-a, a)$ satisfy the equation of the given curve. Hence, the given curve passes through two angular points.

EXERCISE 15 (C)

Find the asymptotes, parallel to the axes, for the following curves:

1. $\dfrac{a^2}{x^2} + \dfrac{b^2}{y^2} = 1$

2. $(x^2 - a^2)\, y^2 = x$

3. $x^2y^2 - y^2 = 2$
4. $xy^3 - x^3 = a(x^2 + y^2)$
5. $x^3 - 2x^2y + xy^2 + x^2 - xy + 2 = 0$
6. $y^4 + x^2y^2 + 2xy^2 - 4x^2 - y + 1 = 0$
7. $x^2y^3 + x^3y^2 = x^3 + y^3$
8. $(x^3 + a^3)\,y = bx^3$
9. $y = \dfrac{x^2 - 1}{x^2 + 1}$

ANSWERS

1. $y = \pm a,\ x = \pm b$
2. $x = \pm a$
3. $y = 0,\ x = \pm 1$
4. $x = 0$
5. $x = 0$
6. $y = \pm 2$
7. $x = \pm 1,\ y = \pm 1$
8. $y = b,\ x = -a$
9. $y = 1$

15.10. Alternative Methods of Finding Asymptotes of Algebraic Curve

FIRST METHOD :

Let the equation of the curve be of degree n.

CASE I. *When $y - m_1 x$ is a non-repeated factor of the* nth *degree terms in the equation of the curve.*

Then the equation of the curve can be written as

$$(y - m_1 x)\,F_{n-1} + P_{n-1} = 0, \qquad ...(1)$$

where, F_{n-1} contains terms of degree $(n-1)$ only and P_{n-1} contains terms of degree not higher than $(n-1)$.

Clearly, m_1 is a root of the equation $\phi_n(m) = 0$ and so there may be an asymptote

$$y - m_1 x = c_1 \qquad ...(2)$$

if a finite value of c_1 can be found out.

Now, $$c_1 = \lim_{x \to \infty} (y - m_1 x),$$

where, (x, y) lies on an infinite branch of the curve given by (1).

But when (x, y) lies on (1), we have

$$y - m_1 x = \frac{P_{n-1}}{F_{n-1}}$$

$$\therefore \qquad c_1 = \lim_{x \to \infty} (y - m_1 x)$$

$$= \lim_{x \to \infty} \frac{P_{n-1}}{F_{n-1}},$$

where, the limit is to be evaluated by using $\lim_{x \to \infty} \frac{y}{x} = m_1$.

Thus, by (2) the required asymptote corresponding to the factor $y - m_1 x$ is,

$$y - m_1 x = \lim_{\substack{x \to \infty \\ y/x \to m_1}} - \frac{P_{n-1}}{F_{n-1}}$$

$$\Rightarrow \qquad y - m_1 x + \lim_{\substack{x \to \infty \\ y/x \to m_1}} \frac{P_{n-1}}{F_{n-1}} = 0$$

If there are a number of non-repeated linear factors of the nth degree terms of the given equation, we can find out the corresponding asymptotes in a similar manner.

COROLLARY : If we have a factor $ax + by + c$ instead of $y - m_1 x$ in the above, i.e., equation (1) is of the form

$$(ax + by + c)\, F_{n-1} + P_{n-1} = 0,$$

then the asymptote corresponding to the factor $(ax + by + c)$ is given by

$$ax + by + c + \lim_{\substack{x \to \infty \\ y/x \to -a/b}} \frac{P_{n-1}}{F_{n-1}} = 0.$$

CASE II. *When $(y - m_1 x)^2$ is a factor of* nth *degree terms of the given equation and $(y - m_1 x)$ is not a factor of the $(n - 1)$ th degree terms.*

Then equation of the curve can be written as

$$(y - m_1 x)^2\, F_{n-2} + P_{n-1} = 0,$$

where, F_{n-2} contains only terms of degree $(n - 2)$ and P_{n-1} contains other terms none of which is of degree higher than $(n - 1)$.

Then proceeding as above in case I,

$$\lim_{\substack{x \to \infty \\ y/x \to m_1}} (y - m_1 x)^2 = \lim_{x \to \infty} - \frac{P_{n-1}}{F_{n-2}} = \infty \text{ or } -\infty.$$

Thus, there is no asymptote parallel to $y - m_1 x = 0$.

CASE III. *When $(y - m_1 x)^2$ is a factor of* nth *degree terms and $(y - m_1 x)$ is also a factor of $(n - 1)$* th *degree terms in the equation of the curve.*

Then the equation of the curve can be written as

$$(y - m_1 x)^2\, F_{n-2} + (y - m_1 x)\, G_{n-2} + P_{n-2} = 0,$$

where, F_{n-2} and G_{n-2} contain only terms of degree $(n - 2)$ and P_{n-1} contains other terms of degree $(n - 2)$ at the most.

Dividing both sides by F_{n-2}, we have

$$(y - m_1x)^2 + (y - m_1x)\frac{G_{n-2}}{F_{n-2}} + \frac{P_{n-2}}{F_{n-2}} = 0$$

Taking limits as $x \to \infty$ and $\frac{y}{x} \to m_1$, we have an equation of the form

$$(y - m_1x)^2 + A\,(y - m_1x) + B = 0$$

which on solving for $y - m_1x$ gives us two parallel asymptotes of the form

$$y - m_1x = c_1 \text{ and } y - m_1x = c_2$$

COROLLARY : If $(ax + by + c)^2$ be a factor of nth degree terms instead of $(y - m_1x)^2$ and $(ax + by + c)$ be a factor of $(n - 1)$ th degree terms, then equation of the curve becomes

$$(ax + by + c)^2 F_{n-2} + (ax + by + c)\, G_{n-2} + P_{n-2} = 0$$

Then the two parallel asymptotes are,

$$(ax + by + c)^2 + (ax + by + c) \lim_{\substack{x \to \infty \\ y/x \to -a/b}} \left(\frac{G_{n-2}}{F_{n-2}}\right) + \lim_{\substack{x \to \infty \\ y/x \to -a/b}} \left(\frac{P_{n-2}}{F_{n-2}}\right) = 0$$

ILLUSTRATIVE EXAMPLES

Example 1. *Find asymptotes for the curve*

$$y^3 - x^2y - 2xy^2 + 2x^3 - 7xy + 3y^2 + 2x^2 + 2x + 2y + 1 = 0$$

Solution : Factorizing the highest degree terms. That is factorizing the third degree terms of the equation,

$$(y - x)(y + x)(y - 2x) - 7xy + 3y^2 + 2x^3 + 2x + 2y + 1 = 0$$

The equation to asymptote corresponding to $y - x$ is,

$$y - x = \lim_{\substack{x \to \infty \\ y/x \to 1}} \frac{7xy - 3y^2 - 2x^2 - 2x - 2y - 1}{(y + x)(y - 2x)}$$

Dividing numerator and denominator by x^2, we get

$$y - x = \lim_{\substack{x \to \infty \\ y/x \to 1}} \frac{7\left(\frac{y}{x}\right) - 3\left(\frac{y}{x}\right)^2 - 2 - \frac{2}{x} - 2\left(\frac{y}{x}\right)\frac{1}{x} - \frac{1}{x^2}}{\left(\frac{y}{x} + 1\right)\left(\frac{y}{x} - 2\right)}$$

$$= \frac{7 - 3 - 2}{(1 + 1)(1 - 2)} = -1.$$

Asymptote is $y - x = -1 \Rightarrow y - x + 1 = 0$.

Equation to asymptote corresponding to $y + x$ is,

$$y + x = \lim_{\substack{y/x \to -1 \\ x \to \infty}} \frac{7xy - 3y^2 - 2x^2 - 2x - 2y - 1}{(y - x)(y - 2x)}$$

$$= \lim_{\substack{y/x \to -1 \\ x \to \infty}} \frac{7\left(\frac{y}{x}\right) - 3\left(\frac{y}{x}\right)^2 - 2 - \frac{2}{x} - 2\frac{y}{x}\cdot\frac{1}{x} - \frac{1}{x^2}}{\left(\frac{y}{x} - 1\right)\left(\frac{y}{x} - 2\right)}$$

$$= \frac{-7 - 3 - 2}{(-1 - 1)(-1 - 2)} = \frac{-12}{6} = -2$$

$\Rightarrow y + x + 2 = 0$

Asymptote corresponding to $y - 2x$ is,

$$y - 2x = \lim_{\substack{x \to \infty \\ y/x \to 2}} \frac{7xy - 3y^2 - 2x^2 - 2x - 2y - 1}{(y - x)(y + x)}$$

$$= \lim_{\substack{x \to \infty \\ y/x \to 2}} \frac{7\left(\frac{y}{x}\right) - 3\left(\frac{y}{x}\right)^2 - 2 - \frac{2}{x} - 2\frac{y}{x}\cdot\frac{1}{x} - \frac{1}{x^2}}{\left(\frac{y}{x} - 1\right)\left(\frac{y}{x} + 1\right)}$$

$$= \frac{7(2) - 3(2)^2 - 2}{(2 - 1)(2 + 1)} = \frac{14 - 12 - 2}{3} = 0$$

$\Rightarrow \quad y - 2x = 0$

The asymptotes of the given curve are—

$$y - x + 1 = 0,\ y + x + 2 = 0 \text{ and } y - 2x = 0$$

Example 2. *Find the asymptotes of the following curves—*

$$y^2 - x^2y + 2y^2 + 4y + x = 0$$

Solution : The equation of the curve is,

$$y^3 - x^2y + 2y^2 + 4y + x = 0 \qquad ...(1)$$

Equating to zero, the coefficient of highest degree term in x, i.e., x^2, the asymptote parallel to X-axis is given by

$$-y = 0 \Rightarrow y = 0$$

Since the coefficient of highest degree term in y, i.e., y^2 is constant (= 1), there is no asymptote parallel to Y-axis.

Factorizing the highest degree term, (1) becomes

$$y(y - x)(y + x) + 2y^2 + 4y + x = 0. \qquad ...(2)$$

$$y = 0,\ y - x = 0 \quad \text{and} \quad y + x = 0$$

The asymptote parallel to $y = 0$, i.e., X-axis has already been found above.

Dividing (2) throughout by $y\ (y + x)$, we get

$$y - x + \frac{2y^2 + 4y + x}{y\ (y + x)} = 0$$

$\therefore$ The asymptote parallel to the line $y - x = 0$ is given by

$$y - x + \lim_{\substack{x \to \infty \\ y/x \to 1}} \frac{2y^2 + 4y + x}{y^2 + yx} = 0$$

$$\Rightarrow \quad y - x + \lim_{\substack{x \to \infty \\ y/x \to 1}} \frac{\frac{2y^2}{x^2} + \frac{4}{x} \cdot \frac{y}{x} + \frac{1}{x}}{\frac{y^2}{x^2} + \frac{y}{x}} = 0$$

[Dividing the numerator and denominator by x^2]

$$\Rightarrow \quad y - x + \frac{2\,(1)^2 + 0\,(1) + 0}{(1)^2 + 1} = 0$$

$$\Rightarrow \quad y - x + 1 = 0$$

Again, dividing (2) throughout by $y\ (y - x)$, we get the asymptote parallel to the line $y + x = 0$ as

$$y + x + \lim_{\substack{x \to \infty \\ y/x \to -1}} \frac{2y^2 + 4y + x}{y\ (y - x)} = 0$$

$$\Rightarrow \quad y + x + \lim_{\substack{x \to \infty \\ y/x \to -1}} \frac{\frac{2y^2}{x^2} + \frac{4}{x} \cdot \frac{y}{x} + \frac{1}{x}}{\frac{y^2}{x^2} - \frac{y}{x}}$$

[On dividing the numerator and denominator by x^2]

$$\Rightarrow \quad y + x + \frac{2\,(-1)^2 + 0\,(-1) + 0}{(-1)^2 - (-1)} = 0$$

$$\Rightarrow \quad y + x + 1 = 0$$

Hence, the asymptotes of the curve are—

$$y = 0,\ y - x + 1 = 0 \quad \text{and} \quad y + x + 1 = 0$$

Example 3. *Find the asymptotes for the curve*

$$(x - 2y)^2\ (x - y) - 4y\ (x - 2y) - (8x + 5y) = 0$$

Solution : Asymptote corresponding to $x + 2y$ is

$$(x - 2y)^2 - 4\ (x - 2y) \lim_{\substack{x \to \infty \\ \frac{y}{x} \to \frac{1}{2}}} \frac{y}{x - y} - \lim_{\substack{x \to \infty \\ \frac{y}{x} \to \frac{1}{2}}} \frac{8x + 5y}{x - y} = 0$$

$$\Rightarrow (x-2y)^2 - 4(x-2y)\lim_{\substack{x\to\infty \\ \frac{y}{x}\to\frac{1}{2}}} \frac{\frac{y}{x}}{1-\frac{y}{x}} - \lim_{\substack{x\to\infty \\ \frac{y}{x}\to\frac{1}{2}}} \frac{8+5\frac{y}{x}}{1-\frac{y}{x}} = 0$$

$$\Rightarrow \quad (x-2y)^2 - 4(x-2y)\frac{\frac{1}{2}}{1-\frac{1}{2}} - \frac{8+\frac{5}{2}}{1-\frac{1}{2}} = 0$$

$$\Rightarrow \quad (x-2y)^2 - 4(x-2y) - 21 = 0$$

$$\Rightarrow \quad x - 2y = \frac{4 \pm \sqrt{16+84}}{2} = \frac{4 \pm 10}{2} = 7, -3$$

$$\Rightarrow \quad x - 2y = 7, \quad x - 2y = -3$$

Asymptote corresponding to $x - y$ is,

$$x - y = \lim_{\substack{x\to\infty \\ y/x\to 1}} \frac{4y(x-2y) - 8x - 5y}{(x-2y)^2}$$

$$= \lim_{\substack{x\to\infty \\ y/x\to 1}} \frac{4\left(\frac{y}{x}\right)\left(1-\frac{2y}{x}\right) - \frac{8}{x} - 5\left(\frac{y}{x}\right)\frac{1}{x}}{\left(1-2\frac{y}{x}\right)^2}$$

$$= \frac{4(1-2)+0+0}{(1-2)^2} = -4$$

$$\Rightarrow \quad x - y = -4$$

The asymptotes are,

$$x - 2y - 7 = 0,\ x - 2y + 3 = 0 \text{ and } y - x - 4 = 0$$

Example 4. *Find the asymptotes for the following curve—*

$$(x-y-1)^2(x^2+y^2+2) + 6(x-y-1)(xy-7) - 8x^2 - 2x - 1 = 0$$

Solution : The equation is,

$$(x-y-1)^2(x^2+y^2+2) + 6(x-y-1)(xy-7) - 8x^2 - 2x - 1 = 0$$

Dividing the equation by $x^2 + y^2 + 2$, we get

$$(x-y-1)^2 + 6(x-y-1)\lim_{\substack{x\to\infty \\ \frac{y}{x}\to 1}} \frac{xy-7}{x^2+y^2-2}$$

$$- \lim_{\substack{x \to \infty \\ \frac{y}{x} \to 1}} \frac{8x^2 + 2x + 1}{x^2 + y^2 - 2} = 0$$

$$\Rightarrow \quad (x - y - 1)^2 + 6\,(x - y - 1) \lim_{\substack{x \to \infty \\ \frac{y}{x} \to 1}} \frac{\frac{y}{x} - \frac{7}{x^2}}{1 + \left(\frac{y}{x}\right)^2 - \frac{2}{x^2}}$$

$$- \lim_{\substack{x \to \infty \\ \frac{y}{x} \to 1}} \frac{8 + \frac{2}{x} + \frac{1}{x^2}}{1 + \left(\frac{y}{x}\right)^2 - \frac{2}{x^2}} = 0$$

$$\Rightarrow (x - y - 1)^2 + 6\,(x - y - 1)\,\frac{1 - 0}{1 + 1 - 0} - \frac{8 + 0 + 0}{1 + 1 + 0} = 0$$

$$\Rightarrow \quad (x - y - 1)^2 + 3\,(x - y - 1) - 4 = 0$$

$$\Rightarrow \quad (x - y - 1) = \frac{-3 \pm \sqrt{9 + 16}}{2} = \frac{-3 \pm 5}{2} = -4,\ +1$$

Asymptotes are,

$$x - y - 1 = -4 \ \Rightarrow\ x - y + 3 = 0$$

$$x - y - 1 = 1 \ \Rightarrow\ x - y - 2 = 0$$

Therefore, there are two asymptotes that is,

$$y - x - 3 = 0 \quad \text{and} \quad y - x + 2 = 0$$

The other two asymptotes are imaginary because linear factors of $x^2 + y^2 + 2$ are imaginary.

SECOND METHOD:

Let $f(x, y) = 0$ be a curve. Collect the highest degree terms in it and factorize them into real linear factors. Then in general there is one asymptote parallel to each such factor. Put each such linear factor equal to k. Substitute either the value of x in terms of y and k or the value of y in terms of x and k from this factor in the equation of the curve. Now, equate to zero the coefficient of the highest power of the resulting equation in order to determine the value of k. This process when repeated for every linear factor gives an asymptote parallel to every factor.

Note : This method may usefully be applied to find the asymptotes of any curve. If the coefficient of the highest power in the resulting equation does not contain k or when equated to zero, does not given real value of k, then we say that there exists no asymptote corresponding to that factor of the highest degree terms in the given equation of the curve.

ILLUSTRATIVE EXAMPLES

Example 1. *Find the asymptotes of the curve*

$$x^3 - xy^2 + x^2 + y^2 + x + y = 0$$

Solution : The equation of the curve is,

$$x^3 - xy^2 + x^2 + y^2 + x + y = 0$$

$$\Rightarrow \quad x(x-y)(x+y) + x^2 + y^2 + x + y = 0$$

[Resolving the highest degree terms into real linear factors]

Clearly, there will be an asymptote parallel to each of the lines

$$x = 0, \; x - y = 0 \text{ and } x + y = 0$$

Asymptote parallel to $x = 0$ **:** Let the asymptote parallel to $x = 0$ be $x = k$.

Put $x = k$ in the equation of the curve, we get

$$k^3 - ky^2 + k^2 + y^2 + k + y = 0$$

$$\Rightarrow \quad y^2(1-k) + y + k^3 + k^2 + k = 0$$

Equating to zero, the coefficient of the highest power of y, i.e., y^2, we get

$$1 - k = 0 \;\Rightarrow\; k = 1$$

Hence, the asymptote parallel to $x = 0$ is $x = 1$.

Asymptote parallel to $x - y = 0$ **:** Let the asymptote parallel to $x - y = 0$ be

$$x - y = k \;\Rightarrow\; x = y + k$$

Put this value of x in the equation of the curve, we get

$$(y+k)^3 - (y+k)y^2 + (y+k)^2 + y^2 + y + k + y = 0$$

$$\Rightarrow \quad y^3 + k^3 + 3k^2y + 3ky^2 - y^3 - ky^2 + y^2 + k^2$$
$$+ 2ky + y^2 + y + k + y = 0$$

$$\Rightarrow \quad y^3(2 + 2k) + y(3k^2 + 2k + 2) + k^2 + k^2 + k = 0$$

Equating to zero, the coefficient of the highest power of y, i.e., y^3, we get

$$2 + 2k = 0 \;\Rightarrow\; k = -1$$

Hence, the asymptote parallel to $x - y$ is $x - y = -1$.

Asymptote parallel to $x + y = 0$ **:** Let the asymptote parallel to $x + y = 0$ be

$$x + y = k \;\Rightarrow\; y = k - x$$

Put this value of y in the equation of the given curve, we get

$$x^3 - x(k-x)^2 + x^2 + (k-x)^2 + x + k - x = 0$$

$$\Rightarrow \quad x^3 - k^2 x - x^3 + 2kx^2 + x^2 + k^2 + x^2 - 2kx + k = 0$$

$$\Rightarrow \quad x^2 (2 + 2k) + x (-2k - k^2) + k + k^2 = 0$$

Equating to zero the coefficient of the highest power of x, i.e., x^2, we get

$$2 + 2k = 0 \Rightarrow k = -1$$

Hence, the asymptote parallel to $x + y = 0$ is $x + y = -1$.

Hence, the asymptotes of the curve are,

$$x = 1, \; x - y + 1 = 0 \text{ and } x + y + 1 = 0$$

Example 2. *Find the asymptotes for the following curve—*

$$y^3 - 5xy^2 + 8x^2 y - 4x^3 - 3y^2 + 9xy - 6x^2 + 2y - 2x + 1 = 0$$

Solution : The terms of highest degree in the given equation are,

$$= y^3 - 5xy^2 + 8x^2y - 4x^3$$

$$= y^3 - xy^2 - 4xy^2 + 4x^2y + 4x^2y - 4x^3$$

$$= y^2 (y - x) - 4xy (y - x) + 4x^2 (y - x)$$

$$= (y - x) (y^2 - 4xy + 4x^2)$$

$$= (y - x) (y - 2x)^2$$

The given equation can be written as

$$(y - x) (y - 2x)^2 - 3 (y - x) (y - 2x) + 2y - 2x + 1 = 0 \quad ...(1)$$

Let corresponding asymptote to $y - x$ be $y - x = k$, then

$$y = x + k$$

Putting the value of y in (1), we get

$$k (k - x)^2 - 3k (k - x) + 2 (x + k) - 2x + 1 = 0$$

Equating the coefficient of highest power of x to zero, we get $k = 0$.

Therefore, the asymptote corresponding to $y - x$ is,

$$y - x = 0$$

Now, let equation to asymptote corresponding to $y - 2x$ be $y - 2x = k$ that is $y = 2x + k$.

Putting the value of y in (1), we get

$$(x + k) k^2 - 3 (x + k) k + 2 (2x + k) - 2x + 1 = 0$$

Equating to zero the coefficient of highest power of x, we get

$$k^2 - 3k - 2 = 0 \Rightarrow (k - 1) (k - 2) = 0 \Rightarrow k = 1, 2.$$

Thus, the equation to asymptote corresponding to $y - 2x$ are—

$$y - 2x = 1 \text{ and } y - 2x = 2$$

The asymptotes will be $y = x$, $y - 2x = 1$ and $y - 2x = 2$.

THIRD METHOD (Asymptotes by Inspection):

If the equation of a curve is of the form $F_n + F_{n-2} = 0$, where F_n is of degree n (i.e., contains terms of degree n and may also contain terms of lower degrees) and F_{n-2} is of degree $(n-2)$ [i.e., contains terms of degree $(n-2)$ and may also contain terms of lower degrees] and if $F_n = 0$ can be broken up into real linear factors none of which is repeated, then all the asymptotes are given by $F_n = 0$.

ILLUSTRATIVE EXAMPLES

Example 1. *Find the asymptotes to the hyporbola*

$$\frac{x^2}{a^2} - \frac{y^2}{b^2} = 1$$

Solution : The given curve is in the form

$$F_2 + F_0 = 0, \quad \text{and} \quad F_2 = 0 \text{ that is}$$

$$\frac{x^2}{a^2} - \frac{y^2}{b^2} = 0$$

$$\Rightarrow \qquad \left(\frac{x}{a} - \frac{y}{b}\right)\left(\frac{x}{a} + \frac{y}{b}\right) = 0$$

This represents two straight lines which are neither parallel nor coincident. Therefore, those two will be the required asymptotes of the given curve.

That, $$\frac{x}{a} - \frac{y}{b} = 0 \quad \text{and} \quad \frac{x}{a} + \frac{y}{b} = 0$$

$$bx - ay = 0 \quad \text{and} \quad bx + ay = 0$$

Example 2. *Find the asymptotes of the curve*

$$x^3 + 2x^2y + xy^2 - x^2 - xy + 2 = 0$$

Solution : The equation of the curve, is

$$x^3 + 2x^2y + xy^2 - x^2 - xy + 2 = 0$$

This equation can be written as,

$$\Rightarrow \qquad x(x + y)^2 - x(x + y) + 2 = 0$$

$$\Rightarrow \qquad x(x + y)(x + y - 1) + 2 = 0$$

Hence, by inspection method, the asymptotes are given by

$$x = 0, \; x + y = 0 \quad \text{and} \quad x + y - 1 = 0.$$

15.11. Intersection of the Curve and Its Asymptotes

We know that all the n asymptotes (non-repeated) of a curve $F_n + F_{n-2} = 0$ of degree n are given by $F_n = 2$. Now, by analytical geometry, it is easy to see that the points common, the curve. $F_n = 0$ and $F_n + F_{n-2} = 0$ lie on curve $F_{n-2} = 0$. Thus, n asymptotes intersect the curve again in points which lie on the curve $F_{n-2} = 0$. Further a straight line cuts a curve of nth degree in n points, so will do an asymptote. But by definition, an asymptote cuts its curve in two points at infinity. Consequently, each asymptote cuts its nth degree curve in $(n-2)$ other points. Therefore, n asymptotes of a curve cut it in $n(n-2)$ other points which lie on a certain curve of degree $(n-2)$.

ILLUSTRATIVE EXAMPLES

Example 1. *Show that the asymptotes of the cubic*

$$x^3 - 2y^3 + xy(2x - y) + y(x - y) + 1 = 0$$

cut the curve again in three points which lie on the straight line

$$x - y + 1 = 0.$$

Solution : The curve is,

$$x^3 - 2y^3 + 2x^2y - xy^2 + xy - y^2 + 1 = 0 \quad ...(1)$$

$$\therefore \quad \phi_n(m) = 1 - 2m^3 + 2m - m^2$$

$$\phi_{n-1}(m) = m - m^2$$

Now, $\phi_n(m) = 0$ gives

$$1 - 2m^3 + 2m - m^2 = 0$$

$$\Rightarrow \quad 1 - m^2 + 2m(1 - m^2) = 0$$

$$\Rightarrow \quad (1 - m^2)(1 + 2m) = 0$$

$$\Rightarrow \quad (1 - m)(1 + m)(1 + 2m) = 0$$

$$\Rightarrow \quad m = 1, -1, -\frac{1}{2}$$

Now,
$$\phi_n'(m) = -6m^2 + 2 - 2m$$

$$\therefore \quad c = -\frac{\phi_{n-1}(m)}{\phi_n'(m)}$$

$$= -\frac{m - m^2}{-6m^2 + 2 - 2m}$$

$$= \frac{m^2 - m}{-6m^2 + 2 - 2m}$$

When $m = 1, \quad c = 0$

When $m = -1, \quad c = -1$

When $m = -\frac{1}{2}, \quad c = -\frac{1}{2}$

Hence, the three asymptotes of the curve are,

$$y = x,\ x + y + 1 = 0 \text{ and } x + 2y - 1 = 0$$

The combined equation of the asymptotes is,

$$(x - y)(x + y + 1)(x + 2y - 1) = 0$$

i.e., $$x^3 - 2y^3 + 2x^2y - xy^2 + xy - y^2 - x + y = 0 \qquad ...(2)$$

Subtracting (2) from (1), we have

$$x - y + 1 = 0$$

which is the required line on which the points of intersection of the asymptote and the curve lie. Again, since curve is a cubic.

$$\therefore \qquad n = 3$$

$$\therefore \qquad n(n-2) = 3(3-2) = 3$$

Hence, the curve is intersected by its asymptotes at three points.

Example 2. *Find the equation of the cubic which has same asymptotes as the curve*

$$x^3 - 6x^2 + 11xy^2 - 6y^3 + x + y + 1 = 0$$

and which touches the axis of Y at the origin and goes through the point (3, 2).

Solution : The equation of the curve is,

$$x^3 - 6x^2y + 11xy^2 - 6y^3 + x + y + 1 = 0$$

$$\Rightarrow \qquad (x - y)(x - 2y)(x - 3y) + x + y + 1 = 0$$

which is of the form $F_n + F_{n-2} = 0$ and F_n consists of non-repeated linear factors. Hence, by inspection method all the asymptotes are given by $F_n = 0$.

i.e., $$(x - y)(x - 2y)(x - 3y) = 0$$

i.e., $$x - y = 0,\ x - 2y = 0,\ x - 3y = 0$$

The combined equation of the asymptotes is,

$$(x - y)(x - 2y)(x - 3y) = 0$$

$$\Rightarrow \qquad x^3 - 6x^2y + 11xy^2 - 6y^3 = 0$$

Now, equation of any cubic having these asymptotes may be taken as of the form

$$x^3 - 6x^2y + 11xy^2 - 6y^3 + ax + by + c = 0, \qquad ...(1)$$

where, a, b, c are to be determined.

Since (1) passes through (0, 0), we have $c = 0$.

$\therefore$ (1) becomes

$$x^3 - 6x^2y + 11xy^2 - 6y^3 + ax + by = 0 \qquad ...(2)$$

The tangent to (2) at origin is (equating to zero the lowest degree terms)

$$ax + by = 0 \qquad ...(3)$$

But this is given to by Y-axis, i.e., $x = 0$ ...(4)

Hence, (3) and (4) must be identical.

$$b = 0$$

$\therefore$ Equation (2) now reduces to

$$x^3 - 6x^2y + 11xy^2 - 6y^3 + ax = 0 \qquad ...(5)$$

This passes through (3, 2),

$$\therefore \quad 27 - 108 + 132 - 48 + 3a = 0$$

$$\Rightarrow \quad 3a + 3 = 0$$

$$\Rightarrow \quad a = -1$$

Hence, (5) becomes

$$x^3 - 6x^2y + 11xy^2 - 6y^3 - x = 0$$

which is the required equation.

EXERCISE 15 (D)

Find the asymptotes of the following curves—

1. $xy^2 = 4a^2 (2a - x)$
2. $y^3 = x^3 + ax^2$
3. $x^4 - y^4 = a^2 xy$
4. $x^3 + y^3 - 3ax^2 = 0$
5. $y^2 (x^2 - a^2) = x$
6. $y^2 (x - 2a) = x^3 - a^3$
7. $y^2 = \dfrac{x^4}{a^2 - x^2}$
8. $(x^2 - a^2) y^2 = x^3 (x^2 - 4a^2)$
9. $2x (y - 3)^2 = 3y (x - 1)^2$

10. $x^2(x-y)^2 + a^2(x^2-y^2) - a^2xy = 0$

11. $x^2y^2 - x^2y + x + y + 1 = 0$

12. $y^3 + x^2y + 2xy^2 - y + 1 = 0$

13. $x^3 + 3x^2y - 4y^3 - x + y + 3 = 0$

14. $y^4 - 2xy^3 + 2x^3y - x^4 - 3x^3 + 3x^2y + 3xy^2 - 3y^3 - 2x^2 + 2y^2 - 1 = 0$

15. $x^3 + x^2y - xy^2 - y^3 + x^2 - y^2 - 2 = 0$

16. $(x^2-y^2)(y^2-4x^2) - 6x^3 + 5x^2y + 3xy^2 - 2y^3 - x^2 + 3xy = 1$

17. $(x-2y)^2(x-y) - 4y(x-2y) - (8x+7y) = 0$

18. $(x-y)^2(x^2+y^2) - 10(x-y)x^2 + 12y^2 + 2x + y = 0$

19. $(x-y)(x+y)(x+2y-1) = 3x + 4y + 5$

20. $(x^2 - 3x + 2)(x + y - 2) + 1 = 0$

21. $(\alpha_1 x + \beta_1 y + \gamma_1)(\alpha_2 x + \beta_2 y + \gamma_2) + \gamma_3 = 0$

22. $x^2y^2(x^2-y^2)^2 = (x^2+y^2)^3$

23. Find the asymptotes of the curve $x^2y - xy^2 + xy + y^2 + x - y = 0$ and prove that they cut the curve again in three points which lie on the straight line $x + y = 0$.

ANSWERS

1. $x = 0$
2. $3y = 3x + a$
3. $x + y = 0,\ x - y = 0$
4. $x + y - a = 0$
5. $x - a = 0,\ x + a = 0,\ y = 0$
6. $x = 0,\ y + x + a = 0,\ y = x + a$
7. $x = \pm a$
8. $x = \pm a,\ x + y = 0,\ x - y = 0$
9. $x = 0,\ y = 0,\ y = \frac{3}{2}x + 3$
10. $x = a,\ x = -a,\ y = x + a;\ y = x - a$
11. $x = 0,\ x = 1,\ y = 0,\ y = 1$

12. $y = 0,\ y + x = 1,\ y + x = -1$

13. $y = -x,\ 2y = -x + 1,\ 2y = -x - 1$

14. $y = x,\ y = 2x + 1,\ y = 2x + 2$

15. $y = x,\ y + x = 0,\ y + x + 1 = 0$

16. $y = x,\ y = 2x,\ y + x + 1 = 0,\ y + 2x + 1 = 0$

17. $x - y + 4 = 0,\ x - 2y = 2 \pm 3\sqrt{3}$

18. $x - y = 2,\ x - y = 3$

19. $x - y = 0,\ x + y = 0,\ x + 2y - 1 = 0$

20. $x = 1,\ x = 2,\ x + y = 2$

21. $\alpha_1 x + \beta_1 y + \gamma_1 = 0,\ \alpha_2 x + \beta_2 y + \gamma_2 = 0$

22. $x = \pm 1,\ y = \pm 1,\ x - y \pm \sqrt{2} = 0,\ x + y \pm \sqrt{2} = 0$

23. $x = 0,\ x = 1,\ x - y + 2 = 0$

OBJECTIVE EXERCISE

Select the correct answers.

1. The asymptote for the curve $\frac{x^2}{a^2} - \frac{y^2}{b^2} = 1$ are—

   (a) $bx = -ay$

   (b) $bx = ay$

   (c) $bx = \pm ay$

   (d) none

2. The asymptote for the curve $x^2 + y^2 = 3axy$ is—

   (a) $x + y = 0$

   (b) $x + y - a = 0$

   (c) $x + y + a = 0$

   (d) $x = 0$

3. The curve of nth degree cuts the asymptotes at—

   (a) n-point

   (b) $(n - 1)$ points

   (c) $(n - 2)$ points

   (d) $(n - 3)$ points

4. The asymptotes to the curve $\frac{a^2}{x^2} - \frac{b^2}{y^2} = 1$ are—

(a) $x = \pm a,\ y = \pm b$

(b) $x = \pm a$

(c) $y = \pm b$

(d) none.

5. The asymptotes parallel to X-axis for the curve,

$$x^2 y^2 + x^2 y - xy^2 + x + y + 1 = 0 \text{ are—}$$

(a) $x = 0,\ x = -1$

(b) $x = 0,\ x = 1$

(c) $y = 1,\ y = 2$

(d) $y = 0,\ y = 1$

6. The asymptote parallel to X-axis for the curve,

$$y^3 + x^2 y + 22xy^2 - y + 1 = 0 \text{ is—}$$

(a) $y = 0$

(b) $y = 2$

(c) $y = 1$

(d) $yx = 1$

7. The asymptotes parallel to Y-axis for the curve $x^2y^3 + x^3y^2 = x^3 + y^3$ are—

(a) $x = \pm 1$

(b) $y = \pm 1$

(c) $y = \pm x$

(d) $y = \pm 2$

8. The asymptotes parallel to X-axis for the curve,

$$x^2y^2 = a^2 (x^2 + y^2) \text{ are—}$$

(a) $y = \pm a$

(b) $x = \pm a$

(c) $y = a$

(d) $x = a$

9. The asymptotes for the curvey $y = \dfrac{x^2 + 1}{x - 3}$ are—

(a) $x = 1$

(b) $x + 3 = 0$

(c) $x = 3$

(d) none

ANSWERS

1. (c)	**2.** (c)	**3.** (c)
4. (b)	**5.** (d)	**6.** (a)
7. (a)	**8.** (a)	**9.** (c)

16

Singular Points

16.1. Definition

A point on the curve at which the curve shows extraordinary behaviour is called a *singular point.*

There are two types of singular points :

(i) Point of Inflexion

(ii) Multiple Points.

16.2. Point of Inflexion

A point on a curve is said to be a point of inflexion if the curve is concave on one side and convex on the other side of the point P with respect to any line AB.

Obviously, the curve crosses its tangent at P, the point of inflexion. Therefore an alternative definition for the point of inflexion may be given as follows:

A point P is said to be a point of inflexion if the curve crosses its tangent at that point.

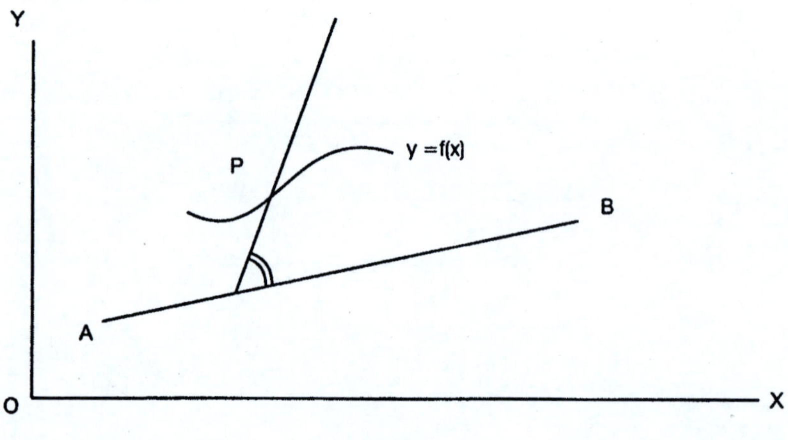

Fig. 16.1

16.3. Concavity and Convexity

Let P be a given point on a curve. Let AB be given straight line which does not pass through P. Then the curve is said to be concave or convex at P with respect to AB, according as a sufficiently small arc containing P lies entirely within or without the acute angle PAB, where AP is a tangent at P.

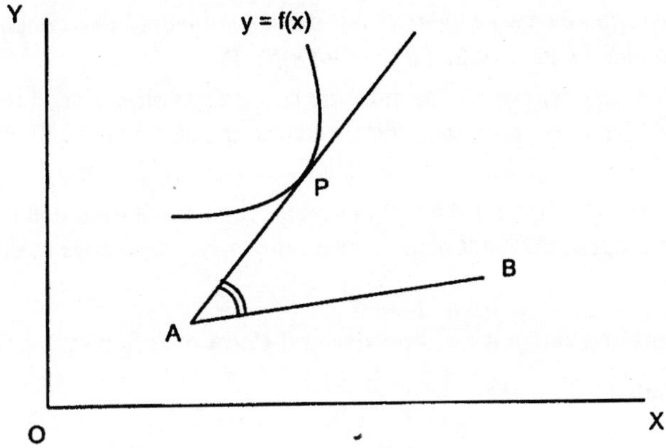

Fig. 16.2

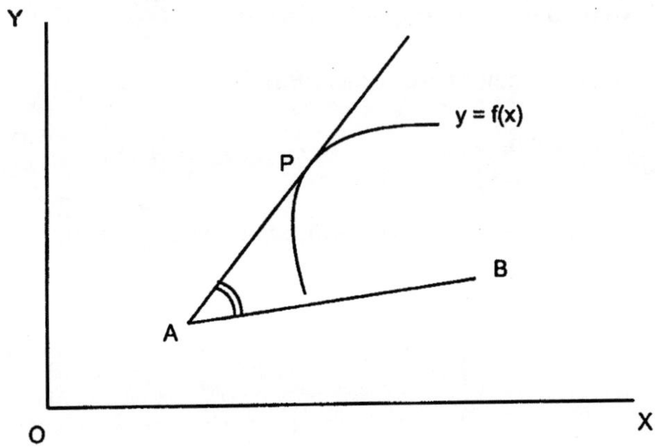

Fig. 16.3

In the Fig. 16.1 given above, the curve is concave at P with respect to AB whereas in the Fig. 16.2, the curve is convex at P with respect to AB.

Alternatively:

The curve is said to be

(i) *Concave upwards (or convex downwards)* at P if in the neighbourhood of P, the curve lies above the tangent at P on both sides. [see Fig. 16.2]

(ii) *Concave downwards (or convex upwards)* at P if in the neighbourhood of P, the curve lies below the tangent at P on both sides. [see Fig. 16.1]

16.4. Another definition of point of inflexion

A point on the curve at which the curve changes from concavity to convexity and vice-versa is called a point of inflexion.

If the curve is concave on one side and convex on other side at the point P w.r.t. line AB or vice-versa, then the tangent to the curve at P will cross the curve at P and then the point P will be called as a point of inflexion.

Note: A point of inflexion is a singular point (i.e., an unusual point) on the curve, for the tangent does not usually cross the curve, as it does at the point of inflexion.

16.5. Criteria for Concavity, Convexity and Point of Inflexion

If $y = f(x)$, then to, show that

(i) the curve is concave upwards at a point P on it if $\frac{d^2y}{dx^2}$ is positive.

(ii) the curve is convex upwards at a point P on it if $\frac{d^2y}{dt^2}$ is negative.

(iii) the curve has a point of inflexion at P if

(a) $\frac{d^2 y}{dx^2} = 0$ and

(b) $\frac{d^2 y}{dx^2}$ changes sign as x passes through P i.e., $\frac{d^3y}{dx^3} \neq 0$

Proof :

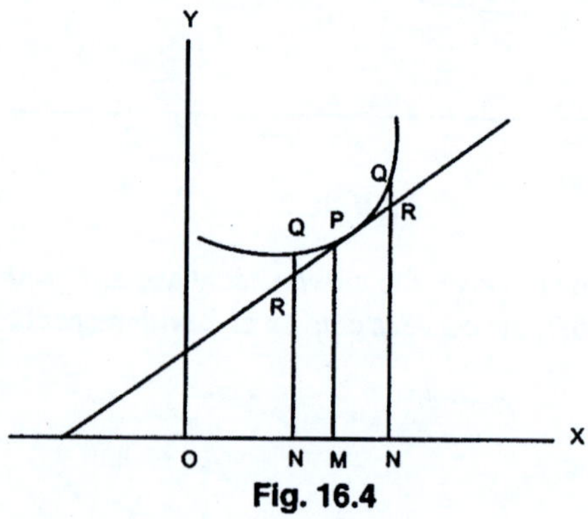

Fig. 16.4

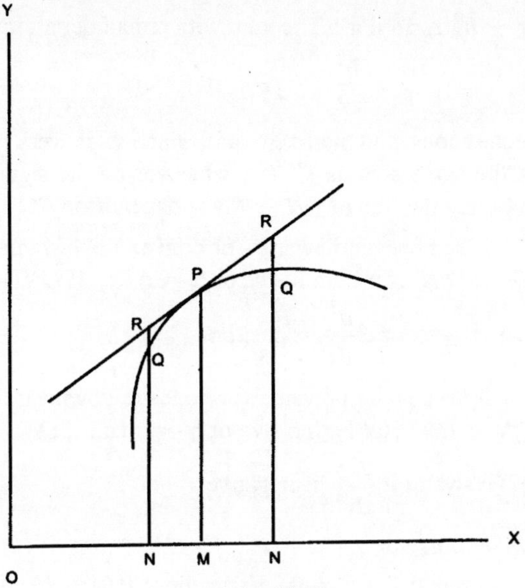

Fig. 16.5

Let $P(x, y)$ be any point on the curve $y = f(x)$. Take a neighbouring point $Q(x + h, \; y + k)$ on both sides of P. From Q draw $QN \perp OX$, meeting the tangent at P to the curve in R.

The equation of the tangent at $P(x, y)$ is

$$Y - y = f'(x) \; (X - x)$$

$$\Rightarrow \qquad Y = y + f'(x) \; (X - x)$$

It meets QN where $X = x + h$ in the point R, so that

$$RN = Y = y + f'(x) \{x + h - x\}$$

$$= y + h f'(x)$$

$$= f(x) + h f'(x)$$

Also QN = Ordinate of Q

= Coordinate corresponding to the abscissa $x + h$

= $f(x + h)$

Now $QN - RN$

$$= f(x + h) - f(x) - h f'(x) \qquad \text{—[A]}$$

$$= \left\{ f(x) + h f'(x) + \frac{h^2}{\lfloor 2} f''(x + \theta h) \right\}$$

$$f(x) - h f'(x)$$

where $0 < \theta < 1$

Expanding $f(x + h)$ by Taylor's Theorem with remainder after two terms

$$= \frac{h^2}{\lfloor 2} f''(x + \theta h) \qquad ...(1)$$

If $f''(x)$ is continuous and non-zero and since h is very small, then $f''(x + \theta h)$ has the same sign as $f''(x)$, whatever be the sign of R. Thus from (1), it follows that the sign of $QN - RN$ depends on $f''(x)$.

(i) The curve will be concave upwards (or convex downwards) at P [see Fig. 1] if $QN > RN$ (for both h +ve or h –ve) i.e., if $QN - RN$ is +ve i.e., if $f''(x)$ is +ve or if $\frac{d^2y}{dx^2}$ is positive.

(ii) The curve will be convex upwards (or concave downwards) at P [see Fig. 2] if $QN < RN$ [for both h +ve or h –ve) i.e., if $QN - RN$ is –ve i.e., if $f''(x)$ is –ve or if $\frac{d^2y}{dx^2}$ is negative.

(iii) *For the point of inflexion* :

Let $f''(x) = 0$ and $f'''(x) \neq 0$. Expanding $f(x + h)$ by Taylor's theorem with remainder after three terms in [A], we get $QN - RN$

$$= \left\{f(x) + h f'(x) + \frac{h^2}{\lfloor 2} f''(x) + \frac{h^3}{\lfloor 3} f'''(x + \theta_1 h)\right\}$$

$$- f(x) - h f'(x), \text{ where } 0 < \theta_1 < 1$$

$$= \frac{h^3}{\lfloor 3} f'''(x + \theta_1 h) \qquad ...(2)$$

$$\because \quad f''(x) = 0 \text{ (given)}$$

$\because$ $f'''(x)$ is a continuous function of x at P and $f'''(x) \neq 0$

$\therefore$ $f'''(x + \theta_1 h)$ has the same sign as $f'''(x)$ in the neighbourhood of P.

But $\frac{h^3}{\lfloor 3}$ changes sign with h

$\therefore$ it follows from (2), that $QN - RN$ changes sign with h as we go from left to right through P.

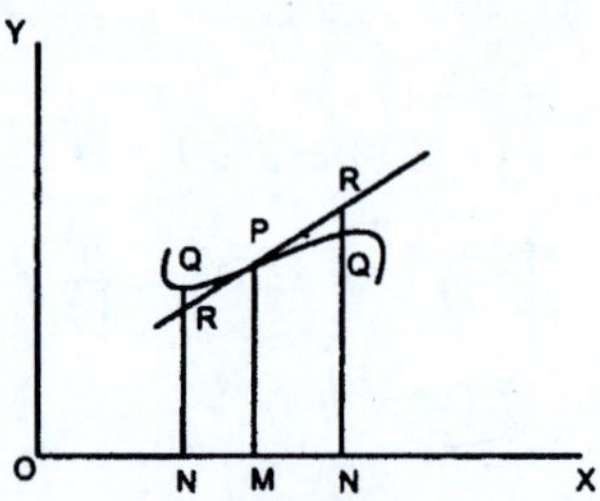

Fig. 16.6

$\therefore$ There is a point of inflexion [see Fig. 3] at P if $f''(x) = 0$ and $f'''(x) \neq 0$.

Thus we have

(i) A curve is concave upwards if $\frac{d^2y}{dx^2}$ is +ve.

(ii) A curve is convex upwards if $\frac{d^2y}{dx^2}$ is – ve and

(iii) At the point of inflexion $\frac{d^2y}{dx^2} = 0$ and $\frac{d^3y}{dx^3} \neq 0$

Generalization

The above result can be generalized.

Thus if $f''(x) = f'''(x) = f^{iv}(x) = \ldots. = f^{n-1}(x) = 0$ and $f^n(x) \neq 0$, then

(i) the curve has a point of inflexion at D if n is odd, and

(ii) the curve is concave upwards or convex upwards according as $f^n(x) > 0$ or < 0, and n is even

Note : In the above investigation, we ahve assumed that $\frac{dy}{dx}$ is finite. If $\frac{dy}{dx}$ becomes infinite, then we must find the points of inflexion by considering $\frac{d^2x}{dy^2}$.

ILLUSTRATIVE EXAMPLES

Example 1. *Show that the curve $y = x^4$ is concave upwards at the origin.*

Solution : The curve is $y = x^4$.

$$\therefore \quad \frac{dy}{dx} = 4x^3, \ \frac{d^2y}{dx^2} = 12x^2, \ \frac{d^3y}{dx^3} = 24x, \text{ and } \frac{d^4y}{dx^4} = 24$$

$\because$ At origin (0, 0)

$$\frac{d^2y}{dx^2} = 0,$$

$$\frac{d^3y}{dx^3} = 0,$$

and $$\frac{d^4y}{dx^4} = 24 = +\text{ve} \neq 0$$

$\therefore$ The curve is concave upwards at the origin.

Example 2. *Find the range of values of x for which the curve $y = 3x^5 - 40x^3 + 3x - 20$ is concave upwards or downwards. Find also the point of inflexion.*

Solution : The equation of the curve is

$$y = 3x^5 - 40x^3 + 3x - 20$$

$$\therefore \quad \frac{dy}{dx} = 15x^4 - 120x^2 + 3$$

and $$\frac{d^2y}{dx^2} = 60x^3 - 240x$$

$$= 60x\,(x - 2)\,(x + 2)$$

In $(-\infty, 2)$, $\dfrac{d^2y}{dx^2}$ is $-$ve

$\therefore$ the curve is concave downwards.

In $(-2, 0)$, $\dfrac{d^2y}{dx^2}$ is +ve

$\therefore$ the curve is concave upwards.

In $(0, 2)$, $\dfrac{d^2y}{dx^2}$ is $-$ve

$\therefore$ the curve is concave downwards.

In $(2, \infty)$, $\dfrac{d^2y}{dx^2}$ is +ve

$\therefore$ the curve is concave upwards.

Putting $\dfrac{d^2y}{dx^2} = 0$, $x = 0, \pm 2$

Also $\dfrac{d^3y}{dx^3} = 180x^2 - 240$ which does not vanish for $x = 0, \pm 2$

$\therefore$ there are points of inflexion at $x = 0, \pm 2$

From (1), when $x = 0$, $y = -20$

when $x = -2$, $y = 198$

and when $x = 2$, $y = -238$

Hence the points of inflexion are

$(0, -20)$, $(-2, 198)$ and $(2, -238)$.

Example 3. *Discuss the concavity and convexity of the curve*

$y = (\sin x + \cos x)\ e^x$ when $0 \le x \le 2\pi$. Find also the points of inflexion.

Solution : The equation of the given curve is

$$y = (\sin x + \cos x)\ e^x \qquad \text{.... (1)}$$

$$\therefore \quad \frac{dy}{dx} = (\sin x - \cos x)\ .\ e^x + e^x\ (\cos x - \sin x)$$

$$= 2e^x \cos x$$

and

$$\frac{d^2y}{dx^2} = 2e^x\ (-\sin x) + \cos x\ .\ 2e^x$$

$$= 2\ (\cos x - \sin x)\ e^x \qquad \text{.... (2)}$$

Now $\frac{d^2y}{dx^2} = 0$ when $\cos x - \sin x = 0$

$[\because\ e^x \ne 0$ for any $x]$

or $\tan x = 1$ or $x = \frac{\pi}{4}, \frac{5\pi}{4}$ in the interval $[0, 2\pi]$

From (2)

$$\frac{d^2y}{dx^2} = 2\ (\cos x - \sin x)\ e^x$$

$$= 2\sqrt{2} = \left(\frac{1}{\sqrt{2}} \cos x - \frac{1}{\sqrt{2}} \sin x\right) e^x$$

$$= 2\sqrt{2} \left(\cos\frac{\pi}{4} \cos x - \sin\frac{\pi}{4} \sin x\right) e^x$$

$$= 2\sqrt{2} \cos\left(x + \frac{\pi}{4}\right) e^x$$

Since e^x is +ve for all values of x,

$\therefore$ when $0 \le x < \frac{\pi}{4}$, $\frac{d^2y}{dx^2} > 0$

$\therefore$ the curve is concave upwards,

when $\frac{\pi}{4} < x < \frac{5\pi}{4}$, $\frac{d^2y}{dx^2} < 0$

$\therefore$ the curve is concave downwards.

and when $\frac{5\pi}{4} < x \le 2\pi$, $\frac{d^3y}{dx^3} > 0$

$\therefore$ the curve is concave upwards.

Now $\frac{d^2 y}{dx^2} = 0$ when $x = \frac{\pi}{4}$ or $\frac{5\pi}{4}$

$\frac{d^2 y}{dx^2}$ changes sign from +ve to –ve at $x = \frac{\pi}{4}$ and therefore gives a point of inflexion.

Also $\frac{d^2 y}{dx^2}$ changes sign from –ve to +ve at $x = \frac{5\pi}{4}$ and therefore gives a point of inflexion.

Now from (1),

$$\text{when } x = \frac{\pi}{4},\ y = \left(\sin\frac{\pi}{4} + \cos\frac{\pi}{4}\right) e^{\frac{\pi}{4}}$$

$$= \left(\frac{1}{\sqrt{2}} + \frac{1}{\sqrt{2}}\right) e^{\frac{\pi}{4}}$$

$$= \sqrt{2}\ e^{\frac{\pi}{4}}$$

$$\text{and when } x = \frac{5\pi}{4},\ y = \left(\sin\frac{5\pi}{4} + \cos\frac{5\pi}{4}\right) e^{\frac{5\pi}{4}}$$

$$= \left(-\frac{1}{\sqrt{2}} - \frac{1}{\sqrt{2}}\right) e^{\frac{5\pi}{4}}$$

$$= -\sqrt{2}\ e^{\frac{5\pi}{4}}$$

Hence the points of inflexion are

$$\left(\frac{\pi}{4}, \sqrt{2}\ e^{\frac{\pi}{4}}\right) \text{ and } \left(\frac{5\pi}{4}, -\sqrt{2}\ e^{\frac{5\pi}{4}}\right)$$

Example 4. *Find the point of inflexion of the curve*

$$y\,(a^2 + x^2) = x^3$$

Solution : The equation of the curve is

$$y\,(a^2 + x^2) = x^3$$

$$\text{or } y = \frac{x^3}{a^2 + x^2}$$

$$\therefore \quad \frac{dy}{dx} = \frac{(a^2 + x^2)\,.\,3x^2 - x^3 \cdot 2x}{(a^2 + x^2)^2}$$

$$= \frac{3a^2x^2 + x^4}{(a^2 + x^2)^2}$$

$$\text{and } \frac{d^2 y}{dx^2} = \frac{(a^2 + x^2)^2\,(6a^2 x + 4x^3) - (3a^2 x^2 + x^4) \cdot 2\,(a^2 + x^2) \cdot 2x}{(a^2 + x^2)^4}$$

$$= \frac{x\,[(a^2 + x^2)^2\,(6a^2 + 4x^2) - 4\,(3a^2x^2 + x^4)]}{(a^2 + x^2)^3}$$

$$= \frac{x\,[6a^4 + 10a^2x^2 + 4x^4 - 12a^2x^2 - 4x^4]}{(a^2 + x^2)^3}$$

$$= \frac{2a^2 x\,(3a^2 - x^2)}{(a^2 + x^2)^3}$$

For the points of inflexion, $\frac{d^2y}{dx^2} = 0$

$\therefore \quad \frac{2a^2x\,(3a^2 - x^2)}{(a^2 + x^2)^3} = 0$

or $2a^2\,x\,(3a^2 - x^2) = 0$

which gives $x = 0$ or $x = \pm\sqrt{3}a$

Now

$$\frac{d^3y}{dx^3} = \frac{2a^2\left[(a^2 + x^2)^3\,(3a^2 - 3x^2) - (3a^2x - x^3)\,.\,3\,(a^2 + x^2)^2\,2x\right]}{(a^2 + x^2)^6}$$

$$= \frac{6a^2\,[(a^2 + x^2)\,(a^2 - x^2) - 2x\,(3a^2x - x^3)]}{(a^2 + x^2)^4}$$

$$= \frac{6a^2\,[a^4 - x^4 - 6a^2x^2 + 2x^4]}{(a^2 + x^2)^4}$$

$$= \frac{6a^2\,(a^4 + x^4 - 6a^2x^2)}{(a^2 + x^2)^4}$$

when $x = 0$, $\frac{d^3y}{dx^3} = \frac{6}{a^2} \neq 0$

when $x = \sqrt{3}\,a$, $\frac{d^3y}{dx^3} = -\frac{3}{4a^2} \neq 0$ and

when $x = -\sqrt{3}\,a$, $\frac{d^3y}{dx^3} = -\frac{3}{4a^2} \neq 0$

Hence $x = 0, \pm\sqrt{3}\,a$ correspond to the points of inflexion.

From (1), when $x = 0$, $y = 0$;

(when $x = \sqrt{3}\,a$, $y = \frac{3\sqrt{3}}{4}a$)

and when $x = -\sqrt{3}\,a$, $y = \frac{-3\sqrt{3}}{4}a$

Hence the points of inflexion are

$$(0,\ 0),\ \left(\sqrt{3a},\ \frac{3\sqrt{3}}{4}\ a\right),\ \left(-\sqrt{3}\ a,\ -\frac{3\ \sqrt{3}}{4}\ a\right)$$

Example 5. *Find the points of inflexion of the following curves :*

$$xy = a^2 \log\left(\frac{y}{a}\right)$$

Solution : Equation of the curve is

$$xy = a^2 \log\left(\frac{y}{a}\right)$$

$$\Rightarrow \qquad x = \frac{a^2}{y} \log\left(\frac{y}{a}\right)$$

Hence y is independent variable and x, the dependent variable.

$$\therefore \quad \frac{dx}{dy} = -\frac{a^2}{y^2} \log\left(\frac{y}{a}\right) + \frac{a^2}{y} \cdot \frac{1}{\frac{y}{a}} \cdot \frac{1}{a}$$

$$= \frac{a^2}{y^2}\left[1 - \log\left(\frac{y}{a}\right)\right]$$

$$\frac{d^2x}{dy^2} = \frac{a^2}{y^2}\left[-\frac{1}{\frac{y}{a}} \cdot \frac{1}{a} + \left(1 - \log\frac{y}{a}\right)\left(\frac{-2a^2}{y^3}\right)\right]$$

$$= -\frac{a^2}{y^3}\left[3 - 2\log\frac{y}{a}\right]$$

$$\frac{d^3x}{dy^3} = -\frac{a^2}{y^3}\left[-2 \cdot \frac{1}{\frac{y}{a}} \cdot \frac{1}{a}\right] + \left(3 - 2\log\frac{y}{a}\right)\left(\frac{3a^2}{y^4}\right)$$

$$= \frac{a^2}{y^4}\left[5 - 2\log\frac{y}{a}\right]$$

Now, for points of inflexion, $\dfrac{d^2x}{dy^2} = 0$ and $\dfrac{d^3x}{dy^3} \neq 0$

Putting $\dfrac{d^2x}{dy^2} = 0$, we get

$$3 - 2\log\left(\frac{y}{a}\right) = 0$$

$$\Rightarrow \quad y = a\ e^{3/2}$$

Putting this value of $y = a\, e^{3/2}$ or $\log \dfrac{y}{a} = \dfrac{3}{2}$ in $\dfrac{d^3 x}{dy^3}$, we get

$$\frac{d^3 x}{dy^3} = \frac{a^2}{a^4 e^6}\left(5 - 2 \cdot \frac{3}{2}\right) = \frac{2}{a^2 e^6} \neq 0$$

Hence $y = a\, e^{3/2}$ gives a point of inflexion.

Putting $y = a\, e^{3/2}$ in equation of curve, we have

$$x = \frac{a^2}{a\, e^{3/2}} \cdot \frac{3}{2} = \frac{3}{2}\, a\, e^{-3/2}$$

Hence the points of inflexion is

$$\left(\frac{3}{2}\, a\, e^{-3/2},\ a\, e^{3/2}\right)$$

Example 6. *Show that every point, in which the same curve*

$$y = c \sin \frac{x}{a}$$

meets the X-axes, is a point of inflexion on the curve.

Solution : Since every point which lies on X-axis as well as on curve, has its ordinate zero, hence putting $y = 0$ in equation of curve, we have

$$\sin \frac{x}{a} = 0$$

$$\Rightarrow \quad \frac{x}{a} = n\pi$$

$\Rightarrow \quad x = an\pi$ where n is any integer.

Now $\dfrac{dy}{dx} = \dfrac{c}{a} \cos \dfrac{x}{a}$

$$\frac{d^2 y}{dx^2} = -\frac{c}{a^2} \sin \frac{x}{a}$$

$$\frac{d^3 y}{dx^3} = -\frac{c}{a^3} \cos \left(\frac{x}{a}\right)$$

For points of inflexion, $\dfrac{d^2 y}{dx^2} = 0$

$$-\frac{c}{a^2} \sin \frac{x}{a} = 0$$

$$\Rightarrow \quad \sin \frac{x}{a} = 0$$

$$\Rightarrow \quad x = an\pi$$

when $x = an\pi$, $\dfrac{d^3y}{dx^3} = -\dfrac{c}{a^3}\cos(n\pi)$

$$= -\frac{c}{a^2}(-1)^n \neq 0$$

$\therefore$ $x = an\pi$ gives the points of inflexion. There are the points in which the curve cuts X-axis.

Example 7. *Show that the points of inflexion of the curve $y^2 = (x-a)^2(x-b)$ lies on the line $3x + a = 4b$.*

Solution : Equation of the curve is

$$y^2 = (x-a)^2(x-b)$$

or

$$y = \pm (x-a)\sqrt{x-b}$$

$$\therefore \quad \frac{dy}{dx} = \left[(x-a)\cdot\frac{1}{2}\frac{1}{\sqrt{x-b}} + \sqrt{x-b}\right]$$

$$= \pm\frac{x-a+2x-2b}{2\sqrt{x-b}}$$

$$= \pm\frac{3x-2b-a}{2\sqrt{x-b}}$$

$$= \pm\frac{1}{2}(3x-2b-a)(x-b)^{-\frac{1}{2}}$$

and $$\frac{d^2y}{dx^2} = \pm\frac{1}{2}\left[(3x-2b-a)\left(-\frac{1}{2}\right)(x-b)^{-3/2} + (x-b)^{-1/2}(3)\right]$$

$$= \pm\frac{1}{2}\left[\frac{a+2b-3x}{2(x-b)^{3/2}} + \frac{3}{(x-b)^{1/2}}\right]$$

$$= \pm\frac{1}{2\sqrt{x-b}}\left[\frac{a+2b-3x}{2(x-b)} + 3\right]$$

$$= \pm\frac{1}{2\sqrt{x-b}}\left[\frac{a-4b+3x}{2(x-b)}\right]$$

For points of inflexion, $\dfrac{d^2y}{dx^2} = 0$

$\therefore$ $3x - 4b + a = 0$

$\therefore$ Points of inflexion of curve lie on the straight line $3x - 4b + a = 0$

16.6. Concavity and Convexity for Polar Curves

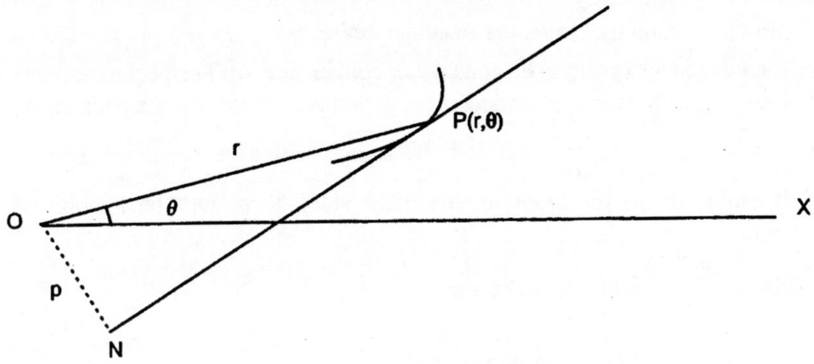

Fig. 16.7 Curve Concave at *P* w.r.t. O

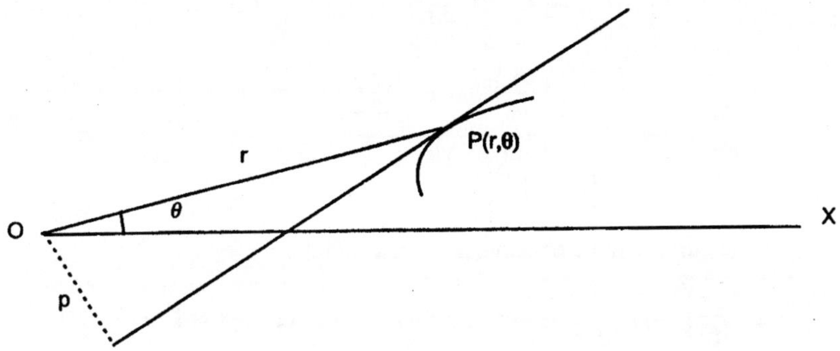

Fig. 16.8 Curve Convex at *P* w.r.t. O

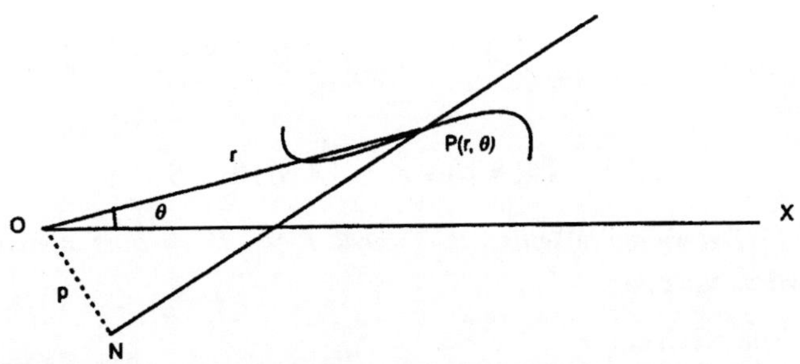

Fig. 16.9 Point of Inflexion at *P*

Let O be the pole and OX be the initial line. Let $P\ (r,\ \theta)$ be any point on the curve $r = f(\theta)$.

Draw $ON \perp$ on the tangent at P and let $ON = p$.

Then the curve is said to be concave or convex at P with respect to the pole O according as p increases or decreases with the increase in r. i.e., according as $\frac{dp}{dr}$ is +ve or –ve. If $\frac{dp}{dr} = 0$ at P, [O being positive for points on one side of P and negative for those on the other side], there will be a point of inflexion at P.

But $r\frac{dr}{dp}$ = radius of curvature

$$= \frac{\left\{r^2 + \left(\frac{dr}{d\theta}\right)^2\right\}^{3/2}}{r^2 + 2\left(\frac{dr}{d\theta}\right)^2 - r\frac{d^2r}{d\theta^2}}$$

$$\frac{dp}{dr} = \frac{r\left\{r^2 + 2\left(\frac{dr}{d\theta}\right)^2 - r\frac{d^2r}{d\theta^2}\right\}}{\left\{r^2 + \left(\frac{dr}{d\theta}\right)^2\right\}^{3/2}}$$

Hence the curve will be concave at P w.r.t. O if

$r^2 + 2\left(\frac{dr}{d\theta}\right)^2 - r\frac{d^2r}{d\theta^2}$ is +ve, the curve will be convex at P w.r.t. O if

$r^2 + 2\left(\frac{dr}{d\theta}\right)^2 - r\frac{d^2r}{d\theta^2}$ is –ve, and there will be a point of inflexion at P,

if $r^2 + 2\left(\frac{dr}{d\theta}\right)^2 - r\frac{d^2r}{d\theta^2} = 0$ or $u + \frac{d^2u}{d\theta^2} = 0$ where $u = \frac{1}{r}$

ILLUSTRATIVE EXAMPLES

Example 1. *Determine whether the spiral $r \cos h\theta = a$ is concave or convex towards the pole.*

Solution : The given curve is

$r \cos h\theta = a$

$\Rightarrow\ r = a \sec h\theta$

$$\therefore \quad \frac{dr}{d\theta} = -a \sec h\theta \tan h\theta$$

$$\frac{d^2r}{d\theta^2} = -a\left[\sec h\,\theta \cdot \sec h^2\,\theta - \tan\theta \cdot \sec h\,\theta \cdot \tan h\,\theta\right]$$

$$= -a\,[\sec h^3\,\theta - \tan h^2\,\theta\,.\,\sec h\,\theta\,]$$

$$= -a \sec h\,\theta\,[\sec h^2\,\theta - \tan h^2\,\theta\,]$$

$$\therefore \quad r^2 + 2\left(\frac{dr}{d\theta}\right)^2 - r\,\frac{d^2r}{d\theta^2}$$

$$= a^2 \sec h^2\theta + 2a^2 \sec h^2\theta \tan h^2\theta + a^2 \sec h^2\theta\,(\sec h^2\,\theta - \tan h^2\theta\,)$$

$$= a^2 \sec h^2\theta + 2a^2 \sec h^2\theta \cdot \tan h^2\theta + a^2 \sec h^4\theta - a^2 \sec h^2\theta \cdot \tan h^2\theta$$

$$= a^2 \sec h^2\theta\,[1 + \tan h^2\theta + \sec h^2\theta\,]$$

$$= 2a^2 \sec h^2\theta \qquad \Big|\because \quad \sec h^2\theta + \tan h^2\,\theta = 1$$

$= +$ve

Hence the curve is concave towards the pole.

Example 2. Show that the curve $r = \dfrac{a\theta^2}{\theta^2 - 1}$ has a point of inflexion at $r = \dfrac{3a}{2}$.

Solution : The given curve is

$$r = \frac{a\theta^2}{\theta^2 - 1} \qquad \text{... (1)}$$

Put $\quad r = \dfrac{1}{u}$, we get

$$\frac{1}{u} = \frac{a\,\theta^2}{\theta^2 - 1}$$

$$\Rightarrow \quad u = \frac{\theta^2 - 1}{a\theta^2}$$

$$\Rightarrow \quad u = \frac{1}{a}\left[1 - \frac{1}{\theta^2}\right] \qquad \text{... (2)}$$

$$\therefore \quad \frac{du}{d\theta} = \frac{1}{a}\left(\frac{2}{\theta^3}\right) = \frac{2}{a\,\theta^3}$$

and $\dfrac{d^2u}{d\theta^2} = -\dfrac{6}{a\theta^4}$

For points of inflexion,

$$\frac{d^2y}{d\theta^2} + u = 0$$

$$\Rightarrow \qquad -\frac{6}{a\theta^4} + \frac{1}{a}\left(1 - \frac{1}{\theta^2}\right) = 0$$

$$\Rightarrow \qquad 1 - \frac{1}{\theta^2} - \frac{1}{\theta^4} = 0$$

$$\Rightarrow \qquad \theta^4 - \theta^2 - 1 = 0$$

$$\Rightarrow \qquad \theta^2 = 3, -2$$

when $\theta^2 = 3, \quad \theta = \pm\sqrt{3}$

when $\theta^2 = -2, \quad \theta$ is imaginary

when $\theta = \pm\sqrt{3}$, then from (1),

$$r = a \cdot \frac{3}{3-1} = \frac{3a}{2}$$

Hence the curve has a point of inflexion at $r = \dfrac{3a}{2}$.

EXERCISE 16 (A)

1. Prove that $y = e^x$ is everywhere concave upwards.
2. Prove that the curve $y = \log x$ is convex upwards everywhere.
3. Find the range of values of x for which the following curves are concave upwards or downwards :

   (a) $y = x^4 - 6x^3 + 12x^2 + 5x + 7$

   (b) $y = (x^2 + 4x + 5)\, e^{-x}$

   Find also the points of inflexion in each case.
4. Test the concavity and convexity of the curve $y = \sin x$ in the interval $[0, 2\pi]$.
5. Find the points of inflexion on the following curves :

   (a) $y = 3x^4 - 4x^3 + 1$

   (b) $y^2 = x(x+1)^2$

   (c) $a^2y^2 = x^2(a^2 - x^2)$

(*d*) $y = (x-2)^6 (x-3)^5$

(*e*) $y(a^2 + x^2) = a^2 x$

(*f*) $x^2 y = a^2 (x - y)$

(*g*) $y = e^{-x^2}$

6. Find the points of inflexion on the following curves :

(*a*) $x = (\log y)^3$

(*b*) $x = (y-1)(y-2)(y-3)$

7. Show that the curve $y = \dfrac{1-x}{1+x}$ has three points of inflexion which lie in a straight line.

8. In the curve $y(a^2 + x^2) = 2x^2$, prove that for varying values of a, the locus of the points of inflexion is the straight line $y = \dfrac{1}{2}$.

9. Show that the line joining the points of inflexion of the curve $y^2 (x - a) = x^2 (x + a)$ subtends an angle of $\dfrac{\pi}{3}$ at the origin.

10. Show that the abscissae of the points of inflexion on the curve $y^2 = f(x)$ satisfy the equation

$[f'(x)]^2 = 2 f(x) f''(x)$.

11. Show that the curve $r\sqrt{\theta} = a$ has a point of inflexion at a distance of $\sqrt{2}\,a$ from the pole.

12. Show that the points of inflexion on the curve $r = a\,\theta^n$ are given by

$$r = a\left[-\frac{n}{n+1}\right]^{n/2}$$

ANSWERS

3. (*a*) Concave upwards in the intervals $(-\infty, 1)$, $(2, \infty)$ and concave downwards in the interval $[1, 2]$; the points of inflexion are $(1, 19)$ and $(2, 33)$.

(*b*) Concave upwards in the intervals $(-\infty, -1)$, $(1, \infty)$ and concave downwards in the interval $[-1, 1]$; the points of inflexion are $(-1, 2e)$ & $(1, \frac{10}{e})$.

4. Concave upwards in $[\pi, 2\pi]$; concave downwards in $[0, \pi]$.

5. *(a)* $(0, 1), \left(\frac{20}{3}, \frac{11}{27}\right)$

(b) $\left(\frac{1}{3}, \pm\frac{4}{3\sqrt{3}}\right)$

(c) $(0, 0)$

(d) at $x = 3, \frac{28 \pm \sqrt{3}}{11}$

(e) at $x = 0, \pm a\sqrt{3}$

(f) at $x = 0, \pm a\sqrt{3}$

(g) at $x = \pm\frac{1}{\sqrt{2}}$

6. *(a)* $(0, 1), (8, e^2)$

(b) $(0, 2)$

16.7. Multiple Point

A point on the curve through which more than one branches of the curve pass is called a *multiple point.*

16.8. Double Point

A point on a curve is called a double point if two branches of the curve pass through it. There are, in general, two tangents at a double point which may be real and distinct or reals and coincident or imaginary.

16.9. Triple Point

A point on a curve is called a triple point if three branches of the curve pass through it.

16.10. Multiple Point of *r*th order

If through a point on the curve, r branches of the curve pass, then the point is called multiple point of rth order. In general, r tangents (real and distinct, coincident or imaginary) can be drawn through a multiple point of order r.

16.11. Classification of double points

There are three kinds of double points.

(a) Node

Def. A node is a point on the curve through which pass two real branches of the curve and two tangents at which are real and distinct. In the following figure, the point P is a node.

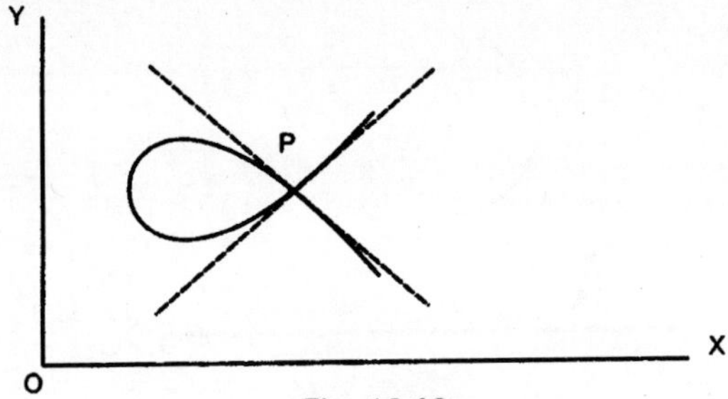

Fig. 16.10

(b) **Cusp**

Def. A double point on the curve through which two real branches of the curve pass and the tangents at which are real and distinct is called a cusp.

In each of the following figures, the point Q is a cusp.

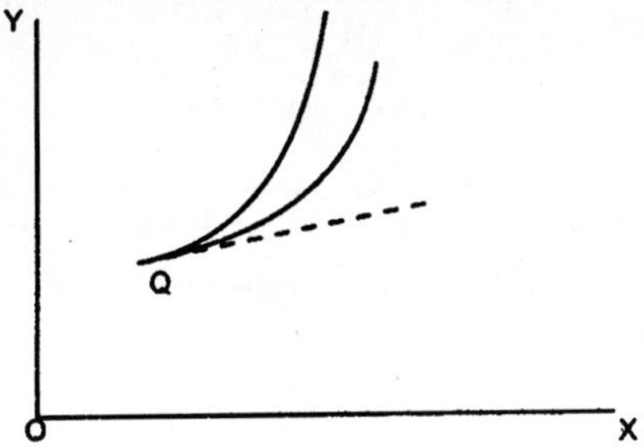

Fig. 16.11

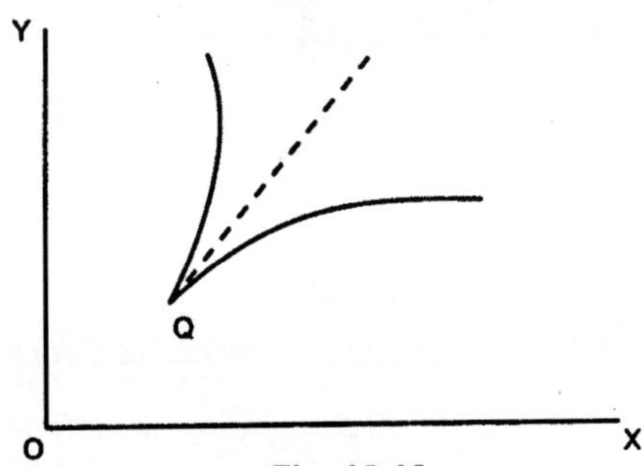

Fig. 16.12

(c) **Conjugate Point or Isolated Point**

Def. A conjugate point or an isolated point on a curve is a point in the neighbourhood of which there are no other real points of the curve.

In the following figure, *R* is a conjugate point. The tangents at a conjugate point are generally imaginary, but sometimes they may be real.

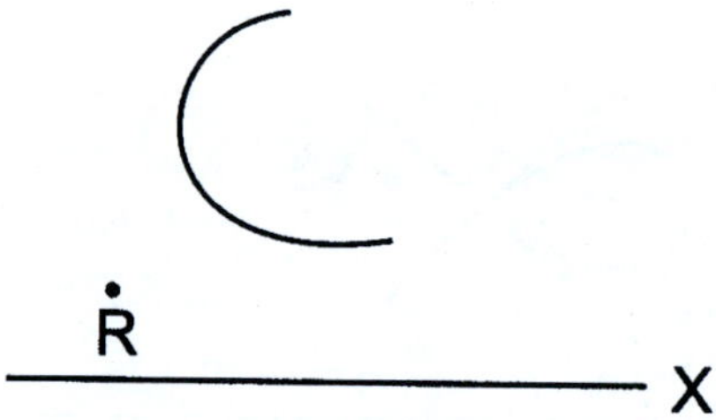

Fig. 16.13

The following example will illustrate the above fact.

Example. Find the nature of the origin on the curve

$$a^4 y^2 = x^4 (x^2 - a^2)$$

Solution : Clearly the curve passes through the origin. The equation of the curve can be written as

$$y = \pm \frac{x^2}{a^2} \sqrt{x^2 - a^2}$$

Thus for small values of $x \neq 0$, +ve or –ve, y is imaginary.

$\therefore$ In the neighbourhood of (0, 0), no other points of the curve lie and hence origin is a conjugate point.

$$\text{Now, } \frac{dy}{dx} = \pm \left[\frac{2x}{a^2} \sqrt{x^2 - a^2} + \frac{x^2}{a^2} \cdot \frac{1}{2} \cdot (x^2 - a^2)^{-1/2} \cdot 2x\right]$$

$$= \pm \left[\frac{2x}{a^2} \sqrt{x^2 - a^2} + \frac{x^3}{a^2 \sqrt{x^2 - a^2}}\right]$$

$$= 0 \text{ at the origin}$$

$\therefore$ Equation of the tangent at the origin (0, 0) is

$$y - 0 = 0\ (x - 0)$$

$$\Rightarrow \quad y = 0$$

which is real, showing that the tangent may be real at a conjugate point.

16.12. *To show that the necessary and sufficient conditions for any point (x, y) and f (x, y) = 0 to be a double point are that*

$$\frac{\partial f}{\partial x} = 0, \quad \frac{\partial f}{\partial y} = 0$$

Proof : The equation of the curve is

$$f\ (x, y) = 0 \qquad \text{...(1)}$$

Let $P(x, y)$ be any point on it. The slope of the tangent to (1) at P is $\frac{dy}{dx}$.

Differentiating (1) w.r.t. x treating y as a function of x, we have

$$\frac{\partial f}{\partial x} + \frac{\partial f}{\partial y}\frac{dy}{dx} = 0 \qquad \text{... (2)}$$

Equation (2) gives $\frac{dy}{dx}$.

Now if P is a double point, then there must be two tangents of the curve at P. These tangents may be real & distinct, real and coincident or imaginary according as P is a node, cusp or an isolated point. Hence $\frac{dy}{dx}$ must have two values at P. But equation (1) gives only one value of $\frac{dy}{dx}$. It can be satisfied by two values of $\frac{dy}{dx}$ if and only if it becomes an identity i.e., if $\frac{\partial f}{\partial x} = 0$ and $\frac{\partial f}{\partial y} = 0$.

Also $P(x, y)$ lies on the curve $f(x, y) = 0$

$\therefore$ the necessary and sufficient condition for any point $P(x, y)$ on the curve $f(x, y) = 0$ to be a multiple point is

$$\frac{\partial f}{\partial x} = 0 \text{ and } \frac{\partial f}{\partial y} = 0 \qquad \text{... (3)}$$

Classification of double points

Thus at a double point

$$\frac{dy}{dx} = -\frac{\partial f | \partial x}{\partial f | \partial y} \qquad \left[\text{Form } \frac{0}{0}\right]$$

$$= -\frac{\frac{d}{dx}\left(\frac{\partial f}{\partial x}\right)}{\frac{d}{dx}\left(\frac{\partial f}{\partial y}\right)} \qquad \text{[By De L' Hospital's rule]}$$

$$= -\frac{\frac{\partial^2 f}{\partial x^2} + \frac{\partial^2 f}{\partial x \partial y}\frac{dy}{dx}}{\frac{\partial^2 f}{\partial x \partial y} + \frac{\partial^2 f}{\partial y^2}\frac{dy}{dx}}$$

$$\Rightarrow \frac{\partial^2 f}{\partial y^2}\left(\frac{dy}{dx}\right)^2 + 2\frac{\partial^2 f}{\partial x \partial y}\frac{dy}{dx} + \frac{\partial^2 f}{\partial x^2} = 0 \qquad \text{[On simplification]}$$

If $\frac{\partial^2 f}{\partial x^2}$, $\frac{\partial^2 f}{\partial x\,\partial y}$ and $\frac{\partial^2 f}{\partial y^2}$ are not all zero at the same time, then this equation, being quadratic in $\frac{dy}{dx}$ will give two values of $\frac{dy}{dx}$ which shows that there shall be two tangents at P (x, y) to the curve. These tangents will real and distinct, real and coincident or imaginary, according as

$$\left(\frac{\partial^2 f}{\partial x\,\partial y}\right)^2 >, = \text{ or } < \left(\frac{\partial^2 f}{\partial x^2}\right)\left(\frac{\partial^2 f}{\partial y^2}\right)$$

Hence the double point P will be a node, cusp or conjugate point, according as

$$\left(\frac{\partial^2 f}{\partial x\,\partial y}\right)^2 >, = \text{ or } < \left(\frac{\partial^2 f}{\partial x^2}\right)\left(\frac{\partial^2 f}{\partial y^2}\right)$$

Alternative Method

Let (h, k) be the double point on the curve $f(x, y) = 0$. Transfer the origin to (h, k) by the substitutions $x = x + h$, $y = y + k$ and let the transformed equation by $F(x, y) = 0$. Since the new origin is a double point, hence the constant terms and the terms of first degree in $F(x, y) = 0$ must be absent.

$\therefore$ Equating to zero the constant term, the coefficient of x and the coefficient of y separately to zero in $F(x, y) = 0$, we get three equations in (h, k). Solve any two of these equations for h and k. If these values of h and k satisfy the third equation, then the point (h, k) is a double point.

Then put the values of h and k in the transformed equation. Now find the tangents at this new origin by equating to zero the lowest degree terms. If the two tangents thus obtained are real and distinct, the point is a node; if they are real and coincident, the point is a cusp and if they are imaginary, then the point is a conjugate point.

16.13. Species of cusps

We know that at a cusp, two branches of a curve have a common tangent and therefore a common normal. Depending upon the positions of the two branches of the curve with respect to the common tangents and normal, the cusps can be of five kinds as shown in the figures given below.

(i) *Single cusp of first species.* If the two branches of the curve lie on the same side of the common normal but on opposite sides of the common tangent, then the cusp is said to be a single cusp of first species.

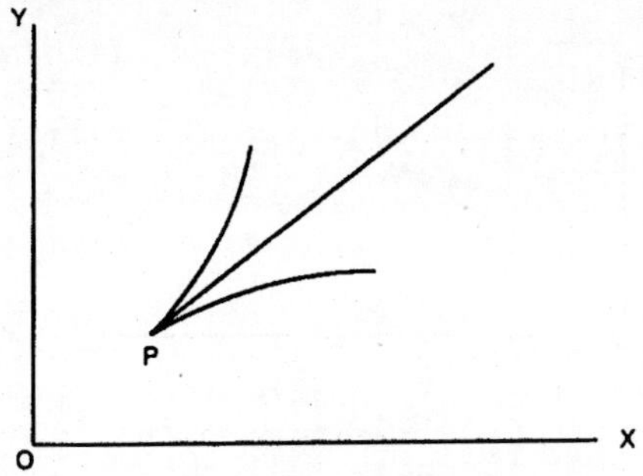

Fig. 16.14

(ii) *Single cusp of second species.* If the two branches of the curve lie on the same side of the common normal and also on the same side of the common tangent, then the cusp is said to be a single cusp of second species.

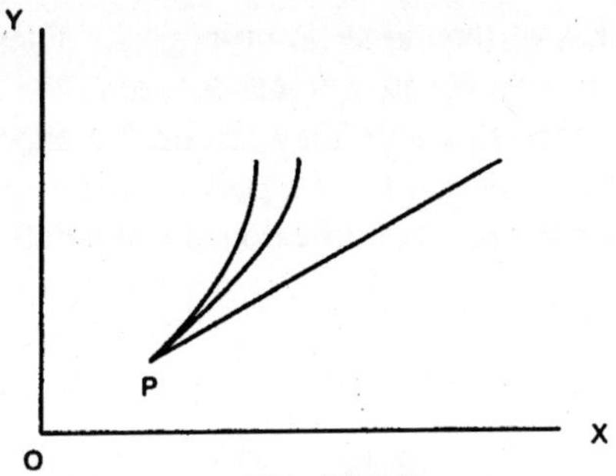

Fig. 16.15

(iii) *Double cusp of first species.* If the two branches of the curve lie on the opposite sides of the common normal and also on the opposite sides of the common tangent, then the cusp is said to be a double cusp of first species.

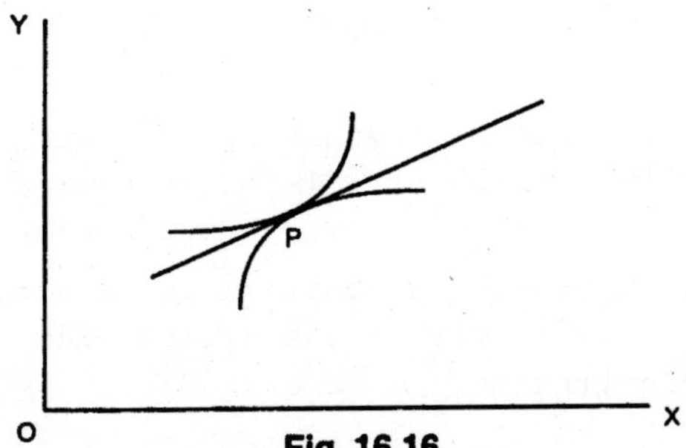

Fig. 16.16

(iv) *Double cusp of second species.* If the two branches of the curve lie on the opposite sides of the common normal but on the same side of the common tangent, then the cusp is said to be a double cusp of second species.

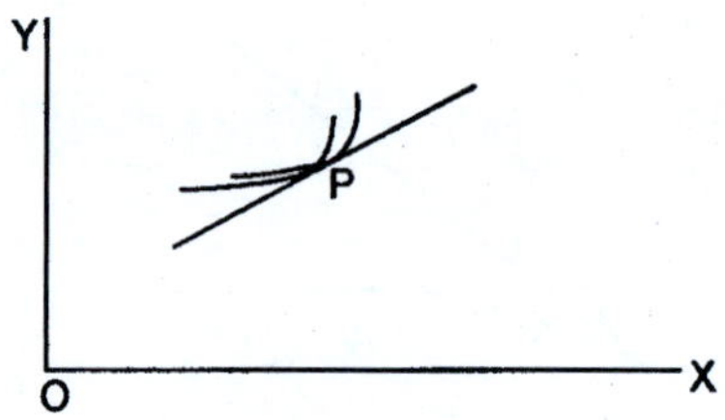

Fig. 16.17

(v) *Point of oscul-inflexion (Double cusp with change of species).* A double cusp which is of the first species on one side of the common normal and of the second species on the other side is called a point of oscul-inflexion. In other words there is a change of species at such a point.

OR

If the two branches of the curve lie on opposite sides of the common normal but on one side the two branches are on the same side of the tangent while on the other side they are on the opposite sides of the tangent, then the cusp is said to be a point of oscul-inflexion. Actually in one branch there is a point of inflexion at that point.

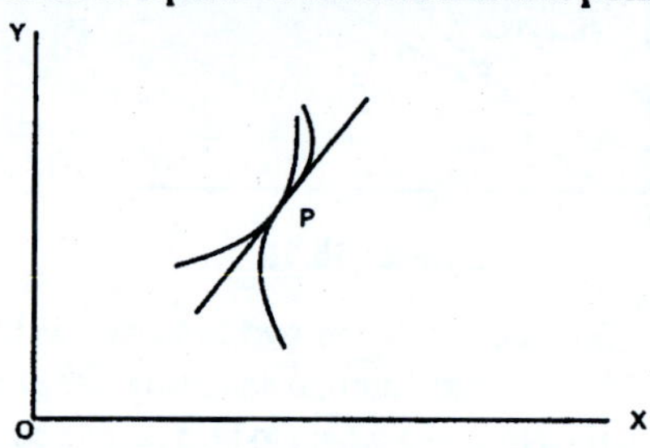

Fig. 16.18

Definition :

A cusp is said to be single or double according as the two branches of the curve lie entirely on the same side or on the opposite sides of the common normal.

Again a cusp (single or double) is said to be of first or second species according as both the branches of the curve lie on the opposite sides or on the same side of the common tangent.

16.14. Nature of cusp at the origin

If the origin is a cusp and the common tangent to the two branches of the curve is the axis of x, then solve the equation of the curve for y. Now the nature of the cusp will be decided as follows :

Find out whether for small values of x (+ve or –ve), y is real or imaginary.

(i) If y is real both for +ve and –ve values of x, there will be a double cusp at the origin.

(ii) If y is imaginary both for +ve and –ve values of x, the origin will be a conjugate point.

(iii) If y is real for one sign of x and imaginary for the other sign of x, in the neighbourhood of origin, there will be single cusp at the origin.

(iv) If the cusp is single and the two values of y are of the same signs, the cusp will be of second species.

(v) If the cusp is single and the two values of y are of the opposite signs, the cusp will be of first species.

(vi) If the origin is a double cusp and the values of y are of the same sign on both sides of the origin, the cusp will be of second species.

(vii) If the origin is a double cusp and the values of y are of the opposite signs on both sides of the origin, the cusp will be of first species.

(viii) If on one side of the origin the value of y are of ther same sign but on the other side they are of opposite sign, there will be a double cusp with change of species at the origin i.e., the origin will be a point of oscul-inflexion.

Again if the origin is a cusp and the common tangent to the two branches of the curve is the axis of y, then solve the equation of the curve for x and decide the nature of the cusp at origin in a similar way as explained above.

Lastly, if the origin is a cusp and the common tangent to the two branches of the curve is the line $ax + by = 0$, then we proceed as follows: Let p_1 be the length of the ⊥ from a point (x, y) on the curve very near to the cusp upon the tangent $ax + by = 0$ then $p_1 = \dfrac{ax + by}{\sqrt{a^2 + b^2}}$

Let $p = ax + by$

Now eliminate x or y (whichever is convenient) from this equation for p and the given equation of the curve. Suppose we eliminate y, then we shall get an equation in p and x. Solve this equation for p (neglecting p^3 and higher powers of p) and discuss the nature of the cusp at the origin exactly in a similar way as explained above.

Note : If we are required to discuss the nature of the cusp at a point other than the origin, then we transfer the origin to that point and proceed exactly as above.

Remember: If a curve passing through the origin be given by a rational integral algebraic equation, then the equation(s) of the tangent(s) at the origin is obtained by equating to zero the lowest degree terms in the given equation of the curve.

ILLUSTRATIVE EXAMPLES

Example 1. *Show that the curve $ay^2 = x^2 y + x^3$ has a cusp of first species at the origin.*

Solution : The given curve is

$$ay^2 = x^2 y + x^3 \qquad \text{... (1)}$$

The curve clearly passes through origin.

Tangents at the origin are obtained by equating to zero the lowest degree terms.

$\therefore \quad y^2 = 0$ is the equation of the tangent at the origin.

Therefore these are two coincident tangents given by $y = 0$ at the origin, hence the origin may be a cusp or a conjugate point.

The equation of the curve can be written as (quadratic in y)

$$ay^2 - x^2 y - x^3 = 0$$

$$\Rightarrow \qquad y = \frac{x^2 \pm \sqrt{x^4 + 4ax^3}}{2a}$$

$$= \frac{x^2 \pm x^2\sqrt{1 + \frac{4a}{x}}}{2a} \qquad \text{... (2)}$$

Now if x is +ve and small, y has two real values one +ve and the other –ve.

Again if x is –ve and small, y is imaginary.

Hence the curve has a cusp of first species at the origin.

Example 2. *Examine the nature of origin on the curve*

$$a^3 y^2 - 2abx^2 y + x^5 + \frac{x^6}{c} = 0$$

Solution : The given curve is

$$a^3 y^2 - 2abx^2 y + x^5 + \frac{x^6}{c} = 0 \qquad \text{... (1)}$$

The curve clearly passes through origin.

Tangents at the origin are obtained by equating to zero the lowest degree terms.

$\therefore \quad y^2 = 0$ is the equation of the tangent at the origin.

Therefore there are two coincident tangents given by $y = 0$ at the origin, hence the origin may be a cusp or a conjugate point.

Neglecting x^6 and solving (1) for y, we get

$$y = \frac{abx^2 \pm \sqrt{a^2 b^2 x^4 - a^3 x^5}}{a^3} \quad \text{... (2)}$$

If x is very small numerically, then $a^2 b^2 x^4 - a^3 x^5$ is the same sign as $a^2 b^2 x^4$ and it is +ve if x is +ve or –ve. It means y is real for small values of x (whether +ve or –ve). Hence there is a double cusp at the origin.

Again when x is +ve and small, $a^2 b^2 x^4 - a^3 x^5 < a^2 b^2 x^4$

$\Rightarrow$ $$\sqrt{a^2 b^2 x^4 - a^3 x^5} < abx^2$$

Thus we see that the small +ve values of x, both the values of y are +ve i.e., on the right hand side of the origin, the curve is of second species.

Again if x is –ve and small,

$$a^2 b^2 x^4 - a^3 x^5 > a^2 b^2 x^4$$

$\Rightarrow$ $$\sqrt{a^2 b^2 x^4 - a^3 x^5} > abx^2 \quad \text{since } a^3 x^5 \text{ is negative.}$$

Hence one value of y given by (2) is now +ve and the other –ve, hence there is a cusp of the first species on the left hand side of the origin. Thus the origin is a point of oscul-inflexion.

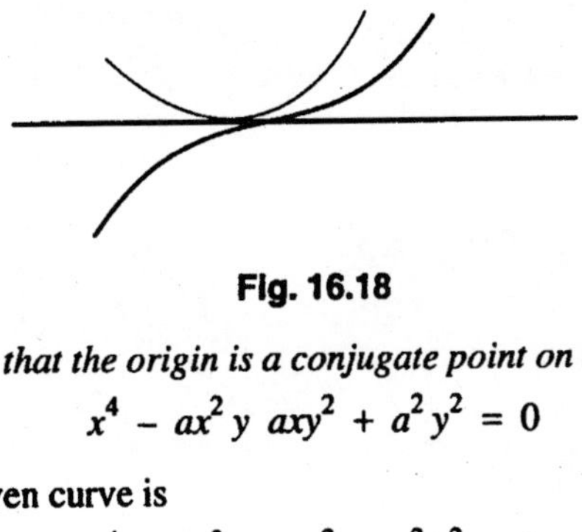

Fig. 16.18

Example 3. *Show that the origin is a conjugate point on the curve*

$$x^4 - ax^2 y \; axy^2 + a^2 y^2 = 0$$

Solution : The given curve is

$$x^4 - ax^2 y \; axy^2 + a^2 y^2 = 0 \quad \text{.... (1)}$$

The curve clearly passes through the origin.

Tangents at the origin are obtained by equating to zero the lowest degree terms.

$\therefore$ $y^2 = 0$ is the equation of the tangent at the origin.

Therefore there are two coincident tangents given by $y = 0$ at the origin, hence the origin may be a cusp or a conjugate point.

Equation (1) can be written as

$$a(x + a)\, y^2 - ax^2 y + x^4 = 0$$

$$\Rightarrow \qquad y = \frac{ax^2 \pm \sqrt{a^2x^4 - 4ax^5 - 4a^2x^4}}{2a(x+a)}$$

$$= \frac{ax^2 \pm x^2\sqrt{-4ax - 3a^2}}{2a(x+a)}$$

For small values of x, the sign of $(-4ax - 3a^2)$ is governed by $-3a^2$, which is –ve.

Hence the value of y near the origin or the branch of the curve near the origin is imaginary. Therefore the origin is a conjugate point.

Example 4. *Show that the curve* $y^2 = bx\sin\frac{x}{a}$ *has a node or a conjugate point at the origin according as a and b have like or unlike signs.*

Solution : The equation of the curve is

$$y^2 = bx\sin(x/a)$$

$$\Rightarrow \; y^2 - bx\sin\left(\frac{x}{a}\right) = 0$$

$\Rightarrow \; f(x, y) = 0$ where

$$f(x, y) = y^2 - bx\sin\left(\frac{x}{a}\right)$$

For double points

$$\frac{\partial f}{\partial x} = -b\sin\frac{x}{a} - \frac{bx}{a}\cos\frac{x}{a} = 0$$

and $\dfrac{\partial f}{\partial y} = 2y = 0$

$\therefore \; y = 0$ and $x = 0$

i.e., (0, 0) may be a double point.

Also, $$\frac{\partial^2 f}{\partial x^2} = -\frac{2b}{a}\cos\frac{x}{a} + \frac{bx}{a^2}\sin\frac{x}{a}$$

$$= -\frac{2b}{a} \text{ at } (0, 0)$$

$$\frac{\partial^2 f}{\partial y^2} = 2 \text{ at } (0, 0)$$

and $\dfrac{\partial f}{\partial x\,\partial y} = 0$ at (0, 0)

Now, $$\frac{\partial^2 f}{\partial x^2}\cdot\frac{\partial^2 f}{\partial y^2} = -\frac{yb}{a},\ \left(\frac{\partial^2 f}{\partial x\,\partial y}\right)^2 = 0$$

$$\therefore \quad \left(\frac{\partial^2 f}{\partial x\,\partial y}\right)^2 > \text{ or } < \left(\frac{\partial^2 f}{\partial x^2}\right)\left(\frac{\partial^2 f}{\partial y^2}\right)$$

according as a and b have like or unlike signs.

Hence the origin will be a node or a conjugate point according as a and b have like or unlike signs.

Example 5. *Determine the position and character of the double points on the curve*

$$y\,(y-6) = x^2\,(x-2)^3 - 9$$

Solution : The equation of the curve is

$$y\,(y-6) = x^2\,(x-2)^3 - 9$$

$$\Rightarrow y\,(y-6) - x^2\,(x-2)^3 + 9 = 0 \quad \text{... (1)}$$

$$\Rightarrow f\,(x,\,y) = 0 \text{ where}$$

$$f\,(x,\,y) = y\,(y-6) - x^2\,(x-2)^3 + 9$$

$$\therefore \quad \frac{\partial f}{\partial x} = 2x\,(x-2)^3 + 3\,(x-2)^2\,.\,x^2$$

$$= x\,(x-2)^2\,[2\,(x-2) + 3x]$$

$$= x\,(x-2)^2\,(5x-4)$$

$$\frac{\partial f}{\partial y} = -2y + 6$$

For the double points

$$\frac{\partial f}{\partial x} = 0, \quad \frac{\partial f}{\partial y} = 0$$

$$\frac{\partial f}{\partial x} = 0 \text{ gives } x\,(x-2)^2\,(5x-4) = 0$$

$$\Rightarrow x = 0,\ 2,\ \frac{4}{5}$$

$$\frac{\partial f}{\partial y} = 0 \text{ gives } -2y + 6 = 0 \Rightarrow y = 3$$

$\therefore$ The possible double points are (0, 3), (2, 3) and $\left(\frac{4}{5},\ 3\right)$.

Out of these only (0, 3) and (2, 3) satisfy (1)

$\therefore$ (0, 3) and (2, 3) are the only double points.

Character of (0, 3)

Shifting the origin to (0, 3), by putting $x = x + 0,\ y = y + 3$, the given equation (1) transforms to

$(y + 3)(y + 3 - 6) = x^2 (x - 2)^3 - 9$

$\Rightarrow y^2 - 9 = x^2 (x - 2)^3 - 9$

$\Rightarrow y^2 = x^2 (x - 2)^3$

Equating to 0, the lowest degree terms, the tangents at the new origin are $y^2 + 8x^2 = 0$ which are imaginary. $\therefore$ new origin (0, 3) is a conjugate point.

Aliter:

at (0, 3), $\dfrac{\partial^2 f}{\partial x^2} = -16,$

$\dfrac{\partial^2 f}{\partial y^2} = -2,$

and $\dfrac{\partial^2 f}{\partial x \partial y} = 0$

$$\therefore \left(\frac{\partial^2 f}{\partial x \partial y}\right)^2 < \frac{\partial^2 f}{\partial x^2} \cdot \frac{\partial^2 f}{\partial y^2}$$

Hence (0, 3) is a conjugate point.

Character of (2, 3)

at (2, 3) $\dfrac{\partial^2 f}{\partial x^2} = 0$

$\dfrac{\partial^2 f}{\partial y^2} = -2$

and $\dfrac{\partial^2 f}{\partial x \partial y} = 0$

$$\therefore \left(\frac{\partial^2 f}{\partial x \partial y}\right)^2 = \frac{\partial^2 f}{\partial x^2} \cdot \frac{\partial^2 f}{\partial y^2}$$

Therefore the point (2, 3) is a cusp. Now to decide the nature of the cusp transfer the origin to the point (2, 3) the equation (1) then becomes

$(y + 3)(y + 3 - 6) = (x + 2)^2 x^3 - 9$

$\Rightarrow y^2 = x^3 (x + 2)^2$

Equating to zero, the lowest degree terms, the tangents at the new origin are $y^2 = 0$ or $y = 0$, $y = 0$ i.e., x-axis is the common tangent.

Solving for y, we get,

$y = \pm (x + 2)\sqrt{x^3}$

$\Rightarrow y = \pm x (x + 2)\sqrt{x}$

This shows that when x is +ve, y has two real values, one +ve and other –ve; and when x is –ve, y is imaginary. Thus near the new origin the curve lies on both sides of x-axis (tangent) but only on one side of the y-axis (normal).

Hence the new origin (2, 3) is a single cusp of first species.

Example 6. *Examine the nature of the origin on the curve.*

$$x^7 + 2x^4 + 2x^3 y + x^2 + 2xy + y^2 = 0$$

Solution : Equating to zero, the lowest degree terms, the tangents at the origin are

$$x^2 + 2xy + y^2 = 0$$

$$\Rightarrow \quad (x + y)^2 = 0$$

from which it is clear that the two tangents at the origin are coincident and therefore (0, 0) is a cusp. Also each of the tangents has the equation as $x + y = 0$.

Let $p = x + y$

$\Rightarrow \quad y = p - x$

Put $y = p - x$ in the given equation of the curve, we get,

$x^7 + 2x^4 + 2x^3 (p - x) + p^2 = 0$

$\Rightarrow \quad p^2 + 2x^3 p + x^7 = 0$

$\Rightarrow \quad p = \dfrac{-2x^3 \pm \sqrt{4x^6 - 4x^7}}{2}$

$\Rightarrow \quad p = -x^3 \pm \sqrt{x^6 - x^7}$ ...(1)

When x is very small, $x^6 - x^7$ has the same sign as x^6 which is positive whether x is positive or negative. Therefore in the neighbourhood of origin, p is real for both positive as well as negative values of x. Hence there is a double cusp at the origin.

When x is +ve and small,

$x^6 - x^7 < x^6$

i.e., $\sqrt{x^6 - x^7} < x^3$

whence it is clear that both the values of p, as given by (1), are negative i.e., both the branches of the curve lie on the same side of the tangent.

Again when x in –ve and small, $-x^7$ becomes +ve and then $x^6 - x^7 > x^6$ i.e., $\sqrt{x^6 - x^7} > x^3$. Therefore the values of p, as given by (1), are of opposite signs i.e., the two branches of the curve lie on the opposite sides of the common tangent.

Thus on one side of the origin, the cusp is of first species and on the other side, it is of second species. Hence there is a point of oscul-inflexion at origin.

Example 7. Determine the existence and nature of the double points on the curve $(x - 2)^2 = y\,(y - 1)^2$.

Solution : Let (h, k) be a double point. Transferring the origin to (h, k), the equation becomes

$$(x + h - 2)^2 = (y + k)\,(y + k - 1)^2 \qquad \text{.... (1)}$$

$\Rightarrow \quad (h - 2)^2 - k\,(k - 1)^2 + 2x\,(h - 2) - y\,\{2k\,(k-1) + (k-1)^2\} +$ terms containing higher powers of x and $y = 0$

(h, k) will be a double point, if

$$(h - 2)^2 - k\,(k - 1)^2 = 0 \qquad \text{.... (2)}$$

$$h - 2 = 0 \qquad \text{.... (3)}$$

and

$$2k\,(k - 1) + (k - 1)^2 = 0 \qquad \text{.... (4)}$$

Equation (3) gives $h = 2$

Equation (4) gives $k = 1$ or $\frac{1}{3}$

Now $h = 2,\ k = 1$ satisfy all the three equations. Therefore (2, 1) is a double point.

Putting $h = 2$ and $k = 1$ in (1), the transformed equation is

$$x^2 = (y + 1)\,y^2$$

Equating to zero the lowest degree terms, the tangents at the new origin (i.e., at the double point) are

$$x^2 = y^2$$

$$\Rightarrow \quad y = \pm\, x$$

Hence these are two real and distinct tangents at (2, 1). Hence (2, 1) is a node of the curve.

EXERCISE 16 (B)

1. Show that the curve $x^3 + y^3 = ax^2$ has a cusp of first species at the origin.
2. Show that the curve $y^3 = (x - a)^2\ (2x - a)$ has a single cusp of first kind at the point $(a, 0)$.

3. Show that the curve $(xy + 1)^2 + (x - 1)^3 (x - 2) = 0$ has a single cusp of the first kind at the point $(1, -1)$.

4. Show that the curve $x^4 - 2a^2 xy - axy^2 + a^2y^2 = 0$ has a cusp of the second kind at the origin.

5. Show that the curve $x^4 - 2x^3 y - xy^2 - 2x^2 - 2xy + y^2 - x + 2y + 1 = 0$ has a single cusp of second species at $(0, -1)$.

6. Find the double points on $x^3 + y^3 - 3axy = 0$.

7. Show that at the origin on the curve $y^2 = ax^2 + ax^3$, there is a node, cusp or a conjugate point according as a is positive, zero or negative.

8. Prove that for the curve $ay^2 = (x - a)^2 (x - b)$ there is a conjugate point at $x = a$ if $a < b$, a cusp if $a > b$ and a node if $a = b$.

9. Find the double points of the curve

$$x^4 - 2ay^2 - 3a^2y^2 - 2a^2x^2 + a^4 = 0$$

10. Find the position and character of the double points on the curve

$$x^4 + y^3 + 2x^2 + 3y^2 = 0$$

11. Determine the position and character of the double points on the curve

$$x^3 - y^2 + 4y - 7x^2 + 15x - 13 = 0$$

12. Examine the following curves for singularities:

(a) $a^4 y^2 = x^4 (2x^2 - 3a^2)$

(b) $x^3 + x^2 + y^2 - x - 4y + 3 = 0$

(c) $(x + y)^3 - \sqrt{2} (y - x + 2)^2 = 0$

(d) $a^2 y^2 = a^2 x^2 - 4x^3$

(e) $y^2 (x + 1) = x^4$

(f) $(x^2 + y^2)^2 = a^2 (x^2 - y^2)$

ANSWERS

6. Node at $(0, 0)$

9. $(a, 0), (-a, 0)$ & $(0, -a)$

10. Conjugate point at $(0, 0)$

11. Node at $(3, 2)$

12. *(a)* $(0, 0)$ is a conjugate point

(b) (– 1, 2) is a node

(c) single cusp of first species at (1, – 1)

(d) node at (0, 0)

(e) origin is a double cusp of first species

(f) (0, 0) is a conjugate point.

17

Curve Tracing

17.1. Curve Tracing

The object of Curve Tracing is to find the approximate shape of a curve without the labour of plotting a large number of points.

If the Cartesian Equation is given, the student will find that he can invariably solve it either for y, or for x, or for r (in terms of θ in the last case), otherwise the curve will be too difficult for him to trace.

Only curve in which we can solve for y need by considered here; because, if the equation cannot be solved for y but can be solved for x we have only to regard y as the independent variable. If the equation can be solved for r, the rules for tracing polar curve will apply.

17.2. Tracing of Cartesian Curves

Procedure :

1. Symmetry : We examine the symmetry of a curve about any line by applying the following rules—

(i) **Symmetry about the X-axis :**—If the equation of a curve remains unaltered when y is changed to $-y$, i.e., if only even powers of y occur in the equation, then the curve is said to be symmetrical about the X-axis. For example, the parabola $y^2 = 4\ ax$ is symmetrical about the X-axis.

(ii) **Symmetry about the Y-axis :**—If the equation of a curve remains unaltered when x is changed to $-x$, i.e., if only even powers of x occur in the equation, then the curve is said to be symmetrical about the Y-axis. For example, the parabola $x^2 = 4ay$ is symmetrical about the Y-axis.

(iii) **Symmetry in opposite quadrants :**—If the equation of the curve remains unaltered when both x and y are changed to $-x$ and $-y$ respectively, the curve is said to be symmetrical in opposite quadrants.

(iv) **Symmetry about the line $y = x$:**—If the equation of the curve is unchanged when x and y are interchanged (i.e., x is changed to y and y to x),

then the curve is said to be symmetrical about the line $y = x$. For example, the curve $x^3 + y^3 = 2axy$ is symmetrical about the line $x = y$.

Note : If the equation of a curve remains unchanged when x and y are changed to $-y$ and $-x$ respectively, then the curve is said to be symmetrical about the line $y = -x$.

2. Origin: *(i)* Find whether the curve passes through the origin or not. It will pass through the origin if the equation of the curve is free from the constant term.

(ii) If the curve passes through the origin, find the equations of the tangents at the origin by equating to zero the lowest degree terms in the equation of the curve.

(iii) Now, we can determine the nature of the origin as to whether origin is an ordinary point or a multiple point. If the origin is found to be a double point, determine whether it is a node, a cusp or a conjugate point.

3. Intersection with coordinate axes : Find the points where the curve cuts the axes of coordinates. Find also the tangents at these points if necessary.

Besides also find out the points whose coordinates clearly appear to satisfy the equation of the curve. Also, find out the tangents at these points if necessary.

If the curve is symmetrical about the line $y = x$, then find out the point of intersection of the curve with this line and the equation of the tangent there at.

4. Asymptotes : Find all the asymptotes of the curve. The curve will not go beyond its asymptotes.

5. Special points :—*(i)* Find the points of the curve where $\frac{dy}{dx} = 0$ or ∞, i.e., the points where the tangent is $\|$ or $\perp$ to the X-axis. At such points, the ordinates or abscissae generally change the character from increasing to decreasing or vice-versa.

(ii) Find some special points on the curve if necessary.

6. Region : *(i)* If possible, solve the equation for y or x. Now, see the behaviour of y or x for different values of the other variable giving particular attention to those values for which y or x becomes infinite or zero or imaginary. Generally, $x = 0$ or $y = 0$ is convenient.

(ii) If the curve is symmetrical either about the X-axis or in the opposite quadrants, only positive values of y need be considered. The curve for negative values of y can be traced out by symmetry.

(iii) Now, observe the variations of y as x decreases from 0 to $-\infty$. If there is symmetry either about Y-axis or in the opposite quadrants, only the values of x need be considered.

(iv) If in the above process, it is observed that y is imaginary for certain range of values of x, say $a < x < b$, then it would imply that there exists no part of the curve between the lines $x = a$ and $x = b$.

(v) Find out the point of inflexion if it exists.

Note : It must be clearly understood here that merely a knowledge of symmetry, double points, asymptotes etc. will not enable us to trace a curve, since the asymptote gives the idea of the curve far removed from the origin, while the tangent at any point gives the nature of the curve in the immediate neighbourhood of that point. The curve can be traced completely if its equation is solved completely for one variable say in terms of another. Now, we should examine how y varies as x varies continuously from $-\infty$ to $+\infty$. However, instead of considering every value of x, we should concentrate on some important values of x such or those values of x for which y is a maximum, or a minimum, or zero or infinite or imaginary. It is not necessary to start with $x = -\infty$ in case of inconvenience. Generally we first consider the values of x from 0 to ∞ and then from 0 to $-\infty$.

ILLUSTRATIVE EXAMPLES

Example 1. *Trace the curve* $y^2 = x^3$ *(semicubical parabola).*

Solution :

1. As only even power of y occurs, so there is symmetry about the X-axis.

X′

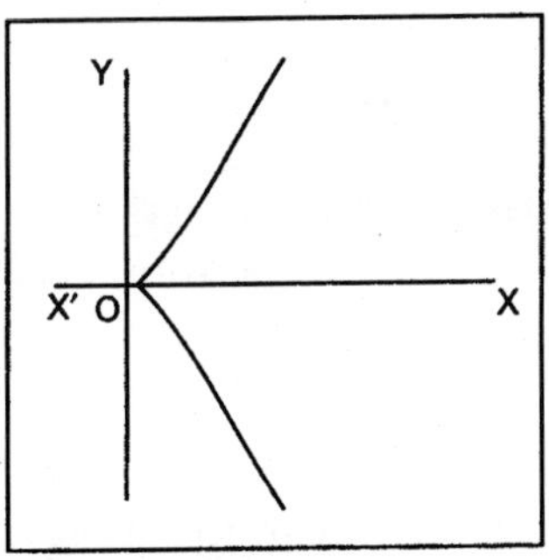

Fig. 17.1

2. Curve passes through the origin. Equating the lowest degree term y^2 to zero, we find the tangents at the origin are $y^2 = 0$, i.e., $y = 0$ and $y = 0$, i.e., X-axis is the tangent at the origin and the tangents being coincident a cusp is expected at the origin.

3. Putting $y = 0$ or $x = 0$ we do not get any new point but the origin.

4. For negative values of x, y^2 is negative, i.e., y is imaginary or the curve does not exist for negative values of x.

5. As x increases from 0 to ∞, y also increases from 0 to ∞.

6. No asymptotes.

7. For $x = 1$, $y = 1$. For $x = 2$, $y = 2\sqrt{2} = 2.8$ nearly. Plot the points (1, 1) and (2, 2.8). These points are on the curve.

With the above data the shape of the curve is as shown in adjoining Fig. 17.1.

Example 2. *Trace the curve $y = x^3$ (cubical parabola).*

Solution :

1. Changing y to $-y$ and x to $-x$ the equation of the curve does not change. Hence, there is symmetry in the opposite quadrants.

2. The curve passes through the origin. The tangent at origin is $y = 0$, on equating the lowest degree term to zero.

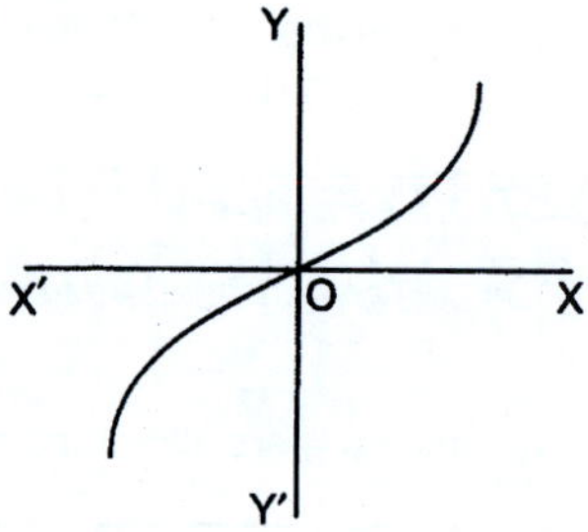

Fig 17.2

3. The curve crosses the axes only at the origin.

4. As x increases from 0 to ∞, y also increases from 0 to ∞.

5. No asymptotes.

6. For $x = 1$, $y = 1$; for $x = 2$ $y = 8$. Plot the points (1, 1) and (2, 8). These points are on the curve.

With the above data the curve is as shown in adjoining Fig. 17.2.

Example 3. *Trace the cissoid of Diocles*

$$y^2 (2a - x) = x^3.$$

Solution : The equation of the curve is

$$y^2 (2a - x) = x^3 \qquad \text{...(i)}$$

(1) Symmetry : Since (1) contains only even powers of y, the curve is symmetrical about X-axis.

(2) Origin : (i) The curve passes through the origin.

(ii) The tangents at the origin are given by $y^2 = 0$. Since the two tangents are real and coincident, the origin is a cusp.

(3) Intersection with axes : The curve meets the axis of X and the axis of Y at the origin only.

(4) Asymptotes : The asymptote parallel to the axis of Y is obtained by equating to zero the coefficient of the highest power, of y *i.e.*, $2a - x = 0$, *i.e.*, $x = 2a$ is the asymptote. There is no other asymptote of the curve since the factorization of the highest degree terms, i.e., $x(x^2 + y^2)$ gives $x(x + iy)(x - iy)$ which shows that two other asymptotes parallel to $x + iy = 0$ and $x - iy = 0$ are imaginary.

(5) Region : From (i),

$$y = x\sqrt{\frac{x}{2a - x}} \quad \text{[Taking +ve root only]}$$

when $x < 0$ or when $x > 2a$, y is imaginary. It means that the curve does not lie to the right of the line $x = 2a$ and to the left of Y-axis.

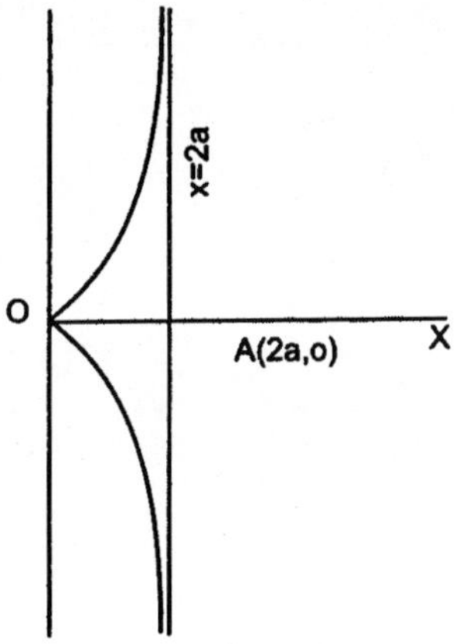

Fig. 17.3

When $x = 0$, $y = 0$, as x increases from 0 to $2a$, y also increases from 0 to ∞.

Hence, the shape of the curve is as shown in the Figure 17.3 given above.

Example 4. *Trace the Folium of Descarte's*

$$x^3 + y^3 = 3axy$$

Solution : The equation of the curve is

$$x^3 + y^3 = 3axy \qquad ...(i)$$

(1) Symmetry : Changing x to y and y to x, equation (i) remains unchanged. Hence, the curve is symmetrical about the line $y = x$.

(2) Origin : The curve passes through the origin. Equating to zero the lowest, degree terms, tangents at the origin are $xy = 0$, *i.e.*, $x = 0$ and $y = 0$. These tangents being real and distinct, origin is a node.

(3) Intersection with axes : Putting $y = 0$ in equation (i), we get $x = 0$. Also putting $x = 0$ in equation (i), we get $y = 0$. Hence, the curve meets the axes only at the origin (0, 0).

(4) Asymptotes : The curve has no asymptote parallel to the axes. Factorising the highest degree terms, we have

$$(x + y)(x^2 + y^2 - xy) = 3axy$$

Clearly, there will be an asymptote || to $x + y = 0$ and rest of the asymptotes are imaginary.

The asymptote || to $x + y = 0$ is,

$$x + y = \lim_{\substack{x \to \infty \\ \frac{y}{x} \to -1}} \frac{3axy}{x^2 + y^2 - xy}$$

$$= \lim_{\substack{x \to \infty \\ \frac{y}{x} \to -1}} \frac{3a\left(\frac{y}{x}\right)}{1 + \frac{y^2}{x^2} - \frac{y}{x}}$$

$$= \frac{-3a}{1 + 1 + 1}$$

$$= -a$$

(5) Special points : The curve (i) meets the line of symmetry $y = x$, where

$$2x^3 = 3\,ax^2$$

or $$x^2(2x - 3a) = 0$$

or $$x = 0 \text{ and } x = \frac{3a}{2}$$

$\therefore$ $$y = 0 \text{ and } y = \frac{3a}{2}$$

Hence, the two points of intersection are (0, 0) and $\left(\frac{3a}{2}, \frac{3a}{2}\right)$.

To find the slope of the tangent at $A\left(\frac{3a}{2}, \frac{3a}{2}\right)$. Differentiating equation (i) w.r.t. x, we get

$$3x^2 + 3y^2 \frac{dy}{dx} = 3a\left(x\frac{dy}{dx} + y\right)$$

or $$(y^2 - ax)\frac{dy}{dx} = ay - x^2$$

or $$\frac{dy}{dx} = \frac{ay - x^2}{y^2 - ax}$$

At the point $A\left(\frac{3a}{2}, \frac{3a}{2}\right)$,

$$\frac{dy}{dx} = -1$$

or $$\tan\psi = -1 = \tan 135°$$

or $$\psi = 135°$$

Hence, the tangent to the curve at A makes an angle of 135° with the axis of X.

(6) Region : x and y both cannot be negative, because that will make L.H.S. of (i) negative and R.H.S. positive. Hence, no portion of the curve lies in the third quadrant.

Hence the curve is as shown in the figure.

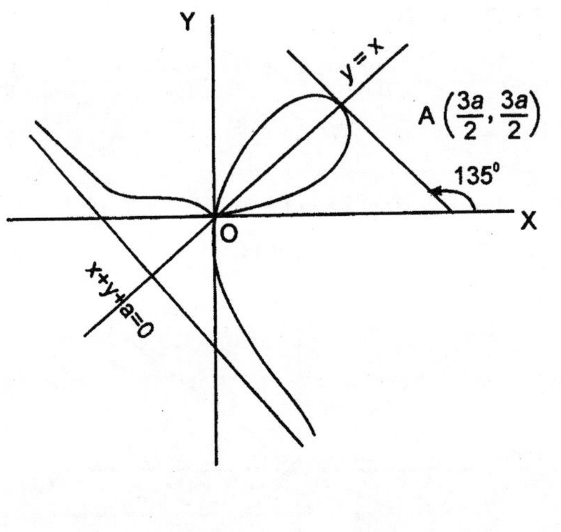

Fig. 17.4

Example 5. *Trace the curve* $y^2(a^2 + x^2) = x^2(a^2 - x^2)$

Solution : The equation of the curve is,

$$y^2 (a^2 + x^2) = x^2 (a^2 - x^2) \qquad \text{...(i)}$$

(1) Symmetry : Since in (i), powers of both x and y are all even, the curve is symmetrical about both the axes. It is obvious that the curve is symmetrical in opposite quadrants.

(2) Origin : The curve passes through the origin and the tangents at the origin are $y^2 = x^2$ or $y = \pm x$, which are real and distinct, showing that origin is a node.

(3) Intersection with axes : The curve meets X-axis at $A\ (a, 0)$ and $A'\ (-a, 0)$ which is seen by putting $y = 0$ in (i).

Transferring the origin to $(a, 0)$, the equation (i) transforms to

$$Y^3 \{2a^2 + 2aX + X^2\} = (X + a)^2 (-2aX - X^2)$$

Equating to zero the lowest degree terms, the tangent at the new origin is $X = 0$, i.e., new Y-axis. Thus, the tangent at $A\ (a, 0)$ is parallel to Y-axis. Similarly, the tangent at $A'\ (-a, 0)$ is also parallel to Y-axis (by symmetry).

(4) Asymptotes : The curve has no asymptotes.

(5) Region : Solving equation (i) for y, we get

$$y = \pm x \sqrt{\frac{a^2 - x^2}{a^2 + x^2}}$$

Now, y is real if and only if

$$a^2 - x^2 \geq 0$$

i.e., $$x^2 \leq a^2$$

i.e., $$|x| \leq a$$

Hence, the whole curve lies between the lines $x = \pm a$

Hence, the curve is as shown in the figure.

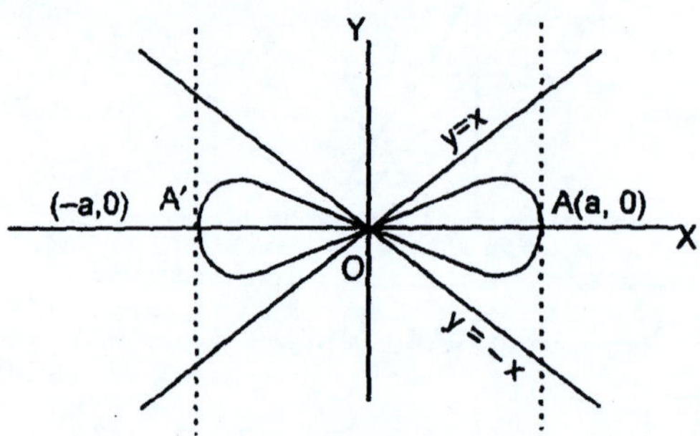

Fig. 17.5

Example 6. *Trace the curve*

$$(x^2 + a^2)\, y = a^2 x$$

Solution : The equation of the curve is,

$$(x^2 + a^2)\, y = a^2 x \qquad \text{...(i)}$$

(1) Symmetry : (i) The curve is not symmetrical about X-axis or Y-axis.

(ii) When x and y are changed to $-x$ and $-y$ respectively, the equation (i) remains unchanged. Hence, the curve is symmetrical in opposite quadrants.

(2) Origin : (i) The curve passes through the origin.

(ii) The tangent at the origin is

$$a^2 (y - x) = 0$$

or

$$y = x$$

(3) Asymptotes : Equating to zero, the coefficient of highest degree term in x, the asymptote parallel to X-axis is $x = 0$. The other asymptotes are imaginary.

(4) Intersection with axes : The curve does not cut the axes at any point other than (0, 0).

(5) Solving (i) for y, we have

$$y = \frac{a^2 x}{a^2 + x^2}$$

Hence, y is +ve when x is +ve and y is –ve when x is –ve. Thus, there is no part of the curve in the second and fourth quadrants.

(6) Differentiating, equation (i) w.r.t. x, we get

$$\frac{dy}{dx} = \frac{a^2 (a^2 - x^2)}{(a^2 + x^2)^2}$$

At maxima or minima,

$$\frac{dy}{dx} = 0$$

$$\therefore \qquad x = \pm a$$

for which

$$y = \pm \frac{a}{2}$$

Hence $A\left(a, \frac{a}{2}\right)$ and $A'\left(-a, -\frac{a}{2}\right)$ are the points of maxima and minima.

Hence, the shape of the curve is as shown in the Figure 17.6.

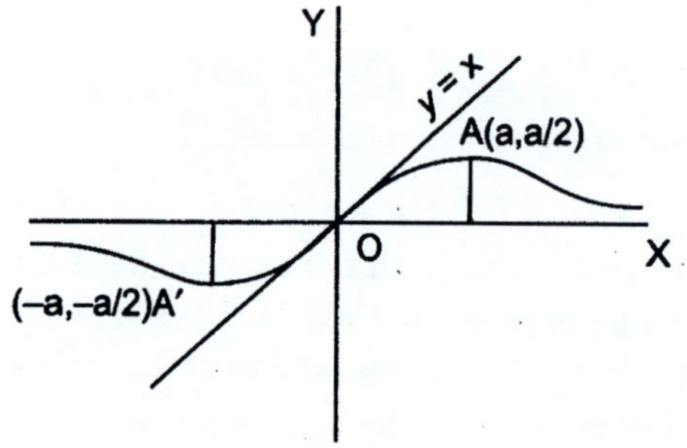

Fig 17.6

Example 7. *Trace the curve*

$$x^2 = y^2 (x + a)^3$$

Solution : The equation of the curve is,

$$x^2 = y^2 (x + a)^3 \qquad ...(i)$$

(1) Symmetry : The curve is symmetrical about X-axis.

(2) Origin : It passes through (0, 0). The tangents at (0, 0) are

$$y = \pm\, xa^{-3/2}$$

Thus, origin is a node.

(3) Asymptotes : The asymptotes are $y = 0$ and $x = -a$.

(4) Region : From (i), $y^2 = \dfrac{x^2}{(x + a)^3}$. Thus, y^2 is – ve when, $x < -a$ i.e., y is imaginary when $x < -a$. Hence, no part of the curve lies beyond $x = -a$.

Hence, the shape of the curve is as shown in the Figure 17.7.

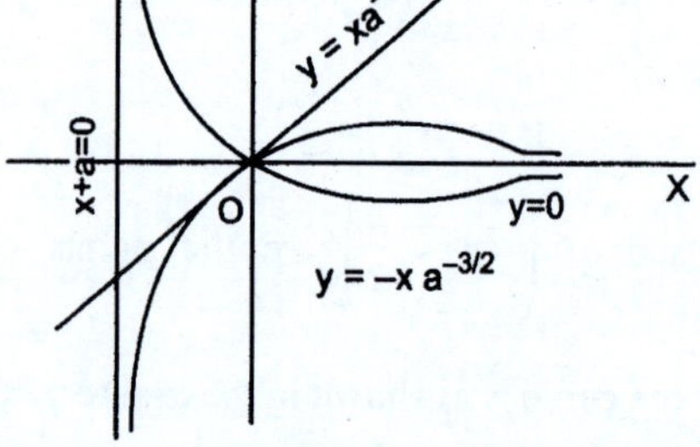

Fig 17.7

Example 8. *Trace the curve*

$$y^2 (x^2 + y^2) + a^2 (x^2 - y^2) = 0$$

or

$$x^2 (a^2 + y^2) = y^2 (a^2 - y^2)$$

Solution : The given curve

$$y^2 (x^2 + y^2) + a^2 (x^2 - y^2) = 0$$

(1) Symmetry : Powers of both x and y are even and the curve remains unchanged when the signs of both x and y are changed. Hence, the curve is symmetrical about both the axes.

(2) Origin : The curve passes through the origin. The tangents at the origin are given by

$$y^2 - x^2 = 0, \quad i.e., \quad y = \pm x$$

which are real and distinct. Hence, origin is a node.

(3) Intersection with axes : Putting $x = 0$, we see that the curve meets the axis of Y in points (0, 0) and (0, $\pm$ a). The curve does not intersect the X-axis at any point other than (0, 0).

(4) Asymptotes : Now, $\frac{dy}{dx}$ at (0, $\pm$ a) is zero. Therefore, the tangents at (0, $\pm a$) are parallel to the X-axis.

(5) Singular point : Solving the given curve for x,

$$x^2 = \frac{y^2 (a^2 - y^2)}{a^2 + y^2}$$

Now, x^2 is positive, i.e., x is real when

$$y^2 < a^2$$

i.e. $\quad |y| < a$ or $-a < y < a$

Thus, the whole curve lies between $y = a$ and $y = -a$.

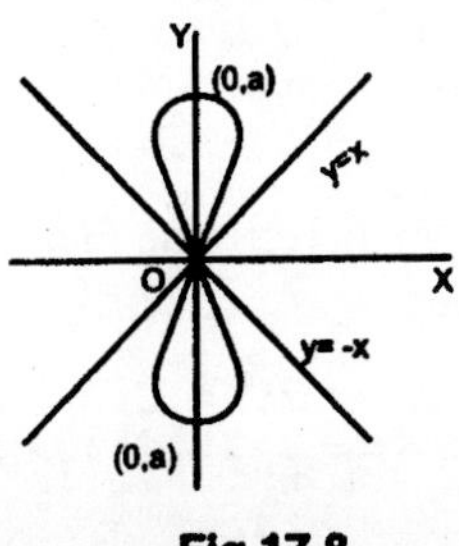

Fig 17.8

(6) Region : Hence, the shape of the curve is as shown in the Figure 17.8.

Example 9. *Trace the curve*

$$xy^2 = 4a^2 (2a - x) \qquad \textit{(Witch of Agensi)}$$

Solution :

(1) Symmetry about X-axis.

(2) The curve does not pass through the origin.

(3) Putting $y = 0$, we get

$$2a - x = 0 \quad \text{or} \quad x = 2a$$

$\therefore$ The curve crosses the x-axis at $(2a, 0)$.

Shifting the origin to $(2a, 0)$ the equation of the curve reduces to

$$(x + 2a)\, y^2 = 4a^2 (2a - x - 2a)$$

or $$y^2x + 2ay^2 + 4a^2x = 0,$$

so that the tangent at the new origin is $x = 0$, i.e., the new Y-axis, i.e., a straight line through $(2a, 0)$ parallel to Y-axis.

(4) The given equation of the curve can be written as

$$y^2 = 4a^2 (2a - x) / x$$

This shows that for $x > 2a$, y^2 is negative or y is imaginary.

$\therefore$ The curve does not exist for values of $x > 2a$. Similarly, for negative values of x or for $x < 0$, the curve does not exist.

(5) $x = 0$ or Y-axis is the asymptote to the given curve.

(6) As x decreases from $2a$ to 0, y increases from 0 to ∞.

The shape of the curve is as shown in adjoining Fig. 17.9.

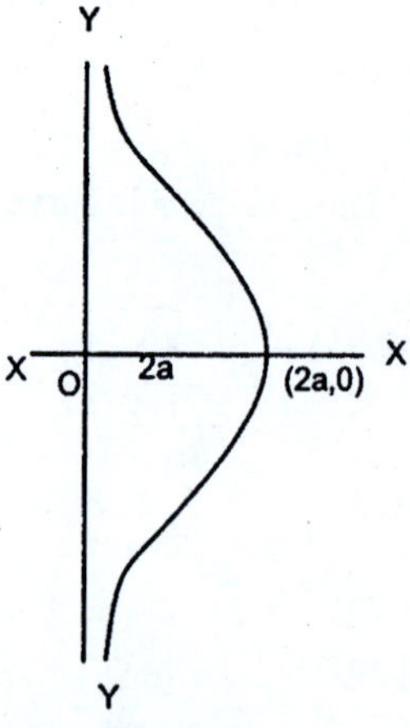

Fig. 17.9

Example 10. Trace the curve

$$y^2 (a + x) = x^2 (a - x), \ a > 0$$

or

$$x (x^2 + y^2) = a (x^2 - y^2)$$

Solution :

(1) Symmetry about x-axis.

(2) The curve passes through the origin. The tangents at the origin are given by

$$ay^2 - ax^2 = 0$$

or

$$y^2 - x^2 = 0 \text{ or } y = \pm x.$$

(3) Putting $y = 0$, we get $x = 0$ and a, i.e., the curve crosses X-axis at $(0, 0)$ and $(a, 0)$. Shifting the origin to $(a, 0)$ the equation of the curve becomes

$$y^2 (a + x + a) = (x + a)^2 (a - x - a)$$

or

$$y^2 (2a + x) = -x (x + a)^2,$$

whence the tangent to the new origin is $x = 0$, i.e., the new Y-axis, i.e., a straight line through $(a, 0)$ parallel to Y-axis.

(4) The given equation of the curve can be written as

$$y^2 = \frac{x^2 (a - x)}{(a + x)}.$$

This shows that for values of $x > a$, y^2 is negative or y is imaginary. The curve does not exist for values of $x > a$. Similarly for values of $x < -a$, the curve does not exist.

Also, as x decreases from 0 to $-a$, y increases from 0 to ∞.

(5) $x + a = 0$ or $x = -a$ is an asymptote.

The shape of the curve is as shown in adjoining Fig. 17.10.

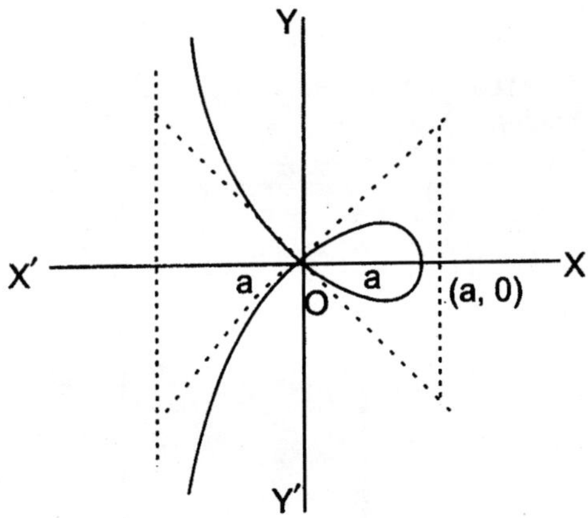

Fig 17.10

Example 10 (a). *Trace the curve*

$$y^2 (a + x) = x^2 (3a - x)$$

Solution : Do as Ex. 10 above.

Example 11. *Trace the curve*

$$(x^2 - a^2)(y^2 - b^2) = a^2 b^2$$

Solution:

(1) Symmetry : The curve is symmetrical about both axis because power on x and y both are even.

(2) The curve passes through given origin because (0, 0) satisfies the equation to the curves.

(3) Tangent at origin : Solving the given equation, we get

$$x^2 y^2 - b^2 x^2 - a^2 y^2 = 0, \qquad \text{...(i)}$$

equating to zero the lowest degree terms.

$$b^2 x^2 + a^2 y^2 = 0 \qquad \text{...(ii)}$$

This equation will give two imaginary straight lines.

The point origin will be a conjugate point.

(4) If put $x = 0$ in equation (ii), we get $y = 0$.

If put $y = 0$, then $x = 0$.

The curve does not meet at any other point.

(5) Asymptote : Equating the coefficients of higher degree of x, then

$$y^2 - b^2 = 0 \Rightarrow y = \pm b$$

There will be two asymptotes. Similarly equating the coefficients of higher degree of x, then

$$x = \pm a$$

will be another two asymptotes.

(6) Solving the equation (ii) for y

$$y = \frac{\pm bx}{\sqrt{x^2 - a^2}}$$

(a) If $x^2 - a^2 < 0$, then the values of y will be imaginary. There will be no branch of the curve in between $x = a$ and $x = -a$.

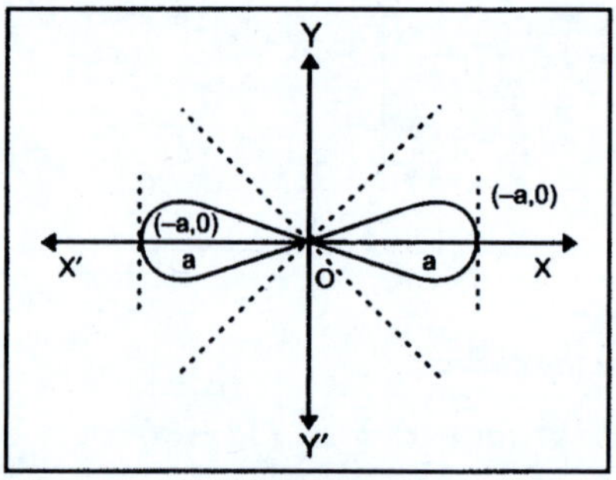

Fig. 17.11

(b) Similarly, there will be no branch of the curve in between $y = a$ and $y = -b$.

(c) If $x \to a$, then $y \to 0$ and if $y \to b$, then $x \to 0$.

Thus, the figure of the given curve will be as shown in Fig. 17.11. There will not be any branch of the curve in the region *ABCD*, except origin.

Example 12. *Trace the curve*

$$ay^2 = x^2 (a - x)$$

Solution : The given equation to the curve is

$$ay^2 = x^2 (a - x) \qquad \text{...(i)}$$

(1) Symmetry : There is symmetry about *X*-axis because power of *y* in (i) is even.

(2) The curve passes through origin.

(3) Tangent at origin : Equating the lowest degree terms to zero. The equation to the tangent at (0, 0) to (i) is

$$y^2 = x^2 \text{ or } y = \pm x.$$

(4) If $y = 0$, then from equation (i)

$$x^2 (a - x) = 0$$

or $$x = 0$$

and $$x = a$$

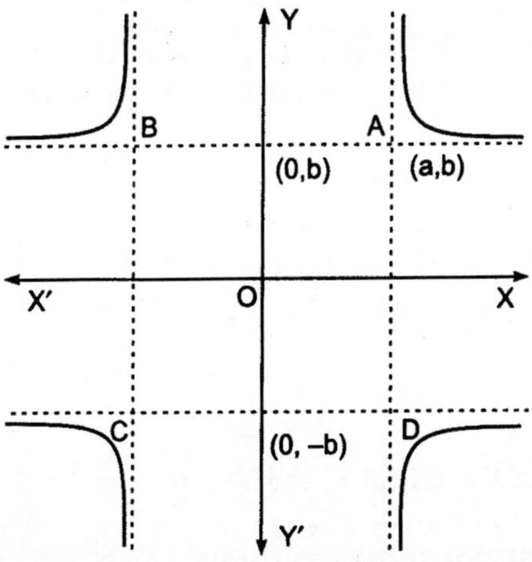

Fig 17.12

The curve meets the X-axis at $(0, 0)$ and $(a, 0)$.

(5) Tangent at $(a, 0)$: Shifting the origin to $(a, 0)$. The new equation will be obtained by writing

$$y = Y + 0$$

and
$$x = X + a$$

$$aY^2 = -X^2 (X^2 + 2aX + a^2)$$

Tangent at new origin,
$$X = 0$$

then
$$X = x - a = 0$$

or
$$x = a.$$

Thus, at $(a, 0)$ the equation to tangent to the given curve will be $x - a = 0$.

(6) Solving the equation (i) for y,

$$y = \pm x \sqrt{\frac{a - x}{a}}$$

If $a - x < 0$, then y will be imaginary.

Thus, there is no branch of the curve on the right of the line $x - a = 0$.

If x changes from 0 to $-\infty$ then y increases rapidly.

Thus, the figure of the given curve will be as in Fig. 17.12.

Example 13. *Trace the curve*

$$y^3 = a^2x - x^3$$

Solution :

(1) Symmetry : The curve is symmetrical in opposite quadrant because when x is replaced by $-x$ and y is replaced by $-y$, there is no change in the equation.

(2) The curve passes through origin.

(3) The tangent at origin is $x = 0$, that is Y-axis.

(4) If $y = 0$, then

$$a^2x - x^3 = 0$$

or
$$x = 0$$

and
$$x = \pm a$$

The curve meets the X-axis at $(0, 0)$ and $(\pm a, 0)$.

(5) Tangent at $(a, 0)$: Put $x = X + a$ and $y = Y + 0$ in the equation of curve for shifting the origin to the point $(a, 0)$.

$$Y^2 = (a + X)\ [a^2 - (a + x)^2]$$

$$= -(a + X)(X^2 + 2aX)$$

$$= -[aX^2 + 2a^2X + X^3 + 2aX^2]$$

Equating to zero the lowest degree terms, we get

$$X = 0, \quad \text{tangent at } (a, 0)$$

that is, $X = x - a = 0$

$$x = a$$

Similarly, tangent at $(-a, 0)$ will be $x + a = 0$

$$\frac{dy}{dx} = 0,$$

$$\frac{d^2y}{dx^2} = 0$$

but $$\frac{d^3y}{dx^3} \neq 0$$

or, the points $(\pm a, 0)$ will be point of inflexion at $(\pm a, 0)$.

(6) **Asymptote :** Putting $y = m$ and $x = 1$, in the equation to curve,

$$\phi_3(m) = m^3 + 1,$$

$$\phi_2(m) = 0$$

From m, $\phi_3(m) = 0$

$$m^3 + 1 = 0$$

$$(m + 1)(m^2 - m + 1) = 0$$

$m = -1$, other roots will be imaginary.

$$c = \frac{-\phi_2(m)}{\phi_3'(m)} = \frac{-0}{3m^2} = 0$$

The asymptote will be $y = -x$ or $y + x = 0$.

(7) Solving the equation for y,

$$y = -x\left[1 - \frac{a^2}{x^3}\right]^{1/3}$$

If $0 < x < a$, then y will be positive.

If $x > a$, then y will be negative.

If $x \to \infty$, then $y \to -\infty$.

Thus, the required figure will be as given in Fig. 17.13.

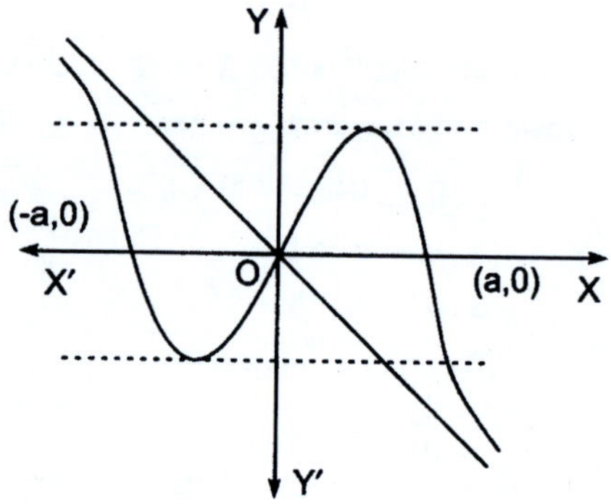

Fig 17.13

Example 14. *Trace the curve*

$$y(x^2 - 1) = x^2 + 1$$

Solution :

(1) The power of x is even in the curve. The curve is symmetric about $y = 0$.

(2) It does not pass through origin because (0, 0) does not satisfy the curve.

(3) If $y = 0$, then $x^2 + 1 = 0$ or $x = \pm i$, therefore, the curve does not meet the X-axis.

If $x = 0$, then $y = 1$

The curve meets the Y-axis at $(0, -1)$.

(4) Tangent at (0, 1): Put $x = X + 0$ and $x = Y + 1$, we get

$$(Y + 1)(X^2 - 1) = X^2 + 1$$

$$Y^2 X^2 - X^2 - Y = X^2$$

$$Y^2 X^2 - 2X^2 - Y^2 = 0$$

The tangent at $(0, -1)$ is $Y = 0$, that is $Y + 1 = 0$.

(5) Asymptote : The equation is,

$$x^2 y - y = x^2 + 1$$

The coefficient of higher power of x is $y - 1$ and coefficient of higher power of y is $x^2 - 1$. The asymptotes are $y - 1 = 0$ and $x^2 - 1 = 0$.

The asymptotes are the lines

$$y = 1, \ x = \pm 1$$

(6) $y = \dfrac{x^2 + 1}{x^2 - 1}$

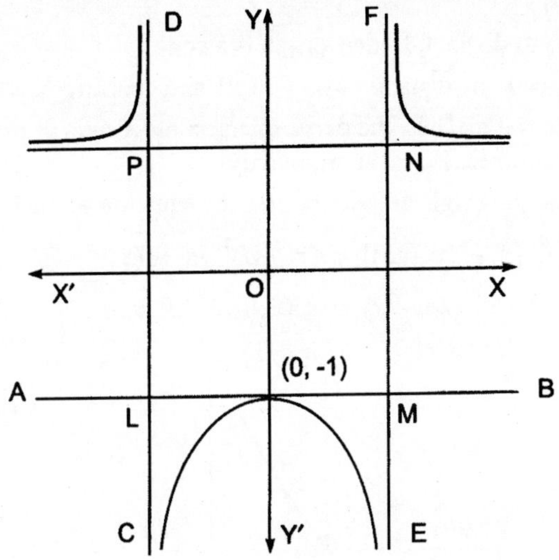

Fig. 17.14

If $x \to \infty$ then $y \to 1$

Again, $$x^2 = \frac{y+1}{y-1} \quad \text{or} \quad x = \pm\sqrt{\frac{y+1}{y-1}}$$

If the value of y lies between 0 and -1, then the value of x is imaginary.

If $|y| < 1$ or $-1 < y < 1$, then the value of y is imaginary.

Therefore, there is no any branch of the curve in the region *LMNP*.

Tracing the curve considering symmetric about *Y*-axis we get the above Figure 17.14.

Example 15. *Trace the curve*

$$y^2 (a - x) = x^2 (a + x)$$

Solution : The given curve is,

$$y^2 (a - x) = x^2 (a + x) \qquad \text{...(i)}$$

(1) Powers of both x and y are even in curve and the curve remains unchanged when the signs of both x and y are changed. Hence, the curve is symmetrical about both the axes.

(2) The curve passes through the origin. The tangents at the origin,

$$ay^2 - ax^2 = 0$$

or $$y^2 = x^2$$

or $$y = \pm x$$

which is real and distinct. Hence, origin is a node.

(3) Putting $x = 0$ in equation (i), $y = 0$ and putting $y = 0$, then $x = 0$ or $x = -a$. Thus, the curve intersect the X-axis at the point (0, 0) and $(-a, 0)$, but intersect Y-axis at origin only.

(4) If substitute the origin at $(-a, 0)$, then the equation of curve will be,

$$y^2\ [a - (x - a)] = (x - a)^2\ [a + (x - a)]$$

or $$y^2\ (2a - x) = x\ (x - a)^2$$

X′

Y′

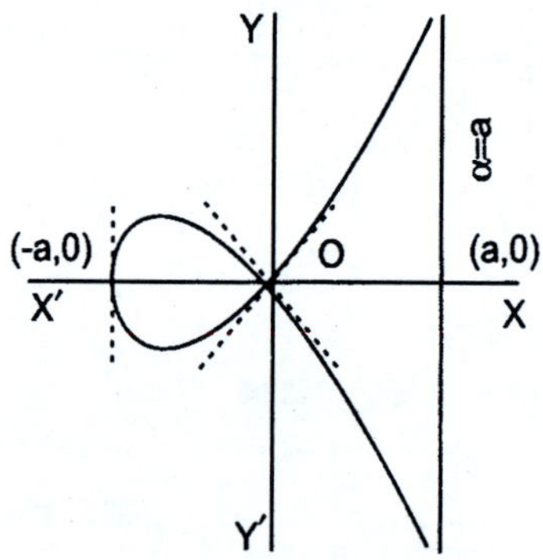

Fig. 17.15

It is clear that the equation of the tangent at origin will be $x = 0$. Hence, the equation of the tangent at $(-a, 0)$ is $x = -a$.

(5) There are no asymptote parallel to X-axis. Hence, the asymptote parallel to Y-axis is,

$$a - x = 0 \quad \text{or} \quad x = a$$

(6) From the equation of curve

$$y^2\ (a - x) = x^2\ (a + x)$$

or $$y = x\sqrt{\frac{a + x}{a - x}}$$

When $x > a$, then y will be imaginary and when $x < -a$, then y will be imaginary.

Example 16. *Trace the curve*

$$y\ (x^2 + 4a^2) = 8a^3$$

Solution :

(1) The curve is symmetric to Y-axis.

(2) The curve does not pass through origin.

(3) Putting $x = 0$ in the equation of curve.

$$4a^2y = 8a^3$$

or

$$y = 2a$$

Hence, the curve intersects Y-axis at $(0, 2a)$.

(4) Replace the origin at $(0, 2a)$, the equation of curve will be

$$(y + 2a)\ (x^2 + 4a^2) = 8a^3$$

or

$$yx^2 + 4a^2 y + 2ax^2 = 0$$

$\therefore$ The equation of tangent at point $(0, 2a)$, $y = 0$.

Hence, $y = 2a$ is the tangent at point $(0, 2a)$.

(5) The asymptote does not parallel to Y-axis. The power of higher coefficient of x is y. Hence, the asymptote parallel to X-axis is $y = 0$.

(6) The equation of curve can be written as

$$x^2 = \frac{8a^3}{y} - 4a^2$$

which shows that the negative values for y is negative for x^2, i.e., x is imaginary. Hence, the curve will not exist $y < 0$. Similarly, the curve will not exist $y > 2a$.

Hence, the figure of the given curve is given as in Fig. 17.16.

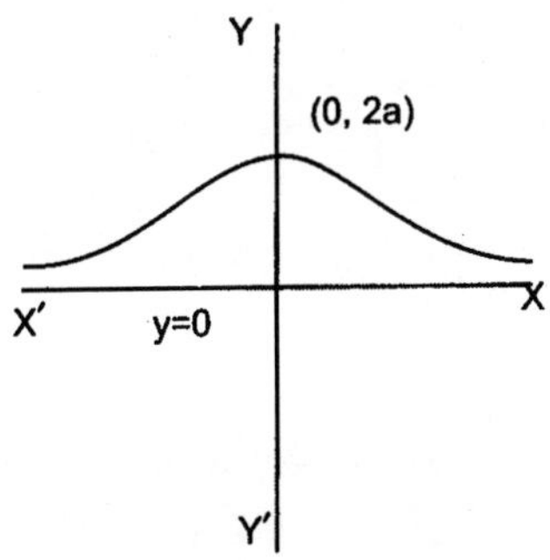

Fig. 17.16

EXERCISE 17 (A)

Trace the following curves—

1. $ay^2 = x^3$ (semi cubical parabola)
2. $a^2y = x^3$ (cubical parabola)

3. $y = a \cosh\left(\dfrac{x}{a}\right)$ (catenary)

4. $xy^2 = 4a^2 (2a - x)$ (with of agnesi)

5. $x^{2/3} + y^{2/3} = a^{2/3}$ (astroid)

6. $(ax)^{2/3} + (by)^{2/3} = (a^2 - b^2)^{2/3}$ (evolute of an ellipse)

7. $y^2 = \dfrac{(1 + x^2)\, x^2}{1 - x^2}$

8. $y^2 (a + x) = (a - x)\, x^2$

9. $x^2 y^2 = (a + y)^2 (b^2 - y^2)$, when $b > a$

10. $y = \dfrac{x^2 + 1}{x^2 - 1}$

11. $x = (y - 1)(y - 2)(y - 3)$

12. $a^2 y^2 = x^3 (2a - x)$

13. $xy^2 + (x + a)^2 (x + 2a) = 0$

14. $x^2 y^2 = a^2 (x^2 + y^2)$

15. $x^3 + y^3 = a^2 x$

16. $x^2 y^2 = (1 + y^2)(4 - y^2)$

17. $x^2 (x^2 - 4a^2) = y (x^2 - a^2)$

18. $y^2 (x^2 - 1) = x$

19. $x^2 y^2 = a^2 (y^2 - x^2)$

ANSWERS

1.

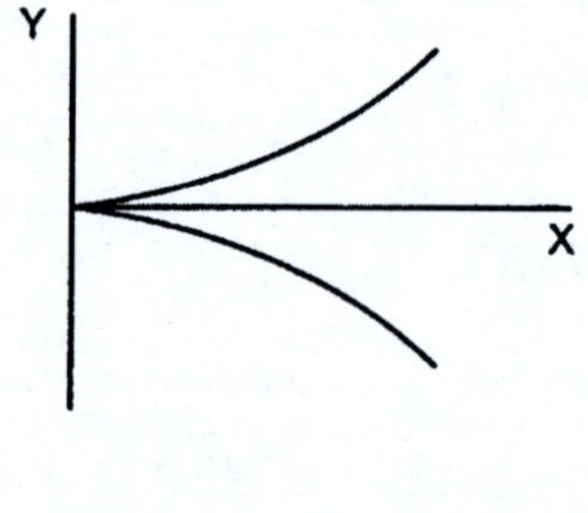

2.

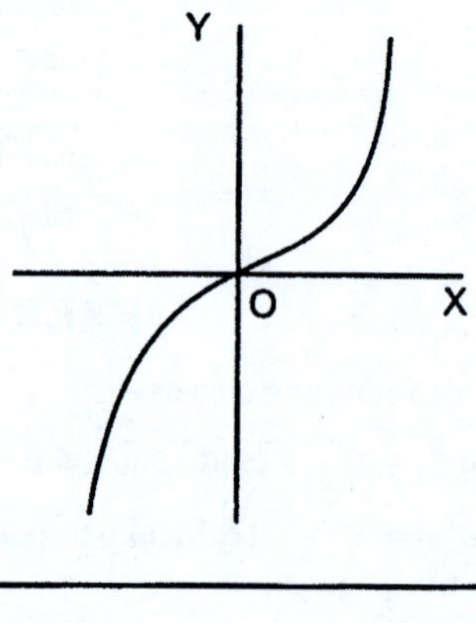

3.

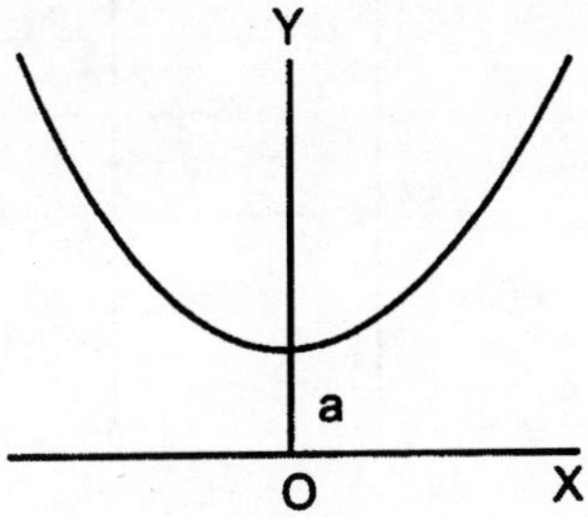
Y
a
O
X

4.

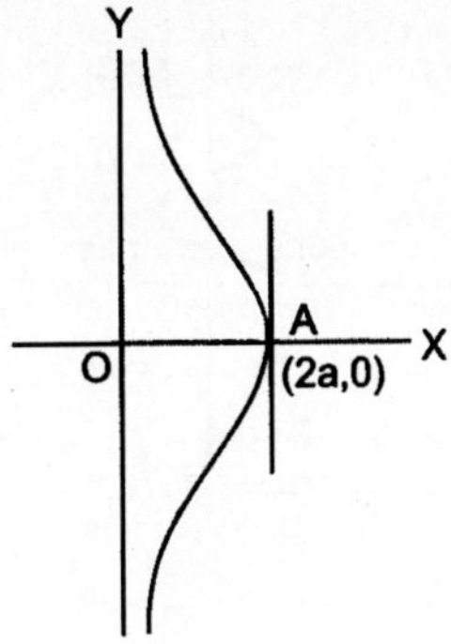
Y
A
X
O
(2a,0)

5.

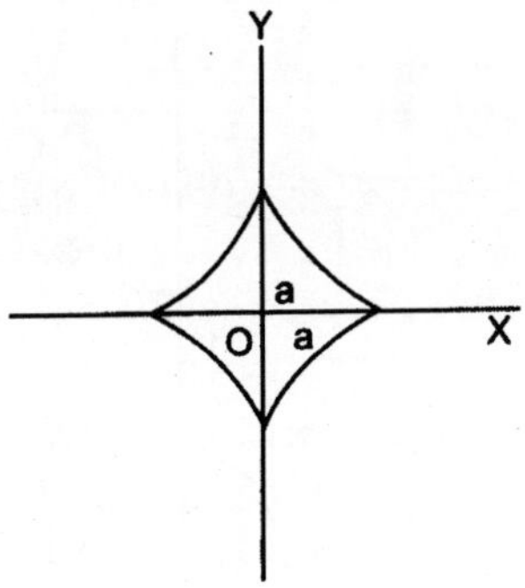
Y
a
O
a
X

6.

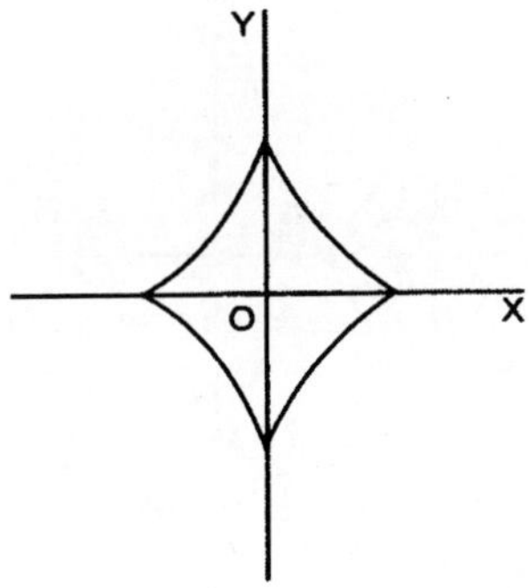
Y
O
X

7.

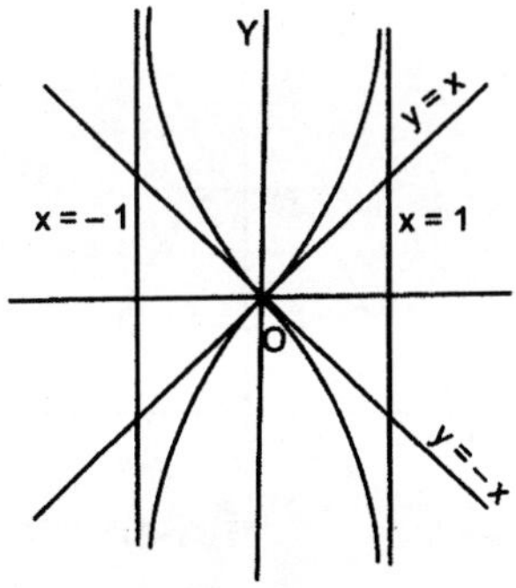
Y
y = x
x = – 1
x = 1
O
y = – x

8.

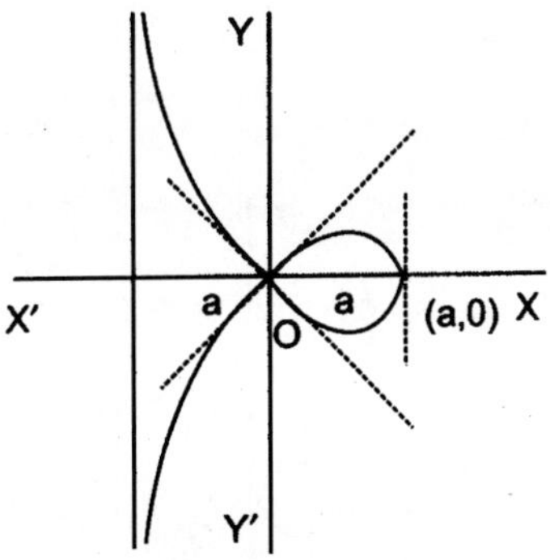
Y
a
a
(a,0)
X
X′
O
Y′

9.

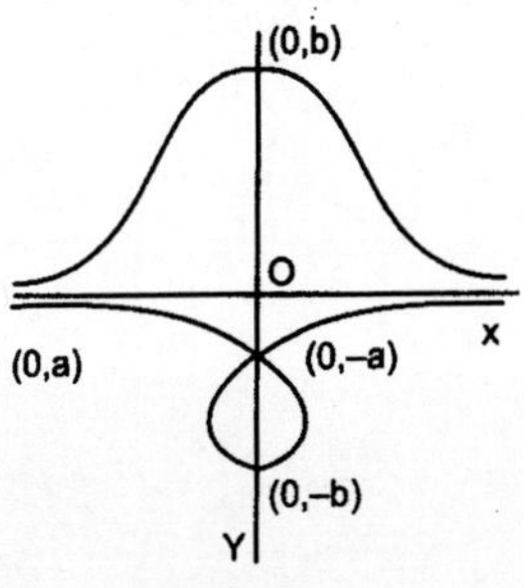
(0,b)
O
x
(0,a)
(0,–a)
(0,–b)
Y

10.

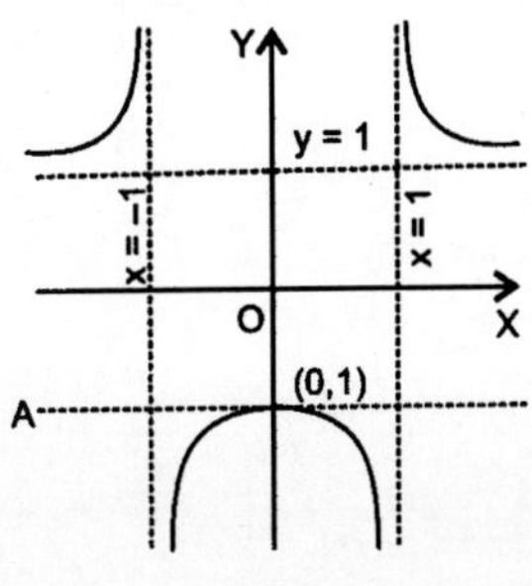
Y
y = 1
x = –1
x = 1
O
X
(0,1)
A

11.

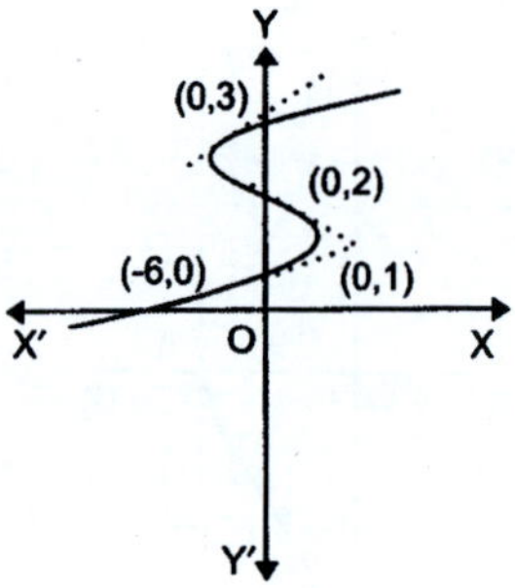

12.

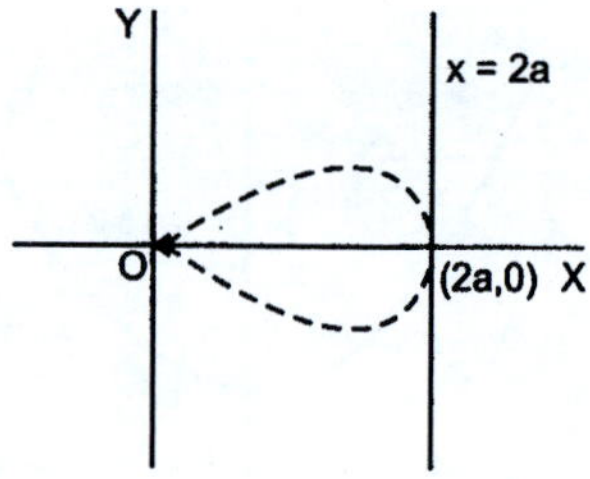

13.

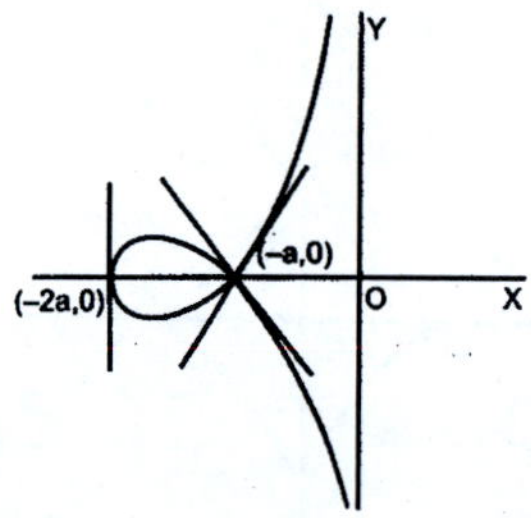

14.

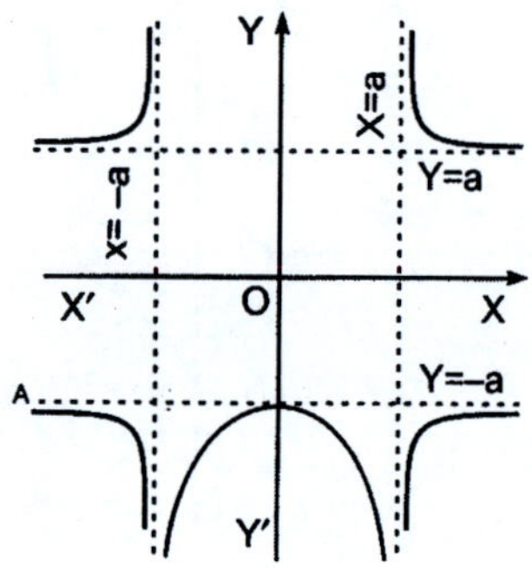

15.

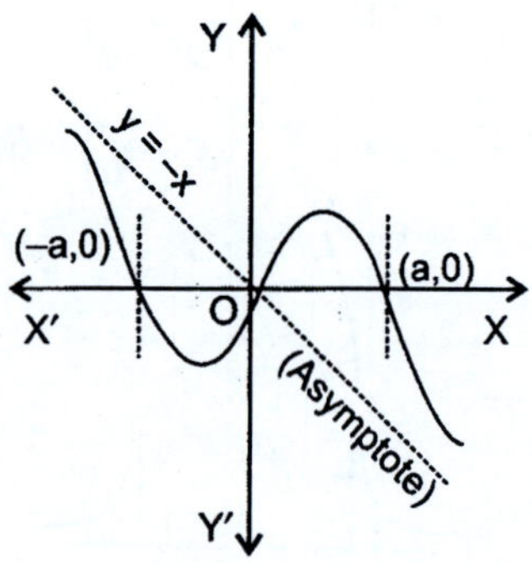

16.

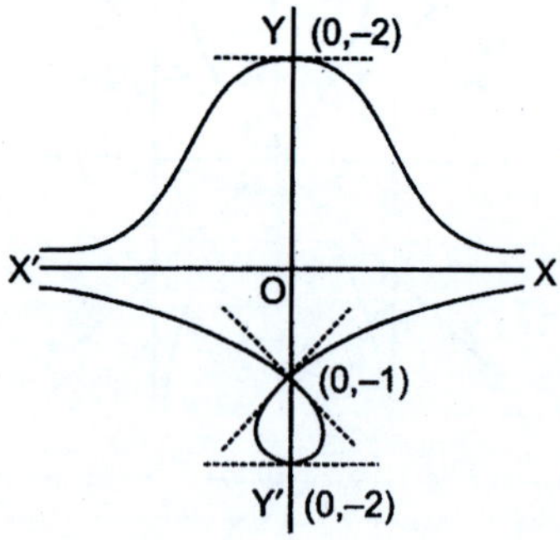

17.

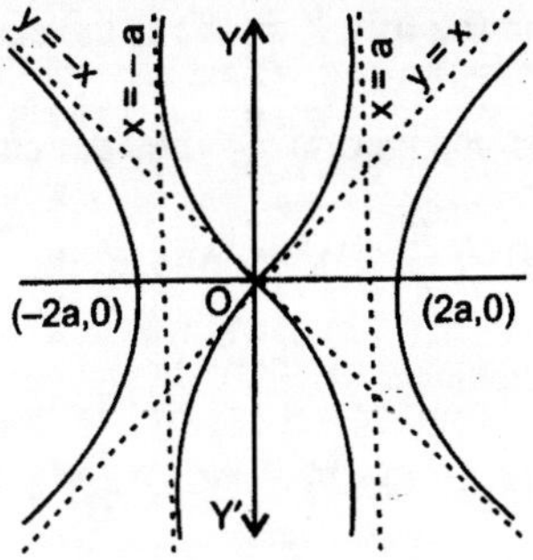

18.

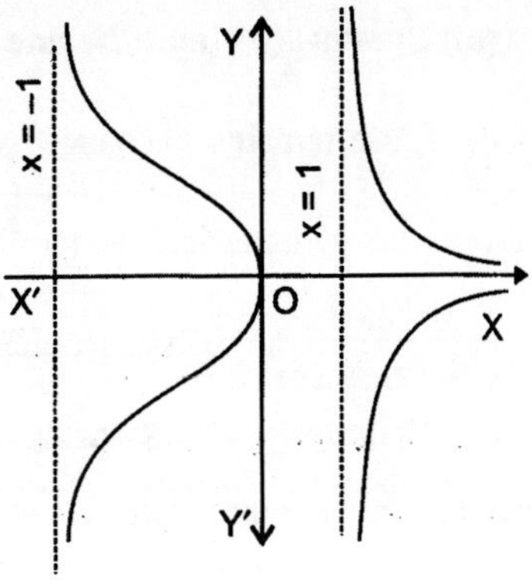

19.

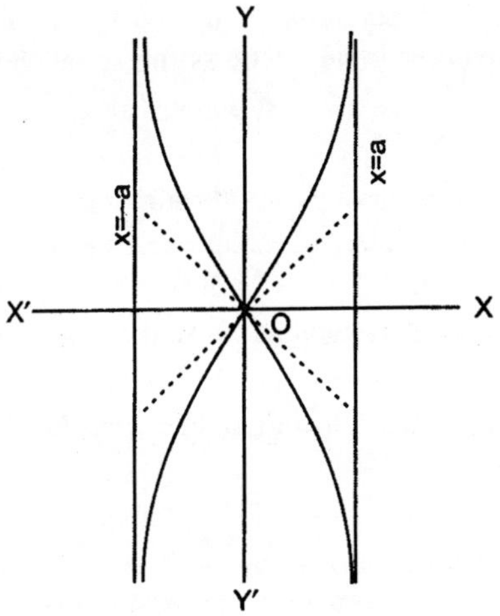

17.3. Tracing of Polar Curves

Procedure : Let the equation of the curve be $f(r, \theta) = 0$.

(1) Symmetry : (i) Symmetry about the initial line or *X*-axis: If the equation of the curve remains unchanged when θ is changed to $-\theta$, the curve is said to be symmetrical about the initial line. For example, the curve $r = a(1 + \cos\theta)$ is symmetrical about the initial line.

(ii) Symmetry about the line $\theta = \frac{\pi}{2}$ or *Y*-axis : If the equation of the curve remains unaltered when θ is changed to $\pi - \theta$ or when θ is changed to $-\theta$ and r to $-r$, the curve is said to be symmetrical about the line $\theta = \frac{\pi}{2}$. For example, the curve $r = a \sin 3\theta$ is symmetrical about the line $\theta = \pi/2$.

(iii) Symmetry about the line $\theta = \frac{\pi}{4}$ or the line $Y = X$: If the equation of the curve remains unchanged when θ is changed to $\frac{\pi}{2} - \theta$, the curve is said to be symmetrical about the line $\theta = \frac{\pi}{4}$. For example, the curve $r = \frac{3a \sin\theta \cos\theta}{\sin^3\theta + \cos^3\theta}$ is symmetrical about the line $\theta = \frac{\pi}{4}$.

(iv) Symmetry about the line $\theta = 3\pi/4$ or the line $Y = -X$: If the equation of the curve remains unaltered when θ is changed to $\frac{3\pi}{2} - \theta$, the curve is said to be symmetrical about the line $\theta = 3\pi/4$. For example, the curve $r = a \sin 2\theta$ is symmetrical about the line $\theta = 3\pi/4$.

(v) Symmetry about the pole : If the equation of the curve remains unchanged when r is changed to $-r$, the curve is said to be symmetrical about the pole. For example, the curve $r^2 = a^2 \cos 2\theta$ is symmetrical about the pole.

(2) Origin or Pole : (i) Find whether the curve passes through the pole or not. The curve will pass through the pole if r is zero for some real value of θ, say $\theta = \alpha$ and then tangent thereat is $\theta = \alpha$.

If it is not possible to find a real value of θ for which $r = 0$, then the curve does not pass through the pole.

(ii) Find the points where the curve meets the initial line and the line $\theta = \frac{\pi}{2}$.

(3) Region : (i) Form a table of corresponding values of r and θ which satisfy the equation of the curve. In polar equations generally periodic functions like $\sin\theta$, $\cos\theta$ occur and so values of θ from 0 to 2π only need be considered in such cases. The remaining values will not give in any new branch of the curve.

(ii) The curve does not lie in the region given by $\theta = \theta_1$ and $\theta = \theta_2$ *i.e.*, $\theta_1 < \theta < \theta_2$ if for every value of θ lying between θ_1 and θ_2 the value of θ is imaginary.

ILLUSTRATIVE EXAMPLES

Example 1. *Trace the cardioid*

$$r = a(1 + \cos\theta)$$

Solution : The equation of the curve is,

$$r = a\ (1 + \cos \theta) \qquad \text{...(i)}$$

(1) Symmetry : The equation of the curve remains unchanged when θ is changed to $-\theta$. Hence, the curve is symmetrical about the initial line.

(2) Pole : When $\theta = \pi$, $r = 0$, hence, the curve passes through the pole. The tangent at pole is $\theta = \pi$.

(3) Again we have

$\theta = 0$	$\pi/2$	π	$3\pi/2$	2π
$r = 2a$	0	0	a	$2a$

Thus, when θ increases from 0 to π, r decreases from $2a$ to 0 and when θ increases from π to 2π, r increases from 0 to $2a$.

Hence, the shape of the curve is shown in the Figure 17.17.

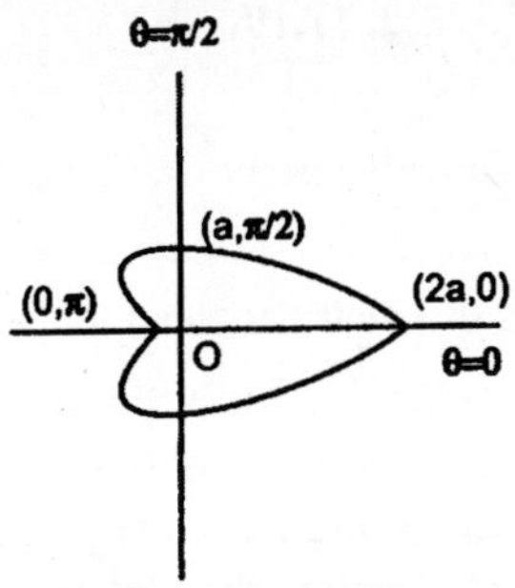

Fig. 17.17

Example 2. *Trace the Limacon*

$$r = a + b \cos \theta.$$

Solution : The equation of the curve is,

$$r = a + b \cos \theta \qquad \text{...(i)}$$

Case I. When $a > b$

(1) Symmetry: (i) The curve remains unchanged when θ is changed to $-\theta$. Hence, the curve is symmetrical about the initial line.

(ii) r is never zero as $a > b$. Hence, pole does not lie on the curve.

(2) Other points on the curve are—

$\theta =$	0	$\pi/2$	π	$3\pi/2$	2π
$r =$	$a + b$	a	$a - b$	a	$a+b$

Thus, when θ increases from 0 to π, r decreases from $a+b$ to $a-b$ and when θ increases from π to 2π, r also increases from $a-b$ to $a+b$. Thus, the shape of the curve is as shown in the Figure 17.18.

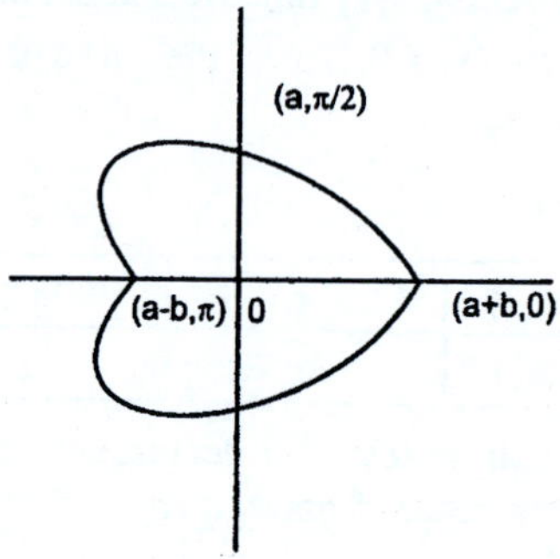

Fig. 17.18

Case II. When $a = b$.

The curve becomes a cardioid which is shown in Example 1 above.

Case III. When $a < b$

(1) Symmetry: The curve is symmetrical about the initial line.

(2) $r = 0$ when $\theta = \cos^{-1}(-a/b)$. Hence, the curve passes through the pole. The values of $\theta = \cos^{-1}(-a/b)$ are two, lying between 0 and 2π; one of them lies in the second quadrant and the other in the third quadrant. Let these values be θ_1 and θ_2. Thus, the tangents at the pole are $\theta = \theta_1$ and $\theta = \theta_2$ which are real and distinct. Hence, the pole is a node.

(3) The other points on the curve are—

$\theta =$	0	$\frac{\pi}{2}$	π	$\frac{3\pi}{2}$	2π
$r =$	$a+b$	a	$-(b-a)$	a	$a+b$

Hence, there is a loop between $\theta = \theta_1$ and $\theta = \theta_2$

Hence, the shape of the curve is as shown in the Figure 17.19.

Example 3. *Trace the Leminscate of Bernoulli*

$$r^2 = a^2 \cos 2\theta$$

Solution : The equation of the curve is,

$$r^2 = a^2 \cos 2\theta \qquad \text{...(i)}$$

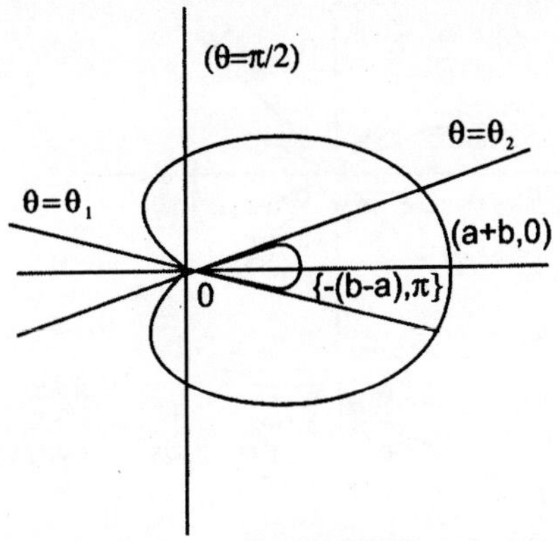

Fig. 17.19

(1) Symmetry : (i) Since equation (i) remains unchanged when θ is changed to $-\theta$, the curve is symmetrical about the initial line.

(ii) Since equation (i) remains unchanged when θ is changed to $\pi - \theta$, the curve is symmetrical about the line $\theta = \pi/2$.

(iii) Since equation (i) remains unchanged when r is changed to $-r$, the curve is symmetrical about the pole.

(2) Pole : (i) r is zero when $\cos 2\theta = 0$, i.e.,

$$2\theta = \pm \pi/2$$

i.e.,

$$\theta = \pm \pi/4$$

Hence, the curve passes through the pole and the tangents thereat are $\theta = \pm \pi/4$.

(ii) When $\theta = 0$, $r^2 = a^2$, *i.e.*, $r = \pm a$. Thus, the curve meets the initial line at the points $(a, 0)$ and $(-a, 0)$.

(3) Region : (i) r^2 decreases from a^2 to 0 as θ increases from 0 to $\pi/4$.

(ii) r^2 is negative when θ lies between $\pi/4$ and $3\pi/4$. Hence, no part of the curve lies between $\theta = \dfrac{\pi}{4}$ and $\theta = \dfrac{3\pi}{4}$.

(iii) Again r^2 increases from 0 to a^2 when θ increases from $3\pi/4$ to π.

Hence, the curve is as shown in the Figure 17.20.

Example 4. *Trace the curve*

$$r = a \sin 2\theta$$

Solution : The equation of the curve is,

$$r = a \sin 2\theta \quad \text{...(i)}$$

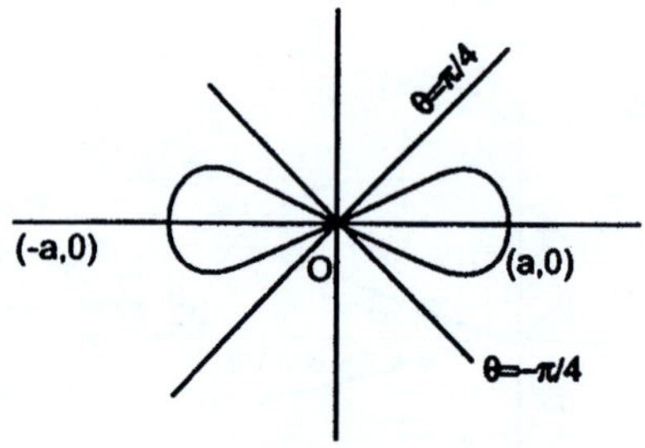

Fig. 17.20

(1) **Symmetry** : The curve is symmetrical about $\theta = \pi/4, \pi/2$ and $3\pi/4$.

(2) **Pole :** r is zero when $\sin 2\theta = 0$,

i.e., $$\theta = 0, \frac{\pi}{2}, \pi, \frac{3\pi}{2} \text{ and } 2\pi.$$

Thus, the curve passes through the pole. The tangents at the pole are $\theta = 0$ and $\theta = \frac{\pi}{2}$. The other values give the same tangents in turn.

(3) The other points on the curve are—

$\theta =$	0	$\frac{\pi}{4}$	$\frac{\pi}{2}$	$\frac{3\pi}{4}$	π	$\frac{5\pi}{4}$	$\frac{3\pi}{2}$	$\frac{7\pi}{4}$	2π
$r =$	0	a	0	$-a$	0	a	0	$-a$	0

We see that the value of r increases from 0 to a when θ increases from 0 to $\pi/4$ and r decreases from a to 0 when θ increases from $\pi/4$ to $\pi/2$. Thus, there is a loop in the first quadrant. Similarly, these are loops each in 2nd, 3rd and 4th quadrants.

Hence, the curve is as shown in the Figure 17.21.

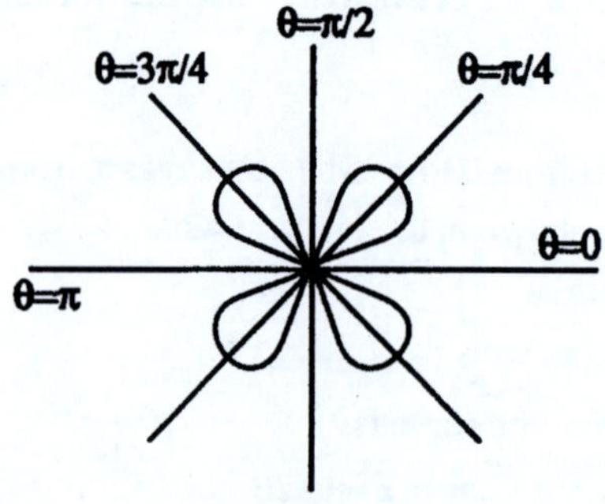

Fig. 17.21

Example 5. *Trace the curve*

$$r = a \sin 3\theta.$$

Solution :

(1) There is no change in the equation to the curve when θ is replaced by $-\theta$ and r is replaced by $-r$. There is no symmetry about $\theta = \frac{\pi}{2}$.

(2) If $r = 0$, then

$$\sin 3\theta = 0 \text{ or } 3\theta = n\pi \text{ or } \theta = \frac{n\pi}{3}, \; n \in I.$$

The curve passes through pole.

The lines $\theta = 0$, $\theta = \pm \frac{\pi}{3}$, $\theta = \pm \frac{2\pi}{3}$ ... etc. are tangents to the curve.

The function $\sin 3\theta$ is a periodic function, we will consider the values of θ in between 0 and π.

(3) The values of r in between 0 and π are given by the following tables—

3θ	0	$\frac{\pi}{6}$	$\frac{\pi}{2}$	$\frac{5\pi}{6}$	π	$\frac{7\pi}{6}$	$\frac{3\pi}{2}$	$\frac{11\pi}{6}$	2π	$\frac{13\pi}{6}$	$\frac{5\pi}{3}$	$\frac{17\pi}{6}$	3π
θ	0	$\frac{\pi}{18}$	$\frac{\pi}{6}$	$\frac{5\pi}{18}$	$\frac{\pi}{3}$	$\frac{7\pi}{18}$	$\frac{\pi}{2}$	$\frac{11\pi}{6}$	$\frac{2\pi}{3}$	$\frac{13\pi}{18}$	$\frac{5\pi}{6}$	$\frac{17\pi}{18}$	π
r	0	$\frac{a}{2}$	a	$\frac{a}{2}$	0	$-\frac{a}{2}$	$-a$	$-\frac{a}{2}$	0	$\frac{a}{2}$	a	$\frac{a}{2}$	0

Then we see that least absolute value of r is zero and maximum value is a.

The value of r is not infinite for any value of θ, therefore, this is no asymptote.

From the table, we get

When θ increases from 0 to $\frac{\pi}{6}$, the value of r increases.

When θ increases from $\frac{\pi}{6}$ to $\frac{\pi}{3}$, the value of r decreases.

Thus, the curve makes a loop in $0 \leq \theta \leq \frac{\pi}{3}$.

When $\frac{\pi}{3} \leq \theta \leq \frac{2\pi}{3}$, we get second loop.

Similarly, the third loop is obtained when $\frac{2\pi}{3} \leq \theta \leq \frac{5\pi}{6}$.

We do not get any branch of the curve when θ changes from π to 2π. Thus, we get the adjoining curve. [Fig. 17.22]

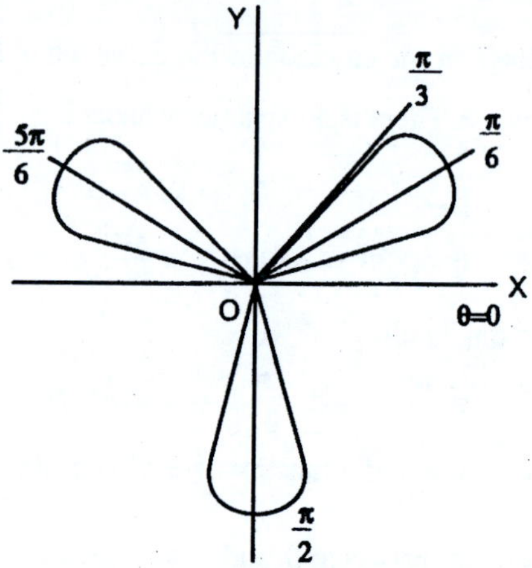

Fig. 17.22

Example 6. *Trace the curve* $r = a\ (\cos\theta + \sec\theta)$.

Solution : We will trace the curve by changing the equation to Cartesian equation. Multiplying the equation by r

$$r^2 = a\left(r\cos\theta + \frac{r}{r\cos\theta}\right)$$

$$x^2 + y^2 = a\left(x + \frac{x^2 + y^2}{x}\right)$$

$$(x^2 + y^2)\,x = a\,(2x^2 + y^2)$$

$$y^2\,(x - a) = x^2\,(2a - x) \qquad \text{...(i)}$$

(1) Symmetry : The power of y is even, therefore, the curve is symmetrical about X-axis.

(2) The curve passes through origin because (0, 0) satisfies the equation.

(3) Tangent at (0, 0) : Equating the lowest defined terms to zero, we get

$$-ay^2 - 2ax^2 = 0$$

$$y = \pm\, ix\sqrt{2}$$

The tangent at (0, 0) will be imaginary. The point (0, 0) will be a conjugate point.

(4) If $y = 0$, then $x = 0$, $x = 2a$. The curve passes through (0, 0) and ($2a$, 0).

(5) Tangent at ($2a$, 0) **:** Put $x = X + 2a$ and $y = Y$, we get

$$Y^2 = \frac{-(X + 2a)^2 X}{X + a}$$

Equating the least degree terms to zero, we get

$$X = 0 \text{ or } x - 2a = 0$$

will be tangent at $(2a, 0)$.

(6) Asymptotes : The $x - a = 0$ will be asymptote to the curve.

(7) Solving the equation for y,

$$y = \pm x \sqrt{\frac{2a - x}{x - a}}$$

If $0 < x < a$, the values of y are imaginary.

If $x > 2a$, the values of y are imaginary.

There will not be any branch of the curve on the left of MN and on the right of PQ.

The curve is symmetrical about, $y = 0$.

Thus, we get the above figure of the curve. [Fig. 17.23]

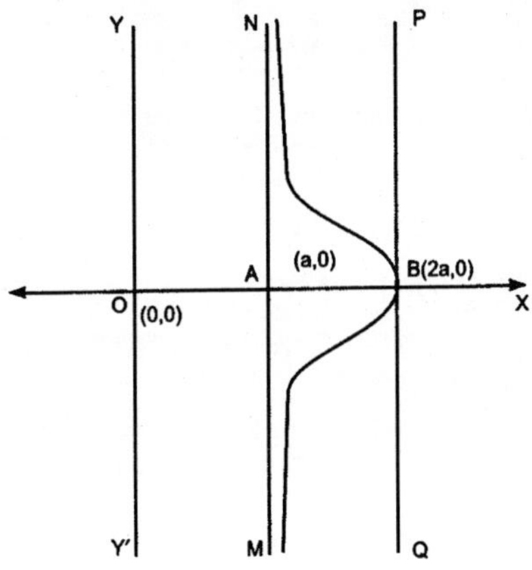

Fig. 17.23

EXERCISE 17 (B)

Trace the following curves—

1. $r = a\ (1 - \cos\ \theta)$
2. $r = a\ (1 + \sin\ \theta)$
3. $r = a\ (1 - \sin\ \theta)$
4. $r = 2a \cos\ \theta$
5. $r = a\ (2 \cos\ \theta + \cos\ 3\theta)$

6. $r = \frac{1}{2} + \cos 2\theta$

7. $r = a \cos 2\theta$

8. $r = a \cos 3\theta$

9. $r^2 \cos 2\theta = a^2$

10. $x^5 + y^5 = 5\, a^2x^2y$

ANSWERS

1.

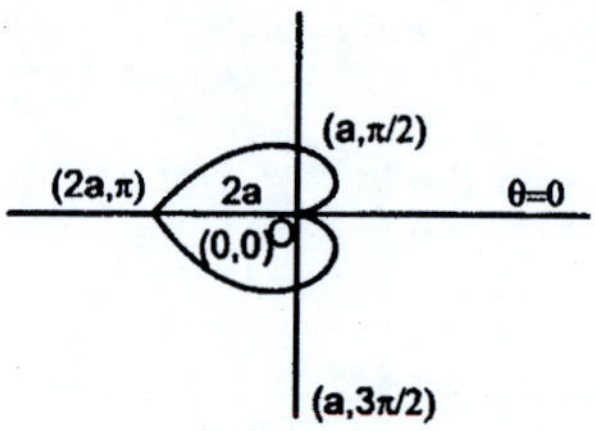

2.

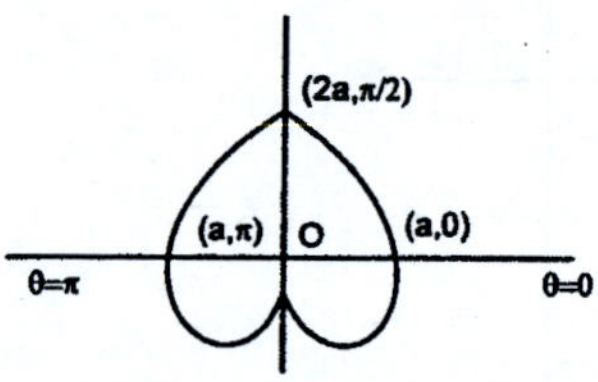

3.

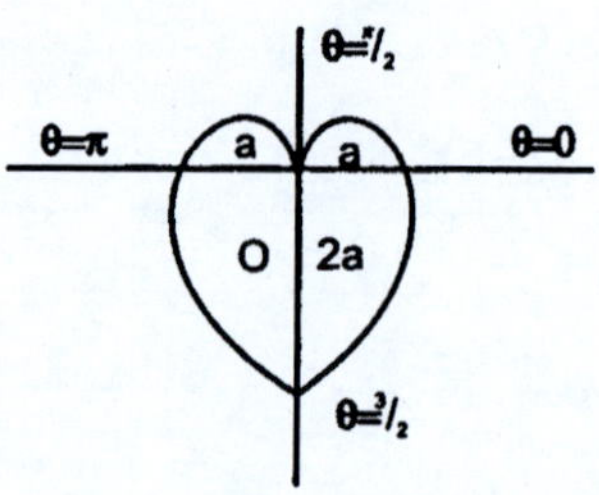

4.

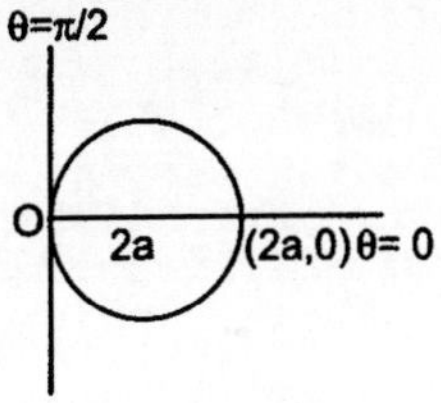
θ=π/2
O
2a
(2a,0) θ= 0

5.

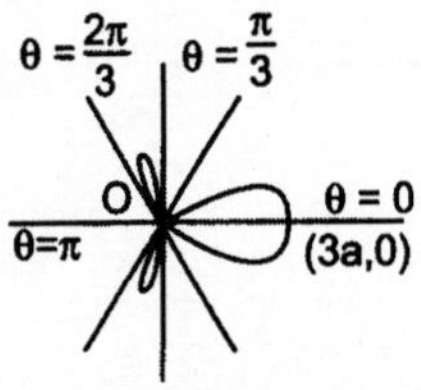
θ = 2π/3
θ = π/3
O
θ = 0
θ=π
(3a,0)

6.

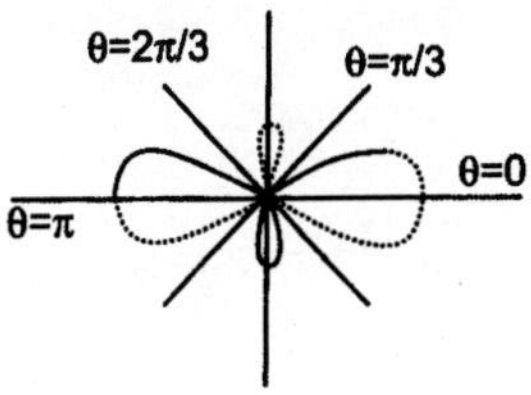
θ=2π/3
θ=π/3
θ=0
θ=π

7.

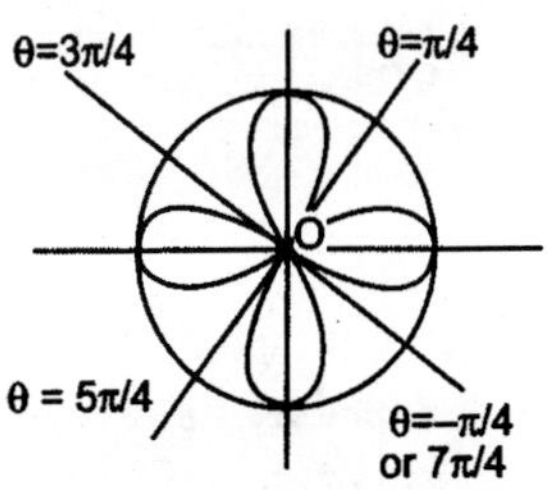
θ=3π/4
θ=π/4
O
θ = 5π/4
θ=−π/4
or 7π/4

8.

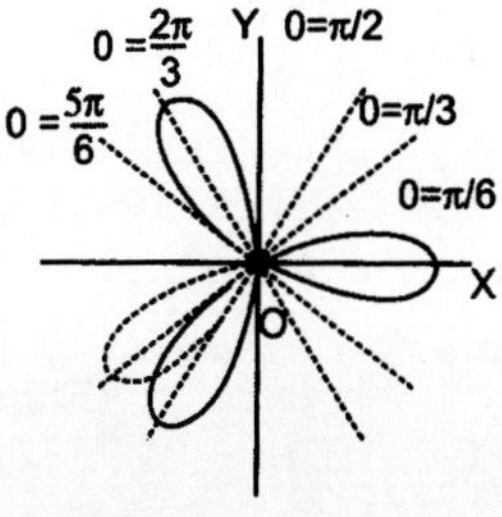
0 = 2π/3
Y
0=π/2
0 = 5π/6
0=π/3
0=π/6
X
O

9.

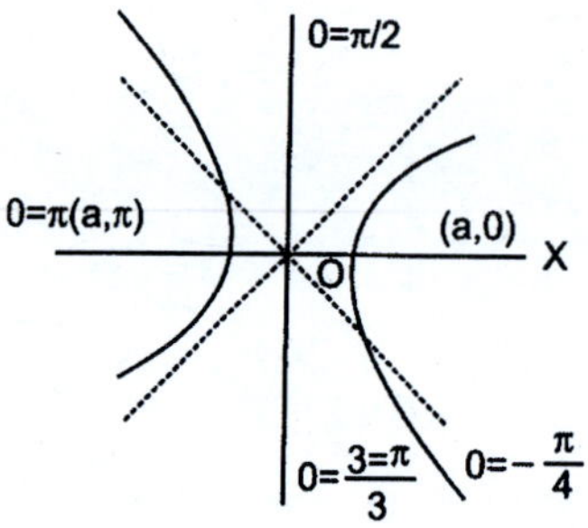

10.

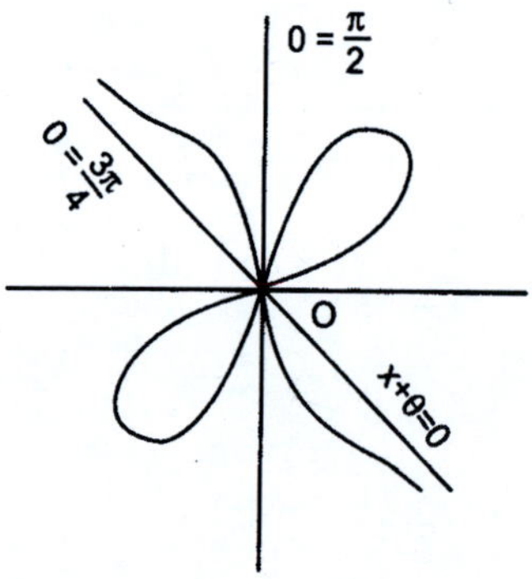

17.4. Tracing of Parametric Curves

Procedure : Let the equations of the curve to be traced be

$$\left.\begin{aligned} x &= \phi\,(t) \\ y &= \psi\,(t) \end{aligned}\right\} \qquad \text{...(1)}$$

Method 1 : First we try to eliminate t. If it is done, the given equations transform to an equation in Cartesian coordinates, the method for which has been given in Art. 8.11. Some important curves are given here.

(i) $$x = a \cos t,\ y = a \sin t$$

or $$x^2 + y^2 = a^2 \qquad \text{(Circle)}$$

(ii) $$x = a \cos t,\ y = b \sin t$$

or $$\frac{x^2}{a^2} + \frac{y^2}{b^2} = 1 \qquad \text{(Ellipse)}$$

(iii) $$x = a\cos^3 t,\ y = b\sin^3 t$$

or $$\left(\frac{x}{a}\right)^{2/3} + \left(\frac{y}{b}\right)^{2/3} = 1$$ (Hypocycloid)

(iv) $$x = a\cos^3 t,\ y = a\sin^3 t$$

or $$x^{2/3} + y^{2/3} = a^{2/3}$$ (Astroid)

(v) $$x = a\sin^2 t,\ y = \frac{a\sin^3 t}{\cos t}$$

or $$y^2(a + x) = x^3$$ (Cissoid)

Method II : If the parameter t cannot be eliminated, then we proceed in the following manner—

(i) Find $\frac{dx}{dt}$ and $\frac{dy}{dt}$. Divide and thus find the value of $\frac{dy}{dx}$.

(ii) Give to the parameter certain values and find the corresponding values of x, y and $\frac{dy}{dx}$.

(iii) Plot the corresponding cartesian points and observe the gradients of the tangents at these points given by the values of $\frac{dy}{dx}$.

ILLUSTRATIVE EXAMPLES

Example 1. *Trace the Tractrix,*

$$x = a\left(\cos t + \frac{1}{2}\log\tan^2 t/2\right),$$

$$y = a\sin t$$

Solution : The equations of the curve are,

$$x = a\left(\cos t + \frac{1}{2}\log\tan^2 t/2\right)$$

$$y = a\sin t. \qquad \text{...(i)}$$

We have, $$\frac{dx}{dt} = \frac{a\cos^2 t}{\sin t}$$

and $$\frac{dy}{dt} = a\cos t$$

$$\frac{dy}{dx} = \frac{dy/dt}{dx/dt} = \tan t$$

Now, we have when

t	0	$\pi/2$	π	$-\pi/2$
x	$-\infty$		∞	
y	0	a		$-a$
$\frac{dy}{dx}$	0	∞		$-\infty$

When $t = 0$, the corresponding point $(-\infty, 0)$ is at infinity on the negative side of X-axis, the slope at which is zero. Hence X-axis is the tangent at this point at infinity. Therefore, X-axis is an asymptote of the curve. A similar case is thereat $t = \pi$. Also, the points $(0, a)$ and $(0, -a)$ are situated on the curve. Hence, the shape of the curve is as shown in the following Figure 17.24.

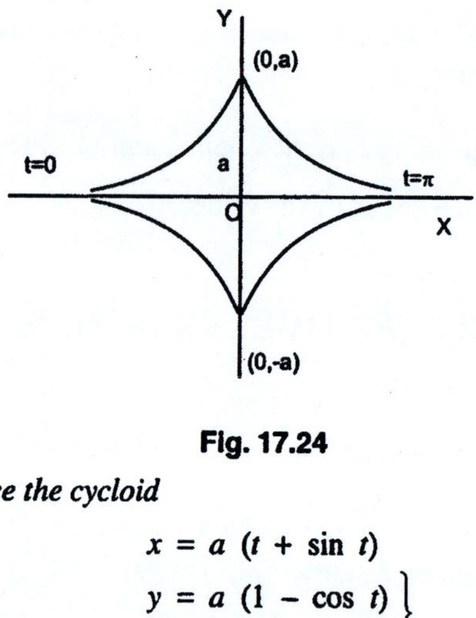

Fig. 17.24

Example 2. *Trace the cycloid*

$$\left.\begin{aligned} x &= a\,(t + \sin t) \\ y &= a\,(1 - \cos t) \end{aligned}\right\}$$

Solution : The equations of the curve are,

$$\left.\begin{aligned} x &= a\,(t + \sin t) \\ y &= a\,(1 - \cos t) \end{aligned}\right\} \quad \text{...(i)}$$

We have, $$\frac{dx}{dt} = a\,(1 + \cos t)$$

and $$\frac{dy}{dt} = a \sin t$$

$\therefore$ $$\frac{dy}{dx} = \frac{dy/dt}{dx/dt} = \tan \frac{t}{2}$$

Now, we have when

$t =$	0	$\frac{\pi}{2}$	π	$\frac{\pi}{2}$	π
$x =$	0	$a\left(\frac{\pi}{2}+1\right)$	$a\pi$	$-a\left(\frac{\pi}{2}+1\right)$	$-a\pi$
$y =$	0	a	$2a$	a	$2a$
$\frac{dy}{dx} =$	0	1	∞	-1	$-\infty$

We see that the curve passes through the point (0, 0) and the slope of the tangent thereat is 0, i.e., X-axis is the tangent at (0, 0).

Again we observe that $\left\{a\left(\frac{\pi}{2}+1\right), a\right\}$ is a point on the curve and the slope of the tangent at it is 1. Hence, the tangent at this point will make an angle 45° with the X-axis. Further $(a\pi, 2a)$ is a point on the curve, the slope at which is ∞. Therefore, the tangent at this point is parallel to Y-axis.

It is quite evident from the table above that the curve is symmetrical about $x = 0$, the Y-axis (for the value of t between $-\pi$ to π).

In a similar manner the other points of the curve can be observed.

Thus, the shape of the curve is as shown in the Figure 17.25. The point O is called the vertex.

Example 3. *Trace the curve*

$$x = a\cos t + \frac{a}{2}\log\tan^2\left(\frac{t}{2}\right),\ y = a\sin t$$

Solution : The equations to curve are,

$$x = a\cos t + \frac{a}{2}\log\left(\tan^2\frac{t}{2}\right) \qquad \text{...(i)}$$

$$y = a\sin t \qquad \text{...(ii)}$$

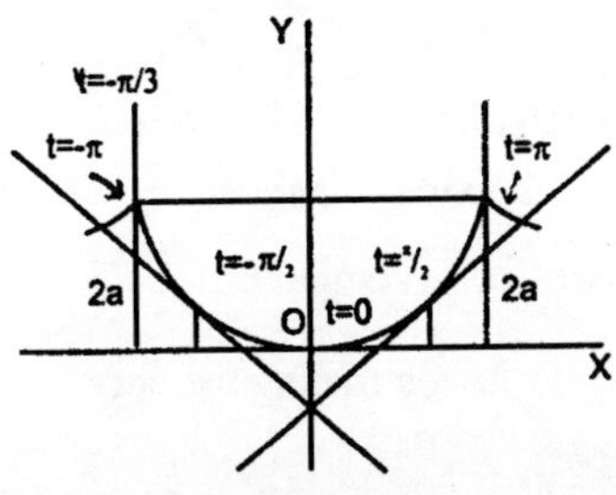

Fig. 17.25

(1) Symmetry : Putting $-t$ for t in the equation to the curve, the value of x remains unchanged but y changes the sign. The curve is symmetric about X-axis.

If t is replaced by $\pi - t$ in (ii), then the value of t remains unchanged, the curve is symmetric about Y-axis.

(2) The different values of x and y with respect to different values of t, we get from the following table—

$t =$	0	$\frac{\pi}{2}$	$\frac{3\pi}{4}$	π
$x =$	$-\infty$	0	$-$ ve	$+\infty$
$y =$	0	a	$+$ ve	0

From table, it is clear that when $t = 0$, then we get the point $(-\infty, 0)$ and for $t = \frac{\pi}{2}$, we get the point $(0, a)$.

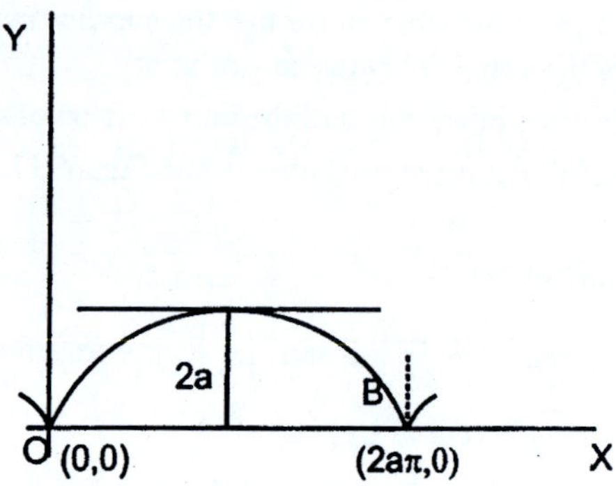

Fig. 17.26

The gradient of tangents at the points

$$m = \frac{dy}{dx} = \frac{dy/dt}{dx/dt} = \frac{a \cos t \sin t}{a \cos^2 t} = \tan t$$

m at the point when $t = 0$, is

$m = \tan 0 = 0$ or X-axis tangent, where $t = 0$, when $t = \frac{\pi}{2}$, $m = \tan \frac{\pi}{2} = \infty$ or Y-axis is tangent.

We get the above Figure 17.26 on tracing the curve.

Example 4. *Trace the curve*

$$x = a\,(\theta - \sin\theta),\; y = a\,(1 - \cos\theta)$$

Solution : The given equations are,

$$x = a\ (\theta - \sin\ \theta) \qquad ...(i)$$

$$y = a\ (1 - \cos\theta) \qquad ...(ii)$$

(1) Symmetric: The value of y remains unchanged when θ is placed by $-\theta$ in equation (ii), then the curve is symmetric about Y-axis.

When θ is replaced by $-\theta$ in equation (i), then the value of x changes.

Thus, the curve is symetrical about Y-axis.

(2) The value of x and y with respect to θ are obtained from the following table—

$\theta =$	0	$\frac{\pi}{2}$	π	$\frac{3\pi}{2}$	2π
$x =$	0	$a\left(\frac{\pi}{2} - 1\right)$	$a\,\pi$	$a\left(\frac{3\pi}{2} + 1\right)$	$2\pi\, a$
$y =$	0	a	$2\,a$	a	0

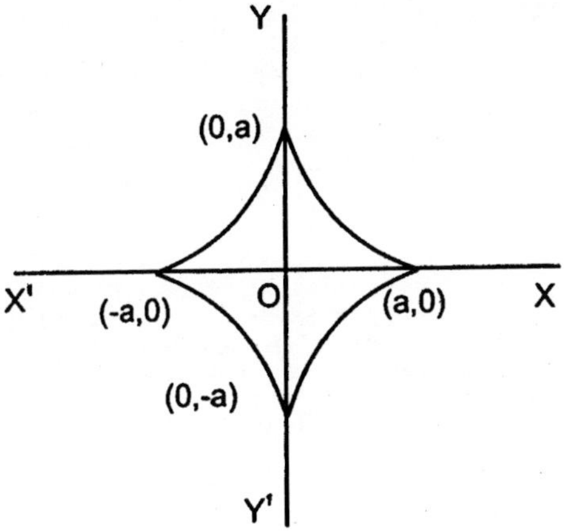

Fig. 17.27

The gradient of the tangent at different points,

$$\frac{dy}{dx} = \frac{a \sin\ \theta}{a\ (1 - \cos\ \theta)} = \cos\ \frac{\theta}{2}$$

At (0, 0), $m = \infty$ or Y-axis. Tangent at (0, 0) when θ changes from 0 to π, then the value of x changes from 0 to $a\pi$ and y changes from 0 to $2a$.

When θ increases from π to 2π, then x increases from $a\pi$ and $2\pi\, a$ and the value of θ decreases from $2a$ to 0.

EXERCISE 17 (C)

Trace the following curves—

1. $x = a\cos^3 t,\ y = b\sin^3 t$
2. $x = a\,(t + \sin\theta),\ y = a\,(1 + \cos t)$
3. $x = a\,(t - \sin t),\ y = a\,(1 + \cos t)$
4. $y = \cosh\left(\dfrac{x}{c}\right)$

ANSWERS

1.

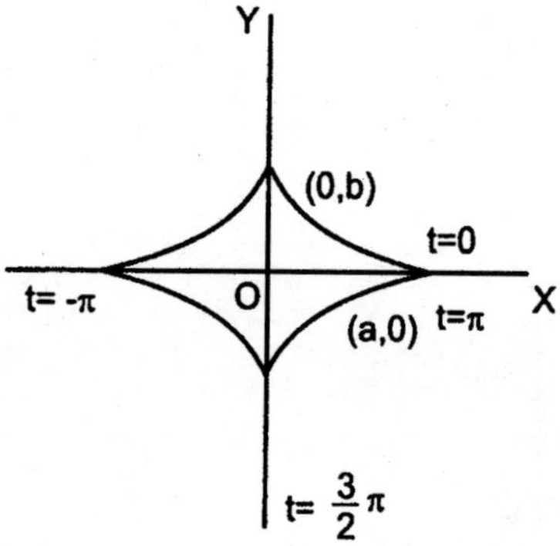

2.

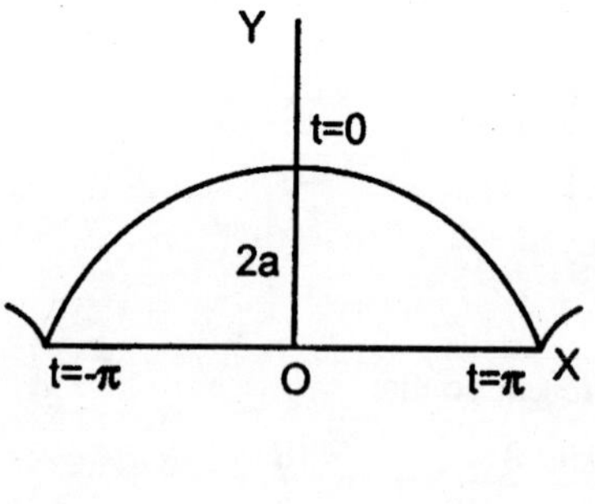